Use of Water Stable Isotopes in Hydrological Process

Use of Water Stable Isotopes in Hydrological Process

Editors

Polona Vreča
Zoltán Kern

MDPI • Basel • Beijing • Wuhan • Barcelona • Belgrade • Manchester • Tokyo • Cluj • Tianjin

Editors
Polona Vreča
Jožef Stefan Institute
Slovenia

Zoltán Kern
Research Centre for Astronomy and Earth Sciences
Hungary

Editorial Office
MDPI
St. Alban-Anlage 66
4052 Basel, Switzerland

This is a reprint of articles from the Special Issue published online in the open access journal *Water* (ISSN 2073-4441) (available at: https://www.mdpi.com/journal/water/special_issues/Isotopes_Hydrological_Process).

For citation purposes, cite each article independently as indicated on the article page online and as indicated below:

LastName, A.A.; LastName, B.B.; LastName, C.C. Article Title. *Journal Name* **Year**, *Article Number*, Page Range.

ISBN 978-3-03943-266-0 (Hbk)
ISBN 978-3-03943-267-7 (PDF)

Contents

About the Editors

Polona Vreča received her Ph.D. degree in Geology from the University of Ljubljana, Ljubljana, Slovenia, in 2000. She is an isotope geochemist working as a senior researcher in the Department of Environmental Sciences of the Jožef Stefan Institute (IJS), Ljubljana, Slovenia. Her expertise comprises the use of stable isotopes of C, N, O and H as tracers of environmental changes, particularly in the geochemical investigations of lake sediments and water cycle. Her current research focus is mainly on the use of H and O isotopes of water molecules in investigations of the water cycle with an emphasis on precipitation–surface water–groundwater interactions. She is actively involved in the development of the Slovenian Network of Isotopes in Precipitation—SLONIP (https://slonip.ijs.si/) and cooperates with different researchers worldwide within the framework of national and international projects.

Zoltán Kern graduated as a teacher of geography and mathematics at the Eötvös University, Budapest, Hungary, in 2001, and completed his Ph.D. in Earth Sciences in 2010. His research interest is focused on climate and environmental changes in the recent geological past. He has research experience in dendroclimatology, dendrochemistry, environmental isotopes, cave ice, time series analysis, and geomorphology. He has been working for the Institute for Geological and Geochemical Research, Research Centre for Astronomy and Earth Sciences (and its predecessors under the aegis of the Hungarian Academy of Sciences) since 2007 and leading the 2ka Paleoclimate Research Group (https://paleoclimate2ka.hu/?lang=en) there since 2012.

Preface to "Use of Water Stable Isotopes in Hydrological Process"

Stable and radioactive isotopes in water are powerful tools in the tracking of the path of water molecules through the whole water cycle. In the last decade, a considerable number of studies have been published on the use of water isotopes, and their number is ever-growing. The main reason is the development of new measurement techniques (i.e., laser absorption spectroscopy) that allow measurements of stable isotope ratios at ever-higher resolutions. Therefore, this compilation of papers has been published to address the current state-of-the-art water isotope methods, applications, and interpretations of hydrological processes, and to contribute to the rapidly growing repository of isotope data, which is important for future water resource management. We are pleased to present here a book with new findings in thirteen original research papers and one review paper issued in the Water MDPI Special Issue (SI) "Use of Water Isotopes in Hydrological Processes". The authors report the use of water isotopes in hydrological processes worldwide, including studies at both local and regional scales related to either precipitation dynamics or to different applications of water isotopes in combination with other hydrochemical parameters in investigations of surface water, snowmelt, soil water, groundwater and xylem water to identify the hydrological and geochemical processes. All of the presentations report on isotopes of precipitation, either in detail, or as a part of the more complex case studies of water resources. Some authors also discuss the multi parameter approach and combine the isotope and other geochemical parameters to address the specific questions. The best example of the latter is the self-organizing map approach presented by Tsuchihara et al. (2020).

A few issues remain as challenges for the future, e.g., the use of the appropriate terminology, improving the traceability of water isotope data to provide results comparable in space and time, the reporting of the "raw" data and detailed explanations in line with FAIR data practice and the precise reporting of the sources of reused isotope data.

Polona Vreča, Zoltán Kern

Editors

Editorial

Use of Water Isotopes in Hydrological Processes

Polona Vreča [1,*] and Zoltán Kern [2,*]

[1] Department of Environmental Sciences, Jožef Stefan Institute, Jamova cesta 39, 1000 Ljubljana, Slovenia
[2] Institute for Geological and Geochemical Research, Research Centre for Astronomy and Earth Sciences, MTA Centre for Excellence, Budaörsi út 45, H-1112 Budapest, Hungary
* Correspondence: polona.vreca@ijs.si (P.V.); zoltan.kern@gmail.com (Z.K.)

Received: 25 July 2020; Accepted: 6 August 2020; Published: 7 August 2020

Abstract: Stable (^{16}O, ^{17}O, ^{18}O, ^{1}H, ^{2}H) and radioactive (^{3}H) isotopes in water are powerful tools in the tracking of the path of water molecules in the whole water cycle. In the last decade, a considerable number of studies have been published on the use of water isotopes, and the number continues to grow due to the development of new measurement techniques (i.e., laser absorption spectroscopy) that allow measurements of stable isotope ratios at ever-higher resolutions. Therefore, this Special Issue (SI) has been compiled to address current state-of-the-art water isotope methods, applications, and hydrological process interpretations and to contribute to the rapidly growing repository of isotope data important for future water resource management. We are pleased to present here a compilation of 14 papers reporting the use of water isotopes in the study of hydrological processes worldwide, including studies on the local and regional scales related either to precipitation dynamics or to different applications of water isotopes in combination with other hydrochemical parameters in investigations of surface water, snowmelt, soil water, groundwater, and xylem water to identify the hydrological and geochemical processes.

Keywords: water cycle; isotope hydrology; measurement traceability; precipitation (rain and snow); surface water; groundwater; water management; networks and data bases; statistical evaluation

---

1. Introduction

Water is vital for all known forms of life and is transported continuously through the different spheres of Earth in the water cycle: evaporation, transpiration, condensation, precipitation, runoff, infiltration, etc. As such, fresh water plays an important role in the world economy as well. Since ancient times, cities have been built around reliable sources of water, and a considerable amount of the total available fresh water is used for irrigation and other agricultural activities to supply humanity with sustenance.

Stable (^{16}O, ^{17}O, ^{18}O, ^{1}H, ^{2}H) and radioactive (^{3}H) isotopes in water molecules are powerful tools for the tracking of the path of water molecules in the water cycle, from precipitation to surface and groundwater, and further, into the drinking water supply. They are commonly used to trace the source of water and its flow pathways, or to quantify exchanges of water, solutes, and particulates between hydrological compartments during different hydrological processes [1].

In the last decade, a considerable number of studies have been published on the use of water isotopes in hydrological processes, and their number is ever-growing. The main reason is due to the development of new measurement techniques (i.e., tunable diode isotope ratio infrared laser absorption spectroscopy—LAS) that have been increasingly employed in hydrological studies since 2007, and allow measurements of stable isotope ratios with much smaller sample sizes [2,3]. In the last few years, *Water* MDPI has published fifty-four papers related to the application of water isotopes, and an exponentially increasing trend can be observed with time: 2016—4 papers; 2017—6; 2018—12; 2019—24. Most of the papers were published in the framework of the three closed Special Issues

(SI), namely "Isotopes in Hydrology and Hydrogeology" (20 papers), "Advances in Isotope Tracer Techniques for Tracing and Quantifying Hydrological Processes" (4 papers), and "Radioactive Isotopes in Hydrosphere" (4 papers).

To continue, the SI "Use of Water Isotopes in Hydrological Processes" addresses the current state-of-the-art methods, applications, and hydrological process interpretations using stable and radioactive water isotopes in the whole water cycle. In addition, accurate and precise measurements are required [3] to provide new data comparable in space and time, or comparable with data obtained with classical isotope ratio mass spectrometry (IRMS) [4]. Therefore, contributions related to some specific topics, such as measurement traceability (comparison of different measurement techniques), conceptual network development, and long-term maintenance of networks on the local to regional scale, as well as papers on different statistical data evaluation approaches, were also given a warm welcome in this SI. The SI was first announced in January 2019, and an invitation was sent to researchers through different channels. The first paper was submitted in March 2019 and the last in April 2020. In total, 18 submissions were received, and following a rigorous peer-review process carried out by qualified reviewers, one review and 13 research papers were selected for publication.

2. Contributions

In the following, we briefly present an overview of fourteen papers published in the period April 2019 to June 2020 [5–18]. The overview shows that contributions can be divided into two groups: the first related to isotopes in precipitation, and the second related to more complex hydrological and hydrogeological investigations. It is interesting that all contributions report on isotopes in precipitation, and five of them [6,7,11,14,18] focus in greater detail on isotopes in precipitation monitoring networks with longer and continuous observations. The importance of isotopes in precipitation data is understandable, as precipitation represents an important part of the water cycle and is the ultimate source of water to catchments. Other papers report on the use of precipitation isotope data as part of more complex hydrological and hydrogeological investigations [5,8–10,12,13,15–17]. Those papers report on different isotope sampling networks, including the collection and evaluation of precipitation, surface, and groundwater data. In some studies, also other samples like snow, ice, or snowmelt [10,12], soil water [13,17], or xylem water [13] were also collected.

The compilation of papers in this SI represents a variety of studies with a global scope. The only review paper in the SI discusses the combined use of stable isotopes of water and nitrate, and focuses on the fundamental background, analytical methods, and hydrograph separation, and reports perspectives related to quantification of relative nitrate contributions to stream water contamination [5].

Other papers present case studies with sampling performed either at a particular location [14,17], or within a research area from a few m^2 [13] to a few tens of km^2 [8,10,15,16], or on a larger regional scale of more than 1000 km^2 [9,12,18]. A few papers focus on the evaluation of published data sets for particular countries, e.g., Chile [6], China [7], or Iran [11].

Further, the isotope measurements discussed herein were performed using either IRMS [8,9] or LAS [10,12,13,15–17]. Some papers combine or evaluate data obtained using different techniques [5–7, 11,14,18]. Although authors usually report details of the instruments used and the degree of analytical precision, it must be admitted that it is not always clear how the calibration of the instrument and normalization of the data are performed [19], nor always how the reported degree of precision was determined. Terminology that was often not in line with the IUPAC recommendations is also an occasional issue [20], though this deficiency was improved during the peer-review process.

An important and valuable contribution of some papers is the supplementary material [6,9–11,14, 17,18], with "raw" data and detailed explanations that enable better use of isotope data in future water resources research and management, in line with FAIR data practice [21].

2.1. Isotopes in Precipitation

All papers in this SI report information on the isotope composition of precipitation. In this section, we focus on those dedicated only to precipitation monitoring, the establishment of local meteoric water lines (LMWL) with the use of different regression models, and those related to precipitation dynamics [6,7,11,14,18].

Boschetti et al. [6] collected isotope data from 32 stations in Chile from different sources, and calculated the LMWL with a confidence interval for this region arrived at by introducing a technique known as error-in variables (EIV) regression. The slopes obtained either using EIV or ordinary last square regression (OLSR) are similar. However, according to authors' conclusions, the new EIV-LMWL is more accurate and suitable than the OLRS or other types of regression models where the measurement errors in the x-axis are violated.

Kong et al. [7] present a meta-analysis of precipitation isotopes by using data from 68 stations from in and around China from different sources. They divided the entire country into five regions according to their major moisture sources, and determined the spatial distribution and seasonal variations of the isotope composition of precipitation. The LMWLs were then calculated for each region by applying the OLRS model on the basis of monthly data. The δ^{18}O-temperature and the δ^{18}O-precipitation amount gradients for the regions were also determined. The conclusion was that the findings could serve as a reference for isotopic application in hydrological and paleo-climate research.

Heydarizad et al. [11] collected isotope data from different sources for 32 stations in Iran and 4 from Iraq and developed three different LMWLs for Iran by applying the OLRS and considering air masses and dominant moisture sources in the particular region. They validated the LMWL by comparison with fresh karstic spring and surface water isotope data from across Iran and determined the main moisture sources important for those water resources.

In contrast to previous papers, Krajcar Bronić et al. [14] focused on a long-term isotope record of precipitation at one station, namely Zagreb in Croatia. The history of isotope monitoring performed since 1976 is described in detail, and in addition to stable isotope data, tritium activity concentration together with air temperature and precipitation amount are reported. A statistical evaluation of the data showed a significant increase in the annual air temperature and larger variations in that of the amount of precipitation, in addition to an increase in the annual δ^{18}O and the δ^2H values and a slight decrease in tritium activity concentration over the last two decades. The LMWL was calculated using different regression models proposed by Crawford et al. [22], and did not result in significant differences.

The last paper in this group evaluates the results of a three-year program monitoring the stable isotope composition of precipitation in the Adriatic-Pannonian region; it focuses on the determination of the isotopic "altitude" and "continental" effects in this particular region [18]. Isotope data were collected from 15 stations (5 in Hungary, 7 in Slovenia, 2 in Austria, and 1 in Croatia). On the basis of the isotope data obtained, an "altitude" effect of −1.2‰/km for δ^{18}O and −7.9‰/km for δ^2H was calculated—and it is recommended that this figure be used for modern precipitation in the region. The estimated mean isotopic "continental" effect could, however, only be determined for the winter precipitation, and figures of −2.4‰/100 km in δ^{18}O and −20‰/100 km in δ^2H were arrived at. The authors conclude that the isotopic "altitude" and "continental" effects thus determined can be used in future isotope hydrological or paleoclimatological applications in the Adriatic–Pannonian region.

2.2. Use of Water Isotopes in Complex Investigations

In this section, we briefly summarize the most important outcomes of other papers reporting on the use of water isotopes in complex hydrological and hydrogeological investigations [8–10,12,13,15–17].

Porowski et al. [8] studied the sources and biogeochemical processes in an aquifer affected by peatland in Kampinos National Park in Poland. Using a combination of chemical and isotopic data, they were able to determine the main sources of the sulfates dissolved in groundwater. The atmospheric sulfates, that is, sulfates formed by the dissolution of evaporate sulfate minerals and sulfates formed by oxidation of pyrite, were also identified. In this paper, water isotopes were used to identify the atmospheric inputs and the groundwater recharge.

Van Lam et al. [9] used different isotope techniques in investigations of groundwater resources of the delta plain of the Red River in Vietnam. The authors carried out short-term observations of local precipitation and used the precipitation water isotope data to determine the LMWL by applying the OLSR. Further, they determined the isotope composition of surface water, groundwater, and sea water, and studied the genesis of and processes controlling the quality of water resources in the region. They also determined the influence of salt water intrusions and studied the connection between different aquifers.

Sing et al. [10] applied water isotope investigations to the Sutri Dhaka glacier basin in the Himalayas, where they established a short-term observation site with the sampling of rain, snow, ice, and snowmelt water. The authors used the precipitation data to determine the moisture sources and LMWL, though they do not report the regression method used for the calculation of lines. Further, a three-component hydrograph separation based on stable isotope methods was applied to estimate the contribution of ice-melt and snowmelt to runoff.

Chen et al. [12] used a similar approach to Sing et al. [10] in the Naqu River basin on the Qinghai-Tibet Plateau. They established a short-term sampling site at which they collected precipitation, snowmelt, and groundwater for isotope analysis to explore the sources of runoff and the spatial variation of the discharge. The authors determined the LMWL and the river water line using the linear relationship, albeit without a more specific explanation of the regression method used. Finally, they used the water isotope data for the hydrograph separation and estimated the proportion of rainwater, snowmelt, and groundwater to the runoff.

Tsuchihara et al. [15] investigated the spatial distribution of the isotopic and hydrochemical composition of groundwater in selected alluvial fans under cultivation as rice paddies in Japan. They used an unsupervised training algorithm of an artificial neural network model developed by Kohonen [23] and applied a self-organizing map (SOM) to characterize the spatial pattern of groundwater hydrochemistry. With the use of the SOM, the groundwater could clearly be classified into four groups, and the different origins of water identified. Their findings represent an important contribution to the proper management of groundwater resources. The systematic, well-organized sampling approach, the data evaluation used, and the outcomes of Tsuchihara et al. [15] are to be recommended as an interesting approach in the future management of water resources.

Marković et al. [16] focused on the application of stable isotopes in combination with hydrochemical parameters for the improvement of the conceptual model of an alluvial aquifer in Croatia, which represents an important source of drinking water in the area studied. The authors used different regression models proposed by Crawford et al. [22] to calculate the LMWL for the area and, as was also the case with Krajcar Bronić et al. [14], no significant differences were observed. The authors also used the water isotope data in combination with water head data to determine the groundwater flow direction.

Jiménez-Rodríguez et al. [13] applied water isotopes to study the effect of willow plantations on soil water conditions in an arid environment in northern China. The authors discuss the influence of afforestation using different species, namely willow bushes and willow trees, and use water isotopes monitored in precipitation, soil water, groundwater, and xylem water to investigate the redistribution of water in the soil. They conclude that both species reduced the effects of soil evaporation after summer. However, the willow bush is capable of extracting soil and groundwater in different proportions according to water availability, while the willow tree is able to extract water and groundwater in specific proportions. Therefore, the selection of species for afforestation programs has to be carried out carefully, not to endanger the scarce water resources in arid regions.

The last paper in this section was contributed by Mattei et al. [17], who investigated the potential of pore water isotope information to parameterize soil water transport models. By conducting a Morris and Sobol sensitivity analysis, they were able to highlight the value of combining water content and pore water isotope composition data in a multi-objective calibration approach to constrain soil hydraulic property parameterization. They tested the model with pore water data collected from sandy soil at two sampling sites close to Montreal, Canada. After employing the model in order to estimate the annual groundwater recharge, they conclude that it is possible to calibrate the model without continuous monitoring data, using only water content and pore water isotopic data from a single sampling. The results presented imply the possibility of reducing long-term monitoring observations and to adapt the sampling according to the objectives of the groundwater recharge investigations.

3. Conclusions

Understanding water exchange within geospheres is vital if we are to address environmental issues linked to water resources management effectively. Isotopes in the water molecule determined in different compartments of the water cycle serve as natural fingerprints and allow us to trace the molecule through the water cycle. This Special Issue is dedicated to the use of water isotopes in hydrological processes and compiles new findings in selected thirteen original research papers and one review paper. All presentations report on isotopes of precipitation, either in detail or as a part of more complex case studies of water resources. Some authors discuss also the multi-parameter approach and combine the isotope and other geochemical parameters to address the specific questions. The best example of the latter is the self-organizing map approach presented by Tsuchihara et al. [15].

Few issues remain challenging for the future, e.g., the use of the appropriate terminology, the improvement of the traceability of water isotope data to provide results comparable in space and time, the reporting of the "raw" data and detailed explanations in line with FAIR data practice, and greater precision in the reporting of the sources of reused isotope data. These issues comprise, suitably, the themes of forthcoming *Water* MDPI Special Issues, "Application of Stable Isotopes and Tritium in Hydrology", "Isotope Hydrology", "Application of Isotopic Data to Water Resource Management", "Geochemistry of Groundwater", and "Applying Artificial and Environmental Tracing Techniques in Hydrogeology".

The Guest Editors of this SI are also considering a sequel to follow up the achievements of this SI, to be entitled "Use of Water Isotopes in Hydrological Processes, part II" in 2021. In addition, such an SI could cover the focus areas of a recently started COST Action "WATer isotopeS in the critical zONe: from groundwater recharge to plant transpiration (WATSON)", CA19120, a compilation in the recent SI which may be considered an inspiration to the working groups in the Action.

Author Contributions: Conceptualization, P.V.; formal analysis, P.V. and Z.K.; writing, reviewing and editing, P.V. and Z.K.; funding acquisition, P.V. and Z.K. All authors have read and agreed to the published version of the manuscript.

Funding: The Guest Editors activities for this Special Issue were supported by the National Research, Development and Innovation Office under Grants SNN118205 and by the Slovenian Research Agency ARRS under Grants N1-0054, J4-8216 and P1-0143.

Acknowledgments: The Guest Editors express their gratitude to the *Water* MDPI in-house editors for their valuable help during the processing of the manuscripts. We also thank the researchers for contributing their findings and all the reviewers for their valuable comments that have greatly improved the outcome of the SI.

Conflicts of Interest: The authors declare no conflict of interest.

References

1. Aggarwal, P.K.; Froehlich, K.F.; Gat, J.R. *Isotopes in the Water Cycle*; Springer: Dordrecht, The Netherlands, 2005; p. 382.
2. Lis, G.; Wassenaar, L.I.; Hendry, M.J. High-Precision Laser Spectroscopy D/H and ^{18}O/^{16}O Measurements of Microliter Natural Water Samples. *Anal. Chem.* **2008**, *80*, 287–293. [CrossRef] [PubMed]
3. Wassenaar, L.I.; Coplen, T.B.; Aggarwal, P.K. Approaches for Achieving Long-Term Accuracy and Precision of δ^{18}O and δ^2H for Waters Analyzed using Laser Absorption Spectrometers. *Environ. Sci. Technol.* **2014**, *48*, 1123–1131. [CrossRef] [PubMed]

4. Wassenaar, L.I.; Terzer-Wassmuth, S.; Douence, C.; Araguas-Araguas, L.; Aggarwal, P.K.; Coplen, T.B. Seeking excellence: An evaluation of 235 international laboratories conducting water isotope analyses by isotope-ratio and laser-absorption spectrometry. *Rapid. Commun. Mass. Spectrom.* **2018**, *32*, 393–406. [CrossRef]

5. Jung, H.; Koh, D.-C.; Kim, Y.S.; Jeen, S.-W.; Lee, J. Stable Isotopes of Water and Nitrate for the Identification of Groundwater Flowpaths: A Review. *Water* **2020**, *12*, 138. [CrossRef]

6. Boschetti, T.; Cifuentes, J.; Iacumin, P.; Selmo, E. Local Meteoric Water Line of Northern Chile (18° S–30° S): An Application of Error-in-Variables Regression to the Oxygen and Hydrogen Stable Isotope Ratio of Precipitation. *Water* **2019**, *11*, 791. [CrossRef]

7. Kong, Y.; Wang, K.; Li, J.; Pang, Z. Stable Isotopes of Precipitation in China: A Consideration of Moisture Sources. *Water* **2019**, *11*, 1239. [CrossRef]

8. Porowski, A.; Porowska, D.; Halas, S. Identification of Sulfate Sources and Biogeochemical Processes in an Aquifer Affected by Peatland: Insights from Monitoring the Isotopic Composition of Groundwater Sulfate in Kampinos National Park, Poland. *Water* **2019**, *11*, 1388. [CrossRef]

9. Van Lam, N.; Van Hoan, H.; Duc Nhan, D. Investigation into Groundwater Resources in Southern Part of the Red River's Delta Plain, Vietnam by the Use of Isotopic Techniques. *Water* **2019**, *11*, 2120. [CrossRef]

10. Singh, A.T.; Rahaman, W.; Sharma, P.; Laluraj, C.M.; Patel, L.K.; Pratap, B.; Gaddam, V.K.; Thamban, M. Moisture Sources for Precipitation and Hydrograph Components of the Sutri Dhaka Glacier Basin, Western Himalayas. *Water* **2019**, *11*, 2242. [CrossRef]

11. Heydarizad, M.; Raeisi, E.; Sorí, R.; Gimeno, L. Developing Meteoric Water Lines for Iran Based on Air Masses and Moisture Sources. *Water* **2019**, *11*, 2359. [CrossRef]

12. Chen, X.; Wang, G.; Wang, F.; Yan, D.; Zhao, H. Characteristics of Water Isotopes and Water Source Identification During the Wet Season in Naqu River Basin, Qinghai–Tibet Plateau. *Water* **2019**, *11*, 2418. [CrossRef]

13. Jiménez-Rodríguez, C.D.; Coenders-Gerrits, M.; Uhlenbrook, S.; Wenninger, J. What Do Plants Leave after Summer on the Ground?—The Effect of Afforested Plants in Arid Environments. *Water* **2019**, *11*, 2559. [CrossRef]

14. Krajcar Bronić, I.; Barešić, J.; Borković, D.; Sironić, A.; Mikelić, I.L.; Vreča, P. Long-Term Isotope Records of Precipitation in Zagreb, Croatia. *Water* **2020**, *12*, 226. [CrossRef]

15. Tsuchihara, T.; Shirahata, K.; Ishida, S.; Yoshimoto, S. Application of a Self-Organizing Map of Isotopic and Chemical Data for the Identification of Groundwater Recharge Sources in Nasunogahara Alluvial Fan, Japan. *Water* **2020**, *12*, 278. [CrossRef]

16. Marković, T.; Karlović, I.; Perčec Tadić, M.; Larva, O. Application of Stable Water Isotopes to Improve Conceptual Model of Alluvial Aquifer in the Varaždin Area. *Water* **2020**, *12*, 379. [CrossRef]

17. Mattei, A.; Goblet, P.; Barbecot, F.; Guillon, S.; Coquet, Y.; Wang, S. Can Soil Hydraulic Parameters Be Estimated from the Stable Isotope Composition of Pore Water from a Single Soil Profile? *Water* **2020**, *12*, 393. [CrossRef]

18. Kern, Z.; Hatvani, I.G.; Czuppon, G.; Fórizs, I.; Erdélyi, D.; Kanduč, T.; Palcsu, L.; Vreča, P. Isotopic 'Altitude' and 'Continental' Effects in Modern Precipitation across the Adriatic–Pannonian Region. *Water* **2020**, *12*, 1797. [CrossRef]

19. Paul, D.; Skrzypek, G.; Fórizs, I. Normalization of measured stable isotopic compositions to isotope reference scales –a review. *Rapid Commun. Mass Spectrom.* **2007**, *21*, 3006–3014. [CrossRef]

20. Coplen, T.B. Guidelines and recommended terms for expression of stable-isotope-ratio and gas-ratio measurement results: Guidelines and recommended terms for expressing stable isotope results. *Rapid Commun. Mass Spectrom.* **2011**, *25*, 2538–2560. [CrossRef]

21. Stall, S.; Yarmey, L.; Cutcher-Gershenfeld, J.; Hanson, B.; Lehnert, K.; Nosek, B.; Parsons, M.; Robinson, E.; Wyborn, L. Make scientific data FAIR. *Nature* **2019**, *570*, 27–29. [CrossRef]

22. Crawford, J.; Hughes, C.E.; Lykoudis, S. Alternative least squares methods for determining the meteoric water line, demonstrated using GNIP data. *J. Hydrol.* **2014**, *519*, 2331–2340. [CrossRef]

23. Kohonen, T. Self-organized formation of topologically correct feature maps. *Biol. Cybern.* **1982**, *43*, 59–69. [CrossRef]

Review

Stable Isotopes of Water and Nitrate for the Identification of Groundwater Flowpaths: A Review

Hyejung Jung [1], Dong-Chan Koh [2], Yun S. Kim [3], Sung-Wook Jeen [4] and Jeonghoon Lee [1,*]

[1] Department of Science Education, Ewha Womans University, Seoul 03760, Korea;
    hyejeong.chung@gmail.com
[2] Groundwater Department, Geologic Environment Division, Korea Institute of Geoscience and Mineral
    Resources, Daejeon 34132, Korea; chankoh@kigam.re.kr
[3] Water Quality Research Center, K-Water Convergence Institute, Daejeon 34350, Korea; yunskim@kwater.or.kr
[4] Department of Earth and Environmental Sciences & The Earth and Environmental Science System Research
    Center, Jeonbuk National University, Jeonju 54896, Korea; sjeen@jbnu.ac.kr
* Correspondence: jeonghoon.lee@ewha.ac.kr; Tel.: +82-2-3277-3794

Received: 1 December 2019; Accepted: 30 December 2019; Published: 1 January 2020

Abstract: Nitrate contamination in stream water and groundwater is a serious environmental problem that arises in areas of high agricultural activities or high population density. It is therefore important to identify the source and flowpath of nitrate in water bodies. In recent decades, the dual isotope analysis (δ^{15}N and δ^{18}O) of nitrate has been widely applied to track contamination sources by taking advantage of the difference in nitrogen and oxygen isotope ratios for different sources. However, transformation processes of nitrogen compounds can change the isotopic composition of nitrate due to the various redox processes in the environment, which often makes it difficult to identify contaminant sources. To compensate for this, the stable water isotope of the H_2O itself can be used to interpret the complex hydrological and hydrochemical processes for the movement of nitrate contaminants. Therefore, the present study aims at understanding the fundamental background of stable water and nitrate isotope analysis, including isotope fractionation, analytical methods such as nitrate concentration from samples, instrumentation, and the typical ranges of δ^{15}N and δ^{18}O from various nitrate sources. In addition, we discuss hydrograph separation using the oxygen and hydrogen isotopes of water in combination with the nitrogen and oxygen isotopes of nitrate to understand the relative contributions of precipitation and groundwater to stream water. This study will assist in understanding the groundwater flowpaths as well as tracking the sources of nitrate contamination using the stable isotope analysis in combination with nitrate and water.

Keywords: groundwater; isotope hydrology; stable water isotopes; stable nitrate isotopes

---

1. Introduction

Identifying groundwater flowpaths can provide important information regarding the movements of water itself and of contaminants therein via interaction with surface water. For example, contaminants can be discharged directly into the stream water but, if they are recharged into groundwater that then passes indirectly into stream water, the groundwater can contribute significantly to the water quality of the stream [1]. In particular, since nitrate is highly mobile and primarily originates from nonpoint source pollution, it is distributed across a wide area through various groundwater flowpaths and it can be difficult to trace the source [2]. In order to effectively control the spread of contaminants, and to clean up the contaminated stream water, it is therefore important to understand the flowpath of groundwater [3].

While concentration-based chemical analyses such as total nitrogen (TN), total phosphorus (TP), total organic carbon (TOC), biochemical oxygen demand (BOD), and chemical oxygen demand (COD)

have traditionally been used to effectively trace mixed contamination, this approach does not easily track contaminant movement and physical processes [4]. By contrast, stable isotope analysis is an effective tool for identifying sources, inferring processes, and determining the contributions of various inputs [5]. In particular, stable water isotopes (δ^{18}O and δD) are affected by meteorological processes that provide a characteristic fingerprint of their origin, which is essential for investigating the source of groundwater [6]. The stable isotopes of nitrogen and oxygen in nitrate (δ^{15}N and δ^{18}O) are also fundamental to identifying the sources of nitrate contamination because the isotopic values are distinct from source to source [7].

While the stable water isotopes have been used as tracers in hydrograph separation studies since the pioneering work of Craig [8], the stable nitrate isotopes have been used to identify nitrate sources since nitrate contaminants became an environmental issue in the 1970s. Even now, nitrate is a very common groundwater pollutant, imposing a serious threat to drinking water supplies and contributing to eutrophication of surface waters [9–11]. Nitrate is the dominant nitrogen species in groundwater, which may be derived from soil organic nitrogen, synthetic fertilizer, livestock waste, sewage effluent, and atmospheric precipitation [11]. In some areas, atmospheric deposition of anthropogenic nitrogen exceeds ecosystem nutrient demand and the influence of atmospheric deposition on nitrogen export has not been well-documented for short-term discharge events such as rainfall and snowmelt [12].

Isotopic hydrograph separation using stable isotopes in water and nitrate provides a useful tool for determining the water flowpath and the source of nitrates. This approach has been widely used to understand the proportion of different water sources contributing to stream water, which can be used to infer the flowpath and residence time [13–16]. In particular, distinguishing between nitrate sources such as direct atmospheric deposition or biological assimilation and release in the soil zone may reveal the flowpath of groundwater into stream or river water [12]. Hence, the isotopic analysis of nitrogen and oxygen in nitrates (the dual isotopic technique) has been used to identify the source of nitrate in many studies. For example, Böttcher et al. [17] determined the sources of nitrate in groundwater downgradient from an agricultural area and Durka et al. [18] later determined the sources of nitrate in an undisturbed watershed in Bavaria, Germany. The dual isotope approach can be used to determine the source of nitrate in stream water because of the distinct isotopic signature of nitrate sources such as event water (rainfall or snowmelt), soil water, and groundwater.

To study the hydrograph separation of stable water isotopes, it is important to understand how precipitation infiltrates into soil water or recharges into groundwater and is subsequently released into stream water. To this end, studies on stable water isotopes in the atmospheric source must first be conducted in order to form a basis for understanding and predicting the movement of contaminants in the groundwater flowpath [16]. For the past 40 years, many studies have been conducted using the hydrograph separation technique through stable water isotopes or conservative chemical tracers to investigate the movement of water components such as groundwater, rainfall, snowmelt, and soil water in the stream water [16,19–26]. In particular, Ladouche et al. [20] investigated the streamflow components using hydrograph separation with stable water isotopes, major chemical parameters, and dissolved organic carbon (DOC). Dahlke et al. [22] used the value of the stable oxygen isotope (δ^{18}O) of water to indicate that the majority of storm runoff was dominated by pre-event water in the 30% glaciated sub-arctic catchment of Tarfala, northern Sweden. Later, Rahman et al. [24] conducted an end-member mixing analysis to describe the daily variation of runoff components in the Alpine watershed, and Kim et al. [16] used chemical and isotopic tracers to identify the impact of the pre-event water component of a granitic watershed with a thin soil layer.

Isotope hydrology involves measuring the stable isotopic compositions of precipitation, stream water, and groundwater samples, then interpreting these measurements in order to quantify or conceptualize the groundwater flowpath and velocity profile along with hydrogeochemical and biogeochemical reactions. With more conventional hydrogeological and hydrogeochemical data, such as information on lithology, meteorology, and solute concentrations, isotopic approaches have been helpful in identifying water movement among various reservoirs, e.g., evapotranspiration, groundwater recharge, discharge, and runoff [12,15,27–31]. The present paper is focused on isotope hydrology reviews dealing with methodological advances and their limitations and lessons drawn from decades of research. This review is motivated by the importance of understanding the groundwater flowpath to rivers and/or streams via analysis of isotopes in water and nitrates. After briefly introducing the systematic processes affecting the oxygen and hydrogen isotopes from precipitation to groundwater and the nitrogen and oxygen isotopes of nitrates, the review goes on to examine the commonly applied isotopic technique of hydrograph separation using stable water isotopes. Hence, this study will help to understand the groundwater flowpath and the tracking of nitrate contamination to its source using the stable isotope analysis of nitrate and water.

2. Hydrograph Separation

Hydrograph separation is the separation of streamflow components into two or more different components that contribute to the stream in a small catchment area or watershed. For example, isotopic hydrograph separation using isotopic tracers was first proposed by Dincer et al. [32], was developed by Sklash and Farvolden [19], and has been evaluated in many studies [33]. The isotopic hydrograph separation technique is based on the assumption that two components contribute to the stream after the precipitation occurs, namely: (1) The runoff caused by the rainwater (new water) and (2) the groundwater (old water). To separate the stream water discharge into rainwater and groundwater components, a two-component mixing model was used. The following mass balance equations introduced by Sklash and Farvolden [19] can be used:

$$Q_t = Q_r + Q_g \tag{1}$$

$$C_t Q_t = C_r Q_r + C_g Q_g \tag{2}$$

$$x = \frac{C_t - C_g}{C_r - C_g} \tag{3}$$

where Q indicates the discharge of each component, C is the concentration of an observed tracer or an isotopic composition, the subscripts t, r, and g indicate total discharge, rainwater, and groundwater, respectively, and x_r is the ratio of stream water contributed by rainwater ($x_r = \frac{Q_r}{Q_t}$).

The following four assumptions underlie the application of these mass balance equations: (1) There is a significant difference between the concentration of tracers in groundwater and rainwater; (2) the concentrations or isotopic compositions of the tracers for groundwater and rainwater are constant in space and time; (3) for two-component hydrograph separation, the concentrations of each tracer are equivalent in groundwater and vadose water, or else the contribution of vadose water is negligible; and (4) surface storage contributes minimally during the runoff. If these assumptions are valid, then two-component hydrograph separation can be used to determine the amounts of stream water contributed by rainwater and groundwater. Otherwise, hydrograph separation of three or more components should be carried out. For example, if the amount of vadose water in the saturation zone is not negligible and must be taken into account, then hydrograph separation of the three components of runoff, soil water, and groundwater should be used. In two-component systems, soil water can be interpreted as runoff or groundwater, depending on the geological characteristics. When considering the soil water among the factors contributing to the stream water after rainfall or snowmelt, hydrograph separation of the three components (soil water, rain or snowmelt, and groundwater) should be used. Hydrograph separation of the three components is basically expressed in the form of a three-way linear system of equations, which can be interpreted as follows:

$$Q_t = Q_r + Q_g + Q_s \tag{4}$$

$$C_t = C_t \frac{Q_r}{Q_t} + C_g \frac{Q_g}{Q_t} + C_s \frac{Q_s}{Q_t} \tag{5}$$

$$I_t = I_t \frac{Q_r}{Q_t} + I_g \frac{Q_g}{Q_t} + I_s \frac{Q_s}{Q_t} \tag{6}$$

where Q indicates the discharge of each component, C is the concentration of an observed tracer, I is the isotopic composition of each component, and the subscripts t, r, g, and s indicate the total discharge, rainwater, groundwater, and soil water, respectively. Since solutions for more than three components are difficult to obtain, matrix operation has been applied to the Equations (4)–(6) in the present work as follows:

$$A = \begin{bmatrix} 1 & 1 & 1 \\ C_r & C_g & C_s \\ I_r & I_g & I_s \end{bmatrix}, X = \begin{bmatrix} \frac{Q_r}{Q_t} \\ \frac{Q_g}{Q_t} \\ \frac{Q_s}{Q_t} \end{bmatrix}, B = \begin{bmatrix} 1 \\ C_t \\ I_t \end{bmatrix}. \tag{7}$$

$$AX = B, X = A^{-1}B \tag{8}$$

A system of linear equations is introduced that enables a three-component hydrograph separation using both isotopic and chemical compositions. MATLAB can be used to solve the matrix. These are mathematically underdetermined systems of n equations in n + 1 unknowns for which there is no unique solution. However, even with n isotope systems and >n + 1 sources, recently published studies introduce software (IsoSource model) that calculates multiple source proportions using mass balance conservation requirements. The IsoSource model, based on the principle of stable isotope mass conservation, can be used to partition contaminant sources in wastewater [34–36].

According to the second assumption mentioned above, there should be no temporal or spatial variation in the isotopic compositions of groundwater and rainwater (i.e., no isotopic fractionation), which would otherwise lead to deviation. Thus, if the isotopic composition of rain and groundwater changes over time, a systematic error in the fraction of rainwater contributing to the stream will arise. This systematic error can be determined using Gaussian error propagation [37,38]. The isotope composition of groundwater (old water) is known to be relatively constant. However, rain or snowmelt (new water) is subject to much greater isotopic fractionation, so hydrograph separation using the mean isotope value generates errors. The uncertainty of new water generated from isotopic fractionation can be calculated according to the following equation [38]:

$$\Delta x_r = -\frac{x_r}{C_r - C_g} \Delta c_r \tag{9}$$

where Δx_r is the systematic error when new water (rain or snowmelt) contributes to the stream, and Δc_r is the error in c_r. This is the variation in the tracer concentration or the ratio of stable isotopes in the rain (new water). Therefore, according to Equation (9), the error generated when considering the effect of new water on the stream is inversely proportional to the difference of the tracer concentration between new and old water, and directly proportional to the actual contribution of new water (x_r) to the stream water and the tracer concentration of the new water over time (Δc_r).

3. Stable Water Isotopes

Water evaporates from the ocean and moves into the continents, cools and condenses to form clouds, then falls to the surface as precipitation (rain or snow). In turn, the precipitated water (stream water, groundwater, and runoff) is evaporated again and recycled. As shown in Figure 1, during the transition from ocean to continent, the isotopic composition is changed through the processes of evaporation and rainout within the hydrologic cycle based on the isotope data from Hoefs [39] and Coplen et al. [40]. When water undergoes a change of physical phase, the water molecules containing heavier isotopes (H^2HO and $H_2^{18}O$) are preferentially concentrated in the more condensed phase (i.e., liquid rather than vapor, and solid rather than liquid), while molecules containing the lighter isotope ($H_2^{16}O$) are concentrated in the remaining phase [41]. Consequently, the rainout process causes continual fractionation of heavy isotopes into the precipitation (Rayleigh like distillation) such that the residual vapor becomes progressively more depleted in heavy isotopes [42]. Hence, subsequent precipitations will be depleted in heavy isotopes compared to previous precipitations originating from the same atmospheric water vapor [43]. Moreover, since the isotope composition of water varies among the components of the water cycle, isotope measurement makes it possible to identify the source of water masses and determine their interrelationships [42]. In particular, because stream water has a complicated relationship between rainfall (new water) and groundwater (old water), isotope composition is a useful tool for determining mixing patterns and relative contribution rates via hydrograph separation [26,38].

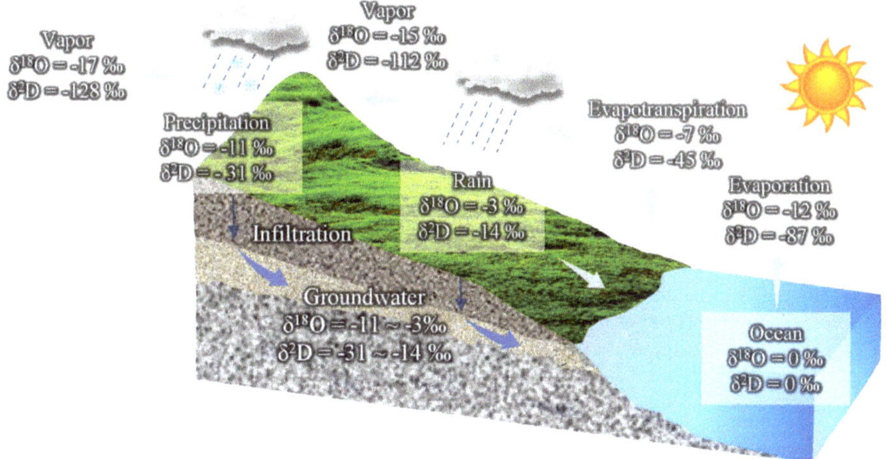

Figure 1. The diagram of isotopic composition change of atmospheric water vapor showing the processes of evaporation and rainout as the air mass proceeds from an ocean to a continent (based on Hoefs [39]; Coplen et al. [40]).

3.1. Isotopes in Precipitation

Unlike other tracers, stable water isotopes are added naturally on the scale of the watershed by precipitation (rain or snowmelt events) and, upon entering the watershed, undergo transport according to the natural movement of the body of water through the watershed. Since the stable isotope compositions of the water only change via the above-mentioned mixing and fractionation processes during evaporation and condensation, these environmental isotopes (supplied by meteoric processes) can be used to trace and identify the different air and water masses contributing precipitation to the watershed [43]. Moreover, since precipitation is a major source of water in the hydrological cycle, an understanding of the processes that control the spatial and temporal isotopic composition distributions of precipitation is essential [43].

In general, the fractionation processes of the stable hydrogen and oxygen isotopes are similar; hence, their behavior in the hydrological cycle is also similar [44]. This similarity causes covariance between the stable hydrogen and oxygen isotope concentrations found in most meteoric water, as first observed by Friedman [45]. This covariance can be explained by the following relationship, which was defined by Craig [8]:

$$\delta^2H = 8\ \delta^{18}O + 10 \tag{10}$$

This linear relationship, termed the meteoric water line (MWL), provides a convenient reference for understanding and tracing the origins of water [43]. In particular, an MWL with an intercept of 10 and a slope of 8 has been defined as the global meteoric water line (GMWL). The GMWL may be explained by the condensation of water vapor under conditions close to equilibrium, producing the slope of 8 [46]. The slope is related to the ratio of the fractionation coefficients and to factors relating to whether the water entering the soil, groundwater, and lakes has experienced evaporation [31,47]. Typically, the evaporation of soil or lake water results in a slope of less than 8 (generally between 4 and 7) for the local meteoric water line [46].

In a plot of Equation (10), the y-intercept is termed the deuterium excess (or d-excess). According to Dansgaard [46] this is defined by the deviation of isotopic equilibrium during evaporation from sourced precipitation and is related to the relative humidity parameter of the vapor source for evaporation. Dansgaard [46] recognized four parameters that determine this depletion in isotope values, namely: Altitude, distance from the shore, latitude, and quantity. Since the affected factors differ regionally, the d-excess is useful for identifying moisture source regions [31,48–50]. More recently, Lee et al. [38] reviewed the results of previous studies on how the fractionation of stable water isotopes significantly differs depending upon the region. In the New Hampshire area of the United States, for example, the difference of stable oxygen isotopic value is 2 to 3‰ [33] and the stable hydrogen isotopic value is 10 to 12‰ [31], while the isotopic values in Incheon, Korea, are 20‰ for oxygen and approximately 60‰ for hydrogen, and in Jeju Island, Korea, the respective isotopic values of oxygen and hydrogen are 7 to 8‰ and 50 to 60‰.

The isotopic compositions of precipitation are dependent upon several factors, including those of its vapor source (typically from nearby oceanic regions) along with the processes of precipitation formation and air mass trajectory (i.e., the influence of vapor source and rainout processes along the pathway of the air mass) [43]. Most of these factors are related to isotopic fractionation through diffusion during physical phase changes such as evaporation, sublimation, condensation, and melting [43]. Further details relating to isotopic fractionation will be discussed in the following sections.

3.2. Isotopic Evolution of Snow

Snowmelt is the largest contributor to groundwater recharge in Alpine environments [51]. Since snow dynamics are highly variable in space and time, an understanding of the hydrological responses of snowmelt contributing to the watershed is crucial for water-resource management [52]. While the isotopic composition of the snowpack profile generally represents the distinct isotopic composition of individual precipitation events, the signal in the snow layers provided by these individual events is attenuated by isotopic exchange, snowpack metamorphism and surface sublimation [53]. The isotopic composition of snowmelt generated from a snowpack results from two major processes, namely: (1) Sublimation and molecular exchange between vapor and the snowpack, and (2) meltwater infiltration and exchange with snow and meltwater within the snowpack [31,33,54,55].

With respect to the first process, Moser and Stichler [56] indicated that the isotopic fractionation associated with sublimation of snow surfaces behaves similarly to that of evaporating water, although Cooper [53] pointed out an exception when the well-mixed conditions of a water body are not present in the snowpack. In the second process, the meltwater is initially depleted in heavy isotopes relative to the remaining snowpack and then becomes gradually enriched in heavy isotopes as the melting proceeds [15,33,54]. This isotopic evolution results from isotopic exchange between liquid water and ice as the liquid water percolates down the snowpack [15,33,54]. Consequently, since the isotopic compositions of snowmelt are generally not the same as those of the bulk snow, hydrograph separations based on the isotope composition of the bulk snow will be erroneous [57]. Since snowmelt is a significant component of groundwater and surface runoff in temperate areas, an understanding of the isotopic evolution of a snowpack is crucial to both climatic and hydrological studies.

Studies of artificial and natural snowpack have demonstrated that complex changes in isotopic compositions can be expected to occur between accumulation and melting [58,59]. The isotopic composition of the upper snow layers is significantly altered by sublimation and exchange with atmospheric water vapor. Enrichment in $\delta^{18}O$ and δ^2H in the snowpack as a result of evaporation is a predictable outcome [60], and theoretical fractionation models developed for evaporation from well-mixed water bodies [61]) are reasonably successful at predicting the effects of simple evaporation once they are modified to account for the less than well-mixed conditions of the natural snowpack. However, isotopic change in the snowpack is more complicated than simple surface evaporation, and is dependent on variable conditions such as soil temperature, soil moisture, relative humidity, air temperature, vegetation cover, and the period of time for which the snowpack is present on the ground. Mast et al. [62] showed that although most of the water in Andrews Creek was new water from snowmelt (based on hydrograph separation using $\delta^{18}O$), much of that water had been transported along subsurface flowpaths prior to reaching the stream, and substantial interaction had occurred with soil or soil-like materials (based on hydrograph separation using dissolved silica). The highest nitrate concentrations in the springs and streams have been found to arise from a combination of the microbial cycle and flushing of nitrates and nitrates directly from rain or snowmelt [12].

3.3. Stable Water Isotope Measurements

The stable isotope composition of water is mainly determined by isotope ratio mass spectrometry (IRMS) [63]. This technique measures the relative isotope ratios of molecular compounds by analyzing mass differences [64]. A spectrum of masses is produced by generating a beam of charged molecules (usually by thermal ionization of gaseous samples) then bending the beam in a magnetic field [6]. In general, stable isotope analysis of water using IRMS requires chemical pretreatment [64]. For example, oxygen isotopes require ion-exchange between H_2O and CO_2 via bicarbonate reactions, and hydrogen isotopes require reduction with metals such as uranium, zinc, platinum and chromium [45,65–70]. Consequently, the oxygen isotope composition is analyzed as CO_2 and the hydrogen isotope composition is analyzed as H_2 [6]. The first dual-inlet mass spectrometer was developed by Alfred Nier in the late 1940s. However, the classical off-line procedures for sample preparation are time consuming and analytical precision depends on the skill of the investigator [6]. These considerations led to the modification of the classic dual inlet technique to create the continuous-flow isotope ratio mass spectrometer in which a trace amount of the gas to be analyzed is delivered in a stream of helium carrier gas [39].

Recently, isotope ratio infrared spectroscopy (IRIS) has been developed to analyze stable water isotopes using laser-based techniques [64]. This technique examines the characteristics of water absorption in the near-infrared wavelength region due to vibration–rotation transitions, which depend upon the ^{18}O and ^2H substitution of H_2O gas molecules [71]. Since these molecular motions are directly related to the proportion of isotopes, the isotope ratio can be measured [72]. The IRIS technique is sub-divided into off-axis integrated cavity output spectroscopy (OA-ICOS) and wavelength-scanned cavity ring-down spectroscopy (WS-CRDS) [64]. Compared to conventional IRMS, the IRIS technique has the advantages of simple preparation and operation, comparative portability for application in the field, and applicability with relatively small amounts of water samples (ppb, ppt) [73–77]. However, the presence of dissolved organic molecules with O–H bonds has the disadvantage of degrading analytical performance due to spectral interferences between the dissolved organics and water molecules [78].

4. Stable Nitrate Isotopes

4.1. Pretreatment Method for Nitrate Isotope Analysis

Dual isotope analysis of nitrates ($\delta^{15}N$ and $\delta^{18}O$) can be a powerful tool for identifying nitrate sources and nitrate cycling mechanisms in stream water because the different sources have isotopically distinct $\delta^{15}N$ and $\delta^{18}O$ compositions [79,80]. Over the past few decades, several pretreatment methods have been developed to concentrate dissolved nitrates for dual isotope analysis. Until recently, almost all nitrates for both $\delta^{15}N$ and $\delta^{18}O$ analysis were prepared using modifications of the silver nitrate method, in which samples are concentrated on anion exchange resins, eluted, and purified to produce $AgNO_3$ [80,81]. The $AgNO_3$ obtained from freeze drying is mixed with a catalyst composed of CuO/Cu wire/CaO and heated to 850 °C in a sealed reactor to generate N_2 gas for $\delta^{15}N$ analysis by IRMS. Meanwhile, $\delta^{18}O$ is analyzed by mixing $AgNO_3$ with graphite (spectroscopic analysis grade) to obtain

CO gas by pyrolysis or CO_2 gas by complete combustion. The combined techniques have been successfully used and published in studies from Alpine, agricultural, and urban environments [3,82]. The ion exchange method described above has the advantages of easy transport and storage, direct applicability in the field, and minimal isotope fractionation of nitrate during ion exchange. However, disadvantages of the ion exchange method include the long time and large cost of sample preparation, and interference due to the presence of other anions (Cl^-, SO_4^{2-}) in the sample. In addition, a relatively large amount of sample is required for precise analysis.

Another nitrate pretreatment method is denitrification by inoculation with a pure culture of denitrifying bacteria that lack the enzyme to reduce nitrate beyond N_2O [83,84]. The gas is then analyzed by IRMS. Microbial denitrification provides a saving in time and cost of sample preparation compared to the silver nitrate method and requires a small amount of sample. Nevertheless, this method involves a long time for culturing the microorganisms and the activity of the microorganisms is affected by toxic substances (antibiotics, heavy metals, pesticides, etc.) in the sample. Moreover, the presence of NO_2^- may distort the composition of the N_2O gas. In order to solve these problems, an advanced method of chemically reducing nitrate to N_2O gas was described by McIlvin and Altabet [85]. In this technique, nitrate (NO_3^-) is converted to nitrite (NO_2^-) using cadmium reduction and then to nitrous oxide (N_2O) using a 1:1 azide and acetic acid solution. The N_2O gas is analyzed in the same manner as in the microbial denitrification method. This chemical reduction method can significantly reduce the time and cost required for sample preparation and requires a small amount of sample for analysis. In addition, unlike microbial denitrification, it is not affected by toxic substances contained in the sample. However, there is a risk of exposure to dangerous chemicals (cadmium, sodium azide) during the sample pretreatment, and inaccurate data can be obtained due to the NO_2^- in the sample, as with microbial denitrification.

More recently, besides IRMS, measurements of $\delta^{15}N$ and $\delta^{18}O$ from the headspace N_2O gas are analyzed in a N_2O triple isotope analyzer (N_2OIA-23e-EP Model 914-0060; Los Gatos Research, Mountain View, CA, USA) using laser absorption spectroscopy after N_2O produced by conversion of NO_3^- by earlier mentioned pretreatment [86,87]. The instrument measures N_2O concentrations (0.3–20 ppm), and $\delta^{15}N_{\alpha}$, $\delta^{15}N_{\beta}$, $\delta^{15}N_{bulk}$, $\delta^{17}O$, $\delta^{18}O$, and H_2O values in air to precisions of 0.03 ppb for N_2O, less than ±1‰(SEM) for N and less than ±2‰ (SEM) for O isotopes over 300 s of measurement integration [86]. However, the laser spectrometry technique is lower precision and accuracy than IRMS technique (less than 0.2‰ for $\delta^{15}N$-NO_3^-, 0.5‰ for $\delta^{18}O$-NO_3^-) [88].

4.2. Identification of Contaminant Source Using Nitrate Isotopes

As previously mentioned, the analyzed nitrate $\delta^{15}N$ and $\delta^{18}O$ isotope ratios provide distinct values for each contaminant source. The value of $\delta^{15}N$ in atmospheric NO_3^- is usually in the range of −15‰ to +15‰ [9,89]. This large range is due to complex chemical reactions of nitrates or related compounds in the atmosphere, seasons, meteorological conditions, types of anthropogenic inputs, proximity to pollution sources, distance from the ocean, etc. [90]. Synthetic nitrogen fertilizers have $\delta^{15}N$ values in the range of −4‰ to +4‰ [79] and the $\delta^{15}N$ value of nitrogen in the soil ranges from −2‰ to +5‰. However, manure and sewage can be more enriched in ^{15}N due to volatilization of ^{15}N-depleted ammonia, and oxidation of much of the residual waste may result in high $\delta^{15}N$ of nitrate [79]. By this process, the $\delta^{15}N$ value becomes significant with a range of +10‰ to +20‰ [91,92]. Hence, the $\delta^{15}N$ is an important indicator of nitrates in the atmosphere, fertilizers, soil, manure and sewage. However, the identification of nitrogen sources and cycles using $\delta^{15}N$ values alone is limited because the ranges of values from precipitation, soil, fertilizer, manure, and sewage show substantial overlap (Figure 2). The analysis is therefore used in combination with $\delta^{18}O$, another indicator for identifying and separating sources of nitrates, in order to reduce the uncertainty of nitrogen isotopes in the identification [3,18,79,93–96].

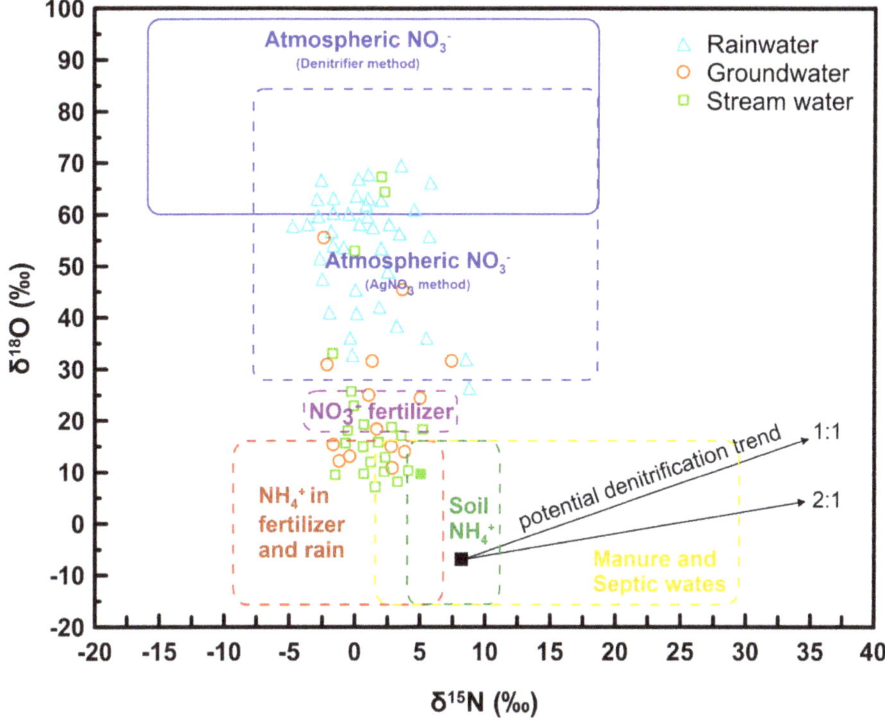

Figure 2. A plot of the δ^{15}N and δ^{18}O values of nitrate from various N sources. The nitrate in stream samples (green squares) was largely derived from groundwater sources. This diagram was modified from Kendall et al. [96] and data from Kendall et al. [79].

The conventional theory asserts that one oxygen atom of newly generated nitrate in soil is derived from dissolved atmospheric oxygen (O_2) and the other two oxygen atoms are from the surrounding water bodies [12,18,79,94,97–100]. If these oxygens are included without any fractionation, and the δ^{18}O values of water and atmospheric sources are known, the δ^{18}O value of microbial nitrate can be calculated as follows:

$$\delta^{18}O_{NO_3} = 2/3\ \delta^{18}O_{H_2O} + 1/3\ \delta^{18}O_{O_2} \tag{11}$$

While the δ^{18}O values of atmospheric-derived nitrates are usually high, between +20‰ and +70‰ [101], the δ^{18}O values of synthetic nitrate fertilizer are 22 ± 3‰; those of soil nitrogen transformed from ammonium via nitrification are between −10 and +10‰; and those of manure and sewage are below 15‰ [79]. As such, nitrate shows distinct isotopic composition of nitrogen and oxygen for each contaminant source, which is useful for contaminant source identification. In addition to identifying contaminant source, it can also be used to identify the contribution of contaminant sources using previously mentioned hydrograph separation.

However, the δ^{15}N and δ^{18}O values of nitrate are altered by isotopic fractionation due to mineralization, absorption/desorption, nitrification, denitrification, volatilization, assimilation (uptake), and leaching from the soil zone [3]. Common microbial organisms preferentially use the lighter isotopes (^{14}N and ^{16}O) over the heavier (^{15}N and ^{18}O), so that the microbial products are isotopically depleted and the residual nitrates are enriched in ^{15}N and ^{18}O [3]. For example, when microbial organisms convert nitrate to nitrogen gases N_2O (denitrification), the formed nitrogen gases are lighter than the remaining nitrates (low δ^{15}N and δ^{18}O). Therefore, denitrification causes increases in the δ^{15}N and δ^{18}O values of the residual nitrates, and the enrichment ratios of δ^{15}N and δ^{18}O are positively correlated by a factor of between 1.3:1 and 2.1:1 [2,17,79,101–104]. This indicates that, even if isotope fractionation by denitrification occurs, the initial isotope composition can be estimated by knowing the enrichment factor [105].

4.3. Movements of Nitrate from Surface to Stream

After reaching the ground, precipitation moves from the surface to the stream, which gradually alters the water isotope composition [42]. These processes typically involve two flow pathways, which are direct and indirect. The direct pathway is the runoff of surface water from rainfall or melting snowpack into the stream water, while the indirect pathway is the vertical movement of dissolved nitrate through the soil profile into the groundwater, after which the groundwater can be flushed out and contribute to the stream water [16,38]. In these processes, the potential sources of nitrates in stream water are atmospheric via rainfall and snowmelt [3], mineralization of soils under snowpacks [106], groundwater [107–109], nitrification [3,12,93], or a combination of these [110]. As mentioned previously, the δ^{18}O values of nitrates from atmospheric sources differ significantly from those of groundwater nitrates originating from nitrification in the soil. Thus, if rainfall and surface water run off directly to the stream, the δ^{18}O value of nitrate is similar to that of the atmospheric source. However, if precipitation is infiltrated into the soil layer and then recharged to groundwater and released into the stream water, the isotope composition of the stream water is similar to that of the groundwater or soil water. As shown in Figure 2, if the isotopic composition of rainwater and groundwater is determined, the typical isotope values of nitrogen and oxygen can be used to identify the source of stream water and the relative contribution rate.

Many previous studies have shown that groundwater (old water) via indirect pathways is the dominant source for stream water (Table 1). By examining the δ^{15}N and δ^{18}O values of nitrate, Kendall et al. [3] concluded that the main source of nitrates in stream water is groundwater, and that a direct contribution of atmospheric-derived nitrate from the snowpack to the stream is a relatively minor source. Ohte et al. [111] studied the nitrate sources of a headwater stream at the Sleepers River Research Watershed in Vermont, USA, during snowmelt using the δ^{18}O values of nitrate with precipitation, soil water, and groundwater as the three end members. The results indicated that, as the groundwater was recharged by meltwater and precipitation during snowmelt, the input to the groundwater gradually increased to eventually make it the dominant source of nitrate. As shown in Figure 3, Piatek et al. [109] analyzed δ^{15}N and δ^{18}O values of nitrate in the stream and compared them to those of snow and groundwater in the Arbutus Watershed of New York State, NY, USA, to indicate that stream water, atmospherically-derived solutions, and groundwater had overlapping nitrate δ^{15}N values. However, while the δ^{18}O values of nitrates displayed similar ranges in stream water and groundwater, these values were significantly lower than those of atmospheric solutions. In addition to these studies, Barnes et al. [112] demonstrated a seasonal variation in the rate of nitrate contribution from atmospheric sources and calculated that, on average, 1–3% of the summer and 10–18% of the winter/spring exported stream NO_3^- is derived from direct atmospheric deposits. Such information is important to the development of efficient and successful abatement strategies that may include ecosystem management, controls on NO_x emissions and possible correlations of nitrogen exports with climate change [109]. Moreover, δ^{15}N and δ^{18}O values of nitrate are useful for identifying the source of nitrate and flowpath process using hydrograph separation because they have distinct isotope values for each source (precipitation, groundwater, soil, etc.).

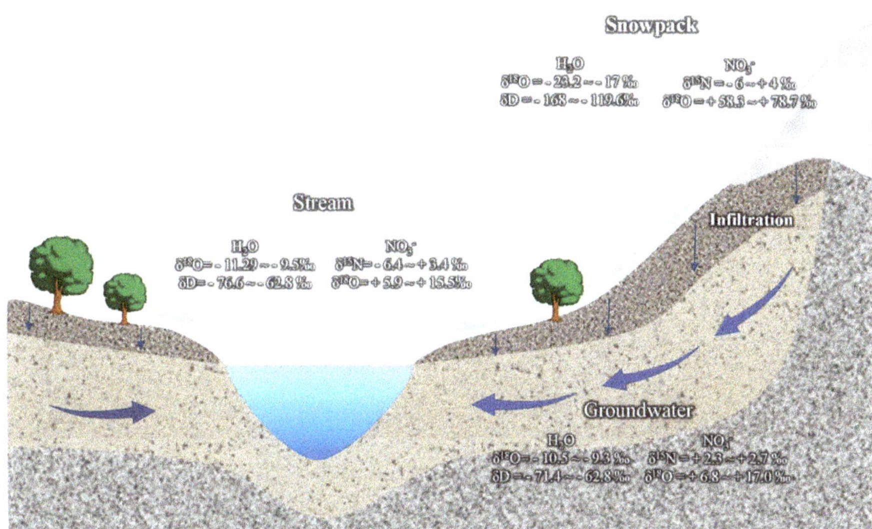

Figure 3. Isotopic compositions of water and nitrate in snowpack, groundwater, stream (data from Kendall [79]; Piatek et al. [109]).

Table 1. Summary of studies that account for more than two different end-members in hydrograph separation using nitrate isotopic tracer.

Location	End-Member (δ^{15}N-NO$_3^-$, δ^{18}O-NO$_3^-$)	Groundwater (Nitrified Sources) Fraction in Stream Water	Reference
Bavaria, Germany	Atmospheric; Nitrification	84–70%	[18]
Catskill Mountains, New York State, USA; Rocky Mountain National Park, Colorado, USA; Danville, Vermont, USA	Snowmelt; Nitrification	Nitrified sources dominant	[3]
Turkey Lakes watershed, Ontario, Canada	Atmospheric; Nitrification	70%	[113]
Catskill Mountains, New York, USA	Precipitation; Snowmelt; Soil water	Soil water dominant	[93]
Loch Vale watershed, Colorado, USA	Nitrification	>75%	[12]
Sleepers River Research Watershed, Vermont, USA	Precipitation; Groundwater; Soil water	Groundwater dominant	[111]
New Hampshire, USA	Precipitation; Nitrification	55–100%	[94]
Arbutus Watershed, New York State, USA	Wet deposition; Groundwater	Groundwater dominant during late winter/early spring	[109]
Green Mountains, Vermont, USA	Precipitation; Soil water	Soil water dominant during snowmelt periods	[114]
Connecticut and Massachusetts, USA	Microbially produced; Atmospheric deposition	Summer 97–99% Winter/Spring 82–90%	[112]
Pennsylvania, USA	Atmospheric sources; Microbial soil nitrification	67%	[115]

Table 1. *Cont.*

Location	End-Member (δ^{15}N-NO$_3^-$, δ^{18}O-NO$_3^-$)	Groundwater (Nitrified Sources) Fraction in Stream Water	Reference
Hubbard Brook Experimental Forest, New Hampshire, USA	Precipitation; Nitrification	66–71% during summer rainfall event	[116]
NMR above-ground streams, Pittsburgh, USA	Atmospheric; Sewage (δ^{15}N: 0‰ to +20‰; δ^{18}O: −15‰ to +15‰)	<66% sewage-derived	[117]
Savannah River, South Carolina, USA	Throughfall; Trench (soil) water; Groundwater	Groundwater predominant	[118]
Savannah River, South Carolina, USA	Atmospheric; Groundwater	Watershed B: 72% Watershed R and C: 90%	[119]

4.4. Implications of the Flowpath of Water and Nitrates

The stable isotope of nitrate (δ^{15}N and δ^{18}O) can be used to trace the nitrate sources in water bodies because nitrate contaminants usually have distinct isotope compositions [7]. In order to increase the reliability of contaminant tracking, there is a need for a multilateral investigation of precipitation, land-use type and area utilization rates, synthetic fertilizers, animal wastes, the presence of point sources (septic tanks and landfills), and the presence of sewer systems. In addition, hydrogeological data such as groundwater flow rate and direction, aquifer geometry, matrix characteristics, nitrate concentrations, electrical conductivity (EC), redox potential (Eh), and dissolved oxygen (DO) can be used to assess variations in the level of contaminants as well as for tracking contaminant sources.

While isotope analysis is a useful tool for tracking nitrate contaminants, isotope fractionation by nitrification, denitrification, and the presence of multiple contaminants continue to make this difficult. Hence, the use of water stable isotope analysis in combination with the isotopic composition of nitrates may improve the reliability of source identification.

5. Summary and Perspectives

Nitrate contamination of stream water has become an environmental problem of global concern [101]. To identify the nitrate source is an effective approach to controlling discharge and emissions of nitrate contamination of stream water. In recent decades, dual nitrate isotope analysis (δ^{15}N and δ^{18}O) has been used as a useful tool for identifying the source and flowpath of nitrate contaminants in water bodies. We have tried to demonstrate in this paper an understanding of the identification water sources and flowpaths process, and the proportion of various sources contributing to stream water via water and nitrate stable isotope technique. However, the application of this method has some limitations due to the multiple nitrogen sources and the influence of isotopic fractionation [101]. In details, nitrates are subjected to multiple physical, chemical, and biological fractionation processes during transport from the original nitrate source to water bodies, and these reactions are influenced by such factors as land-use types, climate, and hydrogeological conditions. Besides, the stable isotope values of nitrate vary according to country or region due to the various regional conditions. To enable the quick and accurate analysis of nitrogen contaminant sources for water bodies, it is therefore suggested that data on the stable isotope values of nitrate from various contaminant sources should be collected in order to establish a global and regional isotope database. For identifying the contaminant sources and tracing the flowpath, it is therefore of great significance to study the influencing factors and transformation processes of nitrates.

More recently, quantification of the relative contributions of nitrate can be improved if other isotope (B, Sr, S, C, Li, U) or chemical tracers [96]. The isotopic signature of boron (δ^{11}B) in association with the nitrates has been demonstrated [120–125]. Strontium and sulphate isotopes give additional information on the sources of contaminant [126,127]. In particular, combined use of boron isotopes with nitrate (δ^{15}N and δ^{18}O) can be a useful tool for nitrate source contributions [120–122,127–129]. Moreover, different nitrate sources can show distinct δ^{11}B values and different processes control the isotopic composition of boron and nitrate [120,127]. Moreover, stable isotopes of dissolved nitrates indicate the absence of denitrification, while the coupled use of boron isotopes evidences, even in rural areas, a contribution from septic effluents [130]. Therefore, the combined use of $\delta^{15}N_{NO3}$, $\delta^{18}O_{NO3}$, and $\delta^{11}B$ is an effective approach to the differentiation of complex NO_3^- sources, assuming that these compounds co-migrate in many environments [7,120,121]. In natural waters, the boron isotopic composition is controlled by the aquifer matrix; the anthropogenic source may be a variable of $\delta^{11}B$ [130]. For example, detergents obtained from evaporites, manure, fertilizers, and organic wastes have high concentrations of boron and distinct $\delta^{11}B$ values [122,128,130–134].

Likewise, there are many effective multi-isotopic toolboxes for identifying the flowpath and the contaminant source of nitrate. In particular, we discuss hydrograph separation using the oxygen and hydrogen isotopes of water in combination with the nitrogen and oxygen isotopes of nitrate to understand the relative contributions of precipitation and groundwater to stream water. While transformation processes of nitrogen compounds can change the isotopic composition of nitrate due to the various redox processes in the environment, the use of the stable water isotopes of the H_2O itself can be used to interpret the multiple hydrological and hydrochemical processes for the movement of nitrate contaminants. Therefore, this study will assist in understanding the groundwater flowpaths as well as tracking the sources of nitrate contamination using the stable isotope analysis in combination with nitrate and water. This suggests that source and process information relating to groundwater and nitrates should be made part of the decision-making process in order to better understand and effectively manage the hydrological and nitrogen cycles.

Author Contributions: H.J. was responsible for the implementation of the reference collection, processing and writing of the manuscript. D.-C.K. provided constructive comments and funding. Y.S.K. and S.-W.J. provided constructive comments for this manuscript. J.L. was responsible for the designing and analyzing of the work. All authors have read and agreed to the published version of the manuscript.

Funding: This research was supported by a grant from the National Research Council of Science & Technology (NST), funded by the Korea government (MSIP) (CAP-17-05-KIGAM). This research was partially funded by the Basic Research Program through the National Foundation of Korea (NRF), which was funded by the Ministry of Education (NRF-2017R1D1A1A09000732).

Acknowledgments: Inputs from two anonymous reviewers improved the quality of this paper.

Conflicts of Interest: The authors declare no conflict of interest.

References

1. Modica, E.; Buxton, H.T.; Plummer, L.N. Evaluating the source and residence times of groundwater seepage to streams, New Jersey Coastal Plain. *Water Resour. Res.* **1998**, *34*, 2797–2810. [CrossRef]
2. Kaushal, S.S.; Groffman, P.M.; Brand, L.E.; Elliott, E.M.; Shields, C.A.; Kendall, C. Tracking Nonpoint Source Nitrogen Pollution in Human-Impacted Watersheds. *Environ. Sci. Technol.* **2011**, *45*, 8225–8232. [CrossRef] [PubMed]
3. Kendall, C.; Campbell, D.H.; Burns, D.A.; Shanley, J.B.; Silva, S.R.; Chang, C.C.Y. Tracing sources of nitrate in snowmelt runoff using the oxygen and nitrogen isotopic compositions of nitrate. In *Biogeochemistry of Seasonally Snow-Covered Catchments*; Tonnessen, K., Williams, M., Tranter, M., Eds.; IAHS Publication: Boulder, CO, USA, 1995; pp. 339–347.
4. Alimi, H.; Ertel, T.; Schug, B. Fingerprinting of hydrocarbon fuel contaminants: Literature review. *Environ. Forensics* **2003**, *4*, 25–38. [CrossRef]
5. McGuire, K.; McDonnell, J.J. Stable isotope tracers in watershed hydrology. In *Stable Isotopes in Ecology and Environmental Science*, 2st ed.; Michener, R., Lajtha, K., Eds.; Wiley/Blackwell: Malden, MA, USA, 2007; pp. 334–540.
6. Clark, I.D.; Fritz, P. *Environmental Isotopes in Hydrogeology*, 2st ed.; CRC Press: New York, NY, USA, 1997; pp. 13–168.
7. Xue, D.; Botte, J.; De Baets, B.; Accoe, F.; Nestler, A.; Taylor, P.; Van Cleemput, O.; Berglund, M.; Boeckx, P. Present limitations and future prospects of stable isotope methods for nitrate source identification in surface and groundwater. *Water Res.* **2009**, *43*, 1159–1170. [CrossRef]

8. Craig, H. Isotopic Variations in Meteoric Waters. *Science* **1961**, *133*, 1702–1703. [CrossRef]

9. Heaton, T.H.E. Isotopic studies of nitrogen pollution in the hydrosphere and atmosphere: A review. *Chem. Geol. (Isot. Geosci. Sect.)* **1986**, *59*, 87–102. [CrossRef]

10. Smith, V.H.; Tilman, G.D.; Nekola, J.C. Eutrophication: Impacts of excess nutrient inputs on freshwater, marine, and terrestrial ecosystems. *Environ. Pollut.* **1999**, *100*, 179–196. [CrossRef]

11. Liu, C.Q.; Li, S.L.; Lang, Y.C.; Xiao, H.Y. Using δ^{15}N- and δ^{18}O-Values to Identify Nitrate Sources in Karst Ground Water, Guiyang, Southwest China. *Environ. Sci. Technol.* **2006**, *40*, 6928–6933. [CrossRef]

12. Campbell, D.H.; Kendall, C.; Chang, C.C.Y.; Silva, S.R. Pathways for nitrate release from an alpine watershed: Determination using d^{15}N and d^{18}O. *Water Resour. Res.* **2002**, *38*, 1052. [CrossRef]

13. Buttle, J.M.; Sami, K. Recharge processes during snowmelt: An isotopic and hydrometric investigation. *Hydrol. Process.* **1990**, *4*, 343–360. [CrossRef]

14. Wels, C.; Cornett, R.J.; Lazerte, B.D. Hydrograph separation: A comparison of geochemical and isotopic tracers. *J. Hydrol.* **1991**, *122*, 253–274. [CrossRef]

15. Taylor, S.; Feng, X.; Williams, M.; McNamara, J. How isotopic fractionation of snowmelt affects hydrograph separation. *Hydrol. Process.* **2002**, *16*, 3683–3690. [CrossRef]

16. Kim, H.; Cho, S.H.; Lee, D.; Jung, Y.Y.; Kim, Y.H.; Koh, D.C.; Lee, J. Influence of pre-event water on streamflow in a granitic watershed using hydrograph separation. *Environ. Earth Sci.* **2017**, *76*, 82. [CrossRef]

17. Böttcher, J.; Strebel, O.; Voerkelius, S.; Schmidt, H.L. Using isotope fractionation of nitrate-nitrogen and nitrate-oxygen for evaluation of microbial denitrification in a sandy aquifer. *J. Hydrol.* **1990**, *114*, 413–424. [CrossRef]

18. Durka, W.; Schulze, E.D.; Gebauer, G.; Voerkeliust, S. Effects of forest decline on uptake and leaching of deposited nitrate determined from ^{15}N and ^{18}O measurements. *Nature* **1994**, *372*, 765–767. [CrossRef]

19. Sklash, M.G.; Farvorden, R.N. The role of groundwater in storm runoff. *J. Hydrol.* **1979**, *43*, 45–65. [CrossRef]

20. Ladouche, B.; Probst, A.; Viville, D.; Idir, S.; Baqué, D.; Loubet, M.; Probst, J.; Bariac, T. Hydrograph separation using isotopic, chemical and hydrological approaches (Strengbach Catchment, France). *J. Hydrol.* **2001**, *242*, 255–274. [CrossRef]

21. Pu, T.; He, Y.; Zhu, G.; Zhang, N.; Du, J.; Wang, C. Characteristics of water stable isotopes and hydrograph separation in Baishui catchment during the wet season in Mt. Yulong region, south western China. *Hydrol. Process.* **2013**, *27*, 3641–3648. [CrossRef]

22. Dahlke, H.E.; Lyon, S.W.; Jansson, P.; Karlin, T.; Rosqvist, G. Isotopic investigation of runoff generation in a glacierized catchment in northern Sweden. *Hydrol. Process.* **2014**, *28*, 11383–11398. [CrossRef]

23. Li, Z.; Feng, Q.; Jianguo, L.; Yanhui, P.; Tingting, W.; Liu, L.; Guo, X.; Gao, Y.; Jia, B.; Guo, R. Environmental significance and hydrochemical processes at a cold alpine basin in the Qilian Mountains. *Environ. Earth Sci.* **2015**, *73*, 4043–4052. [CrossRef]

24. Rahman, K.; Besacier-Monbertrand, A.L.; Castella, E.; Lods-Crozet, B.; Ilg, C.; Beguin, O. Quantification of the daily dynamics of streamflow components in a small alpine watershed in Switzerland using end member mixing analysis. *Environ. Earth Sci.* **2015**, *74*, 4927–4937. [CrossRef]

25. Semenov, M.Y.; Zimnik, E.A. A three-component hydrograph separation based on relationship between organic and inorganic component concentrations: A case study in Eastern Siberia, Russia. *Environ. Earth Sci.* **2015**, *73*, 611–620. [CrossRef]

26. Lee, J.; Lee, H.J.; Ham, J.Y.; Kim, H. Chemical separation of the discharge generated by artificial rain-on-snow experiments in a snowpack. *J. Geol. Soc. Korea* **2016**, *52*, 113–120. [CrossRef]

27. Unnikrishna, P.V.; McDonnell, J.J.; Kendall, C. Isotope variations in a Sierra Nevada snowpack and their relation to meltwater. *J. Hydrol.* **2002**, *260*, 38–57. [CrossRef]

28. Laudon, H.; Hemond, H.F.; Krouse, R.; Bishop, K.H. Oxygen 18 fractionation during snowmelt: Implications for spring flood hydrograph separation. *Water Resour. Res.* **2002**, *38*, 1258. [CrossRef]

29. Worden, J.; Noone, D.; Bowman, K. Importance of rain evaporation and conti-nental convection in the tropical water cycle. *Nature* **2007**, *445*, 528–532. [CrossRef] [PubMed]

30. Yuan, F.; Miyamoto, S. Characteristics of oxygen-18 and deuterium compositionin waters from the Pecos River in America Southwest. *Chem. Geol.* **2008**, *255*, 220–230. [CrossRef]

31. Lee, J.; Feng, X.; Faiia, A.M.; Posmentier, E.S.; Kirchner, J.W.; Osterhuber, R.; Taylor, S. Isotopic evolution of a seasonal snowcover and its melt by isotopic exchange between liquid water and ice. *Chem. Geol.* **2010**, *270*, 126–134. [CrossRef]

32. Dincer, T.; Payne, B.R.; Florkowski, T.; Marthinec, J.; Tongiorgi, E. Snowmelt runoff from measurements of tritium and oxygen-18. *Water Resour. Res.* **1970**, *6*, 110–124. [CrossRef]

33. Taylor, S.; Feng, X.; Kirchner, J.W.; Osterhuber, R.; Klaue, B.; Renshaw, C.E. Isotopic evolution of a seasonal snowpack and its melt. *Water Resour. Res.* **2001**, *37*, 759–769. [CrossRef]

34. Phillips, D.L.; Gregg, J.W. Source partitioning using stable isotopes: Coping with too many sources. *Oecologia* **2003**, *136*, 261–269. [CrossRef] [PubMed]

35. Benstead, J.P.; March, J.G.; Fry, B.; Ewel, K.C.; Pringle, C.M. Testing isosource: Stable isotope analysis of a tropical fishery with diverse organic matter sources. *Ecology* **2006**, *87*, 326–333. [CrossRef]

36. Meng, Z.; Yang, Y.; Qin, Z.; Huang, L. Evaluating Temporal and Spatial Variation in Nitrogen Sources along the Lower Reach of Fenhe River (Shanxi Province, China) Using Stable Isotope and Hydrochemical Tracers. *Water* **2018**, *10*, 231. [CrossRef]

37. Genereux, D. Quantifying uncertainty in tracer-based hydrograph separation. *Water Resour. Res.* **1998**, *34*, 915–919. [CrossRef]

38. Lee, J.; Koh, D.C.; Choo, M.K. Influences of Fractionation of Stable Isotopic Composition of Rain and Snowmelt on Isotopic Hydrograph Separation. *J. Korean Earth Sci. Soc.* **2014**, *35*, 97–103. [CrossRef]

39. Hoefs, J. *Stable Isotope Geochemistry*, 4th ed.; Springer: Berlin, Germany, 1997.

40. Coplen, T.B.; Herczeg, A.L.; Barnes, C. Isotope engineering: Using stable isotopes of the water molecule to solve practical problems. In *Environmental Tracers in Subsurface Hydrology*; Cook, P.G., Herczeg, A.L., Eds.; Kluwer Academic Publishers: Boston, MA, USA, 2000.

41. Ingraham, N.L. Isotopic variation in precipitation. In *Isotope Tracers in Catchment Hydrology*; Kendall, C., McDonnell, J.J., Eds.; Elsevier: Amsterdam, The Netherlands, 1998; pp. 87–118.

42. Gat, J.R. Oxygen and Hydrogen Isotopes in the Hydrologic Cycle. *Annu. Rev. Earth Planet. Sci.* **1996**, *24*, 225–262. [CrossRef]

43. McGuire, K.; McDonnell, J.J. Stable isotope tracers in watershed hydrology. In *Stable Isotopes in Ecology and Environmental Science*, 2st ed.; Michener, R., Lajtha, K., Eds.; Wiley/Blackwell: Malden, MA, USA, 2007; pp. 334–540.

44. Ingraham, N.L.; Zukosky, K.; Kreamer, D.K. Application of Stable Isotopes to Identify Problems in Large-Scale Water Transfer in Grand Canyon National Park. *Environ. Sci. Technol.* **2001**, *35*, 1299–1302. [CrossRef]

45. Friedman, I. Deuterium content of natural waters and other substances. *Geochim. Cosmochim. Acta* **1953**, *4*, 89–103. [CrossRef]

46. Dansgaard, W. Stable isotopes in precipitation. *Tellus* **1964**, *16*, 436–468. [CrossRef]

47. Blasch, K.W.; Bryson, J.R. Distinguishing Sources of Ground Water Recharge by Using δ^2H and $\delta^{18}O$. *Groundwater* **2007**, *45*, 294–308. [CrossRef]

48. Merlivat, L.; Jouzel, J. Global climatic interpretation of the deuterium–oxygen 18 relationship for precipitation. *J. Geophys. Res.* **1979**, *84*, 5029–5033. [CrossRef]

49. Johnsen, S.J.; Dansgaard, W.; White, J.W.C. The origin of Arctic precipitation under present and glacier conditions. *Tellus* **1989**, *41*, 452–468. [CrossRef]

50. Feng, X.; Faiia, A.M.; Posmentier, E.S. Seasonality of isotopes in precipitation: A global perspective. *J. Geophys. Res.* **2009**, *114*, D08116. [CrossRef]

51. Koeniger, P.; Hubbart, J.A.; Link, T.; Marshall, J.D. Isotopic variation of snow cover and streamflow in response to changes in canopy structure in a snow-dominated mountain catchment. *Hydrol. Process.* **2008**, *22*, 557–566. [CrossRef]

52. Lundquist, J.D.; Cayan, D.R. Seasonal and spatial patterns in diurnal cycles in streamflow in the western united states. *J. Hydrometeorol.* **2002**, *3*, 591–603. [CrossRef]

53. Cooper, L.W. Isotopic fractionation in snow cover. In *Isotope Tracers in Catchment Hydrology*; Kendall, C., McDonnell, J.J., Eds.; Elsevier: Amsterdam, The Netherlands, 1998; pp. 119–136.

54. Lee, J.; Feng, X.; Posmentier, E.S.; Faiia, A.M.; Taylor, S. Stable isotopic exchange rate constant between snow and liquid water. *Chem. Geol.* **2009**, *260*, 57–62. [CrossRef]

55. Lee, J. A review on stable isotopic variations of a seasonal snowpack and meltwater. *J. Geol. Soc. Korea* **2014**, *50*, 671–679. [CrossRef]

56. Moser, H.; Stichler, W. Use of environmental isotope methods as a reconnaissance tool in groundwater exploration near San Antonio de Pichincha, Ecuador. *Water Resour. Res.* **1975**, *11*, 501–505. [CrossRef]

57. Talyor, S.; Feng, X.; Kirchner, J.W.; Osterhuber, R.; Klaue, B.; Renshaw, C.E. Isotopic evolution of a seasonal snowpack and its melt. *Water Resour. Res.* **2001**, *37*, 759–769. [CrossRef]

58. Sommerfield, R.A.; Judy, C.; Friedman, I. Isotopic changes during the formation of depth hoar in experimental snowpacks. In *Stable Isotope Geochemistry: A Tribute to Sam Epstein*; Taylor, H.P., O'Neil, J.R., Kaplan, I.R., Eds.; The Geochemical Society: San Antonio, TX, USA, 1991; pp. 205–210.

59. Friedman, I.; Benson, C.; Gleason, J. Isotopic changes during snow metamorphism. In *Stable Isotope Geochemistry: A Tribute to Sam Epstein*; Taylor, H.P., O'Neil, J.R., Kaplan, I.R., Eds.; The Geochemical Society: San Antonio, TX, USA, 1991; pp. 211–221.

60. Rodhe, A. Spring flood: Meltwater or groundwater? *Nord. Hydrol.* **1981**, *12*, 21–30. [CrossRef]

61. Craig, H.; Gordon, L.I. Deuterium and oxygen-18 variations in the ocean and the marine atmosphere. In *Proceedings of the Conference on Stable Isotopes in Oceanographic Studies and Paleotemperatures. Spoleto, Italy*; Tongiorgi, E., Ed.; Consiglio nazionale delle richerche, Laboratorio de geologia nucleare: Pisa, Italy, 1965; pp. 9–130.

62. Mast, M.A.; Kendall, C.; Campbell, D.H.; Clow, D.W.; Back, J. Determination of hydrologic pathways in an alpine-subalpine basin using isotopic and chemical tracers. In *Biogeochemistry of Seasonally Snow Covered Catchments*; IAHS Publ. 228; Tonnessen, K., Williams, M., Trantner, M., Eds.; IAHS Publication: Boulder, CO, USA, 1995; pp. 263–270.

63. Kendall, C.; Coldwell, E.A. Fundamentals of Isotope Geochemistry. In *Isotope Tracers in Catchment Hydrology*; Kendall, C., McDonnell, J.J., Eds.; Elsevier: Amsterdam, The Netherlands, 1998; pp. 51–86.

64. Jung, Y.Y.; Koh, D.C.; Ko, K.S.; Lee, J. Applications of Isotope Ratio Infrared Spectroscopy (IRIS) to Analysis of Stable Isotopic Compositions of Liquid Water. *Econ. Environ. Geol.* **2013**, *46*, 495–508. [CrossRef]

65. Bigeleisen, J.; Perlman, M.J.; Prosser, H. Conversion of hydrogenic materials for hydrogen to isotopic analysis. *Anal. Chem.* **1952**, *24*, 1356–1357. [CrossRef]

66. Epstein, S.; Mayada, T.K. Variations of O-18 content of waters from natural sources. *Geochim. Cosmochim. Ac.* **1953**, *4*, 213–224. [CrossRef]

67. Horita, J. Hydrogen isotope analysis of natural waters using an H_2-water equilibration method: A special implication to brines. *Chem. Geol.* **1988**, *72*, 89–94. [CrossRef]

68. Gehre, M.; Höfling, R.; Kowski, P.; Strauch, G. Sample preparation device for quantitative hydrogen isotope analysis using chromium metal. *Anal. Chem.* **1996**, *68*, 4414–4417. [CrossRef]

69. Coplen, T.B.; Wildman, J.D.; Chen, J. Improvements in the gaseous hydrogen-water equilibration technique for hydrogen isotope-ratio analysis. *Anal. Chem.* **1991**, *63*, 910–912. [CrossRef]

70. Socki, R.A.; Romanek, C.S.; Gibson, E.K. Online technique for measuring stable oxygen and hydrogen isotopes from microliter quantities of water. *Anal. Chem.* **1999**, *71*, 2250–2253. [CrossRef]

71. Kerstel, E. *Handbook of Stable Isotope Analytical Techniques*, 1st ed.; Elsevier Science: Amsterdam, The Netherlands, 2004; pp. 759–787.

72. Berden, G.; Peeters, R.; Meijer, G. Cavity ringdown spectroscopy: Exerimental schemes and applications. *Int. Rev. Phys. Chem.* **2000**, *19*, 565–607. [CrossRef]

73. Wassenaar, L.I.; Hendry, M.J.; Chostner, V.L.; Lis, G.P. High resolution pore water δ^2H and δ^{18}O measurements by H_2O(liquid)-H_2O(vapor) equilibration laser spectroscopy. *Environ. Sci. Technol.* **2008**, *42*, 9262–9267. [CrossRef]

74. Berman, E.S.F.; Gupta, M.; Gabrielli, C.; Garland, T.; McDonnell, J.J. High-frequency field deployable isotope analyzer for hydrological applications. *Water Resour. Res.* **2009**, *45*, W10201. [CrossRef]

75. Gupta, P.; Noone, D.; Galewsky, J.; Sweeney, C.; Vaughn, B.H. Demonstration of high-precision continuous measurements of water vapor isotopologues in laboratory and remote field deployments using wavelength-scanned cavity ring-down spectroscopy (WS-CRDS) technology. *Rapid Commun. Mass Spectrom.* **2009**, *23*, 2534–2542. [CrossRef] [PubMed]

76. Hendry, M.J.; Wassenaar, L.I. Inferring heterogeneity in aquitards using high-resolution δD and δ^{18}O profiles. *Ground Water* **2009**, *47*, 639–645. [CrossRef] [PubMed]

77. Penna, D.; Stenni, B.; Sanda, M.; Wrede, S.; Bogaard, T.A.; Gobbi, A.; Borga, M.; Fischer, B.M.C.; Bonazza, M.; Charova, Z. On the reproducibility and repeatability of laser absorption spectroscopy measurements for δ^2H and δ^{18}O isotopic analysis. *Hydrol. Earth Syst. Sci. Discuss.* **2010**, *14*, 1551–1566. [CrossRef]

78. Brand, W.A.; Geilmann, H.; Crosson, E.R.; Rella, C.W. Cavity ring-down spectroscopy versus hightemperature conversion isotope ratio mass spectrometry: A case study on δ^2H and δ^{18}O of pure water samples and alcohol/water mixture. *Rapid Commun. Mass Spectrom.* **2009**, *23*, 1879–1884. [CrossRef]

79. Kendall, C. Tracing nitrogen sources and cycling in catchments. In *Isotope Tracers in Catchment Hydrology*; Kendall, C., McDonnell, J.J., Eds.; Elsevier: Amsterdam, The Netherlands, 1998; pp. 519–576.

80. Silva, S.; Kendall, C.; Wilkison, D.; Ziegler, A.; Chang, C.C.; Avanzino, R. A new method for collection of nitrate from fresh water and the analysis of nitrogen and oxygen isotope ratios. *J. Hydrol.* **2000**, *228*, 22–36. [CrossRef]

81. Chang, C.C.Y.; Langston, J.; Riggs, M.; Campbell, D.H.; Silva, S.R.; Kendall, C. A method for nitrate collection for δ^{15}N and δ^{18}O analysis from waters with low nitrate concentrations. *Can. J. Fish. Aquat. Sci.* **1999**, *56*, 1856–1864. [CrossRef]

82. Wassenaar, L.I. Evaluation of the origin and fate of nitrate in the Abbotsford Aquifer using the isotopes of ^{15}N and ^{18}O in NO$_3^-$. *Appl. Geochem.* **1995**, *10*, 391–405. [CrossRef]

83. Sigman, D.M.; Casciotti, K.L.; Andreani, M.; Barford, C.; Galanter, M.; Böhlke, J.K. A Bacterial Method for the Nitrogen Isotopic Analysis of Nitrate in Seawater and Freshwater. *Anal. Chem.* **2001**, *73*, 4145–4153. [CrossRef]

84. Casciotti, K.L.; Sigman, D.M.; Galanter Hastings, M.; Böhlke, J.K.; Hilkert, A. Measurement of the Oxygen Isotopic Composition of Nitrate in Seawater and Freshwater Using the Denitrifier Method. *Anal. Chem.* **2002**, *74*, 4905–4912. [CrossRef]

85. McIlvin, M.R.; Altabet, M.A. Chemical Conversion of Nitrate and Nitrite to Nitrous Oxide for Nitrogen and Oxygen Isotopic Analysis in Freshwater and Seawater. *Anal. Chem.* **2005**, *77*, 5589–5595. [CrossRef]

86. Wassenaar, L.I.; Douence, C.; Altabet, M.; Aggarwal, P. N and O isotope (δ^{15}N$_\alpha$, δ^{15}N$_\beta$, δ^{18}O, δ^{17}O) analyses of dissolved NO$_3^-$ and NO$_2^-$ by the Cd-azide reduction method and N$_2$O laser spectrometry. *Rapid Commun. Mass Spectrom* **2018**, *32*, 184–194. [CrossRef] [PubMed]

87. Soto, D.X.; Koehler, G.; Wassenaar, L.I.; Hobson, K.A. Spatio-temporal variation of nitrate sources to Lake Winnipeg using N and O isotope (δ^{15}N, δ^{18}O) analyses. *Sci. Total Environ.* **2019**, *647*, 486–493. [CrossRef] [PubMed]

88. Liu, X.Y.; Koba, K.; Koyama, L.A.; Hobbie, S.E.; Weiss, M.S.; Inagaki, Y.; Shaver, G.R.; Giblin, A.E.; Hobara, S.; Nadelhoffer, K.J. Nitrate is an important nitrogen source for Arctic tundra plants. *Proc. Natl. Acad. Sci. USA* **2018**, *115*, 3398–3403. [CrossRef] [PubMed]

89. Hoering, T. The isotopic composition of the ammonia and the nitrate ion in rain *Geochim. Cosmochim. Ac.* **1957**, *12*, 97–102. [CrossRef]

90. Hübner, H.; Fritz, P.; Fontes, J.-C. *Handbook of Environmental Isotope Geochemistry: The Terrestrial Environment*; Elsevier: Amsterdam, The Netherlands, 1986; Volume 2, pp. 361–425.

91. Kreitler, C.W. *Determining the Source of Nitrate in Ground Water by Nitrogen Isotope Studies*; Report 83; Bureau of Economics and Geology, University of Texas: Austin, TX, USA, 1975; p. 57.

92. Kreitler, C.W. Nitrogen-isotope ratio studies or soil and ground-water nitrate from alluvial fan aquifers in Texas. *J. Hydrol.* **1979**, *42*, 147–170. [CrossRef]

93. Burns, D.A.; Kendall, C. Analysis of δ^{15}N and δ^{18}O to differentiate NO$_3^-$ sources in runoff at two watersheds in the Catskill Mountains of New York. *Water Resour. Res.* **2002**, *38*, WR000292. [CrossRef]

94. Pardo, L.H.; Kendall, C.; Pett-Ridge, J.; Chang, C.C.Y. Evaluating the source of streamwater nitrate using δ^{15}N and δ^{18}O in nitrate in two watersheds in New Hampshire, USA. *Hydrol Process.* **2004**, *18*, 2699–2712. [CrossRef]

95. Savard, M.M.; Somers, G.; Smirnoff, A.; Paradis, D.; van Bochove, E.; Liao, S. Nitrate isotopes unveil distinct seasonal N-sources and the critical role of crop residues in groundwater contamination. *J. Hydrol.* **2010**, *381*, 134–141. [CrossRef]

96. Kendall, C.; Elliott, E.M.; Wankel, S.D. Tracing anthropogenic inputs of nitrogen to ecosystems. In *Stable Isotopes in Ecology and Environmental Science*, 2st ed.; Michener, R., Lajtha, K., Eds.; Wiley/Blackwell: Malden, MA, USA, 2007; pp. 375–449.

97. Andersson, K.K.; Hooper, A.B. O$_2$ and H$_2$O and each the source of one O in NO$_2$ produced from NH$_3$ by *Nitrosomonas*: ^{15}N evidence. *FEBS Lett.* **1983**, *164*, 236–240. [CrossRef]

98. Kumar, S.; Nicholas, D.J.D.; Williams, E.H. Definitive ^{15}N NMR evidence that water serves as a source of 'O' during nitrite oxidation by *Nitrobacter agilis*. *FEBS Lett.* **1983**, *152*, 71–74. [CrossRef]

99. Hollocher, T.C. Source of the oxygen atoms of nitrate in the oxidation of nitrite by *Nitrobacter agilis*: Evidence against a P–O–N anhydride mechanism in oxidative phosphorylation. *Arch. Biochem. Biophys.* **1984**, *233*, 721–727. [CrossRef]

100. Voerkelius, S. Isotopendiskriminierungen bei der Nitrifikation und Denitrifikation: Grundlagen und Anwendungen der Herkunfts-Zuordnung von Nitrat und Disickstoffmonoxid. Ph.D. Thesis, Technische University at Munchen, München, Germany, 1990.

101. Zhang, Y.; Shi, P.; Song, J.; Li, Q. Application of Nitrogen and Oxygen Isotopes for Source and Fate Identification of Nitrate Pollution in Surface Water: A Review. *Appl. Sci.* **2019**, *9*, 18. [CrossRef]

102. Amberger, A.; Schmidt, H.L. Naturliche Isotopen-gehalte von nitrat als indikatoren fur dessen herkunft. *Geochim. Cosmochim. Acta* **1987**, *51*, 2699–2705. [CrossRef]

103. Voerkelius, S.; Schmidt, H.L. Natural oxygen and nitrogen isotope abundance of compounds involved in denitrification: Mitteilungen der Deut. *Bodenkundlichen Gesselschaft* **1990**, *60*, 364–366.

104. Burns, D.A.; Boyer, E.W.; Elliott, E.M.; Kendall, C. Sources and Transformations of Nitrate from Streams Draining Varying Land Uses: Evidence from Dual Isotope Analysis. *J. Environ. Qual.* **2009**, *38*, 1149–1159. [CrossRef]

105. Jeen, S.W.; Lee, H.; Kim, R.H.; Jeong, H.Y. A Review on Nitrate Source Identification using Isotope Analysis. *J. Soil Groundw. Environ.* **2017**, *22*, 1–12.

106. Zak, D.R.; Groffman, P.M.; Pregitzer, K.S.; Christensen, S.; Tiedje, J.M. The vernal dam: Plant-microbe competition for nitrogen in northern hardwood forests. *Ecology* **1990**, *71*, 651–656. [CrossRef]

107. Bottomley, D.J.; Craig, D.; Johnston, L.M. Oxygen-18 studies of snowmelt runoff in a small Precambrian shield watershed: Implications for streamwater acidification in acid-sensitive terrain. *J. Hydrol.* **1986**, *88*, 21–234. [CrossRef]

108. McHale, M.R.; McDonnell, J.J.; Mitchell, M.J.; Cirmo, C.P. A field based study of soil- and groundwater nitrate release in an Adirondack forested watershed. *Water Resour. Res.* **2002**, *38*, 1029–1038. [CrossRef]

109. Piatek, K.B.; Mitchell, M.; Silva, S.R.; Kendall, C. Sources of Nitrate in Snowmelt Discharge: Evidence from Water Chemistry and Stable Isotopes of Nitrate. *Water Air Soil Pollut.* **2005**, *165*, 13–35. [CrossRef]

110. Schleppi, P.; Hagedorn, F.; Providoli, I. Nitrate leaching from a mountain forest ecosystem with Gleysols subjected to experimentally increased N. deposition. *Water Air Soil Pollut.* **2004**, *4*, 453–467. [CrossRef]

111. Ohte, N.; Sebestyen, S.D.; Shanley, J.B.; Doctor, D.H.; Kendall, C.; Wankel, S.D.; Boyer, E.W. Tracing sources of nitrate in snowmelt runoff using a high-resolution isotopic technique. *Geophys. Res.* **2004**, *31*, L21506. [CrossRef]

112. Barnes, R.T.; Raymond, P.A.; Casciotti, K.L. Dual isotope analyses indicate efficient processing of atmospheric nitrate by forested watersheds in the northeastern U.S. *Biogeochemistry* **2008**, *90*, 15–27. [CrossRef]

113. Spoelstra, J.; Schiff, S.L.; Elgood, R.J.; Semkin, R.G.; Jeffries, D.S. Tracing the sources of exported nitrate in the Turkey Lakes Watershed using ^{15}N/^{14}N and ^{18}O/^{16}O isotopic ratios. *Ecosystems* **2001**, *4*, 536–544. [CrossRef]

114. Hales, H.C.; Ross, D.S.; Lini, A. Isotopic signature of nitrate in two contrasting watersheds of Brush Brook, Vermont, USA. *Biogeochemistry* **2007**, *84*, 51–66. [CrossRef]

115. Buda, A.R.; DeWalle, D.R. Dynamics of stream nitrate sources and flow pathways during stormflows on urban, forest and agricultural watersheds in central Pennsylvania, USA. *Hydrol. Process.* **2009**, *23*, 3305–3392. [CrossRef]

116. Wexler, S.K.; Goodalea, C.L.; McGuire, K.J.; Bailey, S.W.; Groffman, P.M. Isotopic signals of summer denitrification in a northern hardwood forested catchment. *Proc. Natl. Acad. Sci. USA* **2014**, *111*, 16413–16418. [CrossRef]

117. Divers, M.A.; Elliott, E.M.; Bain, D.J. Quantification of Nitrate Sources to an Urban Stream Using Dual Nitrate Isotopes. *Environ. Sci. Technol.* **2014**, *48*, 10580–10587. [CrossRef]

118. Klaus, J.; McDonnell, J.J.; Jackson, C.R.; Griffiths, N.A. Where does streamwater come from in low-relief forested watersheds? A dual-isotope approach. *Hydrol. Earth Syst. Sci.* **2015**, *19*, 125–135. [CrossRef]

119. Griffiths, N.A.; Jackson, C.R.; McDonnell, J.J.; Klaus, J.; Du, E.; Bitew, M.M. Dual nitrate isotopes clarify the role of biological processing and hydrologic flowpaths on nitrogen cycling in subtropical low-gradient watersheds. *J. Geophys. Res. Biogeosci.* **2016**, *121*, 422–437. [CrossRef]

120. Seiler, R.L. Combined use of ^{15}N and ^{18}O of nitrate and ^{11}B to evaluate nitrate contamination in groundwater. *Appl. Geochem.* **2005**, *20*, 1626–1636. [CrossRef]

121. Widory, D.; Kloppmann, W.; Chery, L.; Bonnin, J.; Rochdi, H.; Guinamant, J. Nitrate in groundwater: An isotopic multi-tracer approach. *J. Contam. Hydrol.* **2004**, *72*, 165–188. [CrossRef] [PubMed]

122. Widory, D.; Petelet-Giraud, E.; Negrel, P.; Ladouche, B. Tracking the sources of nitrates in groundwater using coupled nitrogen and boron isotopes: A synthesis. *Environ. Sci. Technol.* **2005**, *39*, 539–548. [CrossRef] [PubMed]

123. Saccon, P.; Leis, A.; Marca, A.; Kaiser, J.; Campisi, L.; Bottcher, M.E.; Savarino, J.; Escher, P.; Eisenhauer, A.; Erbland, J. Multi-isotope approach for the identification and characterization of nitrate pollution sources in the Marano lagoon (Italy) and parts of its catchment area. *Appl. Geochem.* **2013**, *34*, 75–89. [CrossRef]

124. Puig, R.; Soler, A.; Widory, D.; Mas-Pla, J.; Domènech, C.; Otero, N. Characterizing sources and natural attenuation of nitrate contamination in the Baix Ter aquifer (NE Spain, using a multi-isotope approach). *Sci. Total Environ.* **2017**, *580*, 518–532. [CrossRef]

125. Martinelli, G.; Dadomo, A.; De Luca, D.A.; Mazzola, M.; Lasagna, M.; Pennisi, M.; Pilla, G.; Sacchi, E.; Saccon, P. Nitrate sources, accumulation and reduction in groundwater from Northern Italy: Insights provided by a nitrate and boron isotopic database. *Appl. Geochem.* **2018**, *91*, 23–35. [CrossRef]

126. Vitòria, L.; Otero, N.; Soler, A.; Canals, A. Fertilizer characterization: Isotopic data (N, S, O, C, and Sr). *Environ. Sci. Technol.* **2004**, *38*, 3254–3262.

127. Nestler, A.; Berglund, M.; Accoe, F.; Duta, S.; Xue, D.; Boeckx, P.; Taylor, P. Isotopes for improved management of nitrate pollution in aqueous resources: Review of surface water field studies. *Environ. Sci. Pollut. Res.* **2011**, *18*, 519–533. [CrossRef]

128. Komor, S.C. Boron content and isotopic composition of hog manure, selected fertilizer, and water in Minnesota. *J. Environ. Qual.* **1997**, *26*, 1212–1222. [CrossRef]

129. Leenhouts, J.M.; Bassett, R.L.; Maddock Iii, T. Utilization of intrinsic boron isotopes as Co-migrating tracers for identifying potential nitrate contamination sources. *Ground Water* **1998**, *36*, 240–250. [CrossRef]

130. Sacchi, E.; Acutis, M.; Bartoli, M.; Brenna, S.; Delconte, C.A.; Laini, A.; Pennisi, M. Origin and fate of nitrates in groundwater from the central Po plain: Insights from isotopic investigations. *Appl. Geochem.* **2013**, *34*, 164–180. [CrossRef]

131. Vengosh, A.; Heumann, K.G.; Juraske, S.; Kasher, R. Boron isotope application for tracing sources of contamination in groundwater. *Environ. Sci. Technol.* **1994**, *28*, 1968–1974. [CrossRef]

132. Eisenhut, S.; Heumann, K.G.; Vengosh, A. Determination of boron isotopic variations in aquatic systems with negative thermal ionization mass spectrometry as a tracer for anthropogenic influences. *Fresenius' J. Anal. Chem.* **1996**, *354*, 903–909. [CrossRef] [PubMed]

133. Barth, S. Application of boron isotopes for tracing sources of anthropogenic contamination in groundwater. *Water Res.* **1998**, *32*, 685–690. [CrossRef]

134. Barth, S.R. Boron isotopic compositions of near-surface fluids: A tracer for identification of natural and anthropogenic contaminant sources. *Water Air Soil Pollut.* **2000**, *124*, 49–60. [CrossRef]

Article

Local Meteoric Water Line of Northern Chile (18° S–30° S): An Application of Error-in-Variables Regression to the Oxygen and Hydrogen Stable Isotope Ratio of Precipitation

Tiziano Boschetti [1,*], José Cifuentes [2], Paola Iacumin [1] and Enricomaria Selmo [1]

[1]   Department of Chemistry, Life Sciences and Environmental Sustainability, University of Parma, Parco Area delle Scienze, 43124 Parma, Italy; paola.iacumin@unipr.it (P.I.); enricomaria.selmo@unipr.it (E.S.)
[2]   Servicio Nacional de Geología y Minería, Avenida Santa María #0104, 7520405 Santiago, Chile; jose.cifuentes@sernageomin.cl
*   Correspondence: tiziano.boschetti@unipr.it; Tel.: +39-0521-905300

Received: 13 March 2019; Accepted: 11 April 2019; Published: 16 April 2019

Abstract: In this study, a revision of the previously published data on hydrogen (^{2}H/^{1}H) and oxygen (^{18}O/^{16}O) stable isotope ratio of precipitation in northern Chile is presented. Using the amount-weighted mean data and the combined standard deviation (related to both the weighted mean calculation and the spectrometric measurement), the equation of the local meteoric line calculated by error-in-variables regression is as follows: Northern Chile EIV-LMWL: δ^2H = [(7.93 ± 0.15) δ^{18}O] + [12.3 ± 2.1]. The slope is similar to that obtained by ordinary least square regression or other types of regression methods, whether weighted or not (e.g., reduced major axis or major axis) by the amount of precipitation. However, the error-in-variables regression is more accurate and suitable than ordinary least square regression (and other types of regression models) where statistical assumptions (i.e., no measurement errors in the x-axis) are violated. A generalized interval of δ^2H = ±13.1‰ is also proposed to be used with the local meteoric line. This combines the confidence and prediction intervals around the regression line and appears to be a valid tool for distinguishing outliers or water samples with an isotope composition significantly different from local precipitation. The applicative examples for the Pampa del Tamarugal aquifer system, snow samples and the local geothermal waters are discussed.

Keywords: precipitation; stable isotope ratios; local meteoric water line; amount-weighted mean; linear regression; confidence; prediction and generalized intervals

---

1. Introduction

To fit a model to data pairs, the linear least squares regression is by far the most widely used modeling method (e.g., [1]). The ordinary least squares regression (OLSR) minimizes the sum of the squared vertical distances between the y data values and the corresponding y values on the fitted line (the predictions). The OLSR design assumes that there is no variation in the independent variable (x) because it is controlled by the researcher. This is also known as 'Model I' regression [2]. In contrast, in 'Model II' regression, both the response and the explanatory variables of the model are random (i.e., not controlled by the researcher), and there are errors associated with the measurements of both x and y [2]. Among the 'Model II' regression, major axis regression (MA) assumes equal variance on both variables and minimizes the orthogonal (perpendicular) distances from the data points to the fitted line. To the best of our knowledge, one of the first suggestions for applying orthogonal regression to the precipitation amount weighted stable isotope ratios of the oxygen (δ^{18}O, the independent variable x) and hydrogen (δ^2H, the dependent variable y) of water molecules in order to determine a better global meteoric water line (GMWL) dates back to the 1980s [3]. Ten years later, the International Atomic Energy Agency (IAEA) proposed the use of the reduced major axis (RMA) regression for calculation of the local meteoric water lines (LMWL) [4]. However, in this latter publication, RMA was wrongly defined as an orthogonal regression. In fact, RMA minimizes the sum of the areas (thus using both vertical and horizontal distances of the data points from the resulting line) rather than the least squares sum of the squared vertical distances, as in OLSR [5]. Meanwhile, the local meteoric water lines in the GNIP dataset of IAEA are calculated through precipitation amount-weighted least squares regression (PWLSR) [6,7]. This regression approach is considered to be the most suitable for representing the isotope composition of groundwater because these are recharged by important rainfall events [6,7]. Alternatively, the precipitation weighted RMA (PWRMA) appears to be most suitable for coastal, island, and Mediterranean sites [6]. Another proposed approach for the calculation of LMWL was the generalized least squares regression model (GENLS), an error-in-variables regression (EIV) that considers the combined standard deviation of the δ^{18}O and δ^2H values [8]. Within the statistical literature, synonyms of this approach are "Deming" [9] or "error-in-variables" [10] regression, the former often distinguished as *simple* (when the data errors are constant among all measurements for each of the two variables) and *general* (different data error at each observation) [11]. It should be noted that in the case where the variance ratio is equal to 1, Deming regression is equivalent to orthogonal regression. From a general statistical and chemometrics point of view, comparisons between the different approaches can be found in the existing literature [12]. In contrast, in the specific case of water isotope geochemistry and the related meteoric water line, a comparison between error-in-variables and other regression methods has never been presented or discussed. Moreover, despite the advantages identified by previous publications of the alternative regression methods [4,6,7,13], most of the studies on the stable isotope ratio of the waters still use OLSR to calculate the LMWLs. While this does not necessarily mean that the use of OLSR is wrong [7], it has been shown, especially in the hydrology study of arid regions, that the differences between the OLSR approach and other regression approaches can be significant, particularly in terms of the slope of the LMWL [7].

In this manuscript, we compare the regression lines obtained from OLSR, RMA, MA, PWLSR, PWRMA, and PWMA with that from the error-in-variables (EIV) approach, which is very similar to the method followed for GENLS [8]. Here, the existing data on δ^{18}O and δ^2H values of precipitation in northern Chile were chosen. This brief study does not attempt to be mathematically or statistically exhaustive and the reader is advised to refer to the referenced literature for more details. In fact, the main purpose here is to provide the results of alternative regression methods applied to stable isotope ratios of precipitation and to provide guidance and advice when the obtained local meteoric water lines are employed for the interpretation of the isotope data of groundwater.

2. Materials and Methods

Northern Chile extends southwards from approximately 18° S to 30° S and includes the two macro-regions of Norte Grande (Regions XV, I and II; Figure 1B) and part of Norte Chico (Regions III and IV; Figure 1B) [14–16].

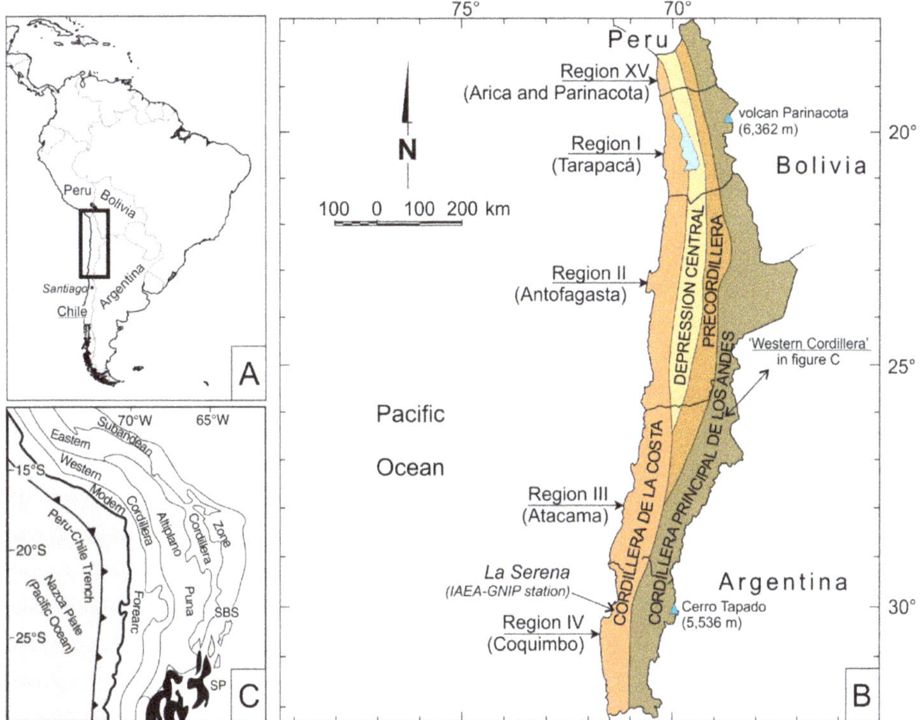

Figure 1. (**A**) Localization of the area of this study. (**B**) Morphostructural domains (shades of brown) and administrative regions (modified from [17]). In blue: Pampa de Tamarugal aquifer. (**C**) Morpho-tectonic units of the central Andes (modified from [18]). SBS: Santa Barbara System; SP: Sierras Pampeanas. The "Modern Forearc" coincides with the Precordillera and the Coastal Cordillera [19].

This presents a narrow strip of land (~300 km) between the Andean ridge and the Pacific coast [17]. The latitudinal range of this northern part of Chile aligns with a major extension of the arid desert climate (BW climate zone [20,21]), followed by a tundra climate (ET climate zone [20,21]), and a semi-arid climate in a minor extension [14,21,22]. The Atacama Desert, the driest place on earth with mean precipitation as low as 10 mm per decade, and the higher elevations in the Andes are striking examples of the BW and the ET climates, respectively. The Amazonian influence in the South American monsoon system, with concentrated summer rainfall or dry winters, affects the climate in this area from the north boundary to Nevado Ojos del Salado [22], a stratovolcano with an altitude of 6893 m, the highest point in Chile and the second highest outside of Asia.

The previous studies on isotope composition of meteoric water revealed a general increase in $\delta^{18}O$ on the Andean plateau (the "Altiplano", Figure 1C) from north to south, concomitant with an increase in aridity and decrease in convective moistening (amount effect; [23]). Stable isotope and seasonal precipitation patterns suggest an eastern provenance of the vast majority of moisture that falls as precipitation across the Andean Plateau and Western Cordillera (Figure 1C), with Pacific-derived moisture contributing a minor amount at low elevations near the coast ([23,24]). However, over most regions, the $\delta^{18}O$ signal of precipitation is influenced by a combination of factors ([24]). The $\delta^{18}O$ -depleted values observed in the high altitude area (i.e., the Andean plateau) were related to processes that affect the air masses that (i) originated over the Atlantic Ocean, (ii) cross the Amazon Basin (continental effect), (iii) ascended the Andes (altitude effect) and (iv) precipitated (convective effect) in the Andean plateau ([25]). In particular, over the eastern Andes, precipitation at low elevations has $\delta^{18}O$ from -2 to $-8‰$, but $\delta^{18}O$ becomes more depleted toward the west as vapor is lifted across the Eastern Cordillera of the Andes ([26–29]). The dominant Altiplano summer rain has $\delta^{18}O$ values from $-8‰$ to $-15‰$, with the expectation that precipitation in the driest places (e.g., Atamaca desert) should be moderately to strongly depleted at all elevations up to approximately $-18‰$ ([28,30]).

During the last forty years, several meteoric water lines calculated through OLSR regression were proposed to describe the best fit for the oxygen and hydrogen stable isotope ratios of the precipitation in northern Chile (Table 1, [31–39]).

Table 1. Slope and intercept values of the previously published northern Chile meteoric water lines [31–39]. All the shown data were calculated by OLSR. The IAEA/GNIP data pairs are referred to as the "La Serena" station (Figure 1B). For this station, other two equations are also proposed on the same database, $\delta^2H = [(7.64 \pm 0.39) \delta^{18}O] + [9.44 \pm 2.32]$ and $\delta^2H = [(8.04 \pm 0.45) \delta^{18}O] + [9.98 \pm 2.59]$ by PWLSR and RMA methods [6,7], respectively, both applied to all the available data ($N = 40$).

	Slope	±	s.e.	Intercept	±	s.e.	N	R^2	Reference	Year	Ref. [#]
	7.8	-		10.3	-		29	-	Fritz et al.	1981	[35]
	8.2	±	0.17	16.6	±	2.2	39	0.98	Chaffaut et al.	1998	[32,33]
	7.8	-		9.7	-		129	-	Aravena et al.	1999	[31]
	7.9	-		14	-		-	-	Herrera et al.	2006	[36]
	7.7	-		9.6	-		-	0.91	Squeo et al.	2006	[38]
	7.95	-		14.9	-		-	0.99	DGA	2015	[34]
	8.07	-		13.5	-		23	0.98	Troncoso et al.	2012	[39]
	7.52	±	0.46	7.18	±	2.64	40	0.87	IAEA/WHO	2015	[37]
mean	7.87			11.97							
std.de.	0.21			3.22							

-: data not calculated or not declared; s.e.: standard error.

However, not all the studies show the oxygen and hydrogen stable isotope ratios of the water molecule along with precipitation amount [31,32,35,40–42]. "La Serena" is the northern Chile station (Region IV, Figure 1B) monitored by the Global Network of Isotopes in Precipitation (GNIP dataset) from 1988 until 2015 [37].

For this study, a dataset of 32 stations has been constructed through collecting previously published data, considering only the isotope values of precipitation with the measured amount and recalculating the amount-weighted mean in order to have only one $\delta^{18}O$ and δ^2H data pair per station (Table 2; Supplementary File S1).

Table 2. Amount-weighted oxygen and hydrogen stable isotope ratios (δ^{18}O and δ^2H in permil versus the standard mean ocean water SMOW, respectively) of the previously published precipitation samples collected in northern Chile [31,32,35,37,40–43]. Calculations are described in the Supplementary File S1. Samples were mainly collected in Region I (from #1 to #27), Region II (from #28 to #31) and Region IV (#32) (Figure 1B).

ID #	Location	Coordinates		Elevation Meters	δ^{18}O (a.w.m. ± c.s.)	δ^2H (a.w.m. ± c.s.)	$N_{a.p.}$	Precipitation Amount (Mean Value) Millimeters	References	Ref. [#]
		Latitude	Longitude		‰ vs. SMOW	‰ vs. SMOW				
1	Apacheta Tapa	−19.5891	−68.9940	4350	−17.48 ± 1.14	−122.3 ± 8.9	4	73.73	Fritz et al. 1981	[35]
2	Chusmiza	−19.7060	−69.1906	3360	−7.51 ± 0.22	−48.9 ± 1.0	3	103.93	Fritz et al. 1981	[35]
3	Alto Mocha	−19.8181	−69.2987	2590	−7.13 ± 0.31	−43.7 ± 1.5	2	33.90	Fritz et al. 1981	[35]
4	Mocha	−19.8436	−69.2840	2200	−4.91 ± 0.90	−29.2 ± 6.7	2	39.65	Fritz et al. 1981	[35]
5	Collacagua	−20.0703	−68.8259	3915	−15.00 ± 1.22	−108.1 ± 8.6	7	82.23	Fritz et al. 1981	[35]
6	Huasco	−20.3144	−68.8065	3800	−17.13 ± 0.65	−127.7 ± 9.3	3	34.63	Fritz et al. 1981	[35]
7	Indio Muerto	−20.3610	−68.9615	4135	−18.27 ± 0.35	−129.9 ± 3.4	4	53.05	Fritz et al. 1981	[35]
8	Tambillos	−20.4672	−69.1415	3300	−8.02 ± 0.27	−50.1 ± 2.1	3	77.13	Fritz et al. 1981	[35]
9	Apacheta Mama *	−20.4734	−69.1580	4115	−19.90 ± 0.15	−146.0 ± 1.0	1	77.00	Fritz et al. 1981	[35]
10	Pumire	−19.0955	−69.1108	4200	−2.25 ± 0.78	−11.5 ± 3.1	3	10.00	Salazar et al. 1998; Aravena et al. 1989	[41,42]
11	Coposa	−20.7089	−68.6942	3460	−16.66 ± 2.01	−119.8 ± 16.2	9	9.17	Salazar et al. 1998; Aravena et al. 1989	[41,42]
12	Collaguasi	−20.9833	−68.7000	4250	−16.96 ± 1.99	−122.3 ± 16.0	5	13.30	Aravena et al. 1999	[31]
13	Ujina	−20.9833	−68.6000	4200	−11.98 ± 1.57	−83.3 ± 12.0	8	18.79	Aravena et al. 1999	[31]
14	Puchuldiza	−19.4000	−68.9500	4150	−14.55 ± 0.93	−98.4 ± 11.0	15	11.88	Aravena et al. 1999	[31]
15	Pampa Lirima	−19.8200	−68.9000	4100	−16.97 ± 1.23	−121.6 ± 9.7	27	9.86	Aravena et al. 1999	[31]
16	Colchane	−19.7003	−68.8833	3965	−14.95 ± 2.28	−106.3 ± 18.2	4	51.90	Aravena et al. 1999	[31]
17	Collacagua	−20.0333	−68.8500	3990	−12.68 ± 1.34	−88.3 ± 10.5	14	15.04	Aravena et al. 1999	[31]
18	Cancosa	−19.8500	−68.6000	3800	−14.00 ± 2.45	−96.4 ± 20.7	5	28.50	Aravena et al. 1999	[31]
19	Huaytane	−19.5433	−68.6042	3720	−14.10 ± 1.48	−100.9 ± 12.1	5	32.20	Aravena et al. 1999	[31]
20	Copaquire	−20.9305	−68.8922	3490	−12.47 ± 0.98	−84.6 ± 7.8	16	5.75	Aravena et al. 1999	[31]
21	Poroma	−19.8667	−69.1833	2880	−5.00 ± 0.66	−28.1 ± 6.0	9	8.01	Aravena et al. 1999	[31]
22	Parca	−20.0167	−68.8500	2570	−6.64 ± 0.90	−40.9 ± 7.8	4	4.88	Aravena et al. 1999	[31]
23	Huatacondo	−20.9167	−69.0500	2460	−8.26 ± 1.26	−53.8 ± 9.8	12	6.21	Aravena et al. 1999	[31]
24	Camina	−19.3116	−69.4299	2380	−7.02 ± 1.00	−42.8 ± 8.1	4	11.50	Aravena et al. 1999	[31]
25	Sillillica	−20.1738	−68.7377	4270	−15.96 ± 1.07	−110.9 ± 7.9	3	95.30	Uribe et al. 2015	[40]
26	Altos del Huasco	−20.3221	−68.9022	3784	−14.82 ± 1.57	−104.8 ± 12.3	3	51.80	Uribe et al. 2015	[40]
27	Diablo Marca	−20.0505	−68.9962	4603	−18.43 ± 1.59	−128.8 ± 11.8	3	78.60	Uribe et al. 2015	[40]
28	Quisquiro	−23.2100	−67.2500	4260	−13.40 ± 2.12	−91.1 ± 13.3	36	14.30	Chaffaut 1998	[32]
29	Tuyajto	−23.9435	−67.5921	4040	−6.23 ± 0.50	−32.4 ± 3.1	2	13.30	Herrera et al. 2016	[43]
30	Pampa Colorada	−23.8588	−67.4774	4426	−8.80 ± 0.86	−51.3 ± 7.6	3	28.50	Herrera et al. 2016	[43]
31	Aguas Calientes-3	−23.9171	−67.6971	3900	−6.00 ± 0.51	−31.9 ± 4.5	3	21.67	Herrera et al. 2016	[43]
32	La Serena	−29.8981	−71.2425	142	−5.78 ± 1.32	−35.9 ± 9.9	34	43.10	IAEA/WHO 2015	[37]

a.w.m. ± c.s.: amount-weighted mean ± combined sigma (see methods paragraph for details); $N_{a.p.}$: Number of accumulation periods of precipitation. This value coincides with the total number of the available isotope data pairs per station (Supplementary File S1); * sample with only one accumulation period ($N_{a.p.}$ = 1). In this case, the combined sigma (c.s.) contain only the standard deviation related to spectrometric measurement (Supplementary File S1). The regression has been calculated both including (Table 3) and excluding (Supplementary File S2) this sample.

Assuming x_i to be the measurement of the stable isotope ratios expressed as $\delta^{18}O$ and δ^2H in ‰ versus SMOW (standard mean ocean water)—determined on a specific precipitation amount p_i (mm) collected over a specific time period—the standard deviation related to the amount weighted mean $\bar{x}_w$ [4] of the meteoric stations with more than one accumulation periods of precipitation ($N_{a.p.} > 1$) was calculated as follows [44]:

$$\sigma_w = \sqrt{\frac{\sum\limits_{i=1}^{N_{a.p.}} p_i \left(x_i - \bar{x}_w\right)^2}{\frac{\left(N_{a.p.}-1\right) \sum\limits_{i=1}^{N_{a.p.}} p_i}{N_{a.p.}}}} \tag{1}$$

It should be noted that the accumulation periods, which coincides with the total number of the isotope data pairs and precipitation amounts available per each station, are not uniformly distributed through the year and not equally long for most of the stations considered (Supplementary File S1).

In the existing literature, there are several definitions of 'combined standard deviation' or 'combined uncertainty.' Here, we deal with two different 'uncertainty' sources [8,45], that is, a standard deviation related to the amount-weighted mean calculation σ_w, as described above, and a standard deviation related to the mass spectrometric measurements σ_{ms}, which differs depending on the publication (Supplementary File S1). One of the easiest ways to calculate the combined standard deviation σ_{csd} involves foreseeing the combination of σ_w and σ_{ms} following the law of propagation of uncertainty, that is, the square root of the sum-of-the-squares [45–47]:

$$\sigma_{csd} = \sqrt{\left(\sigma_w\right)^2 + \left(\sigma_{ms}\right)^2} \tag{2}$$

The combined standard deviations, associated with the $\delta^{18}O$ and δ^2H amount-weighted mean data in 30 of the total 32 stations, were used to calculate the local meteoric water line through EIV regression. In two stations, Quisquiro [32] and La Serena [37], the calculation of the standard deviation of the weighted mean σ_w was performed differently for two reasons. In the case of Quisquiro, the complete dataset is unavailable; therefore, the standard deviation was calculated among the amount-weighted data corresponding to the extreme seasons (winter and summer, [32]). Meanwhile, in the case of La Serena, the σ_w was calculated among the years in which more than 70% of precipitation was analyzed for a given isotope composition [4]. Following this, the σ_{csd} was calculated through using the σ_{ms} declared in the analytical section of the publication dataset [32] or elsewhere [25]. With this, the σ_{csd} on $\delta^{18}O$ and δ^2H is within the mean of the other stations.

The calculation of the EIV regression through different codes employ similar methods based on York and Williamson's works [48–51]. Generally speaking, the process of searching for best straight-line parameters requires an iterative procedure in which each new slope estimate is used to modify a weight term, which combines the reciprocal variance of the x and y variables, in a series of convergent fitting cycles [11,52–55]. The results of this method are compared with those obtained by OLSR, RMA, MA, PWLSR, PWRMA, and PWMA [6]. In accordance with Crawford et al. [6], the Student's t-test was used to check the difference between the two regression slopes. Furthermore, the obtained EIV meteoric water line was also compared with previously published isotope data on precipitation without amount data [28,30,36,56–60], snow [43,58,59,61–68], groundwater [69], and geothermal water [70] samples collected up to this point in northern Chile.

3. Results

The parameters of the northern Chile meteoric water lines obtained through the different software and regression methods are listed in Table 3.

Table 3. Regression results obtained by different methods applied to the data of Table 2. (**a**) Crawford et al.'s non-weighted (OLSR, MA, RMA) and amount weighted (PWLSR, PWRMA, PWMA) regressions [6,7]. (**b**) Ordinary Least Squares Regression (OLSR) and error-in-variables regression (EIV) results by different codes [52,53,55,71]. (**c**) Comparison between regression's slopes obtained by EIV and PWLR, PWRMA, and PWRMA.

(a) regression results according to Crawford et al. 2014

Models	N	Slope		Intercept		rmSSEav	t-Value	p
		Value	s.e.	Value	s.e.			
OLSR	32	7.78	0.10	11.3	1.2	1.0009	-	-
RMA	32	7.80	0.09	11.5	1.2	1.0004	0.1318	0.896
MA	32	7.82	0.10	11.7	1.2	1.0009	0.2593	0.797
PWLSR	32	7.74	0.09	10.7	1.3	1.0054	0.3047	0.763
PWRMA	32	7.76	0.09	10.9	1.3	1.0021	0.1810	0.858
PWMA	32	7.78	0.09	11.1	1.3	1.0000	0.0608	0.952

Local Meteoric Water Line Freeware©—Australian Nuclear Science and Technology Organisation. OLSR: Ordinary Least Square Regression; RMA: Reduced Major Axis; MA: Major Axis; PWLSR: Precipitation Weighted Least Square Regression; PWRMA: Precipitation Weighted Reduced Major Axis. PWMA: Precipitation Weighted Major Axis. s.e.: standard error; rmSSEav: root mean sum of squared error average; Comparison between OLSR and the other regression models listed in (a): t-value: Student's *t*-test value; *p*: probability distribution by Microsoft®Office Excel (d.f. = N − 2).

(b) EIV regression results compared with OLSR

Models	N	Slope		Intercept		95% C.I. Slope		95% C.I. Intercept		r	GOF	rmSSE	t-Value	p
		Value	s.e.	Value	s.e.	Upper	Lower	Upper	Lower					
OLSR	32	7.79	0.10	11.3	1.2	7.98	7.59	13.8	8.7	0.9977	7.1740	2.6784	-	-
EIV-a	32	7.93	0.15	12.3	2.1	8.24	7.63	16.7	8.0	0.9977	0.1457	0.3817	0.8292	0.414
EIV-b	32	7.93	0.15	12.3	2.1	8.24	7.62	16.6	7.9	0.9977	0.1457	0.3817	0.8119	0.423
EIV-c	32	7.93	0.15	12.3	2.1	-	-	-	-	-	0.1457	0.3817	0.8289	0.414
EIV-d	32	7.93	0.06	12.3	0.8	-	-	-	-	-	0.1457	0.3817	1.3185	0.197
EIV-e	32	7.81	0.12	11.5	1.6	8.05	7.56	14.8	8.2	-		-	0.1256	0.901

OLSR: here calculated by Microsoft®Office Excel—2013 (GOF parameter from Origin®Pro). EIV-a: Origin®Pro; EIV-b: SigmaPlot©; EIV-c: BFSL; EIV-d: Cantrell 2008; EIV-e: Real Statistics Using Excel© with a mean variances ratio of λ = Xvar/Yvar = 0.01913. C.I.: Confidence Intervals; r: Pearson's correlation coefficient; GOF: Goodness of Fit; rmSSE: root mean sum of squared error (this parameter is called "standard error" of regression in Microsoft®Excel's OLSR; in other regression codes, it could be easily calculated by the square root of the GOF parameter).

(c) EIV regression results compared with PWLSR, PWRMA, and PWMA

Models	t-Value	p	Models	t-Value	p	Models	t-Value	p
PWLSR	-	-	PWRMA	-	-	PWMA	-	-
EIV-a	1.0843	0.287	EIV-a	0.9883	0.331	EIV-a	0.8946	0.378
EIV-b	1.0662	0.295	EIV-b	0.9705	0.340	EIV-b	0.8770	0.387
EIV-c	1.0840	0.287	EIV-c	0.9881	0.331	EIV-c	0.8944	0.378
EIV-d	1.7481	**0.091**	EIV-d	1.5934	**0.122**	EIV-d	1.4410	**0.160**
EIV-e	0.4141	0.682	EIV-e	0.3029	0.764	EIV-e	0.1949	0.847

The values of slope and intercept through EIV appear to be slightly higher than those obtained by other methods that do not take into account the combined standard deviation of the oxygen and hydrogen measurements. However, there is no difference with other regression methods, aside from the results from Cantrell's bivariate worksheet (EIV-d, Table 3b), which was tested on atmospheric chemical data [54]. The regression results of EIV-d method include: (i) a lower standard error in comparison with those from other codes used for EIV calculation; and (ii) a higher slope when compared to PWLSR, but not significantly different ($p > 0.05$). The difference between the slopes are further smoothed when the sample #9 is removed ($N_{a.p.}$ = 1; Supplementary File S2). It should be noted that lower standard errors on slope and intercept similar to those of Cantrell's bivariate worksheet are obtained in the other EIV models introducing a scale factor in the maximum likelihood method, which is estimated by the square root of the reduced chi-square statistic that results from the fit analysis [11,53]. Lower than half standard error are also obtained by "Deming regression" with a constant variance ratio of $\lambda = \delta^{18}O_{var}/\delta^2H_{var} = 0.019$ (EIV-e, Table 3b), that is the mean ratio of the squared combined sigma (Table 2). Therefore, we consider as most representative of the samples the regression results with the standard errors on slope and intercept calculated without scale factor and with combined sigma for each station (EIV results from "a" to "c" in Table 3b):

$$\text{Northern Chile EIV-LMWL: } \delta^2H = [(7.93 \pm 0.15)\, \delta^{18}O] + [12.3 \pm 2.1] \tag{3}$$

The upper and lower confidence intervals at 95% probability of the slope and intercept in the EIV regressions (results from "a" to "c" in Table 3b) include all the previous published values obtained by OLSR (Table 1) and all the other slopes and intercepts of the meteoric water lines calculated by different regression methods (Table 3a). This is clear and straightforward, in particular looking to the standard errors of slope and intercept, which are higher in EIV regression results.

4. Discussion

In isotope hydrology applications, it is often desirable to know whether a water sample from the surface (river, lake, or lagoon) or from an aquifer is shifted in a statistically significant way in relation to the local meteoric water line. This is particularly true in the following: (i) arid zones, where fractionation due to evaporation produces isotopically enriched residual waters; and (ii) geothermal areas, where hot water samples could be isotopically enriched due to water-rock interaction. Northern Chile offers an ideal location for this kind of study because both of these conditions exist here [65,72].

The calculated confidence intervals obtained through EIV regression represent the scatter of the weighted mean data around the fitted line, disregarding the combined standard deviation (Table 3). In OLSR, prediction intervals for y at a specific x value estimate the bands around the fitted line in which the future observation will fall, with a certain probability (usually 95% as confidence interval), given what has already been observed with the sample dataset employed for the regression (t-distribution multiplied by the standard error of the fit). The OLSR prediction interval can be quickly calculated by using commercial or free codes (e.g., [71,73]). However, OLSR does not take into account the measurement error in x when predicting y, either in the combined standard deviations of the data in the fittings or the prediction interval calculations. Moreover, an officially recognized method for the calculation of prediction intervals related to EIV regression still does not exist.

An approximate prediction area around the EIV local meteoric water line could be traced through using an empirical-graphical method, joining the outer limits of the bars representing the combined standard deviations around the mean of the isotope data of precipitation. However, its irregular shape, related to the non-homogeneous values of the combined standard deviations of the regressed stations, necessitates an ad-hoc correction. Specifically speaking, the lower sector of the line (approximately, $\delta^{18}O < -12‰$ and $\delta^2H < -80‰$) shows numerous monitored stations with wider standard deviations—especially in terms of hydrogen composition δ^2H—in comparison with the higher sector of the line (Figure 2A).

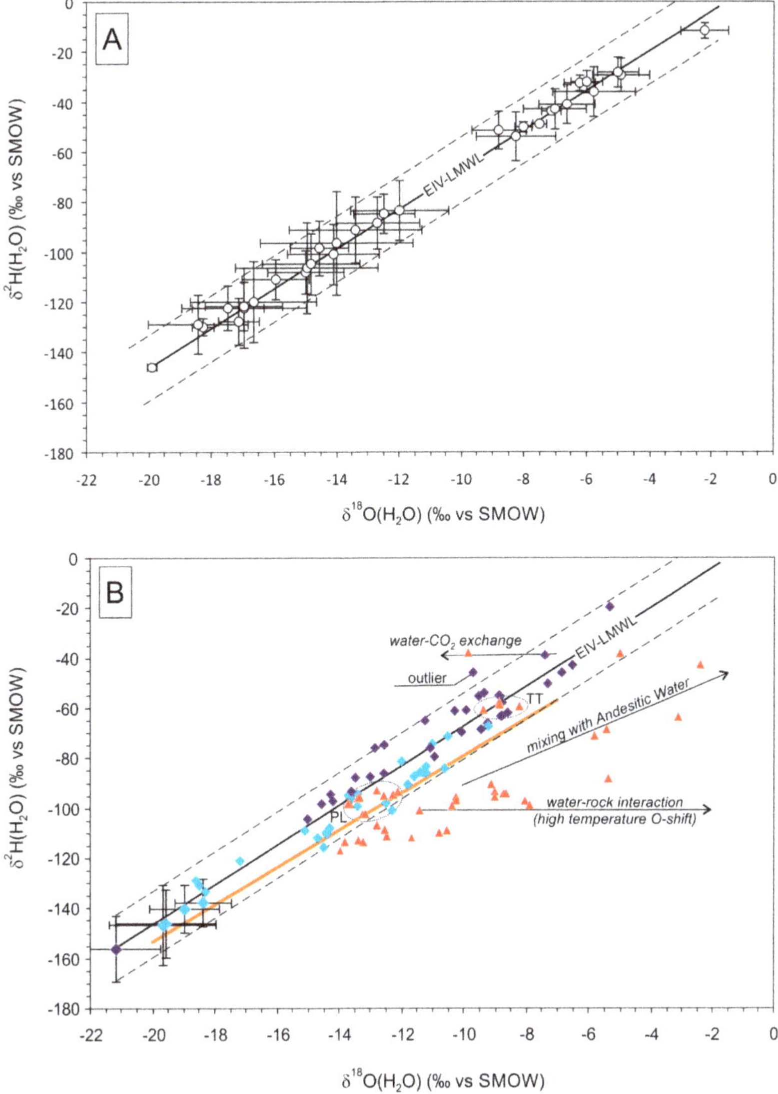

Figure 2. (**A**) δ^2H and δ^{18}O amount-weighted mean of the samples (open circles; Table 2) along with the combined standard deviation on hydrogen and oxygen isotope composition (y and x bars, respectively); error-in-variables local meteoric water line (EIV-LMWL solid line) and the generalized interval (dashed lines; δ^2H = ±13.1‰ up and down the fitted values on the line). (**B**) Pampa de Tamarugal aquifer (thick-orange line, [69]), snow and penitentes (dark and light blue diamonds, respectively [41,43,59,61,63–68]) and hydrothermal groundwater (red triangles, [70]; PL: Pampa Lirima, TT: Torta de Tocorpuri) samples compared with the lines described in 'A'. Arrows explain the isotope effects in hydrothermal groundwater [70,74]. y and x bars in snow and penitentes depict the combined sigma related to the volume weighted mean (samples from Cerro Tapado [67], Supplementary File S5). Light-blue diamonds without error bars are single samples of the penitentes from Parinacota volcano ([64], Supplementary File S5).

The wider combined standard deviation of the samples that fall within the lower sector is mainly due to the irregular/inhomogeneous frequency of the sampling and to the different accumulation periods of precipitation in the same station (mixing of monthly and seasonal data). In contrast, the stations on the higher sector of the line, which are not depleted by the "altitudinal effect" (lower isotope values at a higher elevation, [74,75]), show a minor combined standard deviation. This is probably due to the fact that only four stations for rainwater are located at an altitude of lower than 2500 m, and monitored by a more regular frequency of sampling (e.g., La Serena station).

An innovative method has recently been proposed for the calculation of a "generalized interval," which combines the confidence and predictive concepts in EIV [76–78]. Using this approach, the obtained interval forcing a mean on 12 values per station (qx = qy = 12; i.e., as supposing a sampling of precipitation extended to 12 months) is 8.9 ± 4.2‰ up and down the fitted line, which is fairly similar to the mean combined standard deviation of δ^2H (9.1 ± 4.8‰, excluding the station #9 with $N_{a.p.} = 1$, Table 2; Supplementary File S3). In contrast, the prediction intervals calculated through the jackknife method [71]—a statistical approach which simulate the resampling of the collected data taking into account their variance—and through Deming regression [71]—which necessitate a constant variance ratio (constant λ [71], calculated on the variances ratio of the 32 samples)—give a prediction interval of ±1.53‰, which is narrower than the combined standard deviation (Supplementary File S3).

Considering that the OLSR and EIV meteoric water lines of northern Chile are not different in terms of slope, the stable isotope ratio data on 115 precipitation samples available from the existing literature (Supplementary File S4; [28,30,36,56–60,68,79]) could help with checking the effectiveness of prediction and generalized intervals. While all these isotope data do not report the amount of precipitation and are therefore not included in the EIV regression calculation, most of them are clustered between or extended over the enriched and depleted sector of the meteoric water line, thus completing the station's gap. The upper and lower 95% prediction intervals on the δ^2H values related to the OLSR regression of this latter dataset are shifted ±13.4‰ up and down respectively on the regressed line (Supplementary File S4). This latter value is not significantly different, neither from the upper/lower δ^2H combined standard deviation values of the weighted-amount dataset (±13.8‰), nor from the upper/lower extremes of the generalized interval (±13.1‰). This is particularly true if we take into account that the unweighted rainfall dataset pertains to single rain events, meaning it is quite normal that their predicted δ^2H values are shifted towards the higher value of the combined standard deviation of the weighted rainfall dataset.

To verify the effectiveness of this approach, we compared the EIV regression and the above described generalized interval with the isotope composition of the following samples from northern Chile: (i) the groundwater from Pampa del Tamarugal [69] (Figure 1B); (ii) the snow samples (Supplementary File S5; [43,58,59,61–68]); and (iii) the geothermal waters samples [70].

In terms of the first groundwater sample type (i), in previous studies, the clustering of the $\delta^{18}O$ and δ^2H values of the groundwater from Pampa del Tamarugal falling on the right side of the meteoric water line has led several authors to consider this aquifer system as isotopically different from local precipitation [35,80]. This has been attributed to the following: (i) a recharge in climatic conditions different from the current ones [35]; and (ii) to the evaporation that affects the precipitation in the unsaturated zone of the recharge area [80]. A recent re-evaluation of the isotope data published up to that point shows that the isotope composition of the groundwater from that area is effectively parallel to the meteoric water line, while it was noted that the previous hypothesis on climate variation must be reconsidered [69]. However, the "uncertainty wings" around the meteoric water line have not been traced in any of the above-described cases.

In Figure 2B it is shown how the recent calculated regression line (in orange) representing the groundwater in Pampa del Tamarugal [69] falls within the generalized intervals of the EIV meteoric water line. This demonstrates that the mean composition of the groundwater from this area is not significantly different from the precipitation. This does not exclude the fact that some groundwater samples could be significantly affected by evaporation during recharge. Indeed, the right side of the generalized interval of the EIV regression is the location of the enriched precipitation samples associated with the evaporation during rainfall. Furthermore, the regression line of the groundwater samples should have a proper prediction interval, which certainly does not cover exactly that of the rainwater on that side. However, it is beyond the scope of this study to show also the prediction interval related to groundwater regression.

In terms of the snow samples, all but two of the 61 snow samples from northern Chile fall within the area between the generalized intervals determined in this study, confirming that this area could be approximated as an uncertainty ribbon around the EIV regression (Figure 2B). It should be noted that this dataset includes both samples from precipitation samplers and the Andean snow cover. The two samples out of this area are one outlier (left side of the meteoric water line, Figure 2B) and one isotopically enriched sample from the so-called "penitents" of the Parinacota volcano ([64], Figure 1B—Region I, Supplementary File S5). The penitentes are an annual phenomenon of snow sublimation related to solar radiation and arid climates; consequently, the isotope composition of these samples could be enriched, in particular, if sampled from surface layers [64,67,81]. The volume weighted mean and the related combined standard deviation calculated on the isotope composition of the snow samples from the Cerro Tapado glacier ([67], Figure 1B—Region IV, Supplementary File S5) fall exactly on the EIV-LMWL's extension beyond the most negative value and within the generalized interval (Figure 2B). As expected, the penitentes from the same area are enriched in comparison with the mean isotope composition of the local snow due to sublimation (Figure 2B, [67]). Moreover, although their mean values fall close to EIV-LMWL, the combined sigma of the penitentes can be higher than the generalized interval (up to ±16‰; Figure 2B, Supplementary File S5).

Finally, geothermal water samples from a recent study were plotted with the EIV regression line and the generalized interval, showing more clearly the distinction between the meteoric waters and the hydrothermal water samples that can be isotopically enriched as a result of high-temperature isotopic fractionations between water–rock, mixing with andesitic water or water–gas interactions [70,74] (Figure 2B). Specifically speaking, the Torta de Tocorpuri (TT in Figure 2B) and Pampa Lirima (PL in Figure 2B) samples are not different from meteoric waters. According to the Tassi et al. [70], the limited temperature and involvement of the so-called "andesitic water" does not produce a relevant shift from the meteoric composition in these two sample groups (Figure 2B). In particular, the PL sample group lies precisely between Pampa del Tamarugal line and the EIV local meteoric water line (Figure 2B).

5. Conclusions

The regression analysis based on the error-in-variables approach (EIV) produces a meteoric water line with a slope similar to those obtained with ordinary least squares regression or other regression approaches.

This most likely occurs because the variance ratio (i.e., the square of the combined standard deviations) is high ($\delta^2 H_{var}/\delta^{18}O_{var} >> 3$; [2]).

Further studies that focus on other areas and which use a greater number of amount-weighted data pairs (for example, the GNIP/WMO dataset [37]) could be useful in terms of verifying this.

The present study offers an EIV local meteoric water line that summarizes all the previously published isotope data of precipitation from northern Chile. Moreover, the use of the innovative generalized interval (locally, $\delta^2 H = \pm13.1‰$ up and down the fitted values on the line), which is a good compromise between the confidence and the prediction intervals, appears to be a useful tool for distinguishing significantly different data from precipitation.

The case of the groundwater system of Pampa del Tamarugal, snow samples and the local geothermal waters is emblematic. An efficient application of the generalized interval or prediction band could be extended to waters that have been subjected to evaporative fractionation (surface water or precipitation samples from desert climate areas), while it could be used to distinguish the outliers.

Supplementary Materials: The following are available online at http://www.mdpi.com/2073-4441/11/4/791/s1: Supplementary File S1—Table 2 raw data. Dark blue-highlighted bold values are the data shown in Table 2.

Light blue-highlighted values are the results of partial calculations or additional parameters. Other colors depict parameters from the previously published tables (e.g., yellow-highlighted annual weighted mean in stations 1–9, 10–11, and 12–24) or warnings (grey-highlighted discharged data in stations 29–31 and 32; orange-highlighted measurement error in all the sheets). Supplementary File S2—Regression results as in Table 3 but excluding the sample #9 ($N_{a.p.}$ = 1) from Table 2 dataset. Supplementary File S3—Intervals of the meteoric water line calculated using Table 2 dataset by BivRegBLS [77] and Real Statistics Using Excel© [71]. In the former, code admits only $N_{a.p.}$ > 1 stations, whereas the latter use constant variance ratio ($\lambda = \delta^{18}O_{var}/\delta^2H_{var}$ = 0.019). Supplementary File S4—$\delta^{18}O$ and δ^2H of single precipitation without amount data [28,30,36,56–60,68,79]; OLSR results and prediction band calculated on the isotope composition of the single precipitation dataset. Supplementary File S5—Sheet 1: $\delta^{18}O$ and δ^2H of single snow samples (collected from snow cover or by precipitation sampler; [43,58,59,61–66,68]). Sheet 2: volume weighted mean and combined standard deviation on snow and penitentes [67].

Author Contributions: Study design, data search, data and interpretation, writing the manuscript—original draft preparation, communicating with the journal, T.B.; data search, components for discussion on results and conclusions, writing—review and editing, J.C.; components for discussion on results and conclusions, writing—review and editing, P.I. and E.S.

Funding: This research received no external funding.

Acknowledgments: We are indebted to Bernard Francq, Université Catholique de Louvain—Belgium, who gave a great deal of assistance in the R code and for a critical revision of a first version of the manuscript. Special thanks to Javier Uribe and Jorge Gironás, Pontificia Universidad Católica de Chile, for the raw data on precipitation collected at Salar del Huasco basin. We extend acknowledgments and thanks to all the authors who published isotope data of precipitation in the studied area. Finally, we would like to thank the Editor, Polona Vreča, and two anonymous reviewers for their helpful comments and suggestions.

Conflicts of Interest: The authors declare no conflict of interest.

References

1.	NIST/SEMATECH. NIST/SEMATECH e-Handbook of Statistical Methods. Available online: https://www.itl.nist.gov/div898/handbook/index.htm (accessed on 8 February 2019).

2.	Legendre, P.; Legendre, L. *Numerical Ecology*, 3rd ed.; Elsevier: Amsterdam, The Netherlands, 2012; Volume 24.

3.	Yurtsever, Y.; Gat, J.R. Atmospheric waters. In *Stable Isotope Hydrology: Deuterium and Oxygen-18 in the Water Cycle - Technical Report Series N°210*; Gonfiantini, R., Ed.; International Atomic Energy Agency: Vienna, Austria, 1981; pp. 103–142.

4.	IAEA. *Statistical Treatment of Data on Environmental Isotopes in Precipitation—Technical Report Series n° 331*; International Atomic Energy Agency—IAEA: Vienna, Austria, 1992; p. 781.

5.	Harper, W.V. Reduced Major Axis Regression: Teaching Alternatives to Least Squares. In Proceedings of the Ninth International Conference on Teaching Statistics (ICOTS9), Flagstaff, AZ, USA, 13–18 July 2014.

6.	Crawford, J.; Hughes, C.E.; Lykoudis, S. Alternative least squares methods for determining the meteoric water line, demonstrated using GNIP data. *J. Hydrol.* **2014**, *519*, 2331–2340. [CrossRef]

7.	Hughes, C.E.; Crawford, J. A new precipitation weighted method for determining the meteoric water line for hydrological applications demonstrated using Australian and global GNIP data. *J. Hydrol.* **2012**, *464*, 344–351. [CrossRef]

8.	Argiriou, A.A.; Lykoudis, S. Isotopic composition of precipitation in Greece. *J. Hydrol.* **2006**, *327*, 486–495. [CrossRef]

9.	Deming, W.E. *Statistical Adjustment of Data*; John Wiley & Sons, Inc.: New York, NY, USA, 1943.

10.	Gillard, J. An overview of linear structural models in errors in variables regression. *REVSTAT–Stat. J.* **2010**, *8*, 57–80.

11.	Maind. Sigmaplot di Systat Software. Available online: http://www.maind.it/document/SigmaplotStatistica.pdf (accessed on 27 December 2018).

12.	Francq, B.G.; Govaerts, B.B. Measurement methods comparison with errors-in-variables regressions. From horizontal to vertical OLS regression, review and new perspectives. *Chemom. Intell. Lab. Syst.* **2014**, *134*, 123–139. [CrossRef]

13.	Argiriou, A.A.; Salamalikis, V.; Dotsika, E. A Total Weighted Least Squares Method for the Determination of the Meteoric Water Line of Precipitation for Hydrological Purposes. In *Perspectives on Atmospheric Sciences*; Karacostas, T.S., Bais, A., Nastos, P.T., Eds.; Springer: Cham, Switzerland, 2017; pp. 233–238.

14. Garreaud, R.D. The climate of northern Chile: Mean state, variability and trends. *Rev. Mex. De Astron. Y Astrofísica - Ser. De Conf.* **2011**, *41*, 5–11.

15. Harrison, S.; Glasser, N.F. The Pleistocene glaciations of Chile. In *Developments in Quaternary Sciences*; Elhers, J., Gibbard, P.L., Hughes, C.E., Eds.; Elsevier: Amsterdam, The Netherlands, 2011; Volume 15, pp. 739–756.

16. Yáñez, E.; Lagos, N.A.; Norambuena, R.; Silva, C.; Letelier, J.; Muck, K.P.; San Martin, G.; Benítez, S.; Broitman, B.R.; Contreras, H.; et al. Impacts of climate change on marine fisheries and aquaculture in Chile. In *Climate Change Impacts on Fisheries and Aquaculture*; Phillips, B.F., Pérez-Ramírez, M., Eds.; John Wiley & Sons: Chichester, UK, 2018; pp. 239–332.

17. SERNAGEOMIN. *Mapa geológico de Chile: Versión digital. Escala 1:1'000'000*, 1.0 ed.; Servicio Nacional de Geología y Minería, Subdirección Nacional de Geología: Gobierno de Chile, Santiago, 2003; Publicación Geológica Digital No. 4 (CD-ROM, versión 1, 2003).

18. DeCelles, P.G.; Carrapa, B.; Horton, B.K.; McNabb, J.; Gehrels, G.E.; Boyd, J. The Miocene Arizaro Basin, central Andean hinterland: Response to partial lithosphere removal. In *Geodynamics of a Cordilleran Orogenic System: The Central Andes of Argentina and Northern Chile: Geological Society of America*; DeCelles, P.G., Ducea, M.N., Carrapa, B., Kapp, P.A., Eds.; The Geological Society of America: Boulder, CO, USA, 2015; Volume 212, pp. 359–386.

19. Schilling, F.R.; Trumbull, R.B.; Brasse, H.; Haberland, C.; Asch, G.; Bruhn, D.; Mai, K.; Haak, V.; Giese, P.; Muñoz, M.; et al. Partial melting in the Central Andean crust: A review of geophysical, petrophysical, and petrologic evidence. In *The Andes—Active Subduction Orogeny*; Oncken, O., Chong, G., Franz, G., Giese, P., Götze, H.J., Ramos, V.A., Strecker, M.R., Wigger, P., Eds.; Springer: Berlin/Heidelberg, Germany, 2006; pp. 459–474.

20. Geiger, R. Klassifikation der klimate nach W. Köppen. *Landolt-Börnstein–Zahlenwerte Und Funkt. Aus Phys. Chem. Astron. Geophys. Und Tech.* **1954**, *3*, 603–607.

21. Beck, H.E.; Zimmermann, N.E.; McVicar, T.R.; Vergopolan, N.; Berg, A.; Wood, E.F. Present and future Köppen-Geiger climate classification maps at 1-km resolution. *Sci. Data* **2018**, *5*. [CrossRef]

22. Sarricolea, P.; Herrera-Ossandon, M.; Meseguer-Ruiz, Ó. Climatic regionalisation of continental Chile. *J. Maps* **2017**, *13*, 66–73. [CrossRef]

23. Bershaw, J.; Saylor, J.E.; Garzione, C.N.; Leier, A.; Sundell, K.E. Stable isotope variations (δ^{18}O and δD) in modern waters across the Andean Plateau. *Geochim. Cosmochim. Acta* **2016**, *194*, 310–324. [CrossRef]

24. Vuille, M.; Bradley, R.S.; Werner, M.; Healy, R.; Keimig, F. Modeling δ^{18}O in precipitation over the tropical Americas: 1. Interannual variability and climatic controls. *J. Geophys. Res. Atmos.* **2003**, *108*. [CrossRef]

25. Sánchez-Murillo, R.; Aguirre-Dueñas, E.; Gallardo-Amestica, M.; Moya-Vega, P.; Birkel, C.; Esquivel-Hernández, G.; Boll, J. Isotopic characterization of waters across Chile. In *Andean Hydrology*; Rivera, D.A., Godoy-Faundez, A., Lillo-Saavedra, M., Eds.; CRC Press: New York, NY, USA, 2018; pp. 205–230.

26. Fiorella, R.P.; Poulsen, C.J.; Pillco-Zolá, R.S.; Barnes, J.B.; Tabor, C.R.; Ehlers, T.A. Spatiotemporal variability of modern precipitation δ^{18}O in the central Andes and implications for paleoclimate and paleoaltimetry estimates. *J. Geophys. Res. Atmos.* **2015**, *120*, 4630–4656. [CrossRef]

27. Gonfiantini, R.; Roche, M.A.; Olivry, J.C.; Fontes, J.C.; Zuppi, G.M. The altitude effect on the isotopic composition of tropical rains. *Chem. Geol.* **2001**, *181*, 147–167. [CrossRef]

28. Jordan, T.E.; Herrera, L.C.; Godfrey, L.V.; Colucci, S.J.; Gamboa, P.C.; Urrutia, M.J.; González, L.G.; Jacob, F.P. Isotopic characteristics and paleoclimate implications of the extreme precipitation event of March 2015 in northern Chile. *Andean Geol.* **2019**, *46*, 1–31. [CrossRef]

29. Rohrmann, A.; Strecker, M.R.; Bookhagen, B.; Mulch, A.; Sachse, D.; Pingel, H.; Alonso, R.N.; Schildgen, T.F.; Montero, C. Can stable isotopes ride out the storms? The role of convection for water isotopes in models, records, and paleoaltimetry studies in the central Andes. *Earth Planet. Sci. Lett.* **2014**, *407*, 187–195. [CrossRef]

30. Burgener, L.; Huntington, K.W.; Hoke, G.D.; Schauer, A.; Ringham, M.C.; Latorre, C.; Díaz, F.P. Variations in soil carbonate formation and seasonal bias over >4 km of relief in the western Andes (30° S) revealed by clumped isotope thermometry. *Earth Planet. Sci. Lett.* **2016**, *441*, 188–199. [CrossRef]

31. Aravena, R.; Suzuki, O.; Peña, H.; Pollastri, A.; Fuenzalida, H.; Grilli, A. Isotopic composition and origin of the precipitation in Northern Chile. *Appl. Geochem.* **1999**, *14*, 411–422. [CrossRef]

32. Chaffaut, I. *Précipitations d'altitude, eaux souterraines et changements climatiques de l'altiplano Nord-Chilien*; Universite Paris Sud, U.F.R. Scientifique D'Orsay: Paris, France, 1998.

33. Chaffaut, I.; Coudrain-Ribstein, A.; Michelot, J.L.; Pouyaud, B. Précipitation d'altitude du nord-Chili, origine des sources de vapeur et données isotopiques. *Bull. De L'institut Français D'études Andin.* **1998**, *27*, 367–384.

34. DGA. *Diagnóstico de disponibilidad hídrica en la cuenca del río Lauca, región de Arica y Parinacota*; Ministerio de Obras Públicas, Dirección General de Aguas - Arica y Parinacota, XV Región: Arica, Chile, 2015.

35. Fritz, P.; Suzuki, O.; Silva, C.; Salati, E. Isotope hydrology of groundwaters in the Pampa del Tamarugal, Chile. *J. Hydrol.* **1981**, *53*, 161–184. [CrossRef]

36. Herrera, C.; Pueyo, J.J.; Sáez, A.; Valero-Garcés, B.L. Relación de aguas superficiales y subterráneas en el área del lago Chungará y lagunas de Cotacotani, norte de Chile: Un estudio isotópico. *Rev. Geológica De Chile* **2006**, *33*, 299–325. [CrossRef]

37. IAEA/WMO. Global Network of Isotopes in Precipitation. *The GNIP Database.* Available online: https://nucleus.iaea.org/wiser (accessed on 15 January 2019).

38. Squeo, F.A.; Aravena, R.; Aguirre, E.; Pollastri, A.; Jorquera, C.B.; Ehleringer, J.R. Groundwater dynamics in a coastal aquifer in north-central Chile: Implications for groundwater recharge in an arid ecosystem. *J. Arid Environ.* **2006**, *67*, 240–254. [CrossRef]

39. Troncoso, R.; Castro, R.; Lorca, M.E.; Espinoza, C.; Pérez, Y. Análisis Preliminar de la Composición Isotópica Oxígeno 18 – Deuterio de las Aguas de la Cuenca del Río Copiapó, Región de Atacama: Una Contribución al Conocimiento del Sistema Hidrogeológico. In Proceedings of the XIII Congreso Geológico Chileno, Antofagasta, Chile, 5–9 August 2012; pp. 774–776.

40. Uribe, J.; Muñoz, J.F.; Gironás, J.; Oyarzún, R.; Aguirre, E.; Aravena, R. Assessing groundwater recharge in an Andean closed basin using isotopic characterization and a rainfall-runoff model: Salar del Huasco basin, Chile. *Hydrogeol. J.* **2015**, *23*, 1535–1551. [CrossRef]

41. Salazar, C.M.; Rojas, L.B.; Pollastri, A. *Evaluación de recursos hídricos en el sector de Pica hoya de la Pampa del Tamarugal I region*; Ministerio de Obras Públicas, Dirección General de Aguas, CCHEN: Santiago, Chile, 1998; p. 98.

42. Aravena, R.; Peña, H.; Grilli, A.; Suzuki, O.; Mordeckai, M. Evolución isotópica de las lluvias y origen de las masas de aire en el Altiplano chileno. In *Isotope Hydrology Investigations in Latin America*; International Atomic Energy Agency—IAEA: Vienna, Austria, 1989; pp. 129–142.

43. Herrera, C.; Custodio, E.; Chong, G.; Lambán, L.J.; Riquelme, R.; Wilke, H.; Jódar, J.; Urrutia, J.; Urqueta, H.; Sarmiento, A.; et al. Groundwater flow in a closed basin with a saline shallow lake in a volcanic area: Laguna Tuyajto, northern Chilean Altiplano of the Andes. *Sci. Total Environ.* **2016**, *541*, 303–318. [CrossRef]

44. NIST. DATAPLOT Reference Manual - weighted standard deviation. Available online: https://www.itl.nist.gov/div898/software/dataplot/refman2/ch2/weightsd.pdf (accessed on 28 December 2018).

45. Taylor, B.N.; Kuyatt, C.E. *Guidelines for Evaluating and Expressing the Uncertainty of NIST Measurement Results*; National Institute of Standards and Technology—NIST: Gaithersburg, MD, USA, 1994.

46. Leito, I.; Jalukse, L.; Helm, I. *Estimation of Measurement Uncertainty in Chemical Analysis (Analytical Chemistry) Course*; University of Tartu: Tartu, Estonia, 2018.

47. Bell, S. *A Beginner's Guide to Uncertainty of Measurement*; National Physical Laboratory: Teddington, UK, 1999.

48. Williamson, J.H. Least-squares fitting of a straight line. *Can. J. Phys.* **1968**, *46*, 1845–1847. [CrossRef]

49. York, D. Least-squares fitting of a straight line. *Can. J. Phys.* **1966**, *44*, 1079–1086. [CrossRef]

50. York, D. Unified equations for the slope, intercept, and standard error of the best straight line. *Am. J. Phys.* **2004**, *72*, 367–375. [CrossRef]

51. York, D. Least squares fitting of a straight line with correlated errors. *Earth Planet. Sci. Lett.* **1968**, *5*, 320–324. [CrossRef]

52. Originlab. Algorithms (Fit Linear with X Error). Available online: https://www.originlab.com/doc/Origin-Help/Ref-Linear-XErr#Fit_Parameters (accessed on 27 December 2018).

53. Systat. *Using SigmaStat Statistics in SigmaPlot*; Systat Software: San Jose, CA, USA, 2013; p. 470.

54. Cantrell, C.A. Technical Note: Review of methods for linear least-squares fitting of data and application to atmospheric chemistry problems. *Atmos. Chem. Phys.* **2008**, *8*, 5477–5487. [CrossRef]

55. Sturm, P. bfsl: Best-Fit Straight Line, 0.1.0; CRAN.R-project.org. 2018. Available online: https://cran.r-project.org/web/packages/bfsl/index.html (accessed on 3 February 2019).

56. Aravena, R.; Suzuki, O.; Pollastri, A. Coastal fog and its relation to groundwater in the IV region of northern Chile. *Chem. Geol.* **1989**, *79*, 83–91. [CrossRef]

57. Cifuentes, J.L.; Cervetto, M.M.; López, L.A.; Fuentes, F.C.; Feuker, P.; Espinoza, C. Análisis preliminar de Isótopos Estables en aguas subterráneas, superficiales y lluvia de la Pampa del Tamarugal. In Proceedings of the XIV Congreso Geológico Chileno, La Serena, Chile, 4–8 October 2015; pp. 277–280.

58. DGA. *Levantamiento hidrogeológico para el desarrollo de nuevas fuentes de agua en áreas prioritarias de la zona norte de Chile, Regiones XV, I, II y III, Etapa 2*; Ministerio de Obras Públicas, Dirección General de Aguas, Departamento de Estudios y Planificación; Pontificia Universidad Católica de Chile, Departamento de Ingeniería Hidráulica y Ambiental: Santiago, Chile, 2009.

59. López, L.A.; Cifuentes, J.L.; Fuentes, F.C.; Neira, H.A.; Cervetto, M.M.; Troncoso, R.A.; Feuker, P. *Hidrogeología de la Cuenca de la Pampa del Tamarugal, Región de Tarapacá*; Gobierno de Chile, Servicio Nacional de Geología y Minería: Santiago, Chile, 2017; p. 186.

60. Durán, L.V.L. *Hidrogeoquímica de fuentes termales en ambientes salinos relacionados con salares en los Andes del Norte de Chile*; Universidad de Chile: Santigo, Chile, 2016.

61. Giggenbach, W. The isotopic composition of waters from the El Tatio geothermal field, Northern Chile. *Geochim. Cosmochim. Acta* **1978**, *42*, 979–988. [CrossRef]

62. Godfrey, L.V.; Jordan, T.E.; Lowenstein, T.K.; Alonso, R.L. Stable isotope constraints on the transport of water to the Andes between 22° and 26°S during the last glacial cycle. *Palaeogeogr. Palaeoclimatol. Palaeoecol.* **2003**, *194*, 299–317. [CrossRef]

63. Munoz-Saez, C.; Manga, M.; Hurwitz, S. Hydrothermal discharge from the El Tatio basin, Atacama, Chile. *J. Volcanol. Geotherm. Res.* **2018**, *361*, 25–35. [CrossRef]

64. Peña, H. Mediciones de ^{18}O y ^{2}H en "Penitentes" de Nieve. In *Isotope Hydrology Investigations in Latin America*; International Atomic Energy Agency—IAEA: Vienna, Austria, 1988; Volume IAEA-TECDOC-502, pp. 143–154.

65. Cortecci, G.; Boschetti, T.; Mussi, M.; Herrera Lameli, C.; Mucchino, C.; Barbieri, M. New chemical and isotopic data on waters of El Tatio Geothermal Field, Northern Chile. *Geochem. J.* **2005**, *39*, 547–571. [CrossRef]

66. Alpers, C.N.; Whittemore, D.O. Hydrogeochemistry and stable isotopes of ground and surface waters from two adjacent closed basins, Atacama Desert, northern Chile. *Appl. Geochem.* **1990**, *5*, 719–734. [CrossRef]

67. Sinclair, K.E.; MacDonell, S. Seasonal evolution of penitente glaciochemistry at Tapado Glacier, Northern Chile. *Hydrol. Process.* **2016**, *30*, 176–186. [CrossRef]

68. Cervetto, M.M. *Caracterización hidrogeológica e hidrogeoquímica de las cuencas: Salar de Aguas calientes 2, Puntas negras, Laguna Tuyajto, Pampa Colorada, Pampa Las Tecas y Salar el Laco, II región de Chile*; Universidad de Chile: Santiago, Chile, 2012.

69. Scheihing, K.; Moya, C.; Struck, U.; Lictevout, E.; Tröger, U. Reassessing Hydrological Processes That Control Stable Isotope Tracers in Groundwater of the Atacama Desert (Northern Chile). *Hydrology* **2018**, *5*, 3. [CrossRef]

70. Tassi, F.; Aguilera, F.; Darrah, T.; Vaselli, O.; Capaccioni, B.; Poreda, R.J.; Huertas, A.D. Fluid geochemistry of hydrothermal systems in the Arica-Parinacota, Tarapacá and Antofagasta regions (northern Chile). *J. Volcanol. Geotherm. Res.* **2010**, *192*, 1–15. [CrossRef]

71. Zaiontz, C. Real Statistics Using Excel. Available online: www.real-statistics.com (accessed on 3 February 2019).

72. Boschetti, T.; Cortecci, C.; Barbieri, M.; Mussi, M. New and past geochemical data on fresh to brine waters of the Salar de Atacama and Andean Altiplano, northern Chile. *Geofluids* **2007**, *7*, 35–50. [CrossRef]

73. Verschuuren, G. *Excel 2013 for Scientists*; Holy Macro! Books: Chicago, IL, USA, 2014; p. 250.

74. Clark, I. *Groundwater Geochemistry and Isotopes*; CRC Press, Taylor & Francis Group: Boca Raton, FL, USA, 2015.

75. Gat, J.R. *Isotope Hydrology: A Study of the Water Cycle*; Imperial College Press: London, UK; Singapore, 2010.

76. Francq, B.; Govaerts, B. How to regress and predict in a Bland–Altman plot? Review and contribution based on tolerance intervals and correlated-errors-in-variables models. *Stat. Med.* **2016**, *35*, 2328–2358. [CrossRef] [PubMed]

77. Francq, B.G.; Berger, M. BivRegBLS: Tolerance Intervals and Errors-in-Variables Regressions in Method Comparison Studies, 1.0.0; CRAN.R-project.org. 2017. Available online: https://CRAN.R-project.org/package=BivRegBLS (accessed on 3 February 2019).

78. Berger, M.; Francq, B. BivRegBLS: A new R package in method comparison studies with tolerance intervals and (correlated)-errors-in-variables regressions. In Proceedings of the Chimiométrie XVIII, Paris, France, 30 January–1 February 2017.

79. Lorca, M.E. *Hidrogeología e hidrogeoquímica de la cuenca de la Quebrada Paipote, Región de Atacama*; Universidad de Chile: Santiago, Chile, 2011.
80. Aravena, R. Isotope hydrology and geochemistry of northern Chile groundwaters. *Bull. De L'institut Français D'études Andin.* **1995**, *24*, 495–503.
81. Stichler, W.; Schotterer, U.; Fröhlich, K.; Ginot, P.; Kull, C.; Gäggeler, H.; Pouyaud, B. Influence of sublimation on stable isotope records recovered from high-altitude glaciers in the tropical Andes. *J. Geophys. Res. Atmos.* **2001**, *106*, 22613–22620. [CrossRef]

Article

Stable Isotopes of Precipitation in China: A Consideration of Moisture Sources

Yanlong Kong [1,2,3,*], Ke Wang [1,2,3], Jie Li [4] and Zhonghe Pang [1,2,3,*]

[1] Key Laboratory of Shale Gas and Geoengineering, Institute of Geology and Geophysics, Chinese Academy of Sciences, Beijing 100029, China; wangke@mail.iggcas.ac.cn
[2] Institutions of Earth Science, Chinese Academy of Sciences, Beijing 100029, China
[3] University of Chinese Academy of Sciences, Beijing 100049, China
[4] College of Water Sciences, Beijing Normal University, Beijing 100875, China; lijie_lm@163.com
[*] Correspondence: ylkong@mail.iggcas.ac.cn (Y.K.); z.pang@mail.iggcas.ac.cn (Z.P.); Tel.: +86-10-82998611 (Y.K. and Z.P.)

Received: 17 April 2019; Accepted: 5 June 2019; Published: 13 June 2019

Abstract: An accurate representation of the spatial distribution of stable isotopes in modern precipitation is vital for interpreting hydrological and climatic processes. Considering the dominant impact of moisture sources in controlling water isotopes and deuterium excess, we conducted a meta-analysis of precipitation isotopes using instrumental data from 68 stations around China. The entire country is divided into five regions according to the major moisture sources: Region I (the westerlies domain), Region II (the arctic domain), Region III (the northeast domain), Region IV (the Pacific domain), and Region V (the Tibetan Plateau). Each region has unique features of spatial distribution and seasonal variation for stable precipitation isotopes and deuterium excess. In particular, seasonal variation in Region IV tracks the onset of Asian summer monsoons well. The regional meteoric water lines are presented for each region. A significant temperature effect is found in Regions I and III, with δ^{18}O-temperature gradients of 0.13–0.68‰/°C and 0.13–0.4‰/°C, respectively. However, the reasons for the temperature effects are quite different. In Region I, this effect is caused by the seasonal shift of the westerlies, whereas in Region III, it is caused by the seasonal difference in moisture sources. The precipitation amount effect is most significant in the region along the southeast coast in China, where the δ^{18}O-precipitation amount the gradient is −0.24 to −0.13‰/mm. The findings in our paper could serve as a reference for isotopic application in hydrological and paleo-climatic research.

Keywords: stable isotopes D and ^{18}O; moisture source; temperature effect; precipitation amount effect; regionalization; China

1. Introduction

The stable oxygen and hydrogen isotope compositions of precipitation are powerful tools for studying hydrological and climatic processes. Spatial variations in stable precipitation isotopes are related to various isotope effects, including the continental effect, the precipitation amount effect, and the altitude effect [1]. Generally, atmospheric moisture is derived primarily from low-latitude oceanic regions, where the initial composition is fixed by evaporation and boundary layer diffusion between the ocean and atmosphere. High δ^2H or δ^{18}O values occur near to source regions where vapor is sourced to the atmosphere, and low values occur far from source regions as a result of a progressive rainout history.

Access to data on spatial precipitation isotopes serves as a basis for the widespread application of water isotopes. Since 1961, the International Atomic Energy Agency (IAEA), in cooperation with the World Meteorological Organization (WMO), has been conducting a worldwide survey of the isotopic composition of monthly precipitation [2]. Based on the IAEA and WMO observations, several analyses have been conducted in Southeast Asia, Amazon catchments, Australia, and South Africa [3–8]. In China, systemic work has yielded the following findings (including but not limited to): (1) The moisture source is very important to the spatial variations in precipitation isotopes; (2) the temperature effect is significant in North China; and (3) the precipitation amount effect is significant in South China [6,9–11].

Despite large-scale precipitation isotope monitoring throughout the country, in West China, observations on precipitation isotopes are quite scarce relative to the huge area and its various terrains. Given the vital role of the Tibetan Plateau and Tianshan Mountains in affecting water resources and the economic issues of the billions of people in Asia, it is necessary to conduct long-term observations of water isotopes in West China. Fortunately, considerably more work has been done recently, and increasingly more isotopic data have been obtained in this region [12–18]. The addition of new data from West China lays the foundation for a more precise spatial isotopic pattern for China as a whole.

The most common applications of water isotope data are related to the global meteoric water line (GMWL) for hydrological research and the temperature effect for climatic research [1,19]. The use of the GMWL is based on the assumption that precipitation in the studied region has a meteorological pattern similar to the global average. However, this assumption is not valid on a local scale or even on a regional scale, owing to differences in moisture sources and geographical features. The temperature effect may be significantly altered as a result of the addition of another moisture source, especially recycled continental moisture. Therefore, it is necessary to obtain the regional meteoric water line (RMWL) within a region with similar meteorological conditions.

The derivative parameter "deuterium excess" (which was defined as d excess = δ^2H–8*δ^{18}O by Dansgaard [1]) is evidently inherited from the initial isotopic composition of an air mass and determined by the air–sea interaction regime [1]; it remains relatively invariant during Rayleigh rainout [20]. D excess in precipitation might help to address the moisture sources and reveal processes, including the condensation of atmospheric vapor, evaporation of water, sublimation of ice, and recycling of continental moisture. At least to the authors' best knowledge, there has been no spatial analysis of d excess at the regional scale in China, and such an analysis would be undeniably helpful for improving our understanding of precipitation isotopes and hydro-meteorological processes.

In this paper, we divide China into 5 regions based primarily on the moisture sources. Then, based on the new regionalization, we provide spatial and seasonal distributions of water isotopes and d excess. Explanations of the isotopic patterns are related to moisture sources. The RMWLs are presented for each region, which can be used in comparing with the isotopic composition of surface water and groundwater. The relationship between isotopes and temperature (the temperature effect) and that between isotopes and precipitation amount (the precipitation amount effect) are discussed to shed light on their roles in interpreting paleoclimate proxies.

2. Materials and Methods

2.1. Isotopic Data and Analysis

The precipitation isotopic data available for stations in China from the IAEA-WMO Global Network for Isotopes in Precipitation (GNIP) are taken into account in our analysis. To eliminate the edge effect of interpolation data, isotopic data from the GNIP stations around China are adopted. Although this database contains the most comprehensive observation set for water isotopes in precipitation, data in West China are quite scarce relative to the huge area of this region. Recently, new data have been independently generated and published by Tian et al. [12] and Pang et al. [13] in West China, which improved the spatial integrity of precipitation isotopes in China. Although the data are obtained in different years, they are still comparable in analyzing the impact of moisture sources on it and the isotope–meteorological relationship [21,22]. In total, data from a total of 68 stations are used to characterize the isotopic distribution in China (see Table 1). All the $\delta^{18}O$ and δ^2H values of the precipitation samples are measured by traditional isotope-ratio mass spectrometry or by laser absorption spectroscopy. It should be noted that the data above are from different labs, but all the data were calibrated with respect to Vienna Standard Mean Ocean Water (VSMOW) standards using the same normalization method. Both Tian et al. [12] and Pang et al. [13] tested the same sample in different labs to ensure that the measurement results from different labs were consistent. In total, all the data have a precision within 0.2‰ for $\delta^{18}O$ and 1‰ for δ^2H.

The multi-year annual average isotopic data ($\delta^{18}O_{ave}$) is calculated using precipitation amount as a weighting factor. Taking $\delta^{18}O$ at a station for example,

$$\delta^{18}O_{ave} = \text{sum}(\delta^{18}O_{i,j}*P_{i,j})/\text{sum}(P_{i,j}) \tag{1}$$

where $\delta^{18}O_{i,j}$ and $P_{i,j}$ are the isotopic value and the precipitation amount at month i, year j, respectively.

The regional meteoric water lines (RMWLs) are determined by the least squares regression based on the monthly isotopic data.

2.2. Interpolation

The spatial distribution of precipitation $\delta^{18}O$, δ^2H, and d excess is obtained by adopting the Kriging interpolation method. It takes both the distance and degree of variation among known points into consideration and then estimate the unknown points [23]. Among all the Kriging interpolating methods, the ordinary Kriging is selected due to its merits in providing the best linear unbiased predictions and the best sense of minimum variance [24], which helps us to get a regional overview of isotopes. The spherical model, which is one of the most commonly used models in characterizing precipitation isotopes distribution [25], was chosen to describe the experimental semi-variance in this work. Because we are interested in the impact of moisture sources on the precipitation isotopes (large scale atmospheric circulation, i.e., > 100 km), we chose to do the interpolation without the use of ancillary variables, such as elevation [26], which would introduce high amplitude variability into the interpolated surface over short scale lengths.

In order to make the interpolation more reasonable, all the data at the 68 stations were used to generate a spatial distribution of isotopes to the extent of covering all the stations. Then, we added the extent of China to it. All the interpolation and mapping processes were carried out using the ARCGIS software.

Table 1. Summary of precipitation amount of weighted annual isotopes ($\delta^{18}O$ and $\delta^{2}H$) and d excess with their standard deviations std1, std2, and std3, multi-year average annual precipitation (P), multi-year average annual temperature (T) at the selected stations, the number of months at the GNIP stations, and daily number of samples at stations from Tian et al. [12] and Pang et al. [13] (n).

Sample Site	Longitude (E)	Latitude (N)	Altitude (m)	δ^2H (‰)	Std1 (‰)	$\delta^{18}O$ (‰)	Std2 (‰)	d excess (‰)	Std3 (‰)	P (mm)	T (°C)	n	Observation Period	Source
Omsk	73.38	55.01	94	−98.8	/	−13.5	/	8.9	/	361.0	2.7	8	1990	GNIP
Enisejsk	92.15	58.45	98	−98.4	/	−13.3	/	7.9	/	609.0	0.5	12	1990	GNIP
Novosibirsk	82.90	55.03	162	−104.3	/	−14.6	/	12.8	/	479.0	2.5	12	1990	GNIP
Irkutsk	104.35	52.27	485	−94.5	13.1	−12.2	0.9	3.2	5.7	482.0	2.4	14	1990	GNIP
Habarovsk	135.17	48.52	72	−102.9	/	−14.3	/	11.4	/	911.0	2.2	11	1971	GNIP
Tashkent	69.27	41.27	428	−42.9	/	−7.0	/	13.1	/	316.0	8.6	7	1971	GNIP
Kabul	69.21	34.57	1860	−34.8	25.6	−6.6	3.5	18.3	6.1	348.9	11.8	86	1967–1970, 1973–1974, 1982–1989	GNIP
Karachi	67.13	24.90	23	−24.3	15.0	−4.1	2.0	8.7	4.4	438.8	24.2	39	1961–1968, 1970, 1973	GNIP
New Delihi	77.20	28.58	212	−34.6	18.2	−5.4	2.4	8.3	4.0	899.1	24.8	304	1961–1969, 1973–1996, 2000–2008	GNIP
Allahabad	81.73	25.45	98	−67.8	/	−9.5	/	8.1	/	1531.0	17.8	5	1980	GNIP
Shillong	91.88	25.57	1598	−34.4	30.2	−5.7	3.8	11.1	2.8	1950.0	17.0	30	1969–1970, 1973–1976, 1978	GNIP
Salagiri	79.44	18.19	259	−33.4	/	−5.0	/	7.0	/	640.0	/	5	1977	GNIP
Colombo	79.87	6.91	7	−23.5	1.0	−4.1	0.2	8.9	0.5	2179.3	25.4	54	1983–1987, 1992–1994	GNIP
Hongkong	114.16	22.31	66	−42.1	9.8	−6.6	1.2	10.3	1.9	2342.8	23.0	408	1961–1965, 1973–2007	GNIP
Pohang	129.38	36.03	6	−51.8	17.4	−7.8	2.5	10.3	4.5	1141.0	13.3	110	1961–1966, 1973–1976	GNIP
Ryori	141.81	39.03	260	−55.9	8.1	−8.5	1.2	12.0	4.0	1400.6	10.2	189	1979–1986, 1998–2006	GNIP
Tokyo	139.77	35.68	4	−46.5	7.0	−7.2	0.8	11.4	2.8	1400.2	15.6	183	1961–1969, 1973–1979	GNIP
Yangoon	96.17	16.77	20	−29.4	7.8	−4.5	0.5	6.3	3.5	2434.7	27.3	18	1961–1963	GNIP
Bangkok	100.50	13.73	2	−42.9	7.7	−6.6	3.9	9.5	6.6	1638.8	28.5	347	1968–1970, 1973–2008	GNIP
Ko Sichang	100.80	13.17	26	−38.2	4.9	−6.1	0.6	10.2	1.7	1273.3	28.3	40	1985, 1987–1991	GNIP
Ko Samui	100.05	9.47	7	−28.3	10.4	−4.9	1.3	10.7	0.1	1444.2	27.7	27	1979–1980, 1982–1983	GNIP
Singapore	103.90	1.35	32	−46.5	2.5	−7.5	0.2	13.1	1.4	2345.9	26.2	52	1968–1969, 1973–1975	GNIP
Luang-Prabang	102.13	19.88	305	−53.2	10.5	−7.8	1.3	8.9	2.2	1407.9	26.2	26	1961–1964	GNIP
Qiqihar	123.91	47.38	147	−80.9	6.2	−10.8	0.6	5.8	1.8	570.2	3.9	50	1988–1992	GNIP
Haerbin	126.62	45.68	172	−78.0	4.3	−10.5	0.9	5.8	4.7	623.7	9.1	32	1986–1990, 1996–1997	GNIP
Hetian	79.93	37.13	1375	−34.9	4.8	−5.7	0.6	11.1	1.9	217.8	9.2	46	1988–1992	GNIP
Wulumuqi	87.62	43.78	918	−74.0	15.7	−10.9	2.1	12.9	2.9	336.3	7.3	123	1986–1992, 1996–1998, 2001–2003	GNIP
Zhangye	100.43	38.93	1483	−47.6	8.5	−6.6	1.3	5.4	7.7	150.3	7.8	79	1986–1992, 1995–1996, 2001–2003	GNIP
Lanzhou	103.88	36.05	1517	−41.6	12.0	−6.2	1.5	7.9	6.9	345.8	10.9	39	1985–1987, 1996–1990	GNIP
Yinchuan	106.21	38.48	1112	−43.5	12.3	−6.8	2.2	11.2	6.6	308.4	9.2	30	1988–1992	GNIP
Shijiazhuang	114.41	38.03	80	−56.8	7.7	−8.1	0.8	8.3	3.6	572.4	13.7	129	1985–2003	GNIP
Yantai	121.40	37.53	47	−50.6	4.8	−7.3	0.9	7.7	8.8	575.7	14.1	44	1986–1991	GNIP
Taiyuan	112.55	37.78	778	−62.1	15.1	−8.7	1.8	7.2	1.3	444.2	14.8	20	1986–1988	GNIP
Changchun	125.21	43.90	237	−63.8	9.8	−9.5	0.8	12.5	11.8	448.3	7.4	22	1999–2001	GNIP
Jinzhou	121.10	41.13	66	−54.6	5.2	−7.4	1.1	4.8	4.6	541.0	17.0	12	1986–1989	GNIP
Tianjin	117.16	39.10	3	−50.0	3.6	−7.7	0.5	11.7	1.4	536.5	13.8	64	1988–1992, 2000–2001	GNIP
Baotou	109.85	40.67	1067	−55.9	24.7	−7.9	1.0	7.4	3.4	300.6	7.0	52	1986–1993	GNIP
Lhasa	91.13	29.70	3649	−117.2	35.7	−15.9	4.6	10.3	1.8	427.0	8.4	41	1986–1992	GNIP
Chengdu	104.02	30.67	506	-51.2	11.1	-7.0	1.7	4.7	4.0	854.2	16.3	65	1986–1992, 1996–1999	GNIP

Table 1. *Cont.*

Sample Site	Longitude (E)	Latitude (N)	Altitude (m)	δ^2H (‰)	Std1 (‰)	$\delta^{18}O$ (‰)	Std2 (‰)	d excess (‰)	Std3 (‰)	P (mm)	T (°C)	n	Observation Period	Source
Kunming	102.68	25.01	1892	−69.2	12.1	−10.1	1.7	11.2	5.5	1005.5	15.5	148	1986–1991, 2000–2003	GNIP
Xian	108.93	34.30	397	−50.7	18.9	−7.7	2.7	11.2	3.4	585.5	13.4	48	1985–1993	GNIP
Zhengzhou	113.65	34.72	110	−54.0	15.4	−7.5	2.2	5.8	4.0	642.6	14.1	55	1985–1992	GNIP
Wuhan	114.13	30.62	23	−48.0	19.2	−6.7	2.3	5.4	5.0	1248.7	17.9	49	1986–1998	GNIP
Changqing	106.60	29.62	192	−69.6		−10.4		13.7		493.0	/	5	1992–1993	GNIP
Changsha	113.06	28.20	37	−32.2	0.1	−5.7	0.0	13.0	0.1	1280.7	17.1	57	1988–1992	GNIP
Zunyi	106.88	27.70	844	−55.6	7.0	−8.4	1.0	11.7	1.9	988.1	15.4	70	1986–1992	GNIP
Guiyang	106.71	26.58	1071	−52.0	10.5	−8.3	1.4	14.5	1.5	970.0	15.3	58	1988–1992	GNIP
Guilin	110.08	25.07	170	−34.7	6.1	−6.2	0.9	14.8	1.2	1523.6	19.0	91	1983–1990	GNIP
Nanjing	118.18	32.18	26	−53.0	6.7	−8.2	0.8	12.6	1.0	1212.0	14.4	58	1987–1992	GNIP
Fuzhou	119.28	26.08	16	−43.0	14.0	−6.6	1.5	9.8	3.3	1471.8	20.4	71	1985–1992	GNIP
Liuzhou	109.40	24.35	97	−43.9	10.8	−6.5	1.0	7.9	3.0	1082.5	20.9	45	1988–1992	GNIP
Guangzhou	113.32	23.13	7	−42.8	8.8	−6.1	1.1	6.0	2.4	1830.8	23.0	30	1986–1989	GNIP
Haikou	110.35	20.03	15	−37.6	12.9	−5.7	1.6	8.3	4.7	1923.3	24.9	58	1988–1991, 1996–2000	GNIP
Weathership V	164.00	31.00	/	−24.0	4.0	−6.1	0.6	24.8	3.4	943.4	19.1	81	1962–1969	GNIP
Taguac Guam IS.	144.83	13.55	110	−29.9	12.3	−5.1	1.5	10.7	2.9	2686.1	26.0	103	1962–1966, 1973–1977	GNIP
Wake Island	166.65	19.28	3	−9.8	8.7	−2.2	1.2	7.6	2.6	904.9	26.6	122	1962–1969, 1973–1976	GNIP
Truk	151.85	7.47	2	−32.5	10.8	−5.4	1.4	10.3	1.9	3606.2	27.4	72	1968–1969, 1972–1977	GNIP
Yap	138.09	9.49	23	−34.5	16.2	−5.5	2.5	9.6	4.5	2913.3	27.0	65	1968–1969, 1972–1976	GNIP
Tarawa	172.92	1.35	4	−37.5	/	−6.1	/	11.3	/	3187.0	25.9	17	1991	GNIP
Manila	121.00	14.52	14	−44.2	12.7	−6.8	1.5	9.8	3.8	1975.3	27.0	46	1961–1965	GNIP
Gaoshan	86.83	43.1	3545	−53.6	/	−7.7	/	8.4	/	390	−4.3	59	2003–2004	Pang et al. [13]
Houxia	87.18	43.28	2100	−55.4	/	−7.9	/	7.7	/	424	1.5	88	2003–2004	Pang et al. [13]
Nyalam	85.97	28.18	3811	−85.9	/	−12.1	/	10.6	/	650	/	295	1998–2001	Tian et al [12]
Lhasa	91.03	29.72	3650	−129.7	/	−17.0	/	5.8	/	411	/	152	1998–2001	Tian et al. [12]
Gaize	84.05	32.3	4416	−92.0	/	−12.7	/	9.6	/	170	/	140	1998–2001	Tian et al. [12]
Shiquanhe	80.1	32.5	4279	−76.3	/	−11.2	/	12.9	/	71	/	62	1998–2001	Tian et al. [12]
Yushu	97.02	33.02	3682	−82.0	/	−12.2	/	15.8	/	482	3	339	1998–2001	Tian et al. [12]
Altay	88.08	47.73	737	−100.4	/	−13.8	/	9.9	/	194	/	226	1998–2001	Tian et al. [12]

2.3. Correlation

The coefficient of determination (r^2) between isotopic values, temperature, and precipitation amount is calculated as

$$r^2 = \frac{\left(\sum\limits_{i=1}^{n}(x_i - \bar{x})(y_i - \bar{y})\right)^2}{\sum\limits_{i=1}^{n}(x_i - \bar{x})^2 \sum\limits_{i=1}^{n}(y_i - \bar{y})^2} \tag{2}$$

where n is the number of monthly samples, x is the isotopic value, and y is the temperature or precipitation. Linear regression analysis is employed to compute the gradient between isotopic values, temperature, and precipitation. The t test is employed to test the significance of the correlation. Among all the stations in China, more than 90% stations have at least 19 samples (n = 19). According to the t test (α = 0.05), the correlation is significant $r^2 > 0.3$ (n = 19).

3. Results and Discussion

3.1. Regionalization of China

Based on the climate, vegetation, and topography, Luo [27] divided China into 7 regions (Figure 1a). Following this work, Huang [28] re-divided China into 3 regions by merging Regions 1, 2, 3, 4, and part of 5 as the monsoon-affected region, leaving the rest as the Tibetan Plateau (Region 6 and part of 5) and the arid region (Region 7). Generally, these two regionalization schemes are used for the discussion of climatic and geographic features in China. Araguas-Araguas et al. [7] developed a new regionalization scheme based on the moisture sources dominating the precipitation in China (Figure 1b). There are five major air masses that dominate the pluviometric regime of China [7,9]: (1) the polar air mass originating in the Arctic; (2) the westerlies with recycled continental air mass over central Asia; (3) the tropical-maritime air mass originating in the northern Pacific; (4) the equatorial-marine air mass originating in the western equatorial Pacific; and (5) the equatorial-marine air mass originating in the Indian Ocean. Terzer et al. [29] developed a new model, named the regionalized cluster-based water isotope prediction approach (RCWIP), to predict point and large scale spatial-temporal patterns of the stable precipitation isotopes. In their work, they defined 36 climatic clusters, with 4 of them being related to China, whose representative stations are Chiang Rai, Erenhot, Lhasa, and Shanghai, respectively. Their criterion for the clusters were the differences of the climatic variables of temperature, precipitation amount, and vapor pressure between different stations.

The regionalization by Luo [27] incorporates geographical features in China, but it disregards the factor of moisture sources. As a result, one area that has a similar moisture source throughout is divided into two regions. For example, Regions 3 and 4 have the same moisture source from the South Pacific. However, in some cases, regions that have different moisture sources are treated as a single region. For instance, Region 7 is affected by both the westerlies and the arctic air masses. The regionalization by Huang [28] merges all the regions affected by monsoons but overlooks the different geographic features. Although the regionalization by Araguas-Araguas et al. [7] is based on moisture sources, some regions overlap (Figure 1). Furthermore, the extent of some moisture sources is still controversial. Tian et al. [12] found that the northern limit of summer monsoons from the Indian Ocean is in the middle of the Tibetan Plateau, at approximately 34°–35° N, which is farther north than the extent adopted by Araguas-Araguas et al. [7] and Johnson and Ingram [10]. Li et al. [30], Zhou et al. [31], and Xu et al. [32] found that the East Asian summer monsoons (EASM) can reach the Qaidam Basin (farther west than 100° E), whereas previous studies indicated that the direct influence of EASMs reach as far west as 100° E in China [7,33]. Pang [34] summarized the moisture sources over Northwest China on the basis of a meta-analysis of water isotopes in the region, demonstrating that the Arctic polar air mass could reach as far as the Junggar Basin and that summer monsoons from both the Indian and Pacific Oceans has little impact.

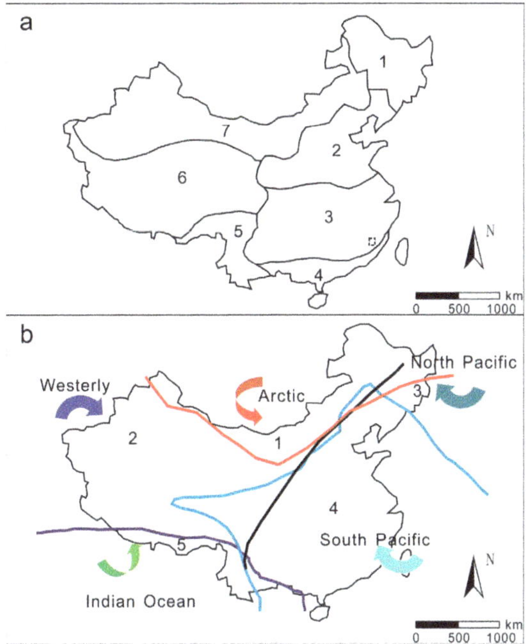

Figure 1. Regionalization of China by Luo [27]. (**a**) Numbers 1, 2, 3, 4, 5, 6, and 7 indicate the Northeast region, the North region, the Central region, the South region, the Sichuan-Yunnan region, the Tibetan region, and the Mongolia-Xinjiang region, as defined by Araguas-Araguas et al. [7]. (**b**) Numbers 1, 2, 3, 4, and 5 indicate the Arctic region, the Westerly region, the North Pacific region, the South Pacific region, and the Indian Ocean region.

When comparing and summarizing from the regionalization schemes of Luo [27], Huang [28], and Araguas-Araguas et al. [7], we find that each region could have its own isotopic patterns affected by moisture sources and other factors. This has also been verified by the recent isotopic studies in China that showed the dominating factor affecting isotopes should be moisture sources and then the local geographical meteorological factors [11,35,36]. To understand the temporal and spatial variation of precipitation isotopes, we devise a new regionalization scheme in this work. The primary factor we consider for this scheme is the moisture source. Thus, we divide the northern continental portion of China into Northwest China (Region I: the westerlies domain) and North China (Region II: the arctic domain). Because the South and North Pacific-dominated regions partially overlap owing to their similar moisture sources, we treat Northeast China as a single region (Region III: the northeast domain) due to its climatological and geographical differences from the rest of the region (Region IV: the Pacific domain). As in the previous regionalization, the Tibetan Plateau is treated as a single region (Region V) because the climatology and tectonic features of the Tibetan Plateau are quite different from those of other regions in China. The Tibetan Plateau plays an important role in affecting regional climatology [7,12,37]. This region impacts both the westerlies and the summer monsoons. The high mountains block the westerlies and split the jet stream, which moves to the south and north of the plateau [7]. The elevated heating by the Tibetan Plateau of the atmosphere plays a fundamental role in the formation and maintenance of the summer circulation, at least over Asia. The onset of the Bay of Bengal monsoon (BOBM) and the EASM can be linked to the thermal and mechanical forces of the Tibetan Plateau [37].

Compared with the regionalization by Araguas-Araguas et al. [7], our regionalization keeps the primary criterion as the moisture sources. However, the improvements can be found in our regionalization by taking into account the geographical and climatic factors. The overlapping regions do not exist as a result of combining the regionalization schemes of Luo [27], Huang [28], and Terzer et al. [29]. The extent of Regions IV and V is ascertained by considering the recent findings of the Indian Ocean and Pacific monsoons mentioned above. The region of the Tibetan Plateau is highlighted in light of its effect on the regional climate.

Collectively, China is divided into 5 regions (Figure 2). All of the results and discussion regarding precipitation isotopes in China will be based on this regionalization.

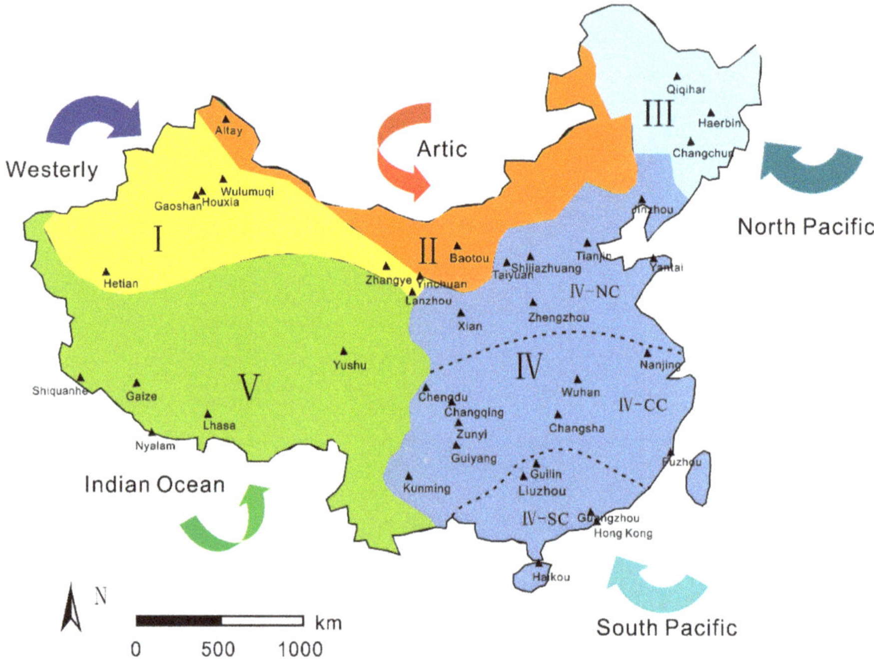

Figure 2. Regionalization of China based on moisture sources and meteorological conditions. The dashed line shows the extent of the sub-regions in Region IV.

3.2. Spatial Distribution of Stable Isotopes and D Excess

A clear pattern of the annual average isotopic distribution can be seen in Figure 3a,b. The minimum $\delta^{18}O$ with a value of −15.9‰ is at Lhasa station, and the maximum value of −5.7‰ is at Haikou station. The minimum δ^2H is −117.2‰ at Lhasa station, and the maximum is −34.7‰ at Guilin station. Isotopes become depleted from south to north overall, which is consistent with the latitude effect reported globally [1]. Though a similar depleted trend can be found from east to west, we could not attribute this trend to the continental or altitude effect because precipitation is formed by different moisture sources from East to West China.

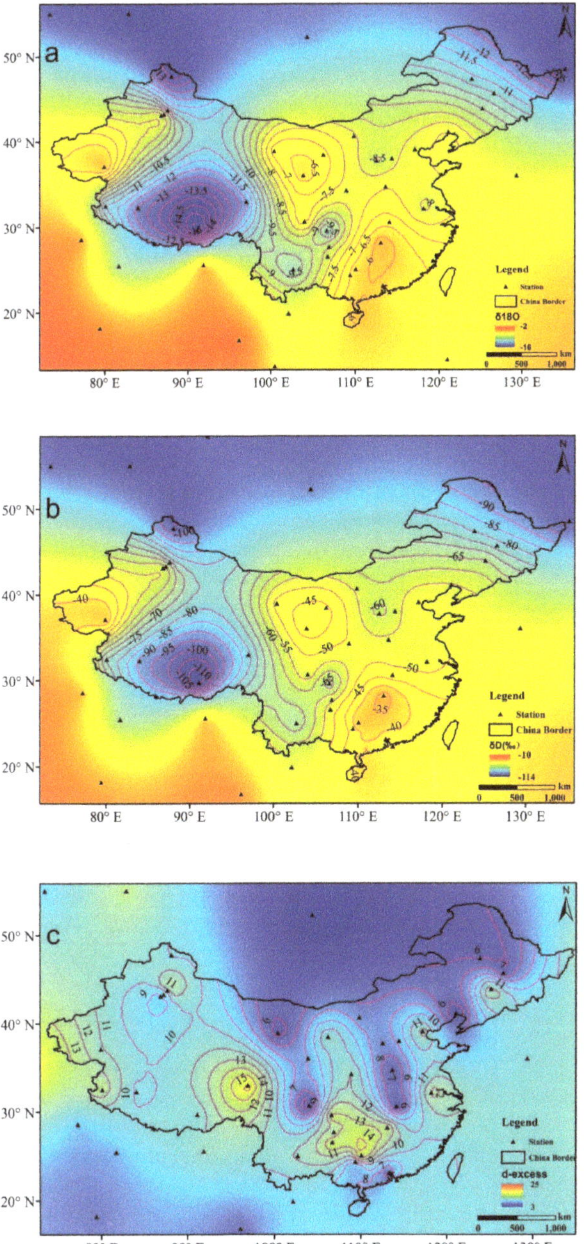

Figure 3. Contours of stable isotopes of precipitation in China: (**a**) long-term annual average $\delta^{18}O$ values; (**b**) long term annual average δ^2H values; (**c**) d excess calculated based on $\delta^{18}O$ and δ^2H values.

The d excess ranges from 4.7‰ at Chengdu station to 14.8‰ at Guilin station (Figure 3c). Most of the values are close to the global average of 10‰ [1]. Comparing the spatial distribution of d excess with that of $\delta^{18}O$ (Figures 2 and 3), we find that in Regions I, III, and V, the contours of $\delta^{18}O$ are intensive, whereas the contours of d excess are scarce; in contrast, in Region IV, the $\delta^{18}O$ values change slightly, whereas the d excess varies as a complex feature. This pattern occurs because in Regions I, III, and V, $\delta^{18}O$ declines gradually along the track of the air masses from the westerlies, the North Pacific, and the Indian Ocean, respectively, whereas in Region IV (the monsoon region), precipitation mainly occurs in summer caused by the monsoons, maintaining consistent $\delta^{18}O$ values [38,39]. This pattern fits well with the dominant role of moisture sources in controlling precipitation isotopes.

3.3. Seasonal Variations of Precipitation Isotopes and D Excess

Stations with a full-year record of stable isotopes are selected to show the seasonal variations and their relationship with moisture sources in each region.

(1)   Region I

The Wulumuqi and Zhangye stations are used to show the seasonal variations in precipitation isotopes (Figure 4). The $\delta^{18}O$ values range from −21.0‰ to −3.7‰, increase gradually from January to July, and then decrease from August to December. The d excess in this region has the largest amplitude (32.0‰) among all the regions. In summer, the d excess is lower than the global average of 10, whereas in winter, it is greater than 10. As noted above, precipitation in this region is mainly formed by moisture from the westerlies. Measuring the isotopic composition of ice core samples taken from a high-altitude glacier in the Tianshan Mountains and the isotopic composition of snow and firn taken from a high mountain of the Siberian Altay, Kreutz et al. [40] and Aizen et al. [41] suggested that seasonal changes in moisture sources and recycling in the Caspian Sea region during transport from the Atlantic to the sampling site in the Tianshan Mountains are responsible for the variability in d excess. Pang et al. [13] and Kong et al. [15] further noted that moisture recycling during summer and autumn amplified the seasonal change in d excess in this region.

(2)   Region II

The Altay and Baotou stations are selected to show the seasonal variations in precipitation isotopes (Figure 4). The isotopic change in Region II is very similar to that in Region I. Nevertheless, the d excess is very different—the d excess has a clear seasonal pattern in Region I, which does not exist in Region II. Comparing the d excess at Wulumuqi and Altay stations, we find they are similar in summer (8.4‰ at Wulumuqi and 7.8‰ at Altay) and different in winter (20.3‰ at Wulumuqi and 9.8‰ at Altay). However, the d excess at Baotou station is closer to that of Wulumuqi in winter, which are both larger than 14.0‰. This finding may show that in winter the only moisture source for Altay is the Arctic Ocean, whereas Baotou has an additional moisture source from the westerlies.

(3)   Region III

The Qiqihar station is selected to show the seasonal variations in precipitation isotopes (Figure 4). The $\delta^{18}O$ values in this region are lower than in other regions. The $\delta^{18}O$ values (−9.6‰) are higher in summer and lower (−25.1‰) in winter. In contrast, the d excess values are lower (3.9) in summer and higher (9.3) in winter. The temporal variations of $\delta^{18}O$ and d excess values in this region are similar to that in Region I. However, Figure 2 shows that moisture sources in Region III are quite different from those in Region I. In Region III, the moisture source in summer is the Pacific Ocean, which has lower d excess and high $\delta^{18}O$ values; however, the source in winter is the Arctic Ocean, which has higher d excess and lower $\delta^{18}O$ values [35,39]. This difference in moisture sources leads to the temporal variation of precipitation isotopes and d excess.

(4)   Region IV

To track the onset of the monsoon season, Region IV is further divided into the sub-regions of South China (SC), Central China (CC), and North China (NC) (Figure 4).

The Haikou, Hong Kong, Guilin, and Liuzhou stations are selected to show the seasonal variations in precipitation isotopes in Region SC. Figure 4 shows that the $\delta^{18}O$ values begin to decrease in April, which is the onset of the monsoons [36,42]. The observed decreasing $\delta^{18}O$ values could be explained by the precipitation amount effect. The d excess values change very little throughout the year, exhibiting features of a marine moisture source.

The Nanjing, Fuzhou, Guiyang, and Wuhan stations are selected to show the seasonal variations in precipitation isotopes in Region CC. The $\delta^{18}O$ values in CC begin to decrease in May, which is a month later than in Region SC. The d excess in winter (15.0‰) is higher than in summer (8.0‰), which is different from Region SC. This finding implies that precipitation isotopes in winter have been affected by moisture from the north.

The Shijiazhuang, Tianjin, Xian, and Zhengzhou stations are selected to show the seasonal variations in precipitation isotopes in Region NC. Both the $\delta^{18}O$ and d excess values in this region vary more significantly than in Regions SC and CC. Obviously, the values are affected by moisture from the north and northwest. However, in summer, the moisture is mainly derived from the Pacific Ocean [38]. The $\delta^{18}O$ values begin to decrease in May. Compared with the $\delta^{18}O$ values in Regions SC and CC, the decreasing $\delta^{18}O$ values reveal the onset of the monsoon season from April to June in different parts of Region IV.

(5) Region V

The Lhasa, Nyalam, and Yushu stations are selected to show the seasonal variations in precipitation isotopes (Figure 4). The seasonal variation in $\delta^{18}O$ values in this region is similar to that in Region SC. However, the $\delta^{18}O$ values are lower than in Region SC because of the higher altitude in Region V. The d excess values in this region are high in winter and low in summer. It should be noted that the d excess values at Yushu station are different from those at Lhasa and Nyalam stations. In summer, the d excess at Yushu station (14‰) is much larger than that at Lhasa and Nyalam stations (10‰ and 5‰) because the moisture is mainly from the Indian Ocean at Lhasa and Nyalam stations in summer, whereas Yushu station is less influenced by Indian Ocean moisture [12].

In summary, it is found that in each region, the moisture source plays a primary role in controlling the precipitation isotopes. What's more, with the long-term observation at the GNIP stations (i.e. Hongkong, Wulumuqi), the seasonality remains robust. Compared with the existing regionalization scheme, such a new regionalization scheme helps to identify the factors affecting isotopic variability. However, when using the regionalization scheme, it should be noted that the similarity of isotopic seasonality at each station is incorporated as a factor for the regionalization scheme, but is not the only factor. For instance, the $\delta^{18}O$ values are higher in summer and lower in winter at both Wulumuqi and Qiqihar stations, but we divide them into different regions due to different moisture sources and geographical factors. Even in one region, the same moisture source might play different roles at different stations, especially for the stations located at the boundary of two regions, such as Baotou in Region II, Shijiazhuang in Region IV, and Yushu in Region V. For example, both stations of Altay and Baotou are located in Region II of the Artic Region, but in winter the Artic moisture source affects the precipitation isotopes at Altay station more significantly than that at the Baotou station, while in summer the effects at both stations are similar. Therefore, the factors influencing the precipitation isotopes are complex. The regionalization scheme helps to illustrate the impact of moisture sources on the precipitation isotopes, but one has to keep in mind that other factors, including local metrological and geographical conditions, could also modify them.

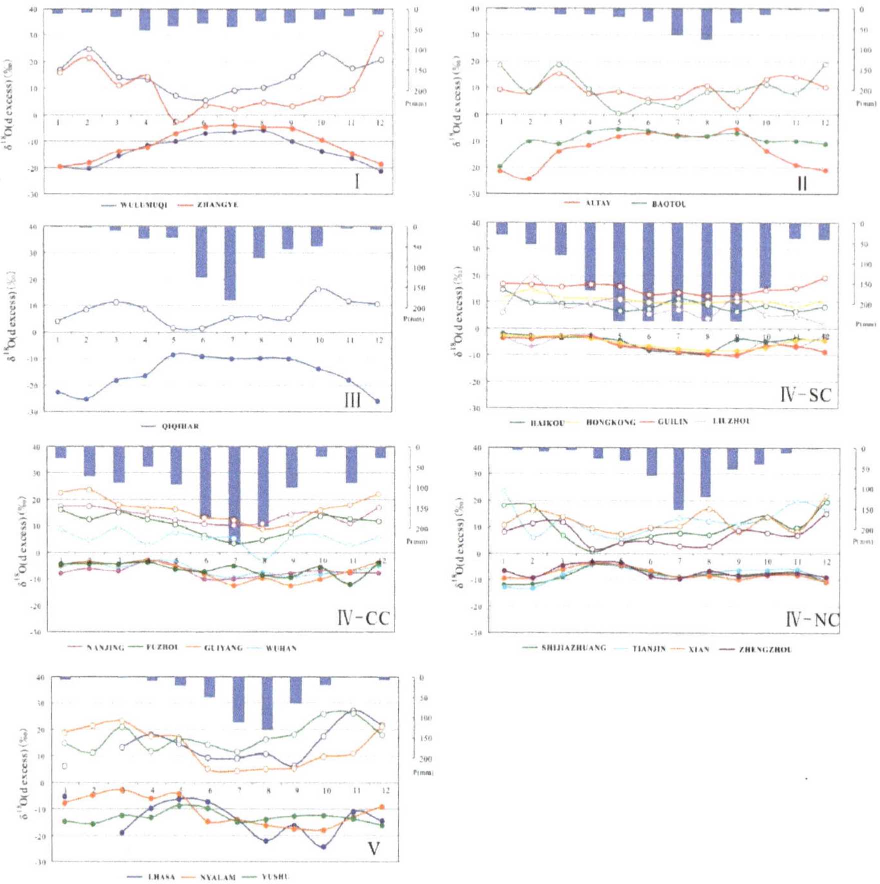

Figure 4. Seasonal variations in precipitation (pillars), $\delta^{18}O$ (solid circles), and d excess (hollow circles) in each region of China.

3.4. Regional Meteoric Water Line (RMWL)

The regional δ^2H-$\delta^{18}O$ relationship, which is well known as the RMWL, provides a reference for interpreting the provenance of surface water and groundwater. The East Asia meteoric water line (EAMWL) of $\delta^2H = 7.92\delta^{18}O + 9.2$ was obtained by Araguas-Araguas et al. [7]. This line is indistinguishable from the GMWL of $\delta^2H = 8\delta^{18}O + 10$. For regions where isotopic data are unavailable, the EAMWL is very useful as a reference for groundwater studies; however, it cannot represent the features of local precipitation for a given region because precipitation isotopes at each region could be modified by local climatic factors, including the origin of the vapor mass, sub-cloud evaporation during rainfall, and the seasonality of precipitation [43]. Thus, in this case, the EAMWL cannot define the groundwater input. Taking these local factors into consideration, we present the meteoric water lines based on the monthly isotopic data for each region (Figure 2) with a dominant single moisture source and similar climatological characteristics (Table 2).

Table 2. Regional meteoric water lines for different regions of the westerlies domain (I), the Artic domain (II), the Northeast China domain (III), the Pacific domain (IV), and the Tibetan Plateau domain (V), respectively.

Regions	RMWL	R^2 (n)	Level of Significance
I	$\delta^2H = (7.24 \pm 0.11)\delta^{18}O + 1.96 \pm 1.38$	0.95 (264)	0.01
II	$\delta^2H = (7.55 \pm 0.10)\,\delta^{18}O + 3.91 \pm 1.22$	0.97 (182)	0.01
III	$\delta^2H = (6.32 \pm 0.12)\,\delta^{18}O - 4.07 \pm 0.97$	0.87 (460)	0.01
IV	$\delta^2H = (7.63 \pm 0.06)\,\delta^{18}O + 8.03 \pm 0.41$	0.93 (1246)	0.01
V	$\delta^2H = (8.41 \pm 0.20)\delta^{18}O + 16.72 \pm 2.98$	0.97 (108)	0.01

The slopes of RMWLs for Regions I–IV, which have an average of 7.19 with a deviation of 0.52, are slightly less than the slope (7.92) of EAMWL and the slope (8) of GMWL, while the slope of RMWL for Region V (the Tibetan Plateau) is 8.41, which is larger than all the other slopes (Table 2). It is known that the sub-cloud evaporation decreases the slope of the meteoric water line, whereas moisture recycling increases it [13,15,43,44]. Although the seasonality of precipitation could also modify the slope of RMWL, there are no significant differences of precipitation patterns between Region V and the other regions. Thus, we deduce that the Tibetan Plateau region (Region V) is affected by moisture recycling, whereas other regions are affected by sub-cloud evaporation. This conclusion maintains consistency with previous studies in the Tibetan Plateau, Northwest, North, South, and Eastern China [15,25,45–47].

3.5. Spatial Extent of the Isotope–Climate Relationship

Given the parallel behavior of $\delta^{18}O$ and δ^2H, we use the $\delta^{18}O$–temperature and $\delta^{18}O$–precipitation amount relationships to reveal the effect of climate on precipitation isotopes. Figure 5 shows the regional variation in correlation coefficients between $\delta^{18}O$, temperature, and precipitation amount. According to the available data, the correlation is significant when the correlation coefficient is larger than 0.3 (pale yellow region in Figure 5a and dark blue region in Figure 5b); otherwise, it is not significant.

Figure 5a demonstrates that the temperature effect is significant in Regions I, II, and III, whereas in the other regions, it is not significant. Rozanski et al. [21] claimed that the temperature effect was mainly observed in regions of middle and high latitudes, which is consistent with our findings. The gradient of the temperature effect in Region I is identified as 0.13–0.68‰/°C (Figure 6a). The westerlies are the dominant moisture source in Region I [13,15]. In the context of an individual moisture source in the region, a shift in moisture source to the North Atlantic Ocean and the alternation of trajectory in summer and winter are considered as the factors leading to enriched isotopes in summer and depleted isotopes in winter [12,13,40,47]. The temperature effect in Region III is reported as increasing in significance from south to north [39,48]. The moisture in Region III is controlled by the westerlies and polar air masses in winter and by the westerlies and the Pacific Ocean in summer (Figures 1 and 2). In winter, when temperatures are low, isotopes incorporated in both westerly and polar air masses are depleted; in contrast, in summer, when temperatures are high, precipitation isotopes from the westerlies and Pacific Ocean are enriched. Therefore, the temperature effect in Region III is significant. From south to north in Region III, the difference in moisture sources becomes increasingly prominent, which leads to a more significant temperature effect. Thus, we can draw the conclusion that the temperature effect in Region I is caused by the seasonal shift of the westerlies, whereas in Region III, the temperature effect is mainly attributed to the seasonal differences in moisture sources.

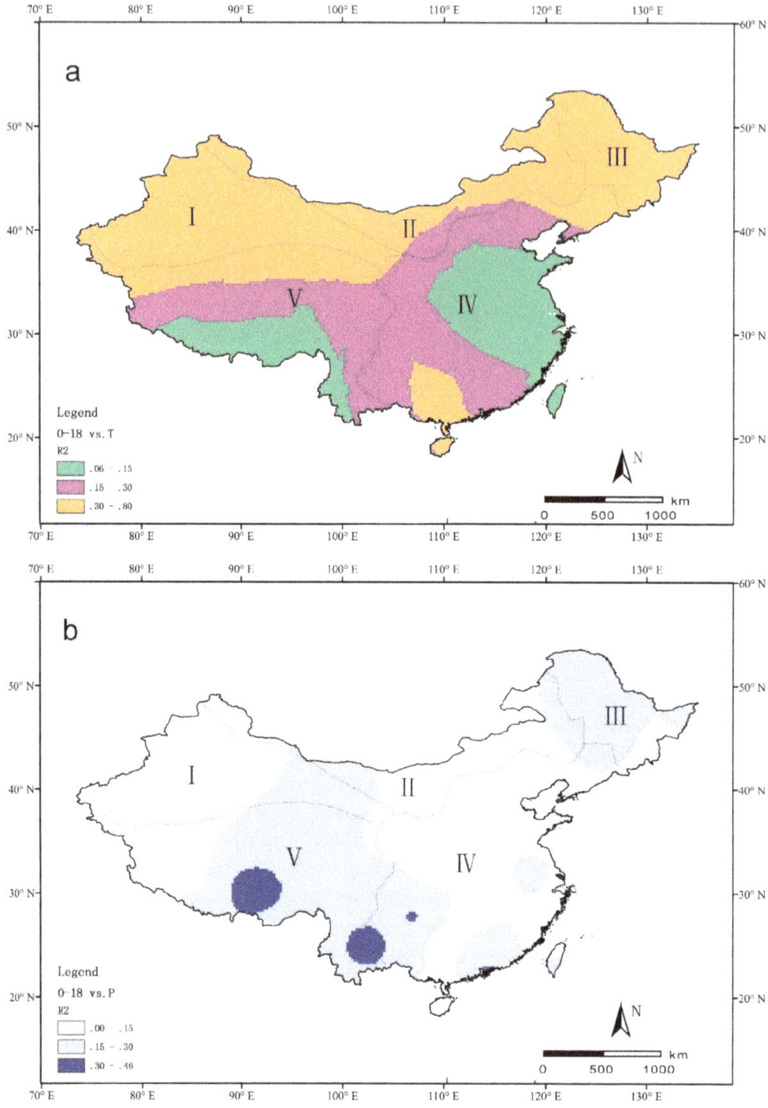

Figure 5. Correlation coefficients between $\delta^{18}O$ and temperature (**a**) and between $\delta^{18}O$ and precipitation amount (**b**). The dotted lines indicate the range of each region.

The correlation coefficients between stable isotopes and precipitation amount demonstrate the precipitation amount effect (Figure 5b). In total, the correlation between isotopes and precipitation amount is not as significant as that between isotopes and temperature. The most significant parts are located in Southwest and Northeast China, but Figure 6b illustrates that in Southwest China, the slope is negative (anti-correlation), whereas in Northeast China, it is positive. Precipitation events in Northeast China occur mainly in summer, when temperatures are high, whereas in winter, both precipitation and temperature are low. Given the significant temperature effect in Northeast China, it is reasonable to consider that precipitation amount has little impact on precipitation isotopes in this region. A significant precipitation amount effect was observed only in Southwest China and in some regions along the southeast coast. This finding implies that precipitation amount has a stronger effect in monsoon-controlled South China. Johnson and Ingram [10] noted that in the monsoon-affected regions, multi-regression analysis may be particularly useful for capturing the effect of precipitation amount on isotopes. Considering the temporal and spatial variation of monsoon intensity, moisture source analysis is indispensable in addressing monsoon isotopic data.

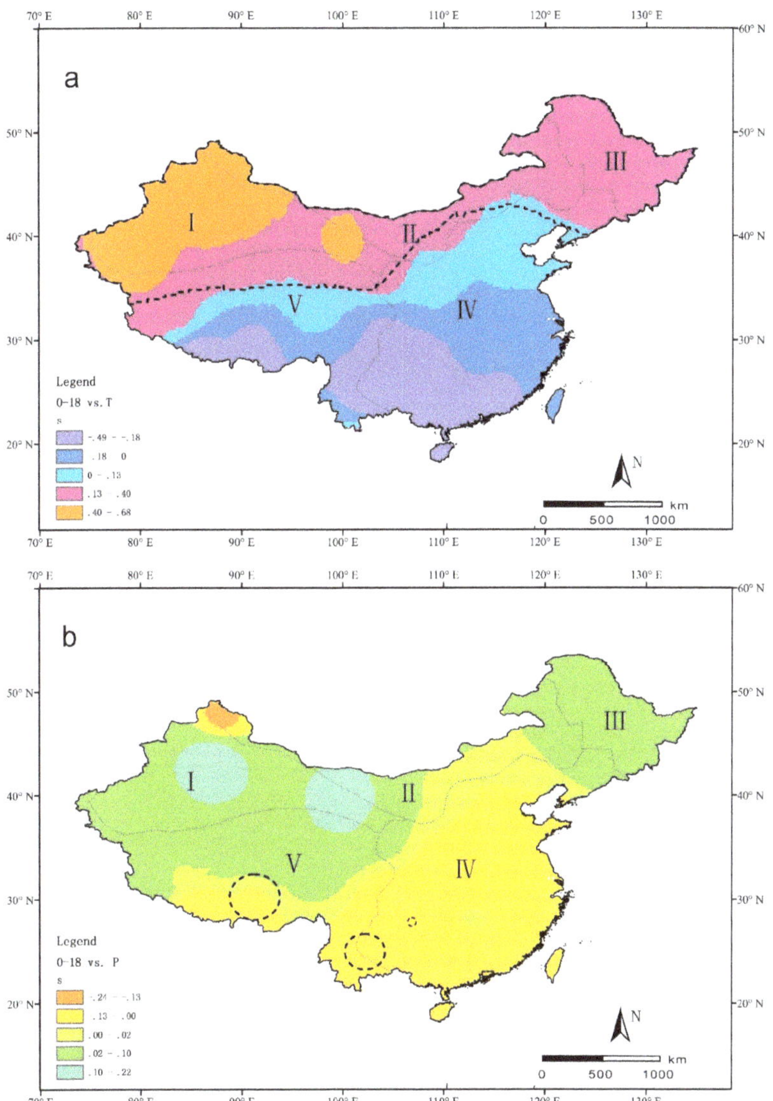

Figure 6. The gradients between (**a**) δ^{18}O and temperature (unit: ‰/°C) and (**b**) between δ^{18}O and precipitation amount (unit: ‰/mm). The dashed lines circle the region where the correlations between δ^{18}O and temperature (a) and between δ^{18}O and precipitation amount (b) are significant (larger than 0.3). The dotted lines indicate the range of each region.

4. Conclusions

The dominant effect of moisture sources on precipitation isotopes is revealed based on long-term observations at 68 stations in and around China. At a regional scale, China is divided into five moisture source regions: Region I (the westerlies domain), Region II (the arctic domain), Region III (the northeast domain), Region IV (the Pacific domain), and Region V (the Tibetan Plateau domain). In Regions I, III, and V, precipitation isotopes exhibit a clear variation trend and d excess almost remains constant; however, in Region IV, both isotopes and d excess show no clear spatial patterns. The seasonal $\delta^{18}O$ and d excess changes in Region IV could be used to track the onset of the EASM.

Regional meteoric water lines (RMWLs) are more useful in comparing with the isotopes of surface water and groundwater. According to the region regionalization above, the RMWLs are $\delta^2H = 7.24\delta^{18}O + 1.96$, $\delta^2H = 7.55\delta^{18}O + 3.91$, $\delta^2H = 6.32\delta^{18}O - 4.07$, $\delta^2H = 7.63\delta^{18}O + 8.03$, and $\delta^2H = 8.41\delta^{18}O + 16.72$ in the westerlies domain, the arctic domain, the northeast domain, the Pacific domain, and the Tibetan Plateau domain, respectively.

The temperature effect in Regions I and III is significant for different reasons: in Region I, the $\delta^{18}O$-temperature gradient ranges from 0.13–0.68‰/°C and is larger in Xinjiang, with a value of 0.40-0.68‰/°C, which is caused by the seasonal shift of the westerlies. In Region III, the temperature effect is mainly attributed to the seasonal differences in moisture sources. The precipitation amount effect is most significant in the region along the southeast coast, where the $\delta^{18}O$-precipitation amount gradient is −0.24 to −0.13‰/mm and is affected by the intensity and extent of the East Asia Summer Monsoons (EASM). Each region has its own features of seasonal variation in $\delta^{18}O$ and d excess. Our study highlights the importance of moisture sources in affecting the spatial and temporal distribution of precipitation isotopes in China. The results show that an analysis with the new regionalization scheme could capture the features of precipitation isotopes, and thus could help in interpreting the hydrological and climatic processes.

Author Contributions: Y.K. and Z.P. designed the framework. Y.K. wrote the paper. Y.K., K.W., and J.L. plotted the figures. All authors revised and approved the manuscript.

Funding: This study is supported by the National Natural Science Foundation of China (Grant 91647101 and U1703122).

Acknowledgments: This paper is dedicated to the memory of our colleague Klaus Froehlich from the IAEA, who was involved in many discussions and provided several suggestions for improving this manuscript. The authors would like to acknowledge Gabriel J. Bowen from the University of Utah, Tao Pu from the Cold and Arid Regions Environmental and Engineering Research Institute, Chinese Academy of Sciences, the academic editor and four anonymous reviewers for their comments, which greatly helped to increase the quality of the paper.

Conflicts of Interest: The authors declare no conflict of interest.

References

1. Dansgaard, W. Stable isotopes in precipitation. *Tellus* **1964**, *16*, 436–468. [CrossRef]
2. IAEA. Isotope Hydrology Information System, the ISOHIS Database. 2006. Available online: http://www.iaea.org/water (accessed on 1 March 2013).
3. Salati, E.; Dall'Olio, A.; Matsui, E.; Gat, J.R. Recycling of water in the Amazon Basin: An isotopic study. *Water Resour. Res.* **1979**, *15*, 1250–1258. [CrossRef]
4. Gonfiantini, R. On the isotopic composition of precipitation in tropical stations. *Acta Amaz.* **1985**, *5*, 121–139. [CrossRef]
5. Joseph, A.; Frangi, J.P.; Aranyossy, J.F. Isotope characteristics of meteoric water and groundwater in the Sahelo-Sudanese Zone. *J. Geophys. Res.* **1992**, *97*, 7543–7551. [CrossRef]
6. Rozanski, K.; Araguas-Araguas, L. Spatial and temporal variability of stable isotope composition of precipitation over the South American continent. *Bulletin De L'Institut Français D'Études Andines* **1995**, *24*, 379–390.
7. Araguas-Araguas, L.; Froehlich, K.; Rozanski, K. Stable isotope composition of precipitation over Southeast Asia. *J. Geophys. Res.* **1998**, *103*, 721–742. [CrossRef]

8. Bowen, G.J.; Revenaugh, J. Interpolating the isotopic composition of modern meteoric precipitation. *Water Resour. Res.* **2003**, *39*, 1299. [CrossRef]

9. Wang, D.; Wang, K. Isotopes in precipitation in China (1986–1999). *Sci. China Ser. E* **2001**, *44*, 48–51. [CrossRef]

10. Johnson, K.; Ingram, B.L. Spatial and temporal variability in the stable isotope systematics of modern precipitation in China: Implications for paleoclimate reconstructions. *Earth Planet. Sci. Lett.* **2004**, *220*, 365–377. [CrossRef]

11. Liu, J.; Song, X.; Yuan, G.; Sun, X.; Yang, L. Stable isotopic compositions of precipitation in China. *Tellus* **2014**, *66*, 22567. [CrossRef]

12. Tian, L.; Yao, T.; MacClune, K.; White, J.W.C.; Schilla, A.; Vaughn, B.; Vachon, R.; Ichiyanagi, K. Stable isotopic variations in west China: A consideration of moisture sources. *J. Geophys. Res.* **2007**, *112*, D10112. [CrossRef]

13. Pang, Z.; Kong, Y.; Froehlich, K.; Huang, T.; Yuan, L.; Li, Z.; Wang, F. Processes affecting isotopes in precipitation of an arid region. *Tellus* **2011**, *63*, 352–359. [CrossRef]

14. Kong, Y.; Pang, Z. Evaluating the Sensitivity of Glacier Rivers to Climate Change based on Hydrograph Separation of Discharge. *J. Hydrol.* **2012**, *434–435*, 121–129. [CrossRef]

15. Kong, Y.; Pang, Z.; Froehlich, K. Quantifying recycled moisture fraction in precipitation of an arid region using deuterium excess. *Tellus* **2013**, *65*, 19251. [CrossRef]

16. Kong, Y.; Pang, Z. Statistical analysis of stream discharge in response to climate change for Urumqi river catchment, Tianshan Mountains, Central Asia. *Quat. Int.* **2014**, *336*, 44–51. [CrossRef]

17. Wang, S.; Zhang, M.; Hughes, C.; Zhu, X.; Dong, L.; Ren, Z.; Chen, F. Factors controlling stable isotope composition of precipitation in arid conditions: An observation network in the Tianshan Mountains, central Asia. *Tellus B* **2016**, *68*, 26206. [CrossRef]

18. Li, Z.X.; Feng, Q.; Li, Z. Climate Background, Facts, and Hydrological Effects of Multiphase Water Transformation in Cold Regions OF the Western China: A review. *Earth-Sci. Rev.* **2018**. [CrossRef]

19. Craig, H. Isotopic variations in meteoric waters. *Science* **1961**, *133*, 1702–1703. [CrossRef]

20. Gat, J. Oxygen and hydrogen isotopes in the hydrological cycle. *Annu. Rev. Earth Planet. Sci.* **1996**, *24*, 225–262. [CrossRef]

21. Rozanski, K.; Araguas-Araguas, L.; Gonfiantini, R. Relation between long-term trends of oxygen-18 isotope composition of precipitation and climate. *Science* **1992**, *258*, 981–985. [CrossRef]

22. Stumpp, C.; Klaus, J.; Stichler, W. Analysis of long-term stable isotopic composition in German precipitation. *J. Hydrol.* **2014**, *517*, 351–361. [CrossRef]

23. Hatvani, I.G.; Leuenberger, M.; Kohan, B.; Kern, Z. Geostatistical analysis and isoscape of ice core derived water stable isotope records in an Antarctic macro region. *Polar Sci.* **2017**, *13*, 23–32. [CrossRef]

24. Oliver, M.A.; Webster, R. A tutorial guide to geostatistics: Computing and modelling variograms and kriging. *Catena* **2014**, *113*, 56–69. [CrossRef]

25. Liu, J.R.; Song, X.F.; Yuan, G.F.; Sun, X.; Liu, X.; Wang, S. Characteristics of δ^{18}O in precipitation over Eastern Monsoon China and the water vapor sources. *Chin. Sci. Bull.* **2010**, *55*, 200–211. [CrossRef]

26. Bowen, G.J.; Wilkinson, B. Spatial distribution of δ^{18}O in meteoric precipitation. *Geology* **2002**, *30*, 315–318. [CrossRef]

27. Luo, K. Draft of natural geography regionalization of China. *Acta Geogr. Sin.* **1954**, *20*, 379–394.

28. Huang, B. Draft of the complex physical geographical division of China. *Chin. Sci. Bull.* **1959**, *18*, 594–602.

29. Terzer, S.; Wassenaar, L.I.; Araguás-Araguás, L.J.; Aggarwal, P.K. Global isoscapes for δ^{18}O and δ^{2}H in precipitation: Improved prediction using regionalized climatic regression models. *Hydrol. Earth Syst. Sci.* **2013**, *17*, 4713–4728. [CrossRef]

30. Li, Z.X.; Yao, T.; Tian, L. Variation of δ^{18}O in precipitation in annual timescale with moisture transport in Delingha region. *Earth Sci. Front.* **2006**, *13*, 330–334.

31. Zhou, S.Q.; Nakawo, M.; Sakai, A.; Matsuda, Y.; Duan, K.Q.; Pu, J.C. Water isotope variations in the snow pack and summer precipitation at July 1 Glacier, Qilian Mountains in northwest China. *Chin. Sci. Bull.* **2007**, *52*, 2963–2972. [CrossRef]

32. Xu, G.; Chen, T.; Liu, X.; An, W.; Wang, W.; Yun, H. Potential linkages between the moisture variability in the northeastern Qaidam Basin, China, since 1800 and the East Asian summer monsoon as reflected by tree ring δ^{18}O. *J. Geophys. Res.* **2011**, *116*, D09111. [CrossRef]

33. Winkler, M.G.; Wang, P.K. The Late-Quaternary vegetation climate of China. In *Global Climates since the Last Glacial Maximum*; University of Minnesota Press: St. Paul, MN, USA, 1993; pp. 221–261.

34. Pang, Z. Mechanism of water cycle changes and implications on water resources regulation in Xinjiang Uygur Autonomous Region. *Quat. Sci.* **2014**, *34*, 907–917.

35. Li, J.; Pang, Z.; Kong, Y.; Zhou, M.; Huang, T. Spatial distributions of stable isotopic composition and deuterium excess in precipitation during the summer and winter seasons in China, Fresenius Environmental Bulletin. *Fresenius Environ. Bull.* **2014**, *23*, 2074–2085.

36. Kong, Y.; Pang, Z. A positive altitude gradient of isotopes in the precipitation over the Tianshan Mountains: Effects of moisture recycling and sub-cloud evaporation. *J. Hydrol.* **2016**, *542*, 222–230. [CrossRef]

37. Wu, G.; Zhang, Y. Tibetan Plateau Forcing and the Timing of the Monsoon Onset over South Asia and the South China Sea. *Mon. Weather Rev.* **1998**, *126*, 913–927. [CrossRef]

38. Yamanaka, T.; Shimada, J.; Hamada, Y.; Tanaka, T.; Yang, Y.; Zhang, W.; Hu, C. Hydrogen and oxygen isotopes in precipitation in the northern part of the North China Plain: Climatology and inter-storm variability. *Hydrol. Process.* **2004**, *18*, 2211–2222. [CrossRef]

39. Liu, Z.; Bowen, G.J.; Welker, J.M. Atmospheric circulation is reflected in precipitation isotope gradients over the conterminous United States. *J. Geophys. Res.* **2010**, *115*, D22120. [CrossRef]

40. Kreutz, K.J.; Wake, C.P.; Aizen, V.B.; Cecil, L.D.; Synal, H.A. Seasonal deuterium excess in a Tien Shan ice core: Influence of moisture transport and recycling in Central Asia. *Geophys. Res. Lett.* **2003**, *30*, 1922. [CrossRef]

41. Aizen, V.; Aizen, E.; Fujita, K.; Nikitin, S.; Kreutz, K. Stable-Isotope Time Series and Precipitation Origin from Firn-Core and Snow Samples, Altai Glaciers, Siberia. *J. Glaciol.* **2005**, *51*, 637–654. [CrossRef]

42. Ding, Y.; Chan, J. The East Asian summer monsoon: An overview. *Meteorol. Atmos. Phys.* **2005**, *89*, 117–142.

43. Clark, I.; Fritz, P. *Environmental Isotopes in Hydrogeology*; Clark, I., Fritz, P., Eds.; Lewis Publishers: New York, NY, USA, 1997.

44. Froehlich, K.; Kralik, M.; Papesch, W.; Rank, D.; Scheifinger, H.; Stichler, W. Deuterium excess in precipitation of Alpine regions-moisture recycling. *Isot. Environ. Health Stud.* **2008**, *44*, 61–70. [CrossRef] [PubMed]

45. Chen, Z.; Nie, Z.; Zhang, G.; Wan, L.; Shen, J. Environmental isotopic study on the recharge and residence time of groundwater in the Heihe River Basin, Northwestern China. *Hydrogeol. J.* **2006**, *14*, 1635–1651. [CrossRef]

46. Tian, L.; Yao, T.; Schuster, P.F.; White, J.W.C.; Ichiyanagi, K.; Pendall, E.; Pu, J.; Yu, W. Oxygen-18 concentrations in recent precipitation and ice cores on the Tibetan Plateau. *J. Geophys. Res.* **2003**, *108*, 4293. [CrossRef]

47. Kong, Y.; Pang, Z.; Li, J.; Huang, T. Seasonal variations of water isotopes in the Kumalak River catchments, Western Tianshan Mountains. Central Asia. *Fresenius Environ. Bull.* **2014**, *23*, 169–174.

48. Li, X.; Zhang, M.; Ma, Q.; Li, Y.; Wang, S.; Wang, B. Characteristics of Stable Isotopes in Precipitation over Northeast China and Its Water Vapor Sources. *Chin. J. Environ. Sci.* **2012**, *33*, 2924–2931.

Article

Identification of Sulfate Sources and Biogeochemical Processes in an Aquifer Affected by Peatland: Insights from Monitoring the Isotopic Composition of Groundwater Sulfate in Kampinos National Park, Poland

Adam Porowski [1,*], Dorota Porowska [2] and Stanislaw Halas [3,†]

[1] Institute of Geological Sciences Polish Academy of Sciences (ING PAN), Ul. Twarda 51/55, 00-818 Warszawa, Poland
[2] Faculty of Geology, University of Warsaw, Ul. Żwirki i Wigury 93, 02-089 Warsaw, Poland
[3] Mass Spectrometry Laboratory, Maria Curie-Skłodowska University, plac M. Curie-Skłodowskiej 1, 20-031 Lublin, Poland
* Correspondence: adamp@twarda.pan.pl
† Stanislaw Halas passed away in 2017.

Received: 28 May 2019; Accepted: 29 June 2019; Published: 5 July 2019

Abstract: Temporal and spatial variations of the concentration and the isotopic composition of groundwater sulfate in an unconfined sandy aquifer covered by peatland have been studied to better understand the sources and biogeochemical processes that affect sulfate distribution in shallow groundwater systems influenced by organic rich sediments. The groundwater monitoring was carried out for one year at hydrogeological station Pożary located within the protected zone of the Kampinos National Park. Sulfur ($\delta^{34}S_{SO4}$) and oxygen ($\delta^{18}O_{SO4}$) isotopic composition of dissolved sulfates were analyzed together with oxygen ($\delta^{18}O_{H2O}$) and hydrogen (δ^2H_{H2O}) isotopic composition of water and major ions concentration at monthly intervals. The research revealed three main sources of sulfates dissolved in groundwater, namely, (a) atmospheric sulfates—supplied to the aquifer by atmospheric deposition (rain and snow melt), (b) sulfates formed by dissolution of evaporite sulfate minerals, mainly gypsum—considerably enriched in ^{34}S and ^{18}O, and (c) sulfate formed during oxidation of reduced inorganic sulfur compounds (RIS), mainly pyrite—depleted in ^{34}S and ^{18}O. The final isotopic composition and concentration of dissolved SO_4^{2-} in groundwater are the result of overlapping processes of dissimilatory sulfate reduction, oxidation of sulfide minerals, and mixing of water in aquifer profile.

Keywords: oxygen isotopes; sulfur isotopes; isotopic composition of water; bacterial sulfate reduction; sulfide oxidation; atmospheric sulfate; peatland; unconfined aquifer; mineralization of organic matter

---

1. Introduction

Organic-rich sediments in hydrogeological profile, their high productivity and complex biogeochemistry have a globally significant influence on the quality of adjacent groundwaters, their chemical composition, trace elements cycling and the composition of atmospheric trace gases [1]. Wetlands, bogs, fens, and peatlands, which cover nearly 5% of the Earth's surface, are such ecosystems where the accumulation of reactive organic matter always implies a series of redox processes which control not only the content of carbon in groundwater but also the distribution of major dissolved compounds, including, for example, nitrogen or sulfur species [1–13]. Peatlands are very heterogeneous environments: inside the peatland bed, macro and micro gradients of redox conditions enable the development of a highly diverse microbial community capable of different redox reactions occurring simultaneously (e.g., nitrification, denitrification, sulfate reduction, oxidation of reduced inorganic sulfides, etc.) on a small spatial scale [7,9,10]. The biotic and abiotic transformations of sulfur compounds are interconnected with other geochemical processes through their common substrates or products. Sulfates reduction in peatlands can result in Eh and pH changes in groundwater, alkalinity generation, and C transformation, and, indirectly, the mobilization and removal of nutrients such as N, P, and C [2,3,6,7]. Moreover, sulfide as a product of dissimilatory sulfate reduction may remove heavy metals contained in waters via the formation of metal sulfides.

Numerous studies concerning wetland systems are focused first of all on the investigation of sulfur compounds transformation, sulfate reduction and generation directly in the organic-reach beds under environmentally relevant conditions in order to better understand the overall sulfur cycle dynamic and net sulfur compounds storage within freshwater peatlands [1,7,9,10,13,14]. Some of these researches are aimed at the direct application of wetlands to natural restoration of water quality e.g., [1,9]. An important part of the sulfur cycle in peatland systems involves the formation and sink of the organic sulfur species, the formation of biogenic minerals, and the interaction between mineral matter and organic matter [4,6,10].

Stable isotopes of O and S are used first of all to investigate the sulfate sources, the sulfate reduction and oxidation zones in the peat, and to compare sulfate sources in various wetland systems often located in different hydrogeological and climatic conditions [4,5,7,14]. Isotopic composition of sulfates is also used to explain the mechanisms and sources of SO_4^{2-} released from peatlands to streams, lakes, or soils [11,13,14].

Still insufficient attention is addressed to investigation the long term, direct influence of the peatland ecosystems on the geochemistry of groundwater masses in adjacent aquifers, especially in changing climate conditions. In this study we utilize an approach involving O and S isotopic composition of sulfates to identify the sulfate sources and biogeochemical processes that affect sulfate distribution in a shallow alluvial aquifer covered by peatland. The research was performed at the Pożary monitoring station, located in an area of Holocene peats of the Vistula River terrace in the Kampinos National Park, Central Poland (52°17′5″ N, 20°29′4″ E). The monitoring station was equipped with four piezometers installed at different depths in a sandy aquifer, namely 2.1, 3.3, 4.75, and 8.35 m (Figure 1). A previous study performed in this area by Porowska and Lesniak [15] focused on the distribution of dissolved inorganic carbon (DIC) and its C isotope composition, as well as on dissolved organic carbon (DOC) and dissolved oxygen (DO) in the aquifer profile. The study indicated that the mineralization of organic matter and the presence of organic carbon in the geological profile may have a significant effect on the biogeochemical processes controlling the chemical composition of groundwater in the aquifer. For example, the concentration of dissolved sulfate showed one of the largest seasonal variations in comparison to other major constituents. Seasonal and spatial variation of sulfates concentration and their stable O and S isotopic compositions is an important tracer of the sulfate sources, the biogeochemical processes in groundwater environment, and the geochemical conditions in the aquifer [4,11,12].

The major forms of sulfur in the hydrogeological subsurface environment include sulfate (SO_4^{2-}) and sulfide (HS^-) minerals, sulfate and sulfide dissolved in water, and hydrogen sulfide gas (H_2S) and sulfur in organic compounds (DOS) [2,3]. These various sources of sulfur may participate in the evolution of the chemical composition of water masses below the ground. Dissolved sulfates (SO_4^{2-}) are usually the dominant form of sulfur, and are one of the most common sulfur species in various natural groundwater environments. The geochemistry of sulfur is complicated by its wide range of oxidation states. However, the key reaction in the global sulfur cycle, i.e., the reduction of sulfate (SO_4^{2-} to hydrogen sulfide (H_2S), may be identified by the distribution of sulfate concentrations and its O and S isotopic composition. When sulfate ions reach the anaerobic zone within the saturated soil, peatland or groundwater, the sulfate reducing bacteria start producing sulfides (S^{2-}); consequently, the concentration of dissolved SO_4^{2-} in water decreases and considerable fractionation of $^{34}S/^{32}S$ and

$^{18}O/^{16}O$ isotope occurs. The residual $SO_4{}^{2-}$ becomes progressively enriched in the heavy isotopes ^{18}O and ^{34}S [5,7,9,11].

This paper gives the results of a thorough one-year monitoring study of the concentration of dissolved sulfate and its sulfur ($\delta^{34}S$) and oxygen ($\delta^{18}O$) isotopic composition together with the chemical and isotopic (δ^2H and $\delta^{18}O$) composition of groundwater in the aquifer vertical profile. The performed study had three principal objectives, namely (i) to distinguish between sulfate sources in the groundwater of the shallow unconfined aquifer covered by peatland, (ii) to identify the biogeochemical processes controlling the distribution and isotopic transformation of sulfates in the shallow groundwater system, and (iii) to evaluate the influence of organic-rich sediments on the geochemical conditions in the groundwater flow system. Explaining these phenomena is essential to obtain a better understanding of the sulfur cycling between the peat and the groundwater, and the influence of organic-rich sediments on the evolution of water chemistry in adjacent unconfined aquifers.

2. Study Area and Hydrogeological Settings

Kampinos National Park, located several kilometers northwest of Warsaw, Poland, is a special protection area where extensive ecological and climatic studies have been conducted for several years. There is one fully equipped meteorological station in this area, as well as several smaller research stations where only piezometers or observation wells are installed to monitor the level and the quality of groundwater for the needs of the Polish National Groundwater Monitoring Network. Our investigation was performed at the Pożary research station, located in an area of Holocene peats on the Vistula River terrace (Figure 1) [16]. The research station was equipped with four piezometers installed in the same aquifer at different depths. The shallowest one (depth: 2.10 m) was installed at a contact between fine- and medium-grained sands. The next two piezometers (depths: 3.30 and 4.75 m) were set into medium-grained sands, and the deepest one (depth: 8.35 m) was set into coarse-grained sands and gravels (Figure 1). The organic matter was accumulated within a peat layer with a thickness of 0.5 m in the uppermost part of the geological profile, within the vadose zone.

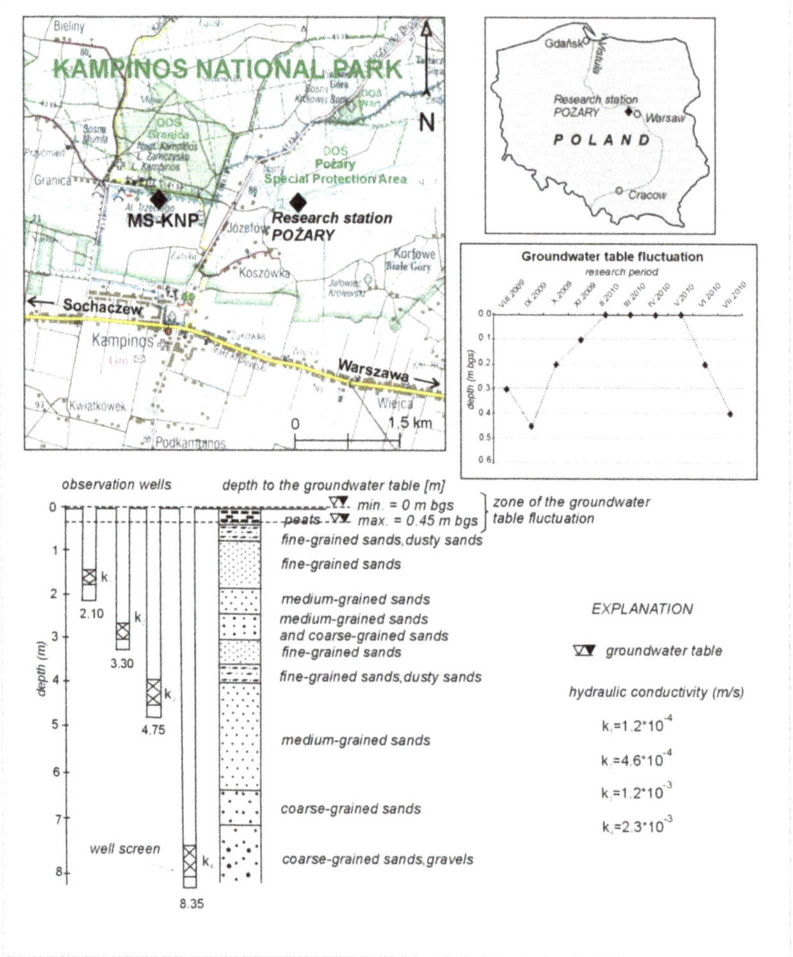

Figure 1. Location of the Pożary research station, geologic profile, and arrangement of piezometers for groundwater sampling. MS-KNP—meteorological station of Integrated Monitoring of the Natural Environment of the Kampinos National Park (KNP). Geological profile after Fic and Wierzbicki [17] modified by the authors.

Depending on the season and position of the groundwater level, the layer of peat may occur within either the saturation or vadose zone. During the one year of monitoring studies, the groundwater table fluctuated in the range from 0.0 to 0.45 m below ground surface (bgs). Generally, the groundwater level was above the peat surface during winter and spring, i.e., from February 2009 to May 2010, and was below the ground surface at depths from 0.10 to 0.45 m during summer and autumn, i.e., from August 2009 to November 2009 and from June 2010 to July 2010. It is important to note that the capillary fringe in the peat can extend to 0.2 m above the water table [18]. The hydraulic conductivity in the aquifer increases with depth, from $1.2 \cdot 10^{-4}$ to $2.3 \cdot 10^{-3}$ m/s (Figure 1)

3. Materials and Methods

Samples of groundwater were collected from four observation piezometers installed close to each other in the same aquifer. The piezometers were simple PVC standpipes slotted in the saturated zone at various depths, i.e., 2.1 (the shallowest one), 3.30, 4.75, and 8.35 m (the deepest one). The samples of groundwater were collected at monthly intervals from August 2009 to July 2010. In December 2009 and January 2010, the field campaign was suspended due to harsh weather and thick snow cover, which caused the piezometers and groundwater to be unavailable for sampling. To collect representative samples of water from the aquifer, sampling procedures were strictly followed [19–22]. Field measurements were made of basic physicochemical water quality parameters such as temperature (T), pH, electrical conductivity (EC), and oxygen/reduction potential (ORP), using a small electrical water pump, an in-line flow-through cell, and portable meters such as a HQ40D multi meter (Hach®, Hach-Lange GMBH, Berlin, Germany) equipped with Intellical™ (Hach-Lange GMBH, Berlin, Germany) pH, EC, and ORP electrodes with temperature sensors. The EC and ORP values were used to determine when formation-quality water was available for sample collection [20].

For chemical analysis, water was filtered through 0.7 μm GF/F (glass microfiber) Whatman's syringe filters and collected in polyethylene bottles—150 mL aliquots for anions and 30 mL aliquots for cations. Then, bottles were put into a refrigerator and delivered to the laboratory within 48 h. A High-Performance Liquid Chromatography (HPLC) method was used for major anion analysis (except bicarbonates and nitrates), and the Inductively Coupled Plasma Absorption Emission Spectrometry (ICP-AES) method was used for cation analysis. Uncertainties in the determination of major ions, as reported by the laboratory, were in the range 5–10%. Bicarbonates were determined by the potentiometric titration method, and concentrations of NO_3^- were determined in the field using a Slandi LF portable spectrophotometer (SLANDI®, SLANDI sp. z o.o., Michalowice, Poland). The anion-cation charge balance method was followed to assess the accuracy of the chemical analyses: for all water samples, the charge balance was less than 5%. The chemical analyses were performed at the Institute of Environmental Protection, National Research Institute in Warsaw, Poland. Table 1 shows a compilation of the physicochemical data obtained for groundwater.

Water samples for the analysis of oxygen stable isotope and hydrogen composition were collected in 30 mL amber-glass bottles and tightly sealed. Routine mass spectrometric techniques were applied for the determination of the isotopic composition of water, namely an off-line technique of CO_2–H_2O equilibration [23,24] to determine $^{18}O/^{16}O$ ratios, and an off-line static batch water reduction on hot zinc to determine $^2H/^1H$ ratios [25–28]. The results were reported using δ notation with respect to the VSMOW international standard. Normalization of measured data was perform according to three international standards, namely: VSMOW (δ^2H = 0.0‰, $\delta^{18}O$ = 0.0‰), VSLAP (δ^2H = −427.5‰, $\delta^{18}O$ = −55.50‰), and GISP (δ^2H = −189.5‰, $\delta^{18}O$ = −24.76‰). The precision of the measurements was ±0.08‰ and ±0.9‰ for $\delta^{18}O$ and δ^2H, respectively. The precision of the isotopic measurements was calculated based on long-term measurements of international reference materials as well as our internal standard water. The analyses were performed at the Institute of Geological Sciences of the Polish Academy of Sciences (ING PAN) in Warsaw, Poland.

Water samples for the analysis of the stable isotopes of sulfur ($^{34}S/^{32}S$ ratio) and oxygen ($^{18}O/^{16}O$ ratio) of dissolved sulfates were collected in HDPE bottles with volumes of 1 or 2 L depending on SO_4^{2-} concentration, which was determined in the field using a Slandi LF portable spectrophotometer. Afterwards, the water samples were acidified to a pH of 2–3 and an appropriate amount of 10% BaCl solution was added in order to precipitate all dissolved sulfates as $BaSO_4$ [29]. A method based on the reduction of $BaSO_4$ with graphite to CO_2 was applied to determine the $^{18}O/^{16}O$ ratio of sulfates [30,31], while a method for the conversion of $BaSO_4$ to SO_2 at 850 °C was applied to determine the $^{34}S/^{32}S$ ratio [32]. The results of the analysis were reported using δ notation: $\delta^{18}O$ with respect to VSMOW and $\delta^{34}S$ with respect to VCDT. Normalization of measured data was perform according to international standards, namely: NBS-127 ($\delta^{34}S$ = +20.3‰ VCDT, $\delta^{18}O$ = +9.3‰ VSMOW) and IAEA-SO6 ($\delta^{34}S$ = −34.1‰ VCDT, $\delta^{18}O$ = −11.35‰ VSMOW). The precision of the measurements was ±0.1 ‰ for both $\delta^{18}O$ and $\delta^{34}S$. The analyses of the isotopic composition of sulfates were performed in the Laboratory of Mass Spectrometry at the Institute of Physics UMCS, Lublin, Poland. Table 2 shows a compilation of the isotopic data obtained for groundwater.

Table 1. Chemical composition of atmospheric precipitation and groundwater from piezometers in the Pożary research station. Rainwater is presented for depth 0.0 m. *nd*—not detected; *na*—not analyzed; <0.5—below detection limit.

Date	D	T	pH	Eh	SC	Ca^{2+}	Mg^{2+}	Na^+	K^+	HCO_3^-	SO_4^{2-}	Cl^-	NO_3^-	$NH4^+$	TDS
	(m)	(°C)	(−)	(mV)	(uS/cm)					(mg/dm^3)					
IX.2009	rain	10	5.87	328	52	3.9	0.4	1.4	2.1	8.6	6.1	3.3	0.6	2.3	28.7
	2.1	10.1	7.06	−28	741	124.0	21.2	11.2	1.2	441.9	54.0	37.6	<0.5	na	691.1
	3.3	10.8	6.66	31	512	93.0	7.2	8.7	1.7	172.0	62.0	58.3	<0.5	na	402.9
	4.75	10.5	6.68	23	484	102.0	12.2	9.6	0.2	241.0	63.0	19.5	<0.5	na	447.5
	8.35	9.4	6.74	25	750	138.0	8.4	9.7	1.0	322.0	105.0	40.8	<0.5	na	624.9
X.2009	rain	9.5	5.74	298	44	3.0	0.4	1.1	0.8	nd	11.5	4.0	0.9	2.0	23.7
	2.1	9.4	7.05	−14	758	125.0	20.6	12.5	2.7	431.8	55.1	35.5	<0.5	na	683.2
	3.3	9.5	6.83	35	521	96.7	11.6	8.9	1.7	222.8	68.0	49.5	<0.5	na	459.2
	4.75	9.5	6.71	26	489	104.6	12.1	9.1	0.2	244.0	65.0	20.3	<0.5	na	455.3
	8.35	8.9	6.97	23	702	116.0	12.7	9.8	0.8	326.0	82.0	40.5	<0.5	na	587.8
XI.2009	rain	10	5.65	278	45	4.1	0.5	1.0	0.4	nd	14.2	5.5	0.79	1.1	27.59
	2.1	10.7	7.16	5	752	122.7	23.2	14.9	3.4	416.2	60.4	50.5	<0.5	na	691.3
	3.3	9.2	6.83	26	528	109.0	9.2	11.6	1.3	199.6	87.2	60.5	<0.5	na	478.4
	4.75	9.6	6.73	14	499	98.9	8.5	9.4	0.2	221.5	62.1	28.0	<0.5	na	428.6
	8.35	9.3	6.93	24	708	135.5	12.3	4.3	0.6	349.7	80.0	43.2	<0.5	na	625.6
III.2010	rain	12	4.44	257	18	0.5	0.1	0.6	0.3	nd	2.1	3.5	0.67	1.0	8.77
	2.1	13.1	7.1	49	780	130.0	25.9	11.9	1.4	423.7	59.1	34.5	<0.5	na	686.5
	3.3	10.9	6.78	73	545	101.0	8.0	8.8	0.3	161.5	113.2	52.9	<0.5	na	445.7
	4.75	11.4	6.56	84	544	103.0	8.3	12.1	0.2	255.7	62.2	22.0	<0.5	na	463.5
	8.35	12.4	6.89	105	749	144.0	15.2	9.6	0.4	374.9	69.5	60.1	<0.5	na	673.7
VI.2010	rain	15	5.22	315	58	1.7	0.7	1.1	3.1	1.0	7.1	6.0	1.27	5.1	27.07
	2.1	15.3	7.09	60	776	129.0	27.5	15.2	0.9	430.9	53.3	33.3	<0.5	na	690.1
	3.3	15.3	6.95	106	560	104.0	8.1	11.5	0.2	186.0	94.2	36.4	<0.5	na	440.4
	4.75	14.5	6.65	131	591	111.0	9.1	11.5	0.3	277.0	58.2	29.7	<0.5	na	496.8
	8.35	15.8	6.79	100	795	149.0	15.9	13.0	0.3	392.0	51.8	39.3	<0.5	na	661.3
VII.2010	rain	16	5.59	292	35	1.3	0.2	0.5	1.3	1.1	4.7	4.0	0.60	1.8	15.5
	2.1	15.7	7.06	55	759	133.0	27.3	12.6	1.5	420.6	50.3	29.3	<0.5	na	674.6
	3.3	17.2	6.95	83	542	106.0	8.0	9.3	0.2	198.6	81.4	39.0	<0.5	na	442.5
	4.75	15.1	6.69	109	511	104.0	8.1	8.3	0.2	245.5	52.0	21.0	<0.5	na	439.1
	8.35	16.1	6.88	86	731	147.0	15.6	10.3	0.2	383.0	67.4	48.4	<0.5	na	671.9

Table 2. The isotopic composition of groundwater and dissolved sulfates in the Pożary aquifer at monitored depths throughout the observation period. *na*—not analyzed.

Date	D	δ^{18}O H$_2$O	δ^2H H$_2$O	δ^{18}O SO$_4$	δ^{34}S SO$_4$	SO$_4^{2-}$	Date	D	δ^{18}O H$_2$O	δ^2H H$_2$O	δ^{18}O SO$_4$	δ^{34}S SO$_4$	SO$_4^{2-}$
	(m)		(‰) vs. VSMOW		(‰) vs. VCDT	(mg/dm^3)		(m)		(‰) vs. VSMOW		(‰) vs. VCDT	(mg/dm^3)
VIII.2009	rain	−7.48	−55.4	*na*	*na*	2.3	II.2010	snow	−18.29	−148.0	13.2	8.8	3.65
	2.1	−8.48	−67.0	13.0	15.8	49.0		2.1	−9.11	−73.7	15.4	19.6	72.0
	3.3	−8.57	−66.5	7.7	4.8	110.0		3.3	−8.45	−68.1	7.8	3.4	115.0
	4.75	−7.96	−66.4	13.5	12.8	61.0		4.75	−7.56	−65.4	15.5	12.8	80.0
	8.35	−7.58	−67.4	10.3	7.9	78.0		8.35	−7.85	−67.7	13.5	12.1	74.0
IX.2009	rain	−8.92	−70.4	*Na*	*na*	6.10	III.2010	rain	−12.04	−91.0	*na*	*na*	2.1
	2.1	−9	−67.2	15.7	19.5	54.0		2.1	−9.08	−73.3	15.4	19.8	59.1
	3.3	−8.45	−65.7	11.0	8.3	62.0		3.3	−8.33	−68.8	6.3	1.4	113.2
	4.75	−8.75	−66.8	15.7	13.1	63.0		4.75	−7.64	−65.0	14.5	13.8	62.2
	8.35	−8.2	−63.4	13.5	6.4	105.0		8.35	−7.7	−66.4	12.8	12.1	69.5
X.2009	rain	−14.75	−93.0	*Na*	*na*	11.5	IV.2010	rain	−10.42	−77.8	*na*	*na*	*na*
	2.1	−8.93	−66.1	15.8	20.9	55.1		2.1	−9.12	−73.0	15.5	19.2	68.0
	3.3	−8.26	−62.6	9.3	6.8	68.0		3.3	−7.84	−69.7	5.1	−2.6	140.0
	4.75	−8.64	−64.6	16.8	15.7	65.0		4.75	−7.47	−64.9	15.6	13.9	80.0
	8.35	−8.4	−65.1	11.5	8.7	82.0		8.35	−6.76	−61.1	14.5	12.3	73.0
XI.2009	rain	−11.03	−87.9	*na*	*na*	14.2	V.2010	rain	−9.1	−68.1	*na*	*na*	2.1
	2.1	−9	−73.0	15.3	19.7	60.4		2.1	−9.03	−73.9	15.0	18.8	70.0
	3.3	−8.29	−68.1	7.7	3.5	87.2		3.3	−7.95	−67.8	6.3	−0.4	115.0
	4.75	−8.6	−70.3	15.8	13.9	62.1		4.75	−7.58	−66.1	15.7	14.6	80.0
	8.35	−8.47	−70.1	12.2	8.7	80.0		8.35	−6.58	−60.1	15.0	13.7	68.0
XII.2009	snow	−15.6	−128.0	9.1	4.3	2.3	VI.2010	rain	−5.64	−40.2	9.8	6.0	7.1
	2.1	*na*	*na*	*na*	*na*	*na*		2.1	−9.11	−65.3	14.4	17.6	53.3
	3.3	*na*	*na*	*na*	*na*	*na*		3.3	−7.93	−58.9	5.9	1.4	94.2
	4.75	*na*	*na*	*na*	*na*	*na*		4.75	−7.98	−61.2	14.6	14.5	58.2
	8.35	*na*	*na*	*na*	*na*	*na*		8.35	−6.94	−55.6	12.99	11.5	51.8
I.2010	snow	−15.67	−124.9	*na*	*na*	1.7	VII.2010	rain	−8.17	−63.5	*na*	*na*	4.7
	2.1	*na*	*na*	*na*	*na*	*na*		2.1	−9.17	−65.7	15.1	16.6	50.3
	3.3	*na*	*na*	*na*	*na*	*na*		3.3	−8.16	−61.1	13.9	10.1	81.4
	4.75	*na*	*na*	*na*	*na*	*na*		4.75	−8.88	−65.3	15.1	15.9	52
	8.35	*na*	*na*	*na*	*na*	*na*		8.35	−7.47	−58.5	12.4	9.7	67.4

The PHREEQC software provided by the USGS (version 2.13.2 with the minteq.dat thermodynamic database), was used to calculate saturation indices (SIs) for selected minerals.

Together with groundwater, rainwater samples were collected and analyzed in order to determine the chemical and isotopic composition of atmospheric precipitation, which is the major input (recharge) component in the studied groundwater system. Rainwater samples were obtained from the meteorological station of Integrated Monitoring of the Natural Environment of Kampinos National Park (52°17′10″ N, 20°27′17″ E), located about 2 km west of the Pożary research station. A standard Hellmann's rain gauge was used to collect and measure atmospheric precipitation. Monthly cumulative rainwater was collected in separate polyethylene bottles with volumes of 1–2 L (one for each month), which were stored in a refrigerator at a temperature of 4 °C and subjected to the same set of chemical and isotopic analyses together with groundwater obtained from the piezometers. Physicochemical parameters of the rainwater were measured in the field during each water collection.

Tables 1 and 2 show data obtained for the rainwater samples. In order to present the entire vertical profile of the chemical and isotopic composition of atmospheric precipitation and groundwater, the data for rainwater are presented for a depth of 0.0 m in plots of Section 4.

4. Results

Chemical analyses of groundwater and rainwater were made six times during an entire monitoring period. The results are presented in Table 1.

Variations of groundwater temperature, pH, ORP, and EC at different depths are presented in Figure 2. The groundwater temperature in the Pożary aquifer varied between 8.9 and 17.2 °C, depending on the seasonal ambient air temperature (Figure 2). The pH of the groundwater ranged from 6.56 to 7.16, indicating slightly acidic and neutral conditions. Taking into account the vertical profile, the lowest pH values in the range of 4.44–5.87 are characteristic of atmospheric precipitation (i.e., rain and snow). In groundwater, the pH is higher than that of precipitation, and clearly varies with depth: It was highest in the upper part of the aquifer (pH > 7.0) and then decreased, reaching the lowest values in the middle part, before again reaching higher values in the deeper part (Figure 2). The EC of groundwater ranged from 484 to 780 µS/cm, and varied with depth, repeating the trend characteristic for pH.

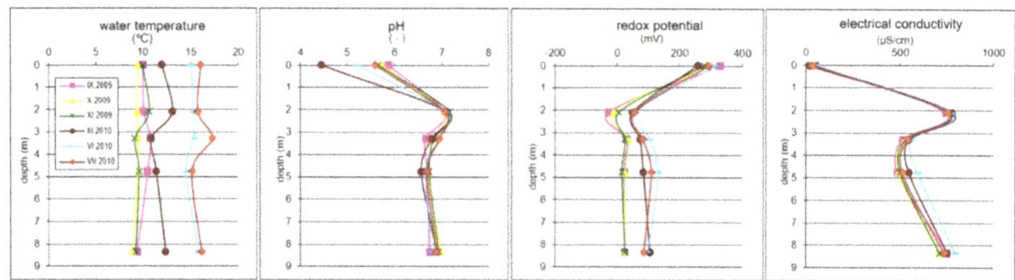

Figure 2. Seasonal variations of physicochemical parameters of groundwater in the Pożary aquifer vertical profile. The values for rainwater are shown at depth 0.0 m.

The ORP varied between −28 and +109 mV, depending on the season and sampling depth. The lowest values occurred at 2.10 m bgs, indicating slightly reducing conditions (from −28 to 60 mV), probably due to the influence of the microbial decomposition of organic matter in the peat layer. In the deeper parts of the aquifer, positive values of redox potential were observed (from 14 to 109 mV), indicating a mildly reducing environment.

The hydrogeochemical type of the water in the aquifer changes with depth as a result of the variation in the concentration of major anions (Figures 3 and 4). In the upper part of the aquifer, water of HCO_3-Ca type dominates. In the middle part, at a depth of 3.3 m, the concentrations of the major anions change abruptly: The content of HCO_3^- reaches the minimum, the contents of SO_4^{2-} and Cl^-

reach the maximum, and the hydrogeochemical type of the water evolves into HCO_3-Cl-SO_4-Ca or HCO_3-SO_4-Cl-Ca. In the deeper parts of the aquifer, the concentration of HCO_3^- gradually increases, whereas the contents of SO_4^{2-} and Cl^- decrease abruptly at a depth of 4.75 m before beginning to rise again with depth; from a depth of 4.75 m, water of the HCO_3-SO_4-Ca type dominates (Figures 3 and 4). On the other hand, at a depth of 3.3 m, the concentrations of major cations (Ca^{2+}, Mg^{2+}, and Na^+) usually exhibit the lowest values, which remain stable or rise slightly in deeper parts of the aquifer. A comparison of archival and actual data of the chemical composition of groundwater shows similar trends of evolution with depth (Figure 3).

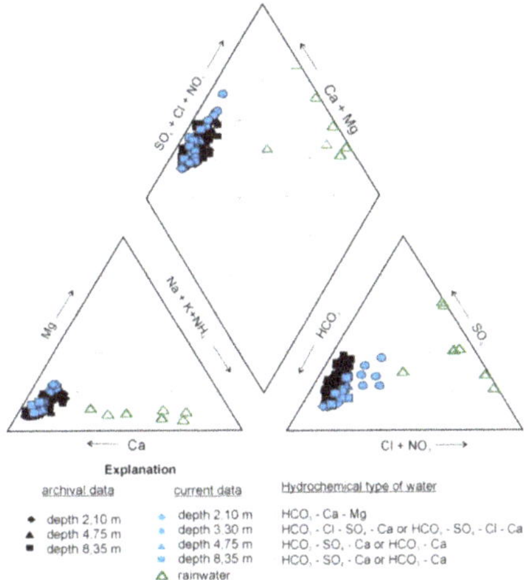

Figure 3. Piper diagram showing chemical composition of groundwater in vertical profile of the Pożary aquifer. Rain water are plotted for reference. Archival records refer to groundwater monitoring period from June 1999 to December 2001 [15,33].

The variation in the stable isotope compositions of groundwater ($\delta^{18}O$ and δ^2H) and dissolved sulfates ($\delta^{34}S$ and $\delta^{18}O$) with depth in the aquifer are summarized in Table 2 and Figure 4.

Typically, the oxygen and hydrogen isotopic composition of groundwater generally reflects the mean weighted annual composition of precipitation. In our case of a shallow unconfined aquifer with a relatively thin unsaturated zone (i.e., about 0.5 m in thickness) composed of a peat layer, the large seasonal variations of the O and H isotopic composition of precipitation are considerably attenuated in the upper part of the saturated zone. For example, the variation coefficient of $\delta^{18}O$ in the precipitation (rain and snow) reached about 35%; while at a depth of 2.1 m, seasonal variations were attenuated to about 2% (Figure 4). The isotopic composition of dissolved sulfates in atmospheric precipitation was measured three times: Two times in rain water, with each value representing rain water collected over a few months (due to a relatively low concentration of dissolved sulfates and the small amount of water available after each month), and one time in snow in December.

The obtained values of isotopic composition (Table 2) are typical for the fallout in rural areas (e.g., [12]). The isotopic composition of SO_4^{2-} dissolved in groundwater ranged from −2.6 to +20.9‰ and from +5.1 to +16.8‰ for $\delta^{34}S_{SO4}$ and $\delta^{18}O_{SO4}$, respectively. As can be seen from Figure 4, both $\delta^{34}S_{SO4}$ and $\delta^{18}O_{SO4}$ demonstrate significant variation along the depth profile. Generally, the observed vertical variation trend of the isotopic composition of SO_4^{2-} is reversed compared to that of SO_4^{2-} concentration. Sulfates were most depleted in ^{34}S and ^{18}O at a depth of 3.30 m in almost all of the observation period (Figure 4). At a depth of 3.30 m, the values of $\delta^{34}S_{SO4}$ gradually decreased from

September 2009 to April 2010, before increasing from April 2010 to July 2010. For $\delta^{18}O_{SO4}$, a seasonal trend similar to that of $\delta^{34}S_{SO4}$ was observed.

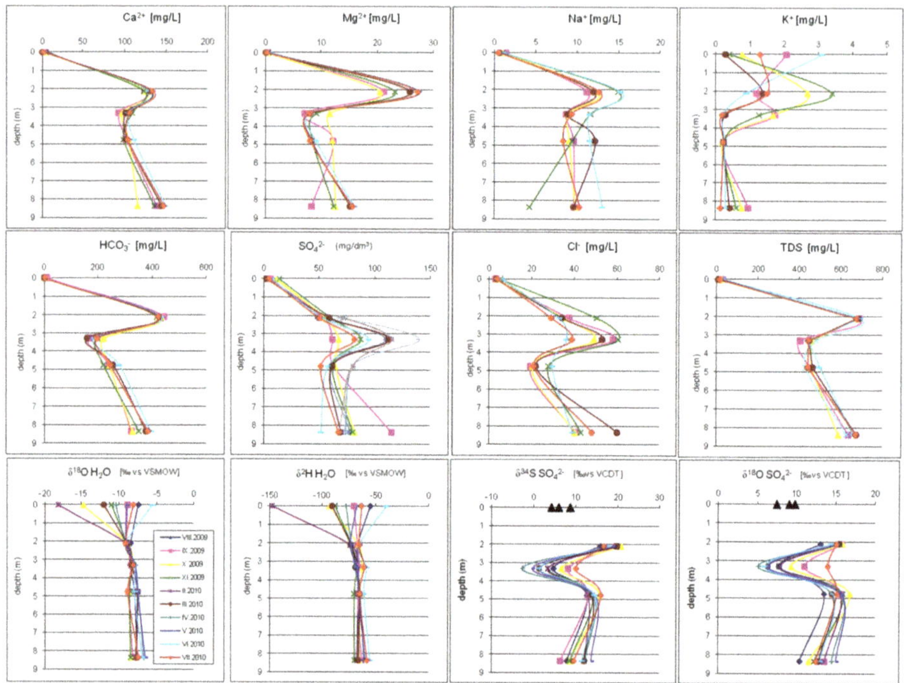

Figure 4. Seasonal variations of chemical and isotopic composition of groundwater in the Pożary aquifer with depth. The values for rainwater are shown at depth 0.0 m. For more explanations see the text.

5. Discussion

The chemical composition of subsurface water is always a complex function of many variables, e.g., the composition of recharge, the mineralogy and composition of the geological environment, and the hydrogeological and hydraulic settings. One of the main factors affecting groundwater chemistry is the presence of organic matter. The mineralization (oxidation) of organic matter in soil and aquifer environments always implies a series of redox reactions which control the distribution of major species dissolved in groundwater, with the organic matter being the major reductant [2–7].

As long as free oxygen is available in the saturated zone, the simplified reaction of organic matter oxidation can be written as follows:

$$CH_2O + O_2 \rightarrow CO_2 + H_2O \tag{1}$$

During this process, firstly carbon is released as CO_2. Depending on the reactivity and composition of organic matter, a combination of organic NO_3^- and HPO_4^{2-} can also be released to groundwater. The concentrations of NO_3^- and HPO_4^{2-} have not been studied in detail in the aquifer of the Pożary research station. However, clear evidence of organic matter decomposition in the aquifer vertical profile was shown by Porowska and Leśniak [15]—namely, an increase in CO_2 partial pressure and DIC content with a carbon isotopic composition typical of organic origin (reported $\delta^{13}C$ from −17 to −25‰ vs. PDB), an increase in organic carbon concentration (DOC), and a decrease in oxygen content (DO, reported concentrations range from 0.48–1.62 mg/dm^3). An additional consequence of the increase in CO_2 concentration in the upper part of the aquifer is an increase in the weathering capacity of groundwater and the induction of carbonate mineral dissolution, which may result, for example, in an increase in Ca and Mg concentrations (see Figure 4).

Water saturation promotes anoxic conditions in soil, vadose zone, or an aquifer. When molecular oxygen is not available (i.e., when it has been used up), the oxidation of organic matter continues and terminal electron acceptors other than O_2 are utilized. Depending on redox potential, pH, and availability of electron acceptors, the following microbial reduction processes can occur after O_2 depletion (i.e. the sequential reduction chain—a series of reactions which represent successively lower Eh levels): Nitrate, manganese, iron, sulfate, and finally CO_2 [7]. The reduction of NO_3^- by organic matter (denitrification) is a bacterially catalyzed process, it starts the sequential reduction chain in the aquifer, and can be written as an overall reaction (after [3]):

$$5CH_2O + 4NO_3^- \rightarrow 2N_2 + 4HCO_3^- + CO_2 + 3H_2O \tag{2}$$

Denitrification predominantly proceeds to the final product of N_2 [8,34–37]. An additional consequence of this process is the increase of HCO_3^- concentration and pH values (e.g., see the upper part of the profile up to 2.1 m in Figures 2 and 4). Different forms of nitrogen in shallow groundwater in wetlands and peat bogs of the Kampinos National Park were studied by Krogulec and Jóźwiak [38]. They typically reported very low concentrations of NO_3^- (from 0 mg/dm^3 to a maximum of 5.3 mg/dm^3) as a result of denitrification. Our research corroborates their conclusions: The concentration of NO_3^- in the groundwater of the Pożary aquifer was below 0.5 mg/dm^3 (see Table 1). Taking into account the aquifer profile below the peat layer, the transition between the denitrification zone and the sulfate reduction zone most likely occurs in the depth range from 2–4 m bgs: Here, the sulfate concentration starts to decrease and the HCO_3^- concentration starts to rise with depth (Figure 3).

The geochemical modeling of sulfur speciation (based on PHREEQC software) shows that sulfate is the main species of sulfur in the groundwater studied. The groundwater sulfates may be derived from various sources, such as atmospheric, pedospheric, lithospheric, and anthropogenic. In shallow aquifers in peatland areas, the sources of sulfate are usually pyrite, gypsum, biogenic sulfur compounds contained in the peat (mostly lithospheric sources), and atmospheric deposition [3,5,6,9,39,40].

However, the concentration, as well as the sulfur and oxygen isotopic composition, of sulfates in groundwater are controlled not only by sulfate sources but also by subsequent biotic and abiotic processes in the aquifer, including hydraulic conditions (e.g., groundwater mixing, fluctuation of ground water level, lateral flow). Hence, four processes may potentially control the distribution and isotope signatures of dissolved sulfate in the Pożary aquifer: (1) Bacterial (dissimilatory) sulfate reduction; (2) the formation and oxidation of reduced inorganic sulfur (RIS); (3) the dissolution of evaporitic sulfates (mainly gypsum); and (4) the mixing of sulfate from different sources. All of these processes result in specific patterns of the concentration and isotopic composition of dissolved SO_4^{2-}, and in favorable hydrogeological settings can be identified by geochemical and isotopic techniques.

5.1. Sulfate Sources in Groundwater Studied

The sources of sulfate in groundwater can have a wide range of $\delta^{34}S$ values. Usually, the local source of sulfate in groundwater is estimated using the relationship between $\delta^{34}S$ and the concentration of dissolved sulfate [12,41–43]. Plotting the $\delta^{34}S$ values of groundwater sulfate versus the inverse of sulfate concentration $(1/[SO_4^{2-}])$ for all water samples yielded a straight line with a correlation factor (R^2) of 0.6 and a y-axis intercept of -7.9‰ (Figure 5A).

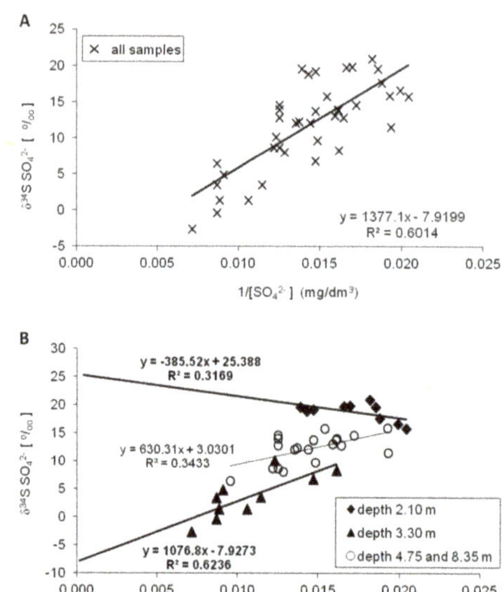

Figure 5. The δ^{34}S values and concentrations of dissolved sulfates in groundwater profile of the Pożary aquifer. For more explanations see the text.

The slope of the line s equals

$$s = C_R \times (\delta^{34}S_R - \delta^{34}S_I) \tag{3}$$

where C_R refers to the concentration of the remaining (i.e., measured) sulfate, and δ_R and δ_I refer to the sulfur isotopic composition of the remaining (i.e., measured) and initial (i.e., source) sulfate, respectively. The y-axis intercept represents δ_I.

The y-axis intercept indicates the sulfur isotopic composition of the potential source of groundwater sulfates. The value of −7.9‰ strongly suggests that the RIS compounds are the dominant lithogenic source responsible for increasing the sulfate concentration in the groundwater of the Pożary aquifer. However, a more detailed picture of the sulfate sources can be obtained by taking into account the entire observation period and the variation of sulfate concentration and sulfur isotopic composition separately in the monitored respective depths of the aquifer (Figure 5B).

The most extreme relationships were observed for groundwater sulfates at depths of 2.1 and 3.3 m. The y-axis intercept for sulfates at the depth closest to the peat layer yielded a $\delta^{34}S_I$ value of +25.4‰, which suggests that the dominant source of dissolved sulfate may be connected with the dissolution of evaporitic sulfates (mainly gypsum) and/or sulfates undergoing bacterial reduction in the peat and aquifer. However, sulfates with different sources may be present at a depth of 2.1 m depending on the season—i.e., mainly depending on the position of the groundwater level, which fluctuates as a result of local climatic conditions, namely precipitation volume, thawing of snow cover, local infiltration or runoff to rivers, etc. When the groundwater level decreases, more of the peat is exposed to air and undergoes decomposition, and more sulfates of evaporitic and biogenic (i.e., the oxidation of organic compounds) origin accumulates in the peat. Rainfall facilitates the dissolution of sulfates (e.g., gypsum) and their transfer to the aquifer. During times of high groundwater level caused by heavy rains in autumn and snow thawing in spring, a peat layer occurs within the saturation zone, and a considerable admixture of additional sulfates with an atmospheric origin or related to gypsum dissolution or dissolution (oxidation) or RIS compounds can reach groundwater, increasing the sulfate concentration in the water in the aquifer below the peat layer [44–46]. At a depth of 3.3 m, the y-axis intercept yielded a $\delta^{34}S_I$ value of −7.9‰, the same as the average for all water samples (Figure 5B).

This strongly suggests that the oxidation of RIS, the most likely being pyrite, is responsible for the increase in SO_4^{2-} concentration in this part of the aquifer profile.

The *y*-axis intercept for sulfates at depths of 4.75 and 8.3 m (which are taken together, as the isotopic composition of sulfates at these depths is quite similar) yielded a $\delta^{34}S_I$ value of +3.0‰. Values of $\delta^{34}S_I$ below +10‰ suggest that the dominant source of dissolved sulfates might be connected with atmospheric sulfates. Indeed, at the Pożary research station, the $\delta^{34}S$ values of sulfates dissolved in rainwater and snow were found to be between +4.25 and +8.79‰ (Table 2). On the other hand, depending on the season and the intensity of the lateral flow of water and direct infiltration, the $\delta^{34}S$ of dissolved sulfates in deeper parts of the aquifer may be a result of the mixing of sulfates from different sources and overlapping biogeochemical processes. At least, the contribution of atmospheric sulfates and sulfates derived from the dissolution of gypsum and pyrite should be taken into account, as well as common processes such as the bacterial reduction of sulfates and the oxidation of sulfides. Anthropogenic sources of sulfur in the groundwater, such as fertilizers, manure, and sewage, are negligible, due to the fact that the Pożary research station is located within the Special Protection Area of the Kampinos National Park, relatively far from arable lands and urban areas.

5.1.1. Atmospheric Sulfates

The sulfur geochemistry of any groundwater in unconfined aquifer is initially controlled by the recharge environment, i.e., atmospheric inputs, namely the sulfate content in snow and rain and the extent of evaporation in the study region. The hydrogen and oxygen isotopic composition of the groundwater indicate a meteoric origin: δ^2H and $\delta^{18}O$ values were located along the Global Meteoric Water Line (GMWL), which is typical for the meteoric recharge of the modern hydrological cycle with some seasonal variation and rainfall effects (Figure 6). There was a loss of seasonal variation of the isotopic composition of groundwater during infiltration through the unsaturated zone (Figure 4): At the piezometer depth of around 2.1 m, the seasonal variation of δ^2H and $\delta^{18}O$ was attenuated to less than 5% of that observed in the precipitation. The sulfur isotopic composition of atmospheric sulfate ($\delta^{34}S_{SO4}$) is usually controlled by emissions from fossil fuel combustion and the biological release of S-bearing compounds, which gives $\delta^{34}S$ values that usually range from slightly negative to about +10‰ CDT [12,42].

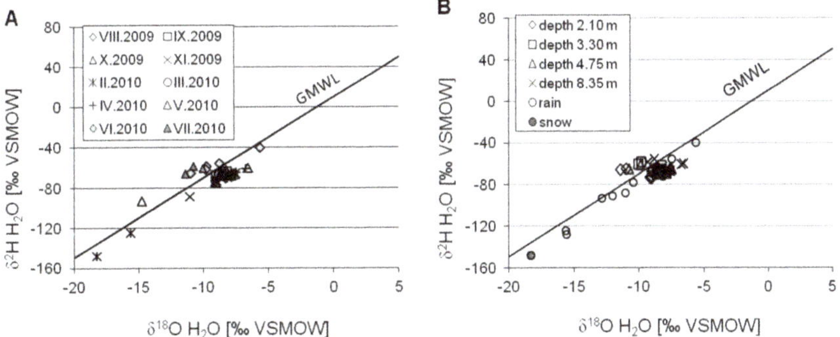

Figure 6. The δ^2H vs. $\delta^{18}O$ relationship in groundwater of the Pożary aquifer. Differentiation of isotopic composition is shown for sampling seasons (**A**) and depth profile (**B**). Global Meteoric Water Line (GMWL) after [47]: $\delta^2H = 8.13 \times \delta^{18}O + 10.8$.

On the other hand, the oxygen isotope composition of freshwater sulfates ($\delta^{18}O_{SO4}$) may be controlled by the oxygen isotope composition of water ($\delta^{18}O_{H2O}$) and dissolved oxygen ($\delta^{18}O_{DO}$). DO comes from the dissolution of atmospheric oxygen which is enriched in heavy isotopes: Values of $\delta^{18}O_{DO}$ are usually around +23.5‰ VSMOW [48,49]. The range of $\delta^{18}O_{SO4}$ for atmospheric deposition in Central Europe varies between +7 and +17‰ VSMOW [41,50], while at the Pożary research station,

$\delta^{18}O_{SO4}$ values in rain and snow were found to be between +9.0 and +13.2‰ VSMOW (based on four measurements, see Table 2). The formation of sulfates in shallow oxygen-rich underground environments produces higher values of $\delta^{18}O_{SO4}$. In the Pożary sandy aquifer, the high $\delta^{18}O$ values of groundwater sulfates (from +10.34 to +16.75‰, except at a depth of 3.30 m) imply that atmospherically derived oxygen may be an important constituent of the total sulfate pool. However, depending on the additional biogeochemical processes which affect the isotopic composition of sulfates, the contribution of atmospheric oxygen to groundwater SO_4^{2-} may vary widely [44,51]. For example, at a depth of 3.30 m, the increase in the SO_4^{2-} concentration is connected with a decrease in the $\delta^{18}O_{SO4}$ and $\delta^{34}S_{SO4}$ values, which strongly suggests that sulfate concentrations are controlled by other processes than only the simple mixing of atmospheric input. Moreover, the groundwater collected from different depths in the studied aquifer revealed a relatively high SO_4^{2-} concentration (from ~52 to 140 mg/dm^3) compared to rain water or snow, whose SO_4^{2-} concentrations range from ~0.1 to 5 mg/dm^3 according to Porowska (2003) [52] and from 1.65 to 14.2 mg/dm^3 according to our measurements. This suggests that, on the scale of the whole aquifer profile, the contribution of sulfates originating directly from atmospheric precipitation may be rather minor, and considerable variations in SO_4^{2-} concentrations and its isotopic composition are controlled by additional sulfate sources and overlapping biogeochemical processes.

5.1.2. Dissolution of Gypsum

Sulfates in the analyzed groundwater may also be derived from soluble minerals such as gypsum ($CaSO_4 \cdot 2H_2O$ or anhydrite $CaSO_4$), which is a common sulfate mineral in freshwater peatland. Similar to pyrite, gypsum is mostly authigenic, and may constitute a considerable source of sulfate dissolved in groundwater affected by peatland [53]. Saturation indices calculated based on our data show undersaturation with respect to gypsum, and the potential dissolution of gypsum along the whole vertical profile of the aquifer, regardless of the hydrogeochemical conditions (Figure 7).

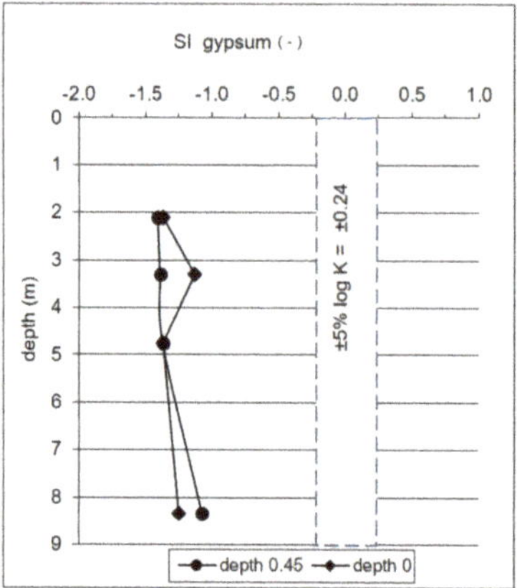

Figure 7. Saturation indices (SI) with respect to gypsum calculated for chemical composition of water in two extreme positions of the groundwater level, namely: −0.45 m bgs in November 2009, and 0.0 m bgs (i.e., the groundwater level reaches the ground surface, the peat bed is fully saturated) in April 2010.

The dissolution of gypsum causes an increase in the sulfate concentration in groundwater, and occurs without measurable isotope fractionation, i.e., dissolved sulfates retain the original isotopic composition of gypsum [40,54–56]. Sulfates in the groundwater of the upper part of the Pożary aquifer—which have high $\delta^{34}S$ and $\delta^{18}O$ values of around +20‰ and +16.7 ‰, respectively (Tab. 1), and have predicted $\delta^{34}S$ values of the source sulfate as high as +25‰ (Figure 5)—may originate from the dissolution of gypsum formed in the peat layer [12].

5.1.3. Mineralization of Carbon-Bonded Sulfur (C-S) Compounds

A large proportion of the sulfates in the studied aquifer might originate from the decomposition of peat during times of low water table. Similar trends were observed in [45,46]. Sulfur compounds may be retained in the peat as both (i) organic forms and (ii) reduced inorganic forms (RIS), depending on the oxygen availability [4]. Several studies have suggested that H_2S, formed as an end product of the dissimilatory sulfate reduction (see reaction in Equation 6) can react rapidly with organic matter, producing carbon-bonded sulfur (C–S) (e.g., [57]). The C–S can be further degraded to biogenic compounds such as CH_3SH (i.e., methanethiol as a volatile form), $(CH_3)_2S$ (i.e., dimethylsulfide as a liquid form), and H_2S gas, which can escape from the soil or can be oxidized back to SO_4^{2-}. In the rainy season, when the groundwater table rises, the sulfates accumulated in the peat are more easily dissolved and transferred into groundwater below the peat layer, thus increasing their concentration. Evidence for the decomposition of organic matter in the Pożary aquifer profile was clearly shown by Porowska and Leśniak [15]. The availability of organic matter and the presence of H_2S (giving a characteristic "rotten egg" odor) from a possible dissimilatory sulfate reduction suggests that C–S compounds can form in the studied peat and groundwater. Their further mineralization and transformation may additionally affect the concentration and isotope signature of sulfates dissolved in groundwater. Processes such as the decay and oxidation of organic compounds do not change the $\delta^{34}S_{SO4}$ significantly [44]. However, C–S compounds were not analyzed in this study, and their role in formation of groundwater sulfates needs to be further studied.

5.2. Precipitation/Oxidation of Reduced Inorganic Sulfur (RIS)

Numerous studies have shown that pyrite (FeS_2) is a commonly occurring species of RIS in freshwater peatlands, while greigite (Fe_3S_4) and mackinawite ($FeS_{0.9}$) are minor components of the RIS pool [1,5,6,57,58]. Two genetic types of pyrite have been identified in peat bogs, namely syngenetic and epigenetic [6,59–61]. Syngenetic pyrite is formed in anoxic environments during bacterial sulfate reduction ($S_{sulfate} \rightarrow S_{pyrite}$), mainly as framboidal pyrite and more rarely as euhedral pyrite. Epigenetic pyrite is produced during the humification of peat, when sulfur from sulphobacteria and carbon from plant respiration bond and are reduced to sulfidic sulfur ($S_{organic} \rightarrow S_{pyrite}$). Berner and Raiswell [62] suggested a greigite (Fe_3S_4) precursor for framboidal pyrite (FeS_2) and amorphous FeS and a mackinawite precursor for euhedral pyrite. Framboidal pyrite (i.e., mostly syngenetic) is clearly the most abundant species of RIS in peat deposits, and was found, for example, in peatland of the Lubartowska Upland, Western Poland [53]. Moreover, Jóźwiak [63] indicated the possibility of the precipitation of pyrite from the groundwater of selected peatland of the Kampinos National Park. According to Berner and Raiswell [62], at the Pożary research station, hydrogeochemical conditions in the aquifer are favorable to the formation of framboidal as well as euhedral pyrite. SIs calculated based on our data show supersaturation with respect to pyrite and the potential possibility of its formation along the entire profile of the aquifer (Figure 8). On the other hand, the negative values of SIs for greigite and mackinawite show that these compounds are not stable and undergo dissolution.

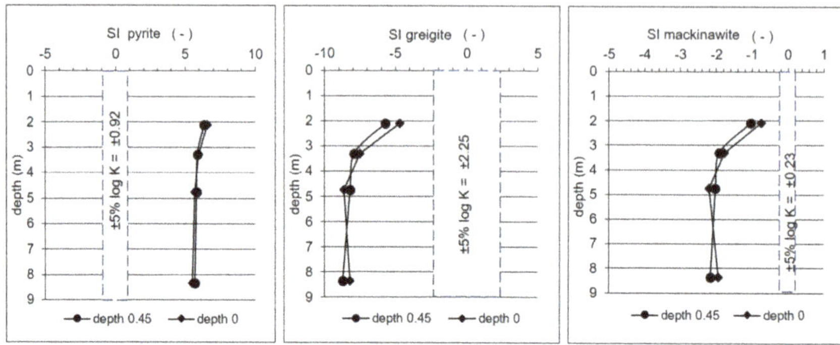

Figure 8. Saturation indices (SI) with respect to pyrite, greigite, and mackinawite. Calculations were made for chemical composition of water in two extreme positions of groundwater level, i.e., −0.45 m bgs in November 2009, and 0.0 m bgs (i.e., the groundwater level reaches the ground surface, the peat bed is fully saturated) in April 2010.

Peatlands are very heterogeneous environments in which SO_4^{2-} reduction and S^{2-} oxidation often coexist [5]. Even if hydrogeochemical conditions temporarily cause an environment which favors sulfate reduction, the short-term products of this reduction, such as pyrite or labile organic sulfur, can be either oxidized or mineralized. In the case of the Pożary sandy aquifer, which is overlain by a peat layer, the formation of RIS (i.e., pyrite) might be connected firstly with the sulfate reduction processes in the peat as well as in the aquifer itself in an anoxic environment. Sulfate reduction results in the formation of sulfides (i.e., H_2S and/or HS^-, depending on the pH), which react with Fe-oxides and form FeS_2, usually in a two-step process [3]:

$$2FeOOH + 3HS^- \rightarrow 2FeS + S^\circ + H_2O + 3OH^- \tag{4}$$

$$FeS + S^\circ \rightarrow FeS_2 \tag{5}$$

In the first step, part of the sulfide reduces Fe(III) and produces S°, while the remainder of the dissolved sulfide precipitates as FeS, which is much less stable than pyrite but forms kinetically very quickly due to the sluggish precipitation of FeS_2. In the second step, FeS transforms to FeS_2, which is an oxidation reaction. The precipitation of pyrite lowers the concentration of sulfide in groundwater. Depending on the environmental conditions, sulfide minerals produced during bacterial sulfate reduction can have $\delta^{34}S$ values which are more than 50% lower than those of the initial sulfate; commonly found $\delta^{34}S$ values for RIS in sedimentary rocks vary between −30 and +5‰ CDT [40]. These sulfur compounds are often finely dispersed in sediments, and under oxidizing conditions they may become a significant source of groundwater sulfates. The oxidation of pyrite or aqueous sulfides is a very common process in surface or near-surface environments and peatlands [11,64]. The process can be bacterially mediated or abiotic [64,65]. Exemplary reactions can be written as follows:

$$FeS_2 + 3\frac{1}{2}O_2 + H_2O \rightarrow Fe^{2+} + 2SO_4^{2-} + 2H^+ \tag{6}$$

$$FeS_2 + 14Fe^{3+} + 8H_2O \rightarrow 15Fe^{2+} + 2SO_4^{2-} + 16H^+ \tag{7}$$

The participating aerobic oxidizers, such as *Thiobacillus*, prefer low pH, the presence of organic matter, and warm temperature [66]. The oxidation of sulfides to SO_4^{2-} generally results in minimal isotopic fractionation, essentially retaining the original isotopic composition of the sulfide minerals [67]. During times of low water table, pyrite can be oxidized, causing an increase in sulfur load with water infiltration into the saturation zone and the aquifer. Even a small drop in the water table is sufficient to promote the oxidation of reduced forms of sulfur [67].

5.3. Bacterial Dissimilatory Sulfate Reduction

Microbial sulfate reduction by organic matter is one of the most important processes responsible for the sulfur cycle in hydrogeological environments. Dissimilatory sulfate reduction is an energy-gaining respiratory process conducted by a specific group of anaerobic prokaryotic bacteria (the genus *Desulfovibrio* and others) through chemical reactions in which organic carbon is oxidized while sulfates are reduced as they serve as terminal electron acceptors [2]. Such a redox reaction can be written as follows:

$$2CH_2O + SO_4^{2-} \rightarrow 2HCO_3^- + H_2S \tag{8}$$

where CH_2O represents a generic form of organic matter with the oxidation state of a carbohydrate. The end-product H_2S gas dominates at low and moderate pH (below 7.0), while at high pH, HS^- and S^{2-} dominate, and concentration is limited only by the solubility of sulfide minerals [3,28]. The geochemical conditions within the studied Pożary aquifer are favorable for bacterial dissimilatory sulfate reduction and H_2S formation. The dissolved organic carbon occurs in the peat and groundwater and provides the source of organic matter for the sulfate reduction process [15]. The surplus of organic carbon promotes the sulfate reduction and stability of the precipitated reduced sulfur compounds [9,68]. The redox potential is favorable for SO_4^{2-} reduction within the whole aquifer profile. In the near-surface zone, the ongoing reduction of SO_4^{2-} can be recognized by the presence of a characteristic "rotten egg" odor of dissolved H_2S, and additionally by the decrease of the organic carbon concentration in the groundwater profile [15], the decrease of the SO_4^{2-} content and the increase of the HCO_3^- content, as well as an increase in pH in comparison to deeper parts of the aquifer.

Typically, bacterial sulfate reduction affects the concentration of dissolved sulfates and is characterized by considerable isotopic fractionation of sulfur and oxygen e.g., [69–72]. As the process proceeds, the heavy sulfur and oxygen isotopes gradually accumulate in the residual sulfate reservoir, i.e., the residual aqueous sulfates are enriched in ^{34}S and ^{18}O isotopes relative to the original SO_4^{2-} due to the preferential consumption of lighter isotopes (^{32}S-O bonds need less energy to break) by bacteria [54,73]. This results in a positive linear (or near-linear) correlation between the $\delta^{34}S$ and $\delta^{18}O$ values of the residual sulfates in the analyzed groundwater (Figure 9A). Additionally, the increase of the $\delta^{34}S$ and $\delta^{18}O$ values of the dissolved sulfates with decreasing SO_4^{2-} concentrations also corroborates the bacterial sulfate reduction process in the groundwater [40,55]. The linear relationship between the $\delta^{34}S$ values and the SO_4^{2-} concentration for all samples from the entire observation period yielded an R^2 value of 0.67 (Figure 9B).

On the other hand, the positive trend between the $\delta^{34}S$ and $\delta^{18}O$ values for all samples yielded an R^2 value of 0.83; the slope of the regression line was 1.68, which falls within the typical range for natural systems of 1.4 to 3.5 [74]. Theoretically, the slope of the $\delta^{34}S$ vs. $\delta^{18}O$ plot represents the ratio of the enrichment factors for sulfur (ε_{34S}) and oxygen (ε_{18O}) during bacterial sulfate reduction. In most groundwater environments (i.e., conditions typical for a closed system, no contact with considerable amounts of sulfate minerals which could replenish their loss), such enrichment in ^{34}S and ^{18}O in the residual sulfate usually follows a Rayleigh fractionation model. The Rayleigh model is an exponential function that, in this case, describes the progressive partitioning of ^{34}S and ^{18}O into the remaining sulfates as the initial sulfate concentrations decrease during the reduction process:

$$\delta^{34}S_R = \delta^{34}S_I + \varepsilon \times \ln f \tag{9}$$

$$\delta^{18}O_R = \delta^{18}O_I + \varepsilon \times \ln f \tag{10}$$

where the subscripts R and I refer to the respective isotope ratios in the remaining (i.e., residual) and initial concentrations of sulfate, ε is the enrichment factor for sulfur and oxygen, and f is the fraction of residual sulfate: $f = [SO_4^{2-}]_R/[SO_4^{2-}]_I$.

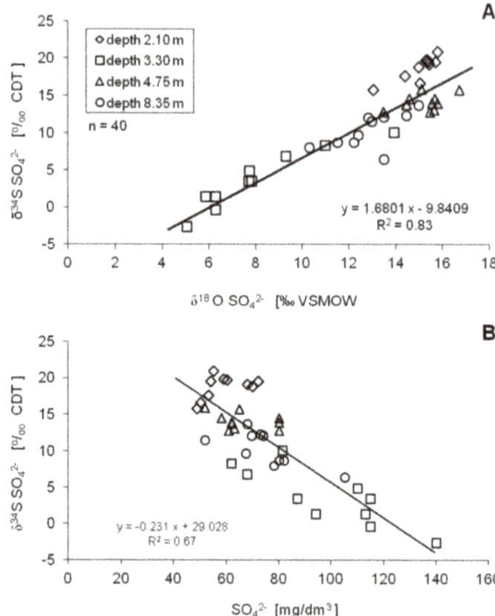

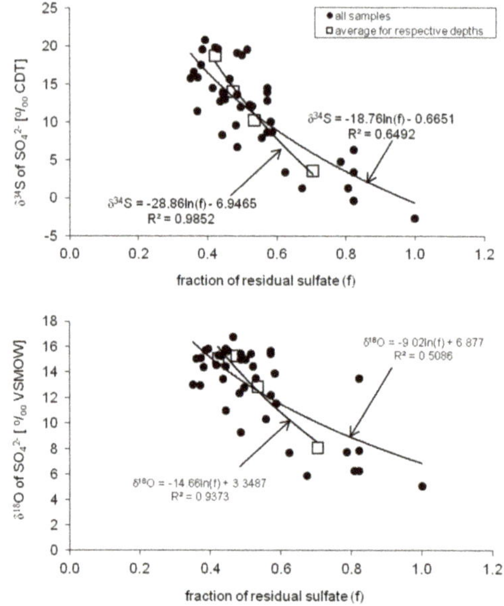

Figure 9. Trends of the δ^{34}S and δ^{18}O values for residual sulfates in groundwater of the Pożary aquifer. Linear correlations strongly suggest the occurrence of the bacterial sulfate reduction.

The relationships between the fraction (f) of the residual sulfate and the δ^{34}S and δ^{18}O values of all groundwater samples from the aquifer of the Pożary research station are shown in Figure 10. In order to calculate the fraction of residual sulfate, the highest measured sulfate concentration in the groundwater was assumed as the initial value, i.e., 140 mg/dm^3, at a depth of 3.3 m.

Figure 10. Relationship between the fraction (f) of the residual sulfate and its δ^{34}S and δ^{18}O values in groundwater samples from the aquifer of the Pożary research station. Note: fraction f was calculated for modeling purpose.

Based on these relationships, the enrichment factors (ε) for sulfur and oxygen were estimated to be $-18.8‰$ and $-9.0‰$, respectively (Figure 10). On the other hand, the logarithmic regressions for the mean values (from the entire observation period) of sulfate concentrations and isotopic compositions at particular depths yielded much better correlation factors (R^2), higher than 0.9. The estimated enrichment factors for sulfur and oxygen were $-28.9‰$ and $-14.7‰$, respectively (Figure 10). The estimated values of the enrichment factors yield $\varepsilon_{34S}/\varepsilon_{18O}$ ratios of about 2.0, which is in quite good agreement with the obtained slope of the $\delta^{34}S$ vs $\delta^{18}O$ enrichment around 1.7 (Figure 9).

The isotopic fractionation during bacterial sulfate reduction, as well as the enrichment factors and the reduction rates, are governed by local environmental and geochemical conditions under which the bacteria function [12,40,75]. The relationships between the fraction of the residual sulfates (f) and the sulfates' $\delta^{34}S$ values in groundwater in the aquifer profile calculated for different seasons reveal that the enrichment factors (ε) vary in a wide range, from about -36.3 to $-13.0‰$. Such variation confirms that local climatic, environmental, and hydrogeological conditions have a strong impact on the bacterial sulfate reduction process in the peat and in the aquifer, and the final isotopic composition of the residual sulfates found in the studied groundwater. It can be seen that the highest absolute values of the enrichment factors (ε) are connected with seasons of decreasing vegetation activity (i.e., late autumn, winter, and early spring), lower ambient air temperature, and high levels of the groundwater table (up to the ground surface). The observed values of the seasonal sulfur isotope enrichment factors for bacterial sulfate reduction in the Pożary aquifer are within the typical range of ε_{34S} reported in the literature for aquifers and wetland peat, that is, from $-60‰$ to $-2.6‰$ [41,76–79].

A general summary of the sulfate sources in groundwater of the unconfined Pożary aquifer deduced from distribution of dissolved sulfates and their O and S isotopic compositions is shown in Figure 11.

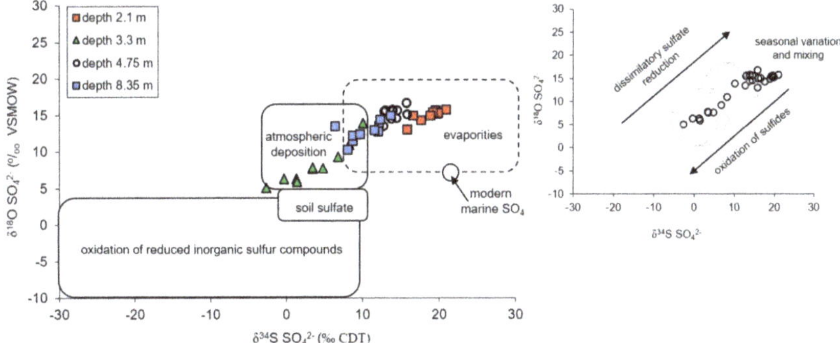

Figure 11. The $\delta^{34}S$ versus $\delta^{18}O$ of dissolved SO_4^{2-} in groundwater of unconfined sandy aquifer against the background of typical sulfate sources and trends of biogeochemical processes.

There are three sources of sulfates that are most likely in the studied groundwater. The sulfates connected with atmospheric deposition (i.e., rain and snowmelt) reach the aquifer by direct infiltration though the peat layer and vadose zone and/or by lateral groundwater inflow from an adjacent area. The second source of sulfates is connected with sulfate flax from the peat layer. These sulfates are enriched in heavy oxygen ^{18}O and sulfur ^{34}S isotopes which is typical for dissolution of evaporitic sulfates (most likely gypsum formed in the peat) or for residual sulfates formed during dissimilatory sulfate reduction in the peat bed. The third source of sulfates is connected with oxidation of reduced inorganic sulfur compounds (i.e., first of all the precipitation/dissolution of pyrite, the dissolution of greigite and mackinawite) in the aquifer.

The bacterial sulfate reduction seems to be the most important biogeochemical process affecting the concentration and the isotopic composition of sulfates dissolved in groundwater across entire profile of the Pożary aquifer. However, the bacterial sulfate reduction as a single process cannot fully explain the distribution of SO_4^{2-} concentration and its isotopic composition in the groundwater profile. The imposition of other processes such as precipitation of sulfide minerals, sulfide oxidation/re-oxidation, groundwater lateral inflow and mixing, must also occur in the studied aquifer.

6. Conclusions

The results of this study show that the peat layer above the unconfined sandy aquifer of the Pożary research station controls the occurrence, origin, and seasonal variation of sulfates and physiochemical composition of groundwater.

The peat layer is firstly the source of organic matter to groundwater in the aquifer. The mineralization of organic matter in the peat and in the aquifer affects in the first place the distribution and concentration of nitrates, sulfates, and carbon species according to the sequential reduction chain.

Two main factors control the concentration and isotopic composition of the SO_4^{2-} dissolved in the groundwater of the studied aquifer, namely, (i) different sources of sulfates, and (ii) the imposition of biogeochemical processes taking place in the peat layer and in the aquifer.

Detailed analysis of the isotopic composition of aqueous sulfates and the distribution of SO_4^{2-} concentrations across the aquifer profile during the entire observation period revealed three main sources of sulfates in the studied groundwater:

(a) The atmospheric deposition—sulfates which are introduce to the aquifer with the recharge water such as rain and snowmelt. Their initial values of $\delta^{34}S_{SO4}$ in the area of the Pożary research station documented for rain and snow varied from +4.3 to +8.8‰ with concentrations ranging from 1.7 to 14.2 mg/dm^3. Such isotopic composition is typical for sulfate of anthropogenic origin in atmospheric precipitation in industrialized regions of the northern hemisphere with $\delta^{34}S$ values ranging from −3 to +9‰. The atmospheric sulfates reach the aquifer by direct infiltration of recharging water though the peat layer and vadose zone and by lateral groundwater inflow from an adjacent areas, where the aquifer is not cover by peat.

(b) The evaporitic sulfate minerals (lithogenic source)—sulfates enriched in heavy ^{18}O and ^{34}S isotopes with $\delta^{34}S_{SO4}$ ranging from +15.8 to +20.9‰ and $\delta^{18}O_{SO4}$ from +13.0 to +15.8‰. Such isotopic composition is typical for sulfates originating most likely from the dissolution of evaporitic gypsum which may be formed in the peat layer or in the aquifer's vadose zone during seasons with low groundwater level. Sulfates of such isotopic composition were observed exclusively in the most upper part of the aquifer profile (monitoring depth of 2.1 m) which is the closest to the overlying peat bed. It is very likely that the source of sulfates is connected with sulfate flax from the peat layer. Calculated initial value of the dominant source of dissolved sulfate at the depth of 2.1 m yielded a $\delta^{34}S_I$ value of +25.4‰, which also strongly suggests the influence of sulfates originating from bacterial sulfate reduction.

(c) The reduced inorganic sulfur compound (RIS; lithogenic source)—sulfates depleted in heavy ^{18}O and ^{34}S isotopes with $\delta^{34}S_{SO4}$ ranging from −2.6 to +10.1‰ and $\delta^{18}O_{SO4}$—from +5.10 to +13.9‰. Sulfates of such isotopic composition were observed in the aquifer exclusively in monitoring depth of 3.3 m. Depletion in heavy isotopes is connected with increasing SO_4^{2-} concentration: Usually increased from about 1.5–2 times (depending on the season) in relation to the uppermost part of the aquifer profile. The source of these sulfates is connected with oxidation/re-oxidation of reduced inorganic sulfur compounds in the aquifer i.e., first of all with the dissolution/precipitation of pyrite, the dissolution of greigite and mackinawite. Calculated initial value of the dominant source of dissolved sulfate at the depth of 3.3 m yielded a $\delta^{34}S_I$ value of −7.9‰ which corroborates that the oxidation of RIS is responsible for the increase in SO_4^{2-} concentration in this part of the aquifer profile.

The observed linear trend of $\delta^{34}S$ vs $\delta^{18}O$ for aqueous sulfates can be attributed to (i) bacterial (dissimilatory) sulfate reduction, (ii) a mixing process in the aquifer, and/or (iii) the precipitation of sulfide minerals. Bacterial sulfate reduction is the most important biogeochemical process affecting the concentration and isotopic evolution of sulfates dissolved in groundwater in the Pożary aquifer.

The sulfur isotope enrichment factor (ε_{34S}—expresses the partitioning magnitude of the ^{34}S isotopes into the residual sulfates during bacterial sulfate reduction undergoing the Rayleigh fractionation model) calculated for each monitoring season revealed wide variation range from about −36.3 to −13.0‰. Such variation of ε_{34S} corroborates that local climatic, environmental, and hydrogeological conditions have a strong impact on the bacterial sulfate reduction rate in the peat and in the aquifer, and on the final isotopic composition of the residual sulfates found in the studied groundwater. The highest absolute values of the enrichment factor ε_{34S} are connected with seasons of decreasing vegetation activity (i.e., late autumn, winter, and early spring), lower ambient air temperature, and high levels of the groundwater table up to the ground surface (i.e., full saturation of the peat layer).

The bacterial sulfate reduction as a single process cannot fully explain the distribution of SO_4^{2-} concentrations and the isotopic compositions of sulfur in the groundwater profile. The imposition of sulfide oxidation/re-oxidation, as well as groundwater mixing processes, must also be taken into account in the studied aquifer.

The biotic and abiotic transformations of sulfur compounds in the studied aquifer are interconnected with other geochemical processes through their common substrates or products. Bacterial sulfate reduction in groundwater profile results in change of isotopic signature of recharge water, a decrease of redox potential (Eh), slight decrease of pH, decrease of SO_4^{2-}, increase of HCO_3^- (alkalinity), and precipitation of sulfide minerals. On the other hand, the imposition of the RIS oxidation causes the increase of water pH, the increase of SO_4^{2-} concentration, and the decreases of alkalinity. The influence of this competitive processes of sulfur species transformations are reflected by the complex variations of the physicochemical parameters of groundwater and the concentration of chemical compounds across the aquifer profile. The monthly monitoring of the aquifer vertical profile allowed us to identify the zones where these different processes are dominant.

There are two issues that need further studies from the perspective of our experience gained during this research, namely, (i) the seasonal monitoring of the sulfate content and its isotopic composition inside the peat layer, and (ii) the identification and analysis of the main forms of the carbon-bonded sulfur compounds (C–S) in order to evaluate their influence on the formation of aqueous sulfates and sulfide minerals in groundwater in contacts with peatlands.

Author Contributions: Conceptualization, A.P. and D.P.; methodology, A.P.; software, A.P.; validation, S.H., D.P. and A.P.; formal analysis, S.H., A.P. and D.P.; investigation, A.P. and D.P.; data curation, D.P.; writing—original draft preparation, A.P. and D.P.; writing—review and editing, A.P.; visualization, D.P.

Funding: This research received funding in the framework of a joint statutory research task supported by the Institute of Geological Sciences of the Polish Academy of Sciences (INGPAN) and the Faculty of the Geology University of Warsaw. Part of the analytical studies was supported by the National Science Center Poland, grant No. UMO-2015/17/B/ST10/03295.

Acknowledgments: The authorities of the meteorological base station of Integrated Monitoring of the Natural Environment of the Kampinos National Park are thanked for their help in the collection of rainwater samples for chemical and isotopic analysis.

Conflicts of Interest: The authors declare no conflict of interest. The funders had no role in the design of the study; in the collection, analyses, or interpretation of data; in the writing of the manuscript, or in the decision to publish the results.

References

1. Johnston, S.G.; Burton, E.D.; Aaso, T.; Tuckerman, G. Sulfur, iron and carbon cycling following hydrological restoration of acidic freshwater wetlands. *Chem. Geol.* **2014**, *371*, 9–26. [CrossRef]
2. Langmuir, D. *Aqueous Environmental Geochemistry*; Prentice Hall: Upper Saddle River, NJ, USA, 1997; 600p.
3. Appelo, C.A.J.; Postma, D. *Geochemistry, Groundwater and Pollution*; A.A. Balkema: Rotterdam, The Netherlands, 1996; 536p.

4. Chapman, S.J. Sulfur forms in open and afforested areas of two Scottish peatlands. *Water Air Soil Pollut.* **2002**, *128*, 23–39. [CrossRef]

5. Novak, M.; Vile, M.A.; Bottrell, S.H.; Stepanova, M.; Jackova, I.; Buzek, F.; Prechova, E.; Newton, R.J. Isotope systematic of sulfate-oxygen and sulfate-sulfur in six European peatlands. *Biogeochemistry* **2005**, *76*, 187–213. [CrossRef]

6. López-Buendia, A.M.; Whateley, M.K.G.; Bastida, J.; Urquiola, M.M. Origins of mineral matter in peat marsh and peat bog deposits, Spain. *Int. J. Coal Geol.* **2007**, *71*, 246–262. [CrossRef]

7. Alewell, C.; Paul, S.; Lischeid, G.; Storck, F.R. Co-regulation of redox processes in freshwater wetlansd as a function of organic matter availability? *Sci. Total Environ.* **2008**, *404*, 335–342. [CrossRef] [PubMed]

8. Schwientek, M.; Einsiedl, F.; Stichler, W.; Stögbauer, A.; Strauss, H.; Maloszewski, P. Evidence for denitrification regulated by pyrite oxidation in heterogeneous porous groundwater system. *Chem. Geol.* **2008**, *255*, 60–67. [CrossRef]

9. Wu, S.; Kuschk, P.; Wiessner, A.; Müller, J.; Sasd, R.A.B.; Dong, R. Sulphur transformations in constructed wetlands for wastewater treatment: A review. *Ecol. Eng.* **2013**, *52*, 278–289. [CrossRef]

10. Wiessner, A.; Kappelmeyer, U.; Kuschk, P.; Kastner, M. Sulphate reduction and the removal of carbon and ammonia in a laboratory-scale constructed wetland. *Water Res.* **2005**, *39*, 4643–4650. [CrossRef]

11. Schiff, S.L.; Spoelstra, J.; Semkin, R.G.; Jeffries, D.S. Drought induced pulses of from a Canadian shield wetland: use of δ^{34}S and δ^{18}O in to determine sources of sulfur. *Appl. Geochem.* **2005**, *20*, 691–700. [CrossRef]

12. Clark, I.; Fritz, P. *Environmental Isotopes in Hydrogeology*; CRC Press LLC: Boca Raton, FL, USA, 1997; 328p.

13. Mandernack, K.W.; Lynch, L.; Krouse, H.R.; Morgan, M.D. Sulfur cycling in wetland peat of the New Jersey Pinelands and its effect on stream water chemistry. *Gechimica Cosmochiica Acta* **2000**, *64*, 3949–3964. [CrossRef]

14. Bottrell, S.H.; Hartfield, D.; Bartlett, R.; Spence, M.J.; Bartle, K.D.; Mortimer, R.J.G. Concentrations, sulfur isotopic compositions and origin of organosulfur compounds in pore waters of a highly polluted raised peatland. *Org. Geochem.* **2010**, *41*, 55–62. [CrossRef]

15. Porowska, D.; Leśniak, P.M. Identification of processes controlling groundwater chemistry below peatland—Pożary, Kampinos Park Narodowy. *Geol. Surv.* **2008**, *56*, 982–990. (In Polish)

16. Słowański, W.; Piechulska-Słowańska, B.; Gogołek, W. *Geological Map of Poland in the Sacle 1:200 000. Warszawa-Zachód Sheet*; PIG: Warszawa, Poland, 1994.

17. Fic, M.; Wierzbicki, A. Organizacja sieci monitoringu wód podziemnych na terenie rezerwatu "Pożary" w Kampinoskim Parku Narodowym. *Geol. Surv.* **1994**, *42*, 1004–1008. (In Polish)

18. Ingram, H.A.P. Hydrology. In *Mires, Swamps, Bog, Fen and Moore*; Gore, A.J.P., Ed.; General Studies; Elsevier: Amsterdam, The Netherlands, 1983; Volume 4A, pp. 67–158.

19. Witczak, S.; Kania, J.; Kmiecik, E. *Katalog Wybranych Fizycznych i Chemicznych Wskaźników Zanieczyszczeń Wód Podziemnych i Metod ich Oznaczania*; IOŚ: Warszawa, Poland, 2013; 648p.

20. Nielsen, D.M.; Nielsen, G.L. *The Essential Handbook of Ground-Water Sampling*; CRC Press: Boca Raton, FL, USA, 2007; 309p.

21. Weight, W.D.; Sonderegger, J.L. *Manual of Applied Field Hydrogeology*; McGraw-Hill: New York, NY, USA, 2000; 608p.

22. Hermanowicz, W.; Dojlido, J.; Dożańska, W.; Koziorowski, B.; Zerbe, J. *Fizyczno-Chemiczne Badanie Wody i ścieków*; Wydawnictwo Arkady: Warszawa, Poland, 1999; 558p. (In Polish)

23. Epstein, S.; Mayeda, T. Variation of ^{18}O content of waters from natural sources. *Geochim. Cosmochim. Acta* **1953**, *4*, 213–224. [CrossRef]

24. Roether, W. Water-CO_2 set-up for routine ^{18}Oxygene assay of natural waters. *Int. J. Appl. Radiat. Isot.* **1970**, *21*, 379–387. [CrossRef]

25. Coleman, M.L.; Shepherd, T.J.; Durham, J.J.; Rouse, J.E.; Moore, G.R. Reduction of water with zinc for hydrogen isotope analysis. *Anal. Chem.* **1982**, *54*, 993–995. [CrossRef]

26. Kendall, C.; Coplen, T.B. Multisample conversion of water to hydrogen by zinc for stable isotope determination. *Anal. Chem.* **1985**, *57*, 1437–1440. [CrossRef]

27. Schimmelmann, A.; DeNiro, M.J. Preparation of organic and water hydrogen for stable isotope analysis: Effects due to reaction vessels and zinc reagent. *Anal. Chem.* **1993**, *65*, 789–792. [CrossRef]

28. Porowski, A.; Kowski, P. Determination of δ^{2}H and δ^{18}O in saline oil-associated waters: the question of simple vacuum distillation of water samples prior to isotopic analyses. *Isot. Environ. Health Stud.* **2008**, *44*, 227–238. [CrossRef]

29. Carmody, R.W.; Plummer, L.N.; Busenberg, E.; Coplen, T.B. *Methods for Collection of Dissolved Sulfate and Sulfide and Analysis of Their Sulfur Isotopic Composition*; U.S. Geological Survey Open-File Report; USGS Publication Warehouse: Reston, VA, USA, 1998; pp. 97–234.

30. Mizutani, Y. An improvement in the carbon-reduction method for the oxygen isotopic analysis of sulphates. *Geochem. J.* **1971**, *5*, 69–77. [CrossRef]

31. Halas, S.; Szaran, J.; Czarnacki, M.; Tanweer, A. Refinements in $BaSO_4$ to CO_2 preparation and $\delta^{18}O$-calibration of the sulphate standards NBS-127, IAEA SO-5 and IAEA SO-6. *Geostand. Geoanal. Res.* **2007**, *31*, 61–68. [CrossRef]

32. Halas, S.; Szaran, J. Use of Cu_2O-$NaPO_3$ mixtures for SO_2 extraction from BaSO4 for sulphur isotope analysis. *Isot. Environ. Health Stud.* **2004**, *40*, 229–231. [CrossRef] [PubMed]

33. Porowska, D. Content of dissolved oxygen and carbon dioxide in groundwaters within the selected hydrogeocheical environment. *Monogr. Kom. Gospod. Wodnej Pan* **2004**, *24*, 137.

34. Trudell, M.L.; Gillham, R.W.; Cherry, J.A. An in-situ study of the occurrence and rate of denitrification in a shallow unconfined sand aquifer. *J. Hydrol.* **1986**, *86*, 251–268. [CrossRef]

35. Smith, R.L.; Duff, J.H. Denitrification in a sand and gravel aquifer. *Appl. Environ. Microbiol.* **1988**, *54*, 1071–1078. [PubMed]

36. Böttcher, J.; Strebel, O.; Voerkelius, S.; Schmidt, H.L. Using isotope fractionation of nitrate nitrogen and nitrate oxygen for evaluation of denitrification in a sandy aquifer. *J. Hydrol.* **1990**, *114*, 413–424.

37. Smith, R.L.; Howes, B.L.; Duff, L.H. Denitrification in nitrate-contaminated groundwater: occurrence in steep vertical geochemical gradients. *Geochim. Cosmochim. Acta* **1991**, *55*, 1815–1825. [CrossRef]

38. Krogulec, E.; Jóźwiak, K. Mineral nitrogen in shallow waters of the Kampinos National Park. In Proceedings of the XII Sympozjum Współczesne Problemy Hydrogeologii, Kraków-Krynica, Poland, 21–23 June 2007; pp. 105–113.

39. Krouse, H.R. Sulfur isotopes in our environment. In *Handbook of Environmental Isotope Geochemistry. Volume 1: The Terrestrial Environment*; Fritz, P., Fontes, J.C., Eds.; Elsevier: Amsterdam, The Netherlands, 1980; pp. 435–472.

40. Krouse, H.R.; Mayer, B. Sulfur and oxygen isotopes in sulfate. In *Environmental Tracers in Subsurface Hydrology*; Cook, P., Herczeg, A.L., Eds.; Kluwer Academic Publishers: Dordrecht, The Netherlands, 2000; pp. 195–231.

41. Knöller, K.; Fauville, A.; Mayer, B.; Strauch, G.; Friese, K.; Veizer, J. Sulfur cycling in an acid mining lake and its vicinity in Lusatia, Germany. *Chem. Geol.* **2004**, *204*, 303–323.

42. Cook, P.G.; Herczeg, A.L. *Environmental Tracers in Subsurface Hydrology*; Kluwer: Boston, MA, USA, 2000; 529p.

43. Porowski, A. Isotop hydrogeology. In *Handbook of Engineering Hydrology. Volume 1. Fundamentals and Applications*; Eslamian, S., Ed.; CRC Press: Boca Raton, FL, USA, 2014; pp. 345–378.

44. Trembaczowski, A. Sulfur and oxygen isotopes behavior in sulfates of atmospheric groundwater system, observation and model. *Nord. Hydrol.* **1991**, *22*, 49–66. [CrossRef]

45. Bottrell, S.H.; Coulson, J.; Spence, M.; Roworth, P.; Novak, M.; Forbes, L. Impacts of pollutant loading, climate variability and site management on the surface water quality of a lowland raised bog, Thorne Moors, E. England, UK. *Appl. Geochem.* **2004**, *19*, 413–422. [CrossRef]

46. Skrzypek, G.; Akagi, T.; Drzewicki, W.; Jędrysek, M.O. Stable isotope studies of moss sulfur and sulfate from bog surface waters. *Geochem. J.* **2008**, *42*, 481–492. [CrossRef]

47. Rozanski, K.; Araguas-Araguas, L.; Gonfiantini, R. Isotopic patterns in modern global precipitation. In *Climate Change in Continental Isotopic Records. Geophysical Monograph*; Stewart, P.K., Lohmann, K.C., McKenzie, J., Savin, S., Eds.; American Geophysical Union: Washington, DC, USA, 1993; pp. 1–36.

48. Kroopnick, P.; Craig, H. Atmospheric oxygen: isotopic composition and solubility fractionation. *Science* **1972**, *175*, 54–55. [CrossRef] [PubMed]

49. Everdingen, R.O.; Krouse, H.R. Isotope composition of sulphates generated by bacterial and abiological oxidation. *Nature* **1985**, *315*, 395–396. [CrossRef]

50. Mayer, B.; Fritz, P.; Prietzel, J.; Krouse, H.R. The use of stable sulfur and oxygen isotope ratios for interpreting the mobility of sulfate in aerobic forest soils. *Appl. Geochem.* **1995**, *10*, 161–173. [CrossRef]

51. Taylor, B.E.; Wheeler, M.C.; Nordstrom, D.K. Stable isotope geochemistry of acid mine drainage: experimental oxidation of pyrite. *Geochim. Cosmochim. Acta* **1984**, *48*, 2669–2678. [CrossRef]

52. Porowska, D. Content of dissolved oxygen and carbon dioxide in rainwaters and groundwaters within the forest reserve of the Kampinos National Park and the urban area of Warsaw, Poland. *Geol. Q.* **2003**, *47*, 187–194.

53. Rydelek, P. Origin and composition of mineral particles of selected peat deposits in Lubartowska Upland. *Woda Śr. Obsz. Wiej.* **2011**, *11*, 135–149.

54. Mayer, B. Assessing sources and transformations of sulphate and nitrate in the hydrosphere using isotope techniques. In *Isotopes in the Water Cycle; Past, Present and Future of a Developing Science*; Aggarawal, P.K., Gat, J., Froehlich, K.F.O., Eds.; Springer: Dordrecht, The Netherlands, 2007; pp. 67–90.

55. Zuber, A. *Metody znacznikowe w badaniach hydrogeologicznych, poradnik metodyczny*; Oficyna Wydawnicza Politechniki Wrocławskiej: Wrocław, Poland, 2007; 402p. (In Polish)

56. Brown, K.A. Sulphur distribution and metabolism in waterlogged peat. *Soil Biol. Biochem.* **1985**, *17*, 39–45. [CrossRef]

57. Rickard, D.; Luther, G.W., III. Chemistry of Iron Sulfides. *Chem. Rev.* **2007**, *107*, 514–562. [CrossRef]

58. Rickard, D. *Sulfidic Sediments and Sedimentary Rocks*; Elsevier Science: Amsterdam, The Netherlands, 2012; Volume 65, 816p.

59. Cohen, A.D.; Spackman, W. Methods in peat petrology and their application to reconstruction of paleoenvironments. *Geol. Soc. Am. Bull.* **1972**, *83*, 129–142. [CrossRef]

60. Casagrande, D.J.; Seiffert, K.; Berschinski, C.; Sutton, N. Sulfur in peat forming systems of the Oekefenokee Swamp and Florida Everglades: origins of sulfur in coal. *Geochim. Cosmochim. Acta* **1977**, *41*, 161–167. [CrossRef]

61. Philips, S.; Bustin, R.M. Sulfur in Changuinola peat deposit, Panama, as an indicator of the environments of deposition of peat and coal. *J. Sediment. Res.* **1996**, *66*, 184–196. [CrossRef]

62. Berner, R.A.; Raiswell, R. Burial of organic carbon and pyrite sulfur ion sediments over Phanerozoic time: A new theory. *Geochim. Cosmochim. Acta* **1983**, *47*, 855–862. [CrossRef]

63. Jóźwiak, K. Bogs iron ore in the marshy ground areas—e.g., Kampinoski National Park. *Biul. Pig* **2011**, *445*, 237–244.

64. Wallace, A.; Wallace, G.A. Factors influencing oxidation of iron pyrite in soil. *J. Plant Nutr.* **1992**, *15*, 1579–1587. [CrossRef]

65. Balci, N.; Shanks, W.C., III; Mayer, B.; Mandernack, K. Oxygen and sulfur isotope systematics of sulfate produced by bacterial and abiotic oxidation of pyrite. *Geochim. Cosmochim. Acta* **2007**, *71*, 3796–3811. [CrossRef]

66. Canifeld, D.E. Isotope fractionation by natural populations of sulfate-reducing bacteria. *Cosmochim. Acta* **2001**, *65*, 1117–1124. [CrossRef]

67. Samborska, K.; Halas, S. ^{34}S and ^{18}O in dissolved sulfate as tracers of hydrogeochemical evolution of the Triassic carbonate aquifer exposed to intense groundwater exploitation (Olkusz–Zawiercie region, southern Poland). *Appl. Geochem.* **2010**, *25*, 1397–1414. [CrossRef]

68. Miao, Z.; Brusseau, M.L.; Carroll, K.C.; Carreón-Diazconti, C.; Johnson, B. Sulfate reduction in groundwater: characterization and applications for remediation. *Environ. Geochem. Health* **2012**, *34*, 539–550. [CrossRef]

69. Kaplan, I.R.; Rittenberg, S.C. Microbiological fractionation of sulfur isotopes. *J. Gen. Microbiol.* **1964**, *34*, 195–212. [CrossRef]

70. Rees, C.E. A steady-state model for sulfur isotope fractionation in bacterial reduction process. *Geochim. Cosmochim. Acta* **1973**, *37*, 1141–1162.

71. Fritz, P.; Basharmal, G.M.; Drimmie, R.J.; Isen, J.; Qureshi, R.M. Oxygen isotope exchange between sulphate and water during bacterial sulfate reduction. *Chem. Geol.* **1989**, *79*, 99–105.

72. Habicht, K.S.; Canfield, D.E. Sulfur isotope fractionation during bacterial sulfate reduction in organic-rich sediments. *Geochim. Cosmochim. Acta* **1997**, *61*, 5351–5361. [CrossRef]

73. Mizutani, Y.; Rafter, T.A. Oxygen isotopic composition of sulphates, part 3: Oxygen isotopic fractionation in the bisulphate ion-water system. *N. Z. J. Sci.* **1969**, *12*, 54–59.

74. Ahron, P.; Fu, B. Microbial sulfate reduction rates and sulfur and oxygen isotope ratios at oil and gas seeps in deepwater Gulf Mexico. *Geochim. Cosmochim. Acta* **2000**, *64*, 233–246. [CrossRef]

75. Hoefs, J. *Stable Isotope Geochemistry*, 6th ed.; Springer: Berlin, Germany, 2009; 286p.

76. Bottrell, S.H.; Smart, P.I.; Whitaker, F.; Raiswell, R. Geochemistry and isotope systematic of sulfur in the mixing zone of Bahamian blue holes. *Appl. Geochem.* **1991**, *6*, 97–103. [CrossRef]

77. Bottrell, S.H.; Hayes, P.J.; Bannon, M.; Williams, G.M. Bacterial sulfate reduction and pyrite formation in a polluted sand aquifer. *Geomicrobiol. J.* **1995**, *12*, 75–90. [CrossRef]
78. Bottrell, S.H.; Moncaster, S.M.; Tellam, J.H.; Lloyd, J.W.; Fisher, Q.J.; Newton, R.J. Control on bacterial sulfate reduction in a dual porosity aquifer system: The Lincoln Limestone aquifer, England. *Chem. Geol.* **2000**, *169*, 461–470. [CrossRef]
79. Asmussen, G.; Strauch, G. Sulfate reduction in a lake and the groundwater of a former lignite mining area studied by stable sulfure and carbon isotopes. *Water Air Soil Pollut.* **1998**, *108*, 271–284. [CrossRef]

 *water*

Article

Investigation into Groundwater Resources in Southern Part of the Red River's Delta Plain, Vietnam by the Use of Isotopic Techniques

Nguyen Van Lam [1], Hoang Van Hoan [2,*] and Dang Duc Nhan [3]

[1]   Department of Hydrogeology, Hanoi University of Mining and Geology, Duc Thang Ward, North Tu Liem District, 1000 Hanoi, Vietnam; lamdctv@gmail.com
[2]   National Center for Water Resources Planning and Investigation, Sai Dong Ward, Long Bien District, 1000 Hanoi, Vietnam
[3]   Institute for Nuclear Science and Technology, Nghia Do Ward, Cau Giay District, 1000 Hanoi, Vietnam; dangducnhan50@gmail.com
*   Correspondence: hoanghoandctv@gmail.com; Tel.: +84-983-653-229

Received: 10 August 2019; Accepted: 8 October 2019; Published: 12 October 2019

Abstract: Groundwater in the Red River's delta plain, North Vietnam, was found in Holocene, Pleistocene, Neogene and Triassic aquifers in fresh, brackish and saline types with a total dissolved solids (TDS) content ranging from less than 1 g L^{-1} to higher than 3 g L^{-1}. Saline water exists inHolocene aquifer, but fresh and brackish water exist in Pleistocene, Neogene and Triassic aquifers. This study aims at the investigation into genesis and processes controlling quality of water resources in the region. For this isotopic, combined with geochemical techniques were applied. The techniques include: (i) measurement of water's isotopic compositions (δ^2H, $\delta^{18}O$) in water; (ii) determination of water's age by the ^3H- and ^{14}C-dating method, and (iii) chemical analyses for main cations and anions in water. Results obtained revealed that saline water in Holocene aquifer was affected by seawater intrusion, fresh water in deeper aquifers originated from meteoric water but with old ages, up to 10,000–14,000 yr. The recharge area of fresh water is from the northwest highland at an altitude of 140–160m above sea level. The recharge water flows northwesterly towards southeasterly to the seacoast at a rate of 2.5m y^{-1}. Chemistry of water resources in the study region is controlled by ferric, sulfate and nitrate reduction with organic matters as well as dissolution of inorganic carbonate minerals present in the sediment deposits. Results of isotopic signatures in water from Neogene, Triassic and Pleistocene aquifers suggested the three aquifers are connected to each other due to the existence of faults and fissures in Mesozoic basement across the delta region in combination with high rate of groundwater mining. Moreover, the high rate of freshwater abstraction from Pleistocene aquifer currently causes sea water to flow backwards to production well field located in the center of the region.

Keywords: isotopic techniques; water isotopic signature; ^3H- and ^{14}C-dating; saltwater intrusion; Red River's delta; Vietnam

---

1. Introduction

The Red River Delta plain (RRDP), covering an area of approximately 15,000 km^2 is situated across the northern part of Vietnam and is one of the two most productive deltas in the country. In RRDP, water is used for irrigation, industrial and domestic purposes, out of which irrigation consumes around 95% of the 83.031 km^3 annual production [1]. In the RRDP, groundwater is the main source of domestic usage water. Reports state that the groundwater storage capacity of the RRDP is over 23 billion m^3, out of which, roughly 20% is located in the southern part, i.e., Thai Binh, Nam Dinh and a part of the Ninh Binh provinces [2,3].

Groundwater in the RRDP was found in four aquifers, namely in Holocene (qh), Pleistocene (qp), in the fissured basement rock Neogene (n) and Triassic (t) aquifers [3,4]. The origin and chemistry of groundwater in entire RRDP were studied by a number of researchers, e.g., Luu et al. [5] and Nguyen et al. [6,7]. Groundwater in the RRDP was classified into three types: fresh, low saline and high saline [5,7]. Vietnamese hydrogeologists have drawn conclusions that fresh groundwater recharges across mountainous areas in the northwest (NW) and discharges in the southeast (SE) into the sea. This hypothesis is based on the topography of the Delta without any conclusive evidence [3].

The aims of this study are to elucidate the genesis of groundwater, to determine recharge area, to estimate flow rate of the recharged water and to provide evidences proving hydro-geochemical processes controlling quality of groundwater in the southern part of the RRDP by isotopic and related techniques.

1.1. Study Site

The study site encompasses the southern RRDP that comprises a part of Thai Binh, Nam Dinh and Ninh Binh provinces (Figure 1).

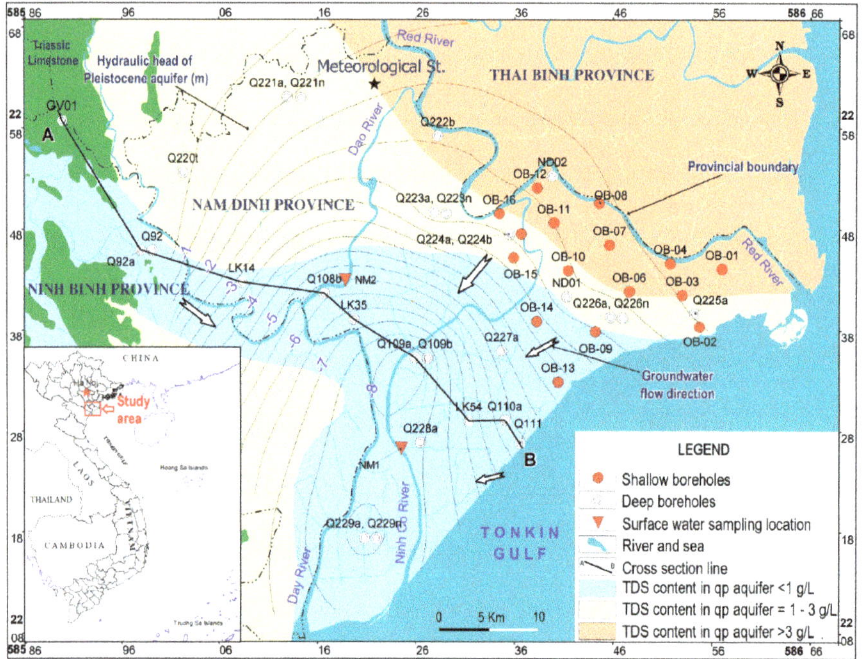

Figure 1. Map showing the study site: southern part of the Red River's Delta plain: Thai Binh, Nam Dinh, and Ninh Binh provinces; sampling locations for groundwater of Holocene aquifer (●points), surface water (▼points); saline-brackish-fresh water boundary; and isolines of water hydraulic head of groundwater in Pleistocene (qp) aquifer.

The study site occupies an area of 4600 km^2 where 4.848 million habitants are residing [8]. The climate in the study region is sub-tropical with a rainy season from May to October and a dry season from November to April. During the rainy season, precipitation ranges from 1400 to 1750 mm and from 150 to 300 mm during the dry season. The potential annual evaporation in the region is in the range of 700–800 mm [9]. The monthly average temperature varies between 16 °C and 35 °C with the lowest temperature in January and the highest in July [9].

1.2. Geology and Hydrogeology of the Study Site

The geology and hydrogeology of the study site drawn for a NW-SE oriented cross-section AB (Figure 1) is shown in Figure 2.

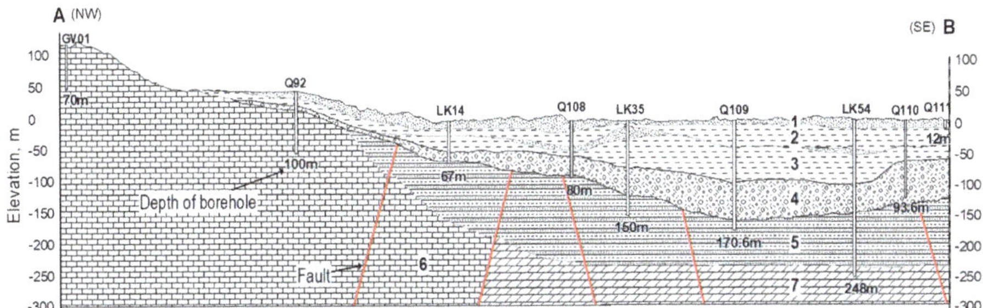

Figure 2. Geological setting of the study region drawn for the transect AB. See explanations for the sequence of the geological formation in the text.

The Red River Delta plain sediments consist of Holocene, Pleistocene and Neogene deposits [3]. Holocene deposits in study region are found from elevation of +2 m to −15 m (layer #1, Figure 2) and contain alluvial sediments and fine sand with disseminated wood debris as well as biogenic calcite such as shells and carbonate mud, etc. [10]. The Holocene sediment layer forms the Holocene unconfined aquifer (qh) where water exists within sediment pores [2,3]. Below Holocene deposit there is a silty-clay layer of Holocene and Pleistocene with a thickness up to 100 m (layers #2 and #3, Figure 2) that forms an aquitard, a low permeable clay layer, to separate the Holocene aquifer from deeper aquifers. Following aquitard, in the NE area there are medium to coarse fluvial and marine sediment deposits, but in the SW area there are silty and fine to medium grained marine sand deposits of Pleistocene with a thickness up to 75 m (layer #4, Figure 2). The NE and SW areas were presumably separated from each other by boundary between blue and light brown color areas in the Nam Dinh province (Figure 1). The NE area comprises the northern part of Nam Dinh and Thai Binh provinces. The SW area comprises the southern part of Nam Dinh province and a part of Ninh Binh province (Figure 1).

The sediment in Pleistocene deposit contains disseminated fragments of organic materials like wood remnants as well as biogenic carbonate such as shells, skeletal debris, etc., [10]. The porous water in Pleistocene sediment layer forms Pleistocene (qp) confined aquifer [2,3]. The underlying Neogene silt and sandstone deposits are located from 120 to 130m to 220–230m below the surface (layer #5, Figure 2). The water existing in the Neogene formation forms the Neogene water bearing layer or Neogene (n) aquifer. The Mesozoic stone basement is located from 230m below the ground surface (layer #7, Figure 2).

Data of drilling works combined with geophysical measurements revealed that the Mesozoic block #6 and #7 comprises of 4 zones: uplifted, moderate depressed, depressed and active depressed ones that created Ninh Binh, Song Hong, Nam Dinh, Song Chay and Vinh Ninh faults oriented from NW towards SE. In each zone, it was observed also uplifted and depressed subzones that created from seaside to inland Hai Hau, Xuan Truong and Yen Khanh faults oriented from SE towards NE. All the faults as well as fissures formed during the tectonic activities had created channels allowing recharge water from mountainous region in the uplifted zone in the SW to leak up to the Neogene and Quaternary formations. The water in fractured Mesozoic rocks basement constitutes Triassic (t) water zone (layers #6 and #7, Figure 2).

1.3. Hydrology in the Study Region

The river network of the study region is very dense. Almost 3% of the land area is covered by rivers and irrigation canals [11]. The main rivers flowing through region are Red River and Day River (Figure 1). Dao and Ninh Co Rivers (Figure 1) are tributaries of Red River that have been man-made during the Tran's Dynasty in the thirteenth century to control floods during the rainy season and for irrigation purposes in the dry season. Red River annually discharges 119,837 km^3 of water into the Ton Kin Gulf, and discharge from Day River is 1848 km^3, annually [12].

The tidal regime in the region is diurnal with a level ranging from 2.6 to 3.6 m as recorded at Quang Phuc hydrological monitoring station, around 50 km NE of the mouth of Red River [13].

2. Methods

In this study, isotopic techniques combined with chemical analyses of the ionic contents of local precipitation, water from the Red River and groundwater samples, are applied. Results of chemical analyses give an insight into geochemical processes controlling quality, while isotopic techniques provide evidences to verify the mechanisms behind geochemical processes as well as genesis and flow pattern of groundwater. The isotopic techniques used in this study include: (1) tritium (^{3}H) and carbon-14 (^{14}C) dating to verify recharge areas, the patterns of groundwater flows and to estimate the flow rate of recharge water; (2) determination of isotope composition of hydrogen (δ^2H) and oxygen (δ^{18}O) in water to delineate the hydraulic interaction between local precipitation and groundwater and the inter-connection between aquifers, as well as to examine the recharge area; (3) determination of isotope composition of carbon (δ^{13}C) of dissolved inorganic carbon (DIC) to evaluate the type of mineralization of groundwater existing during water seepage and recharge conditions. All these approaches were successfully applied in hydrogeological studies worldwide, e.g., in [14–20].

During the water cycle, isotope composition of hydrogen and oxygen in water will be changed due to the isotopic fractionation effect. The extent of the fractionation during phase transformation was expressed in delta notation as follows:

$$\delta^2H = \left(\frac{R_{2H,Sample}}{R_{2H,Std}} - 1 \right) \times 1000 \tag{1}$$

$$\delta^{18}O = \left(\frac{R_{18O,Sample}}{R_{18O,Std}} - 1 \right) \times 1000 \tag{2}$$

where: $R_{2H,Sample}$, $R_{2H,Std}$, $R_{18O,Sample}$, $R_{18O,Std}$ are the isotopic ratios of [^{2}H]/[^{1}H] and [^{18}O]/[^{16}O] in water samples and standard, respectively. The delta notation is expressed in per mil (‰).

The standard used in these analyses is Vienna Standard Mean Ocean Water 2 (VSMOW2) supplied by the Isotopes Hydrology Laboratory of International Atomic Energy Agency (IAEA IHL) based in Vienna.

The fractionation of stable carbon isotope in DIC due to geochemical processes, e.g., isotopic exchange between atmospheric CO_2 and biogenic carbon dioxide released from plants' roots respiration, as well as dissolution of carbonate, etc., which occur during seepage of water in the unsaturated zone was characterized by delta notation as follows:

$$\delta^{13}C = \left(\frac{^{13}R_{DIC,\,sample}}{^{13}R_{std}} - 1 \right) \times 1000 \tag{3}$$

Standard used in this analysis is IAEA-603, which is calcite minerals supplied by the IAEA IHL.

At the global scale the relationship of δ^2H vs. δ^{18}O for precipitation will follow a so-called Global Meteorological Water Line (GMWL) [21]:

$$\delta^2H = 8 \times \delta^{18}O + 10 \tag{4}$$

At the regional scale, the slope and intercept of Equation (4) are much dependent upon the conditions under which the water was yielded, so for assessing the origin of groundwater of a certain region using water isotopic signatures (δ^2H and δ^{18}O), regional meteoric water line (RMWL) is used instead of the GMWL. The approach for assessing the genesis of groundwater based on the water isotopic compositions was described in detail in, e.g., [15,18].

2.1. Sampling Procedure and Field Measurement

Sixteen (16) groundwater samples from Holocene, twelve (12) samples from Pleistocene, five (5) samples from Neogene, three (3) samples from Triassic aquifers and two (2) surface water (NM) samples from Dao and Ninh Co River (Table S1) were taken during rainy (August 2011) and dry (March 2012) seasons using a submersible pump, a Grundfos MP1. Sampling locations for Holocene aquifer were marked with symbol OB and red dots in the region adjacent to the Red River as shown in Figure 1. The sampling locations for Pleistocene, Neogene and Triassic aquifers are shown in Figure 3 along with depth of boreholes, δ^{13}C- and ^{14}C-values in groundwater from those boreholes.

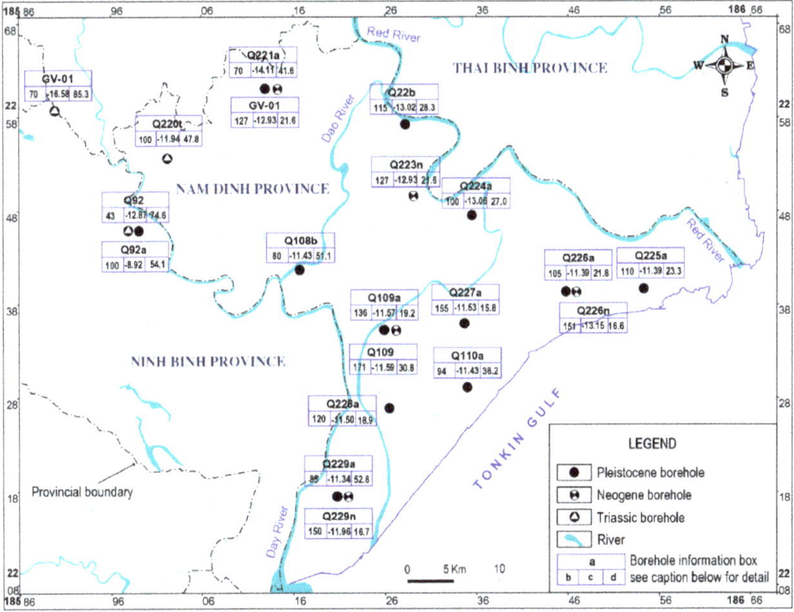

Figure 3. Sampling locations of the groundwater from Pleistocene, Neogene and Triassic aquifers along with information related to boreholes and water in the boreholes. In the borehole information box: *a*: Borehole ID (Q stands for boreholes of the National Groundwater Monitoring Network with a sequence number in the Network; Characters a or b, n and t standing aside the borehole ID indicate the Pleistocene, Neogene and Triassic aquifers, respectively); *b*: borehole depth (meter from the ground surface); *c*: δ^{13}C of DIC (‰); *d*: ^{14}C content in DIC (pMC).

Before taking groundwater samples, water table level in each sampling location was first measured using a logger system then the stagnant water in the wells was completely flushed out by pumping it out till the pH and temperature of water were unchanged. A flow cell equipped with probes for pH, temperature, and electrical conductivity (EC) was mounted directly on sampling tube. The measurements were carried out with a WTW Multi 197i multi-purpose instrument using a WTW Tetracon 96 EC probe and a WTW SenTix 41 pH electrode.

Alkalinity was determined shortly after sampling using the Gran-titration method [22]. Ferrous ion (Fe^{2+}) concentration in the water samples was determined by colorimetry using a Hach DR/2010 instrument and Ferrozine method [23]. The dissolved inorganic carbon (DIC) including bicarbonate, carbonate ions and CO_2 dissolved in water, used to determine $\delta^{13}C$ and ^{14}C content was precipitated in the form of $BaCO_3$ from around 100 L of groundwater at pH 10 using saturated $BaCl_2$ solution and CO_2-free KOH (Merck based in Darmstadt, Germany).

Around 100 mL of groundwater were first filtered through 0.45 µm mesh polycarbonate membranes to remove suspended matters then split into two parts. One part was acidified with HNO_3 (65%, PA grade, Merck) to a pH of about 1–2. These samples were subject to the quantification of cations NH_4^+, Na^+, K^+, Ca^{2+} and Mg^{2+}. Another part was kept without acidification for Cl^-, NO_3^- and SO_4^{2-} determination. All samples were stored in bottles made from high density polyethylene (HDPE) resin and refrigerated until laboratory analysis.

Surface water from Dao and Ninh Co River (NM1 and NM2 sites, Figure 1) was collected the same day of groundwater sampling. The samples were taken at a depth of 0.5 m from surface and around 2 m apart from the bank of the rivers.

Precipitation was collected monthly for the 2011–2012 years using a device constructed following an IAEA's recommendation to establish regional meteoric water line and to determine ^3H-activity [24]. The device was installed on the roof of the premises of the Meteorological Station (marked with a star in Figure 1) in the city of Nam Dinh. Precipitation, surface water and groundwater samples for water stable isotopic composition determination were filtered through 0.45 µm mesh polycarbonate membranes and stored in 50 mL capacity double caps HDPE bottles.

For the tritium determination, in each sampling site 1 L of water was sampled into a HDPE bottle closed with a tight cap to avoid the isotopic exchange with atmospheric moisture. The samples were transferred to laboratory in Ha Noi for further treatment and measurement of tritium activity.

2.2. Samples Treatment and Analytical Procedure

The ionic content of water samples was quantified by ion chromatography method (IC) using a DIONEX 600 at Institute for Nuclear Science and Technology (INST in Ha Noi). A quality control program was applied for the ionic content determination by analyzing standard solutions supplied by the IC supplier (DIONEX). The standard deviation of results derived by the laboratory was within ±5% from the certified value for respective constituents.

The stable isotopes compositions of hydrogen, oxygen and carbon (δ^2H, $\delta^{18}O$ and $\delta^{13}C$) were analyzed at the INST on an Isotope Ratio Mass-Spectrometer (IR MS, IsoPrime, GV, UK) equipped with an Elemental Analyzer (EA 3000, Eurovector based in Milan, Italy). To analyze for $\delta^{13}C$, the barium carbonate precipitate was first washed off the excess of alkali by boiled deionized water till neutral pH was attained then freeze-dried. The dry and alkali-free carbonate barium was stored in HDPE vials till analysis. For the δ^2H and $\delta^{18}O$ analyses, the water samples collected and filtered in the field did not need any additional treatment in the laboratory.

Procedures applied for analyzing the stable isotopic compositions were implemented as per the supplier's manual [25] and as follows.

For the δ^2H analysis, 2 µL of water was injected into the EA where the sample was decomposed on the Ni-catalyst at 1050 °C to form hydrogen gas [25]. A continuous flow of He-carrier gas carried the H_2 gas through a $Mg(ClO_4)_2$ moisture trap to dry then by passing through a chromatographic column to purify from contaminants before entering the ionization chamber of the IR MS where H_2 was ionized. The flow of He-carrier gas carried the H_2^+ ions into the mass separator of the IR MS where masses 2 ($^1H_2^+$) and 3 ($^1H^2H^+$) were separated from each other. The H_2^+ and H_3^+ ions were collected and counted by the respective Faraday cups installed on the exit from the separator.

For the $\delta^{18}O$ analysis, the decomposition of water was carried out in the EA on the glassy carbon catalyst at 1150 °C to form CO_2 gas [25]. A continuous flow of He-carrier gas carried the CO_2 through a $Mg(ClO_4)_2$ moisture trap followed by the chromatographic purification before entering the ionization chamber to ionize the CO_2 gas. The carrier gas carried the ions of masses 44 ($^{12}C^{16}O_2^+$) and 46 ($^{12}C^{16}O^{18}O^+$) into the mass separator of the IR MS where they were separated from each other. Ions $^{12}C^{16}O_2^+$ and $^{12}C^{16}O^{18}O^+$ were collected and counted by the respective Faraday cups installed on the exit from the separator.

For the $\delta^{13}C$ analysis, around 100 µg of the freeze-dried $BaCO_3$ was wrapped in tin capsules and subjected to decomposition at 1050 °C on the chromium oxide catalyst in the EA of the IR MS to form CO gas [25]. The formed CO was carried by a continuous flow of He-carrier gas through a chromatographic column to purify from contaminants before entering the ionization chamber to ionize. A flow of the He-carrier gas carried the ions of masses 28 ($^{12}C^{16}O^+$) and 29 ($^{13}C^{16}O^+$) into the mass separator of the IR MS where they were separated from each other. The ions $^{12}C^{16}O^+$ and $^{13}C^{16}O^+$ were collected and counted by respective Faraday cups installed on the exit from separator.

To calculate the δ^2H, $\delta^{18}O$ and $\delta^{13}C$ the Mass Lynx software supplied by the IR MS supplier was used. Precision of δ^2H determination was ±2‰ and that of $\delta^{18}O$ and $\delta^{13}C$ was ±0.2‰.

For tritium measurement, water samples were first subjected to distillation to remove the minerals dissolved till EC was less than 10 µS cm^{-1}, then tritium content in the samples were enriched by electrolysis at 4 °C till around 10 mL was attained [26,27]. The tritium enriched water samples were mixed with the low tritium Ultima Gold scintillation cocktail (Hewlett-Packard, HP Supplier based in Palo Alto, CA, USA) in vials of 20 mL capacity to count for the ^3H activity on a low background HP Liquid Scintillation Counter TriCarb TR 3700. The ^3H activity in water was expressed in the Tritium Unit (TU, 1 TU = 0.118 Bq L^{-1}). The limit of quantification (LOQ) for ^3H was estimated to be as low as 0.4 TU. The accuracy of the determination was validated by our participation in the inter-comparison exercises of the TRI-2004 and TRI-2008 organized by the IAEA IHL in the years 2004 and 2008. In the 2004 exercise, the Hanoi laboratory (no.74) produced results having Z-scores of −1.25 and 0.59 for the samples of 1.74 TU and 5.43 TU, respectively. In the 2008 exercise, the laboratory (no. 27) produced results with Z- scores of 0.42 and 1.57 for the samples of 4.07 TU and 1.54 TU, respectively [28,29].

The ^{14}C-analysis was conducted at the Center for Nuclear Techniques in the city of Ho Chi Minh (HCM) where $BaCO_3$ was decomposed by concentrated H_3PO_4 (PA grade, Merck supplier) to obtain CO_2 for further benzene synthesis [30,31]. The yield of the benzene synthesis was from 95% to 98%. The benzene obtained was mixed with the Ultima Gold scintillation cocktail (HP Supplier) in vials of 20 mL capacity, and then counted for the ^{14}C activity on the HP LSC TriCarb TR 3770. Counting time for the ^{14}C activity determination was estimated in order to achieve a precision of 5%, usually it took 1000 min but split into 10 cycles, each cycle was set for 100 min to count.

2.3. Estimate Groundwater Age by the ^{14}C-Dating Method

The aim of groundwater age determination is to verify follow pattern and to estimate its flow rate. The age of groundwater was calculated based on the law of radioactive decay [15,18] as follows:

$$^{18}t = 8268 \times ln \; \frac{^{14}a_{in}^0}{^{14}a_{sample}} \; (aBP) \tag{5}$$

where ^{14}t denotes the age in years Before Present (aBP) of a groundwater sample estimated by the ^{14}C-content in the DIC; the number 8268 is quotient of half-life of ^{14}C-isotope (5730 a) to ln2; $^{14}a_{in}^0$ is the initial ^{14}C-content in the DIC before entering the saturated zone (pMC), and $^{14}a_{sample}$ is the ^{14}C-content (pMC) in the DIC of the sample to be measured. The term "Before Present" implies the age relative to the time when the standard was produced. This means that the $^{14}a_{sample}$ must be measured relative to a standard to reduce uncertainty of the ^{14}C-activity measurements [15,18] and it was expressed by the formula:

$$^{13}a_{sample} = \frac{^{14}A_{sample}}{^{14}A_{std}} 1000 \; (pMC) \tag{6}$$

where $^{14}a_{sample}$ is the relative ^{14}C-activity or the ^{14}C-content in a sample; $^{14}A_{sample}$ and $^{14}A_{std}$ are the absolute activities (Bq g^{-1} C) of ^{14}C in the sample and standard, respectively.

The standard used in this study was oxalic acid II (ox-II) made from French beet molasses planted in 1977 and supplied by the National Institute of Standards and Technology (NIST based in Gaithersburg, MD, USA), which has a ^{14}C-activity of 0.2147 Bq g^{-1} C and δ^{13}C = −25‰ [32].

The $^{14}a^0_{in}$ in Equation (5) was calculated using an isotope mixing model referred to as the model of complete exchange with CO_2 in the unsaturated zone proposed by Gonfantini [33] for a closed/confined system was applied as follows:

$$^{14}a^0_{in} = \frac{\delta^{13}C_{DIC} - \delta^{13}C_{cc}}{\delta^{13}C_{CO_{2,org}} - \delta^{13}C_{cc} + \varepsilon_{CO_2/DIC}} \tag{7}$$

where $\delta^{13}C_{DIC}$, $\delta^{13}C_{cc}$ and $\delta^{13}C_{CO_{2,org}}$ are carbon-13 compositions, respectively, in DIC of a groundwater sample, in calcareous materials in soil/sediment and in biogenic dioxide originated from decomposition of organic matters; $\varepsilon_{CO_2/DIC}$ is fractionation coefficient for ^{13}C in the isotopic exchange reaction between biogenic carbon dioxide and DIC, which is temperature dependent [34,35]:

$$\varepsilon_{CO_2/DIC} = \left(\frac{9483}{T} + 23.89\right)‰ \tag{8}$$

where T is the temperature of the groundwater sample, in Kelvin.

In the study area the $\delta^{13}C_{cc}$ in soil was found to be ranging from 1 to 2‰ with an average value of 1.5‰ was taken to calculate for $^{14}a^0_{in}$. The $\delta^{13}C_{CO2,org}$ in Equation (7) was taken as high as −23‰, as it is characterized for carbon dioxide generated from mineralization of remnants of C3 plants in tropical areas followed by its diffusion to aquifer [36].

Detail of procedure for $^{14}a^0_{in}$ calculation could be found in Fontes and Garnier [37] and Fontes [38]. The computer code NETPATH [39] was used for this correction. The data of ^{3}H activity measured for water samples were used to confirm whether it has old or modern age.

3. Results

3.1. Regional Meteoric Water Line

Results of the stable isotopic compositions of precipitation collected in Nam Dinh city for one meteorological year, beginning in August 2011 to the end of July 2012 allowed one to construct a regional meteoric water line (RMWL), a graph of δ^2H vs. δ^{18}O of local precipitation. This relationship was deduced based on the ordinary least square regression method. The RMWL follows a model with a correlation coefficient R^2 = 0.99 as shown by Equation (9):

$$\delta^2\text{H} = 8.48 \times \delta^{18}\text{O} + 15.88 \tag{9}$$

The slope and intercept of the RMWL in region (Equation (9)) was a little higher than those of the Global Meteoric Water Line of, respectively, 8 and 10, probably due to the kinetic effect during rain fall in the tropical regions where atmospheric temperatures are usually high [15].

The RMWL was inserted in Figure 4a,b to compare with water stable isotopic compositions of groundwater samples. In Figure 4a,b the mean weighted isotopic composition of precipitation was taken from [40] which was calculated based on the precipitation depth and δ^2H and δ^{18}O of each event over whole the Red River's delta plain for a time period from March 2001 till March 2011.

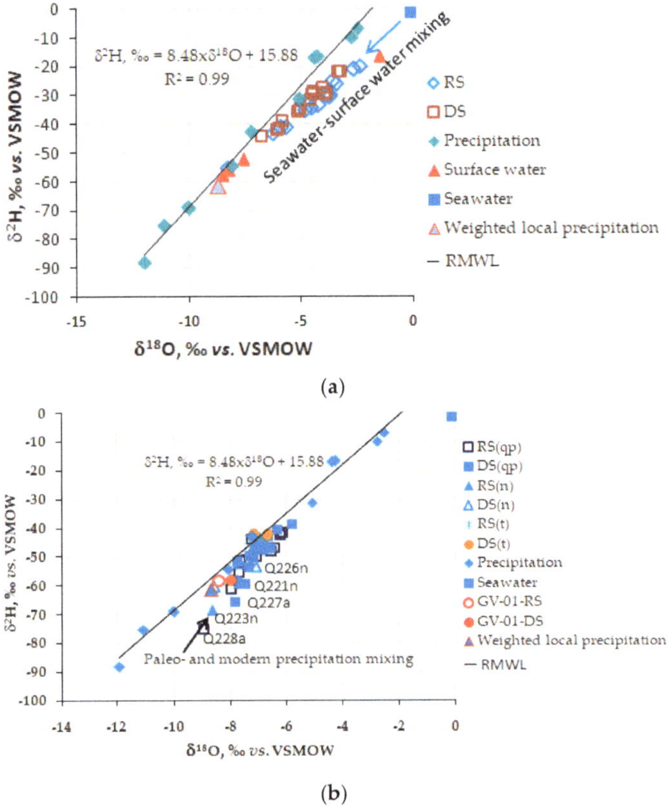

Figure 4. Isotopic composition in regional precipitation along with that in water from Holocene (**a**) and Pleistocene (qp), Neogene (n) and Triassic (t) aquifers (**b**) of the study area during dry (DS) and rainy (RS) seasons.

3.2. Water Level

Table 1 shows elevation, in meter above mean sea level (m asl), of water table in boreholes installed in different aquifers from those groundwater samples were taken.

Table 1. Elevation of groundwater table in boreholes from those water samples were taken.

Borehole	Depth, m bgs	Elevation of Water Table, m asl	Borehole	Depth, m bgs	Elevation of Water Table, m asl
Shallow Holocene Aquifer in NE Area			Deep Aquifers in NE Area		
OB-01	7.6	0.38	Q223n	138	−1.94
OB-02	8.5	0.34	Q224a	100	−2.46
OB-04	8.3	0.25	Q225a	110	−1.45
OB-06	6.7	0.26	Q226a	105	−2.27
OB-07	7.3	0.42	Q226n	151.5	−2.25
OB-08	8.1	0.40	Q227a	105.5	−4.57
OB-09	6.1	0.45	Deep Aquifers in SW Area		
OB-10	8.0	0.50	GV01	70	94.5
OB-11	7.8	0.57	Q220t	100	0.96
OB-12	7.6	0.74	Q92a	43	0.50

Table 1. *Cont.*

Borehole	Depth, m bgs	Elevation of Water Table, m asl	Borehole	Depth, m bgs	Elevation of Water Table, m asl
OB-13	6.7	0.52	Q92t	100	1.03
OB-14	8.4	0.60	Q108b	80	−5.89
OB-15	8.8	0.62	Q109a	136	−8.51
OB-16	9.6	0.78	Q109n	171	−7.45
Deep Aquifers in NE Area			Q228a	120	−7.57
Q221a	70	−0.97	Q110a	94	−5.12
Q221n	127	−0.96	Q229a	85	−8.18
Q222b	115	−2.05	Q229n	150	−6.89

Note: m bgs: meter below ground surface; m asl: meter above mean sea level; a, n, and t stand for Pleistocene, Neogene and Triassic aquifers, respectively.

As seen from Table 1, elevations of water table of Holocene aquifer were all above the mean sea level. Elevations of water table in deep aquifers are descending from northeasterly towards southwesterly, from northwesterly towards the seaside and from seaside towards production well field located in proximity to borehole 109a. These data were used to draw isolines of hydraulic heads of groundwater as it was depicted in Figure 1.

3.3. Groundwater Isotopic Composition

Figure 4a,b depicts the isotopic compositions in groundwater collected during the rainy (RS) and dry (DS) seasons from Holocene (Figure 4a), Pleistocene, Neogene and Triassic aquifers (Figure 4b), respectively, along with the RMWL.

The open and solid red color circles in Figure 4b represent isotopic composition in water collected in the RS and DS from borehole GV01 installed in Gia Vien district, Ninh Binh province. Though the borehole was made into Triassic limestone bedrock (Figure 2) water in it characterizes the local and modern precipitation with its isotopic composition positioned close to the RMWL (Figure 4b), it contains (85.30 ± 0.87) pMC and 2.03 TU (Table S1), i.e., comparable with the average tritium activity found year round in precipitation in the Red River Delta region [40].

3.4. The Age of Groundwater in the Aquifers

Tritium activity in groundwater from Holocene aquifer was comparable with the surface water ranging from 2.0 to 3.0 TU (Table S1). However, the tritium content in groundwater from Pleistocene, Neogene and Triassic aquifers was lower than LOQ of determination of 0.4 TU, excepted for water in borehole GV01. This finding reflects the fact that groundwater in Holocene aquifer is modern and that it is continuously recharged from surface water. This was confirmed by the ^{14}C-age of water from wells OB-10 and OB-12 for which the ^{14}C-content was 101.2 pMC and 103.5 pMC, respectively (Table S1). In contrast, groundwater in Pleistocene, Neogene and Triassic aquifers did not receive recent recharge, therefore it was of an older age, e.g., water in borehole Q223n made into Neogene aquifer has a ^{14}C-age as old as 14.5 ka BP (Table S1).

Results of ^{14}C-dating revealed that along transect AB (Figure 1) groundwater in borehole GV01 made into Triassic aquifer is modern (i.e., recent precipitation), while in Pleistocene aquifer it was 1.1 ka old at borehole Q92, 3.3 ka old in Q108a and 11.3 ka old in borehole Q109a (Figure 5). However, the age of groundwater found in Q110a borehole installed behind borehole Q109 at the sea coast decreased down to 6 ka (Figure 5 and Table S1).

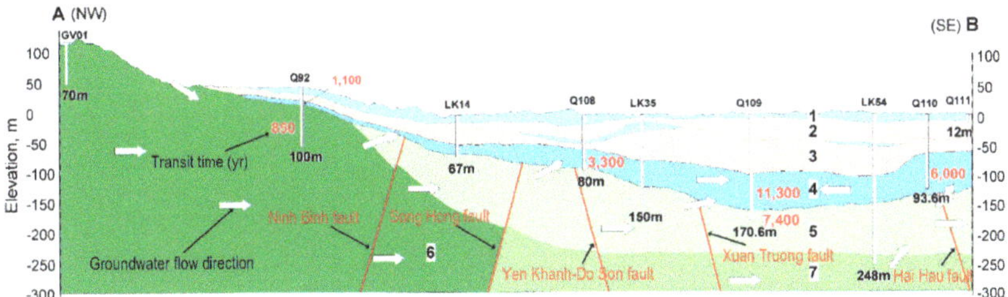

Figure 5. ^{14}C-ages and conceptual model indicating flow directions of groundwater in Pleistocene, Neogene and Triassic aquifers in the study region.

3.5. Groundwater Chemistry

The chemistry of the water samples studied were presented in Table S2. Sodium and chloride were the dominant cation and anion in water from Holocene aquifer, excepted for water tapped from borehole OB-09 (Table S2). Concentration of Na^+ ion ranged from 3.0 to 5961.5 mg L^{-1} in the RS, and from 3.5 to 5768.5 mg L^{-1} in the DS. Concentration of Cl$^-$ ranged from 3.0 to 11,699.5 mg L^{-1} in the RS and from 3.6 to 11,699.5 mg L^{-1} in the DS.Molar [Ca^{2+}] to [Mg^{2+}] ratio ranged from 0.1 to 4.99 in both rainy and dry seasons. The lowest concentrations of [Na+] and [Cl$^-$] ions and the highest value of molar [Ca^{2+}] to [Mg^{2+}] ratiowas found in water tapped from borehole OB-09 (see Figure 1 for location). It seems that water in Holocene aquifer of the study region is affected by seawater.

Concentration of Na^+ ion in water tapped from deep aquifers of the study region varied from 3.5 to 1003.3 mg L^{-1} in the RS and from 4.6 to 1146.3 mg L^{-1} in the DS, whilst concentration of Cl$^-$ ion in water varied from 7.4 to 2216.0 mg L^{-1} in the RS and from 6.8 to 2247.0 in the DS. Molar [Ca^{2+}] to [Mg^{2+}] ratio ranged from 0.1 to 5.1 in the RS and from 0.32 to 6.8 in the DS (Table S2). The highest value of the [Ca^{2+}] to [Mg^{2+}] ratio was found in water tapped from borehole GV01 located in a mountainous site of Ninh Binh province where it was expected to be the recharge area for groundwater in the study region. The source of salinity in water of deep aquifers seems to be not from the sea.

Groundwater in the study region could be divided into three types based on its total dissolved solids (TDS) content. In the NE area, groundwater was found of two types. in Thai Binh province and northernmost part of Nam Dinh province (Figure 1), groundwater in all Holocene, Pleistocene, Neogene and Triassic aquifers is saline with TDS > 3 g L^{-1} but in a part of northern Nam Dinh province lying along the Red River (Figure 1), groundwater in these aquifers is brackish with TDS ranging from 1 to 3 g L^{-1}. In the SW area, the southern part of Nam Dinh and northern part of Ninh Binh province (Figure 1), groundwater in the deep aquifers is fresh containing TDS < 1 g L^{-1}. The fresh-brackish and brackish-saline water boundaries drawn based on the TDS content in the deep aquifers groundwater of the study region is presented in Figure 1. The blue, light brown and dark brown color in Figure 1 indicate the areas of fresh, brackish and saline waters, respectively. Brackish water in the deep aquifers in the NE area seems to be resulting from the migration of saline water from the NE or/and from the diffusion downwards of saline water from the upper aquitard due to the high rate of fresh water mining in the SW and this will be discussed later on.

4. Discussion

4.1. Genesis of Groundwater Resources in the Southern Part of Red River's Delta Plain

Groundwater in Holocene aquifer of the study region is affected by seawater intrusion. This thought was evident from Figure 4a, where the trend of water isotopic signatures appeared to extend to the point characterizing for seawater (Figure 4a). Moreover, as it was shown in Figure 6a, that the molar concentrations of sodium ion in water of Holocene aquifer are strongly correlated (R^2 = 0.986) with concentrations of chloride for both rainy and dry seasons (Figure 6a). The slope of 0.76 of the correlation is very close to the value of 0.8 representing for seawater.

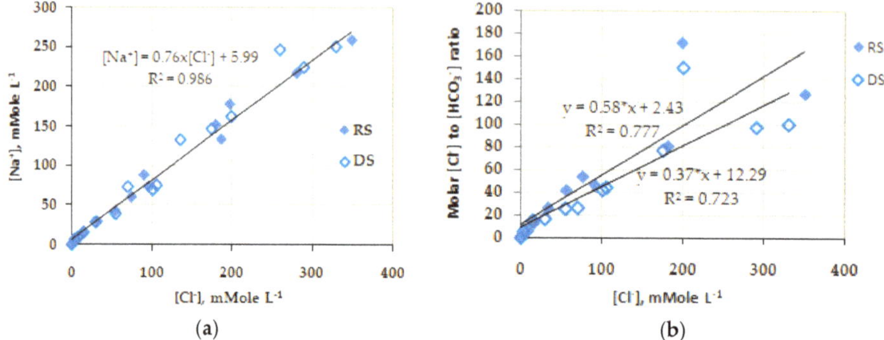

(a)   (b)

Figure 6. Relationship between molar concentrations of Na^+ and Cl^- ions (**a**), and relationship between molar $[Cl^-]$ to $[HCO_3^-]$ ratio and $[Cl^-]$ (**b**) for water tapped from Holocene aquifer indicating the influence of seawater to water in the shallow aquifer (the data for constructing graphs (**a**,**b**) were taken from Table S2).

Data presented in Table S2 showed molar ratio of $[Cl^-]$ to $[HCO_3^-]$ of water from Holocene aquifer to range from 0.21 to 170.2 and it is positively correlated with concentrations of chloride ions ($R^2 > 0.7$) for both RS and DS, as seen in Figure 6b. Among 16 samples from Holocene aquifer, the only sample taken from borehole OB-09 (see Figure 1 for location) has the lowest value of $[Cl^-]/[HCO_3^-]$ ratio of 0.21, less than 0.5. This reflects a fact that water in borehole OB-09 was not affected by seawater like Mohan Babu et al. have shown [41].

Groundwater in Holocene aquifer is a mixture of three end-members, namely: local precipitation, water from Red River and seawater. The three end-members mixing character of water in Holocene aquifer is clearly demonstrated in Figure 7, where the relationship of $\delta^{18}O$ vs. chloride concentration ($[Cl^-]$) in water was arranged within a triangle with three apexes representing local precipitation, Red River's water and seawater (Figure 7).

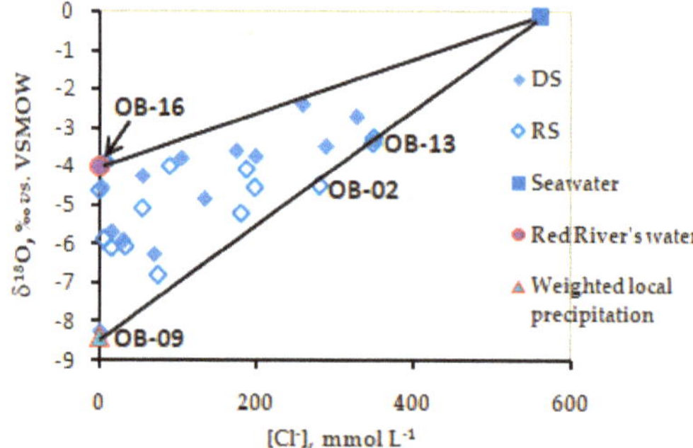

Figure 7. The $\delta^{18}O$ vs. molar $[Cl^-]$ relationship for groundwater in Holocene aquifer in the study region representing a mixture of three end-members: local precipitation, Red River's water and seawater (see Figure 1 for OB-09, OB-16, OB-02 and OB-13 locations).

Let the contribution of precipitation's water to groundwater in Holocene aquifer be x, the River's be y and seawater be z, then one could compute x, y and z for each water sample based on $\delta^{18}O$, and $[Cl^-]$ using three end-members model [42] as follows:

$$x + y + z = 1 \tag{10}$$

$$x \times \delta^{18}O_p + y \times \delta^{18}O_r + z \times \delta^{18}O_s = \delta^{18}O_{gw} \tag{11}$$

$$x \times [\text{Cl}^-]_p + y \times [\text{Cl}^-]_r + z \times [\text{Cl}^-]_s = [\text{Cl}^-]_{gw} \tag{12}$$

where subscripts p, r, s, and gw denote the precipitation, river's water, seawater and groundwater, respectively.

Table 2 shows results of estimated contribution (in percent) by local precipitation, river's water and seawater to groundwater in different boreholes made into Holocene aquifer in the study region during the RS and DS. This estimate was made based on the average $\delta^{18}O_p$, $\delta^{18}O_r$ and $\delta^{18}O_s$ of −8.40‰, −4.0‰ and −0.1‰ (vs. VSMOW), and $[\text{Cl}^-]_p$, $[\text{Cl}^-]_r$, and $[\text{Cl}^-]_s$ of 0.10 mmol L^{-1}; 0.11 mmol L^{-1} and 566 mmol L^{-1}, respectively. The data of $\delta^{18}O_p$, $\delta^{18}O_r$, $[\text{Cl}^-]_p$, and $[\text{Cl}^-]_r$ for precipitation and Red River's water in the RRDP were from [43]. The $\delta^{18}O_{gw}$ and $[\text{Cl}^-]_{gw}$ were taken from Tables S1 and S2, respectively. Here, the river's water must be included as well as water from irrigation canals in the region because this system is directly connected to Red River via Dao and Ninh Co tributaries (Figure 1).

Table 2. Contribution (%) of seawater (SW), Red River's water (RW) and local precipitation (LP) to groundwater in Holocene aquifer in study region during rainy and dry seasons.

	Rainy Season				Dry Season			
Borehole	SW	RW	LP	Total	SW	RW	LP	Total
OB-01	36.97	28.49	34.53	100	36.63	43.25	20.12	100
OB-02	53.62	1.34	45.03	100	51.96	22.00	26.03	100
OB-04	61.02	7.98	31.00	100	64.56	14.29	21.15	100
OB-06	12.93	29.91	57.16	100	13.85	37.11	49.04	100
OB-07	0.60	66.59	32.81	100	1.08	88.73	10.19	100
OB-08	2.75	59.12	38.13	100	2.80	65.87	31.33	100
OB-09	0.00	1.67	98.33	100	0.00	5.58	94.43	100
OB-10	4.95	37.17	57.88	100	33.38	29.46	37.16	100
OB-11	19.39	67.71	12.90	100	16.62	76.06	7.32	100
OB-12	5.52	55.15	39.32	100	6.08	56.50	37.43	100
OB-13	48.07	32.23	19.70	100	64.72	19.45	15.83	100
OB-14	32.35	44.51	23.14	100	34.66	49.03	16.32	100
OB-15	10.15	64.36	25.49	100	10.18	78.66	11.16	100
OB-16	0.45	99.62	0.00	100	1.56	98.43	0.00	100

As seen from Table 2, groundwater from boreholes located deep inland, e.g., OB-07, OB-08, OB-12, and OB-16 (see Figure 1 for locations) has low contribution of seawater. This contribution was only from 0.4 to 5.5% in the RS and from 1.1 to 6.1% in DS. However, waters from boreholes nearby the River's mouth and coastline do have a high contribution of seawater. Water from the OB-01 and OB-04 boreholes installed close to the Red River's mouth contains, respectively, 37% and 61% seawater in the RS and 37% and 65% seawater in the DS (Table 2). Borehole OB-01 is about 500-m from the River's bank, while OB-04 is installed right on the River's bank. Apparently, seawater intruded Holocene aquifer via the River. Active sand excavation (for construction purposes) from the River's bed combined with construction of upstream hydroelectric reservoirs, reduce downstream sediments deposits; the combination of these factors could much facilitate seawater to intrude deep inland and to diffuse to the aquifers. Boreholes OB-02 and OB-13 are located at 50 and 100m from the sea shore (Figure 1) so apparently groundwater around these locations could be directly affected by seawater intrusion.

From Table 2, one can see that groundwater from borehole OB-09 was fresh with no seawater detected. $\delta^{18}O$ and chloride concentration in groundwater from borehole OB-09 were −8.25‰ and 0.10 mmol L^{-1}, respectively (Tables S1 and S2) that almost coincided with the heavy oxygen isotope signature and $[\text{Cl}^-]$ found in local precipitation (see the point of isotopic signature weighted for local precipitation in Figure 4a). This supports the rainwater origin of groundwater around OB-09 borehole. In fact, borehole OB-09 is located in a lower area of Giao Thuy district that is capable of storing rain water year-round making rain water to be the most dominant source recharging shallow aquifer. On the other hand, the aquifer sediment around borehole OB-09 is purely coarse sand so chemistry of groundwater in this area is mainly dependent upon the chemistry of the local precipitation.

Groundwater in Pleistocene and Neogene aquifers in the NE area is brackish and saline. The EC and [Cl$^-$] in water from boreholes Q221a, Q222b, Q224a, Q225a, Q226a, Q221n, Q223n and Q226n range from 2540 to 6980 μS cm^{-1} and from 22 to 63 mmol L^{-1}, respectively, in the RS and DS (Table S2). The high salinity and chloride concentration in groundwater of Pleistocene and Neogene aquifers in the NE area could not be explained by recharge of saline water from Holocene aquifer or salt intrusion neither through the River's bank nor the Ghyben-Herzberg mechanism like in the case of Holocene aquifer aforesaid. It was revealed that the relationships of the δ^{18}O vs. [Cl$^-$] as well as the molar [Na$^+$] to [Cl$^-$] ratio vs. [Cl$^-$] for groundwater in Pleistocene and Neogene aquifers within entire study region do not reflect a conservative mixing character of fresh and seawater (Figure 8a,b).

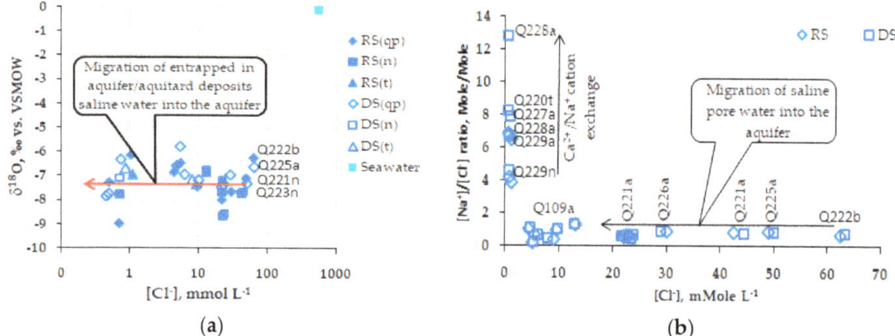

Figure 8. Relationships of δ^{18}O vs. molar [Cl$^-$] (**a**) and molar [Na$^+$] to [Cl$^-$] ratios vs. [Cl$^-$ (**b**) for groundwater in Pleistocene and Neogene aquifers within the study region showing the migration of entrapped in aquifer and/or aquitard deposits saline pore water to the recharge-fresh water (**a**) as well as the Ca^{2+}/Na$^+$ cation exchange that proceeds in the areas around the Q227a, Q228a and Q229a boreholes (**b**). (**a**) was drawn in semi-logarithmic scale (see Figure 3 for the boreholes locations).

Non-parametric z-test was applied to examine hydraulic interaction between groundwater in Holocene and Pleistocene aquifers in the study region. For the test, the mean δ^{18}O and its standard deviation in water from Holocene and Pleistocene aquifers were used as variables. It was revealed that mean of δ^{18}O in water from Holocene aquifer, $-(4.78 \pm 1.08)$‰, n =15, was very significant different from that in water of Pleistocene aquifer $-(7.21 \pm 0.62)$‰, n = 12, as z = 7.05 (in RS) and 8.49 (in DS) > z-critical = 2.58 at α = 0.01. This indicates a fact that water from Holocene aquifer does not recharge Pleistocene aquifer.

As seen from Figure 8a, the δ^{18}O in groundwater of Pleistocene, Neogene and Triassic aquifers vary within a narrow range of $-(7.26 \pm 0.76)$‰, n =17,whilst chloride concentration in water does within a wide range, from 22 mmol L^{-1} (Q221a) up to 63 mmol L^{-1} (Q222b). By definition (Equation (2)), δ^{18}O for seawater should be close to 0‰, therefore if seawater intruded to Pleistocene and Neogene aquifers then the heavy oxygen isotope content in water of those aquifers must be enriched more than $-(7.26 \pm 0.76)$‰.On the other hand, results presented in Figure 4b show the mixing character of young age precipitation with paleo-water in Q221n, Q223n, Q222b, Q224a, Q226n, Q227n having enriched oxygen-18 signature (Figure 4b). The trend of water line for deep aquifers (Figure 4b) is similar to those that was observed for groundwater in Qatar city [44]. Here, the paleo-water must be understood as groundwater having an old age, up to several thousand years. In fact, it was found that the age of water in Pleistocene and Neogene aquifers, e.g., water in Q221a, Q222b, Q224a, Q225a, Q228a, Q221n, Q223n and Q226n boreholes, was from 6 to 14.5 ka old (Table S1, see Figure 3 for locations), i.e., the age of the middle and early Holocene.

Data in Table S2 showed molar $[Ca^{2+}]$ to $[Mg^{2+}]$ ratio in water of Pleistocene and Neogene aquifers in NE area (Q221a, Q222b, Q224a, Q226a, Q227a, Q223n boreholes, see Figure 3 for locations), in both RS and DS, to vary within a range of (0.60 ± 0.05), n = 9, which is 3 times higher than that ratio characterizing for seawater of around 0.2. It seems that groundwater in deep aquifers in that area is freshening. This freshening process would cause dead marine flora and fauna in the aquifer deposits to mineralize making molar $[Br^-]$ to $[Cl^-]$ ratio in water of Pleistocene and Neogene aquifers in NE area to be higher than those ratio in seawater. Calculation showed that the molar $[Br^-]$ to $[Cl^-]$ ratio for water in Pleistocene and Neogene aquifers in NE area was $(2.4 \pm 0.5) \times 10^{-3}$ (n = 9) that was higher than that ratio representing for seawater of around 1.5×10^{-3}.

All results presented suggest that brackish and saline waters in Pleistocene and Neogene aquifers in NE area originated from entrapped in aquifer and/or aquitard deposits saline pore water which existed since Holocene transgression as it was also pointed out by [3,5–7]. High salinity in groundwater of deep aquifers in NE area (Q221a, Q222b, Q225a, Q226a) associated with molar $[Na^+]$ to $[Cl^-]$ ratios varying within the range of seawater (0.73 ± 0.14) (Figure 8b) probably indicates that the migration rate of saline water from the aquifer/aquitard deposits was over the rate of submarine discharge. The rate of the two processes apparently depends upon hydraulic conductivity of both aquifer and aquitard deposits.

Recently, it has been documented that the hydraulic properties and thickness of the aquitard in the Quaternary Red River Delta plain control recharge and leaching of saline water into Pleistocene aquifer [45]. Results from the simulation work of Larsen et al., [45] allow one to assume that permeability of clayey aquitard in the NE area was higher than that in the SW area. This made the diffusion rate of saline water from the aquitard to Pleistocene aquifer in latter area to be insignificant compared to the rate of the submarine discharge flow, hence groundwater in Pleistocene aquifer of the SW area became fresh today.

Groundwater of Pleistocene and Neogene aquifers in SW area have low EC and $[Cl^-]$, except for water from borehole Q109b. Groundwater in Q92, Q108b, Q110a, Q227a, Q228a and Q229a boreholes (see Figure 3 for the locations) has EC < 1000 μS cm^{-1} and $[Cl^-]$ < 7 mmol L^{-1}. The salinity of groundwater around borehole Q109bresults from brackish water in the NE flowing to production well field located in proximity of the borehole, due to the high mining rate of fresh water from Pleistocene aquifer. This was evident from Figure 1 and Table 1 that the isolines of water hydraulic head in Pleistocene aquifer were lowdown from −2 m above sea level (asl) in the area close to the Red River to −8 masl in area around Q109b borehole.

4.2. Recharge Area, Flow Direction and Flow Rate of Groundwater in the Southern Part of the Red River's Delta Plain

The recharge to aquifers in the southern part of the Red River's Delta plain is from mountainous areas around GV01 point (Figure 1), where the water isotopic composition is close to that of recent precipitation, ^{14}C-content is modern and tritium activity is in the range of 2–3 TU that prevails in the RRDP surface water (Table S1). There is no recharge to Pleistocene aquifer through the thick Holocene clays, as indicated by high salinity profiles in these clays as well as z-test for mean $\delta^{18}O$ in groundwater of Pleistocene and Holocene aquifers that was discussed in previous section.

Mean value of $\delta^{18}O$ in groundwater of Pleistocene and Neogene aquifers in study region was −7.25‰ vs. VSMOW (Table S1), whereas average $\delta^{18}O$ in precipitation over the RRDP was −8.40‰ vs. VSMOW ([40] and Figure 4a, b). Thus, the difference of mean $\delta^{18}O$ in groundwater of Pleistocene and Neogene aquifers and those in precipitation over the region was −1.15‰ vs. VSMOW. Considering an altitude gradient of −0.3‰ per each 100m rise [14,18], the recharge area to Pleistocene and Neogene aquifers in study region was supposed to be a region at about 380m higher than aquifer altitude or around 150 m above sea level. This area should correspond to the northwest extension of the region and outcrops in Ninh Binh province, i.e., an area around the GV01 point, where the elevation is from 140 to 160m above sea level (Figures 1 and 5).

A non-parametric, Mann-Whitney or z-test using $\delta^{18}O$ as variable was revealed that mean of $\delta^{18}O$ in water from Pleistocene aquifer was not different from that in water in Neogene and Triassic aquifers, $z = 0.75 < z$-critical $= 1.96$ at $\alpha = 0.05$. This implies that water in Pleistocene aquifer connected with those in Neogene and Triassic aquifers and vice versa. This was termed as the inter-aquifer leakage. The inter-aquifer leakage of water between deep aquifers in this case was proven by the water line depicted in Figure 4b where the isotopic compositions of groundwater from Pleistocene, Neogene and Triassic aquifers positioned along a line characterized for mixing paleo-water of old age with recent precipitation. In addition, it was observed that hydraulic head of water in Neogene aquifer was always higher than that in Pleistocene aquifer (Table 1). As an example, Figure 9 shows the comparison of hydraulic heads of water in Q109b and Q109a boreholes installed, respectively, in Neogene and Pleistocene aquifer. Water levels in boreholes 109b and 109a were consecutively monitored during the time from December 1994 to December 2018. It was clear that water level in Neogene aquifer all time was higher than that in Pleistocene aquifer, though production well field is located in proximity to those boreholes (Figure 9). This provides another evidence for Pleistocene-Neogene inter-aquifer connection. The inter-aquifer leakage of groundwater in the study region is possible because of the existence of faults and fissures in the basement rock as it was mentioned early.

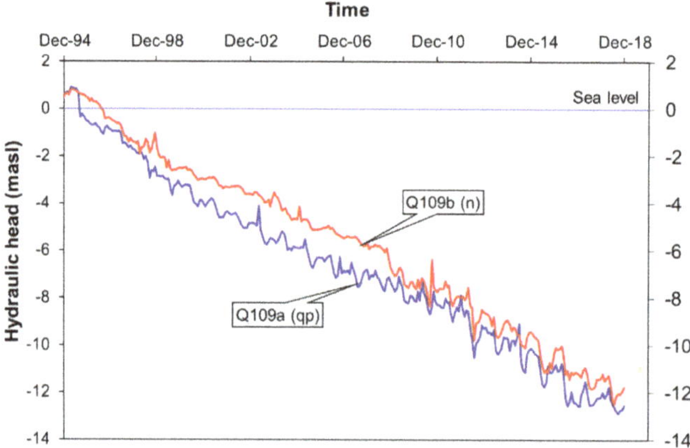

Figure 9. The hydraulic heads of groundwater in Pleistocene and Neogene that have been consecutively monitored in boreholes Q109a and Q109b since late 1994 indicating inter-aquifer leakage between the two aquifers due to over rate of freshwater mining from Pleistocene aquifer.

Based on the results of the ^{14}C-dating, a conceptual model of groundwater flow in Pleistocene and Neogene aquifers was suggested, as depicted in Figure 5. As seen in Figure 5, recharge water in study region flows from northwesterly towards southeasterly to the sea, which is in phase with [3] statement. From Figure 5 one can see also that groundwater from Neogene and Triassic aquifers is leaking upwards to Pleistocene aquifer making the age of water in the upper aquifer to be older than that of water in the lower aquifers because the leakage needs time to travel from one to another aquifer.

The yield of fresh water mining in recent years in the center of region, in Nam Dinh city, was reportedly to be as high as 95,000 m^3 a day [46] that seems to be over rate of recharge that caused not only inter-aquifer leakage but also could cause backwards flowing of sea water from seaside to well field. Water in borehole Q110 in seaside tapped from Pleistocene aquifer has a ^{14}C-age of 6 ka, almost two times younger than the age (11.3 ka) of water in borehole Q109a installed in proximity of well field (Figure 3). Apparently, in SW area seawater is currently intruding into deeper aquifers according to the Ghyben-Herzberg rule making the age of water in boreholes installed at proximity to sea coast to be younger than those inland. Modeling with the use of a MODFLOW model confirmed the scenario [47].

It should be noted, that EC of water in borehole Q110 was 825 μS cm^{-1}, which was lower than EC found in the water from boreholes Q109a and Q92a(see Figure 3 for locations) being around 1100 μS cm^{-1} (Table S2). This implies that saline water in boreholes Q109a and Q92a was not only from the sea but also coming from NE. Results from water head measurements confirmed three flow directions of groundwater in study region. The isolines of water head measured for Pleistocene aquifer are depicted in Figure 1. The descent of hydraulic water head delineates direction of water flow. As seen from Figure 1, groundwater in study region flows from northwesterly towards southeasterly to the sea, northeasterly towards southwesterly and from the seaside to the production well field in proximity to borehole Q109 as shown by the white arrows (Figure 1).

The ineluctable consequence from current over-extraction rate of fresh groundwater in region is that good quality water in the deep aquifers could be under the threat of salt intrusion in future.

Assuming the recharge area is in the highlands around the GV01 point then one can estimate the mean flow rate based on the transit time and distance between recharge area and sampling point. The recharge rate to Neogene aquifer of borehole Q221n was estimated to be as high as 2.5 m a year with a 30 km distance from the recharge area and a water age of 11.3 ka.

4.3. Chemistry of Groundwater in the Southern Part of the Red River's Delta Plain

Sulphate and nitrate ions were found to be in low concentrations in groundwater of Holocene aquifer whilst concentrations of ferrous and ammonia ions were in an elevated range, of up to 0.85 mmol L^{-1} and 5.0 mmol L^{-1}, respectively (in borehole OB-02 in the RS, Table S2). The [Fe^{2+}] ion found in groundwater tapped from Holocene aquifer in both RS and DS seems to correlate with concentrations of NH$_4^+$ ion, though R^2 ~ 0.4 (figure not shown here), suggesting that reduction of iron-oxyhydroxide by organic matters is on-going in Holocene aquifer as shown by reaction (13).

$$CH_2O + 4FeOOH + 7H^+ \rightarrow 4Fe^{2+} + HCO_3^- + 6H_2O \qquad (13)$$

The relatively weak correlation between [Fe^{2+}] and [NH$_4^+$] is probably due to the influence of an additional amount of ammonia generated from biological reduction of nitrate in the aquifer that proceeds following reaction (14) making concentration of nitrate in water to be reduced:

$$NO_3^- + 10H^+ + enzyme \rightarrow NH_4^+ + 3H_2O \qquad (14)$$

In addition, it was found that [Fe^{2+}] ions in groundwater of Holocene aquifer were not correlated with [HCO$_3^-$], i.e., reaction (13) seemingly does not proceed in aquifer. However, there would be several processes which could make the correlation of [Fe^{2+}] vs. [HCO$_3^-$] in water to be pure. These could be ferrous ions formed are either adsorbing onto aquifer deposits or precipitating as siderite or pyrite. Table 3 presents saturation indices (SI) of several minerals possibly existed in aquifer deposits of Holocene age for RS. In DS absolute values of SI for all the minerals were a little bit changed, but the trend of dissolution/precipitation of the minerals in groundwater was the same. Unfortunately, in this study concentrations of sulphide in groundwater were not analyzed so that SI for pyrite was not be able to calculate, however water from Holocene aquifer strongly smelled with H$_2$S. It was believed that pyrite is precipitating also.

As seen in Table 3, gypsum in all sampling sites is undersaturated suggesting dissolution of the mineral is on-going. Calcite and aragonite in 8 out of 16 sampling points are dissolving (Si$_{cc}$ < 0 and SI$_{arag}$ < 0), dolomite and siderite in most sampling points are precipitating (SI$_{dol}$ > 0, SI$_{sid}$ > 0), excepted for the case of borehole OB-09 in which deposit was pure sandy as it was discussed previously. Thus, siderite's precipitation is one of the reasons why the concentrations of ferrous ions in groundwater of Holocene aquifer werepurely correlated with those of bicarbonate.

Depletion of δ^{13}Cin DIC as it was checked for DICin water of OB-08 (-17.27‰ vs. VPDB, Table S1) partly proved the mineralization of organic matters by goethite (reaction 13) is occurring in Holocene aquifer, as it was well known that organic matters originated from C3 plants in the tropical regions like Vietnam do have δ^{13}C ranging from −20‰ to −27‰ vs. VPDB [36].

Table 3. Saturation indices (SI) of several minerals possibly existed in Holocene deposits in rainy season.

Boreholes	Calcite	Aragonite	Dolomite	Siderite	Gypsum
OB-01	−0.23	−0.36	0.72	1.64	−1.89
OB-02	1.19	1.06	3.01	2.64	−1.47
OB-04	1.24	1.11	3.89	2.57	−2.24
OB-06	−0.02	−0.15	1.10	1.16	−2.63
OB-07	−0.06	−0.19	0.16	1.48	−1.53
OB-08	−0.48	−0.61	−0.46	1.31	−2.27
OB-09	−1.06	−1.19	−2.37	−0.47	−2.79
OB-10	0.34	0.21	1.86	1.88	−2.22
OB-11	0.41	0.28	1.71	1.17	−1.90
OB-12	0.01	−0.12	0.51	1.54	−1.99
OB-13	0.72	0.59	2.61	1.85	−1.67
OB-14	0.05	−0.08	1.35	1.38	−2.24
OB-15	−0.30	−0.43	−0.34	1.36	−1.71
OB-16	−0.30	−0.43	0.09	0.80	−2.15

Low concentration of sulphate in groundwater samples could be explained by its reduction with organic matters following reaction (15):

$$2CH_2O + SO_4^{2-} \rightarrow H_2S + 2HCO_3^- \rightarrow H^+ + HS^- + 2HCO_3^- \rightarrow 2H^+ + S^{2-} + 2HCO_3^- \tag{15}$$

The reduction of sulphate occurred not only in Holocene aquifer but also in Pleistocene one. It was revealed that $[SO_4^{2-}]$ correlated with $[HCO_3^-]$ for both aquifers as shown in Figure 10. As seen from Figure 10 the higher $[HCO_3^-]$ the lower $[SO_4^{2-}]$ in groundwater. Here, one can see that the gained $[HCO_3^-]$ were not equivalent to the lost $[SO_4^{2-}]$, but it is understood because bicarbonate could be formed from other routes, e.g., from reaction (13) or dissolution of calcite, aragonite and dolomite minerals presented in aquifers' deposits.

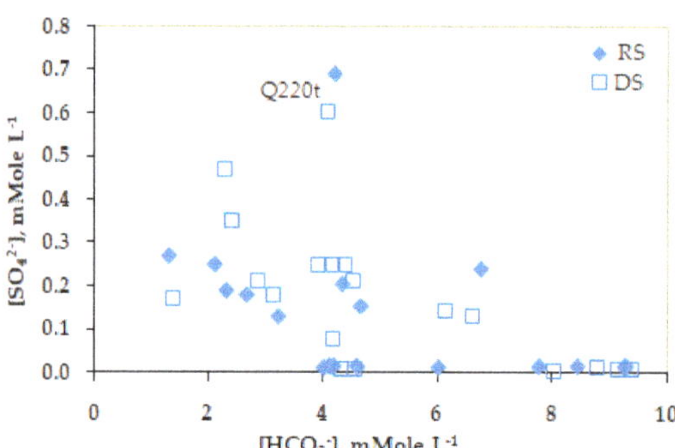

Figure 10. Scatter plot of molar $[SO_4^{2-}]$ vs. $[HCO_3^-]$ for groundwater in both Holocene and Pleistocene aquifers showing the sulphate reduction led to lowering SO_4^{2-} concentration in water.

Results presented in Table S2 show that the $[Ca^{2+}$, meq $L^{-1}]$ to $[Na^+$, meq $L^{-1}]$ ratio for water in Pleistocene and Neogene aquifers in study region ranged from 0.01 to 0.57, which is much lower compared to that ratio for water in borehole GV01 where groundwater is purely fresh and of Ca-HCO$_3$ type for which the equivalent $[Ca^{2+}]$ to $[Na^+]$ ratio was 28. This finding combined with the results presented in Figure 8b suggests that cation exchange between calcium ions in groundwater and sodium ions adsorbed on sediment's surface [36]. Figure 8b shows the molar $[Na^+]$ to $[Cl^-]$ ratio in water from boreholes Q227a, Q228a, Q229a and Q229n be much higher than 0.8, the value representing seawater. This indicates the excess of sodium over chloride concentration that is due to the release of sodium from aquifers sediment by Ca^{2+}/Na^+ exchange as it was explained by otherresearchers, e.g., [48]. However, it could be happened, that the weathering of silicate minerals releasesboth sodium and calcium but calcium ions would tie up in calciteand other precipitating carbonate minerals in the system, so that $[Na^+]$ in water increased whilst $[Ca^{2+}]$ decreased resulting in increase of $[Na^+]/[Ca^{2+}]$ ratio in water. Calculation with the results presented in Table S2 showed that all calcite, aragonite and dolomite in deposits around boreholesQ227a, Q228a, Q229a and Q229n are dissolving as it was shown in Table 4. This is evident for the fact, that Ca^{2+} ions are not lost from the systemof those boreholes.

Table 4. Saturation indices (SI) of calcite, aragonite and dolomite in groundwater of Pleistocene, Neogene and Triassic in the study region.

Borehole	Rainy Season			Dry Season		
	SI_{cc}	SI_{arag}	SI_{dol}	SI_{cc}	SI_{arag}	SI_{dol}
Q221a	0.44	0.31	1.16	0.48	0.34	1.15
Q222b	−0.39	−0.52	−0.30	−0.37	−0.51	−0.32
Q224a	0.01	−0.12	0.26	0.00	−0.14	0.25
Q225a	0.68	0.55	1.87	0.70	0.56	1.84
Q226a	0.46	0.33	1.39	0.53	0.39	1.37
Q227a	−0.39	−0.52	−0.03	−0.16	−0.30	−0.31
Q228a	−1.31	−1.44	−1.24	−0.84	−0.98	−0.91
Q229a	−0.16	0.03	−0.21	−0.11	−0.03	−0.11
Q110a	0.07	−0.06	0.48	0.17	0.03	0.60
Q109a	−0.45	−0.58	0.09	−0.24	−0.38	0.19
Q108b	−0.90	−1.03	−1.07	−0.91	−1.05	−1.13
Q92	0.50	0.37	1.41	0.45	0.31	1.33
Q229n	−0.62	−0.75	−0.55	−0.70	−0.84	−0.55
Q109	0.51	0.38	1.48	0.51	0.37	1.51
Q226n	0.42	0.29	1.22	0.55	0.41	1.26
Q221n	0.40	0.27	1.24	0.47	0.33	1.27
Q223n	−0.53	−0.66	−0.84	−0.52	−0.66	−0.76
Q220t	−1.51	−1.64	−1.93	−1.19	−1.33	−1.96
Q92a	0.46	0.33	1.39	0.54	0.40	1.47

It was revealed that concentrations of ferrous ions in water of Pleistocene, Neogene aquifers are inversely correlated with $\delta^{13}C$ in DIC as shown in Figure 11. This suggests that reduction of goethite by organic matters (reaction 13) is also occurring in these aquifers.

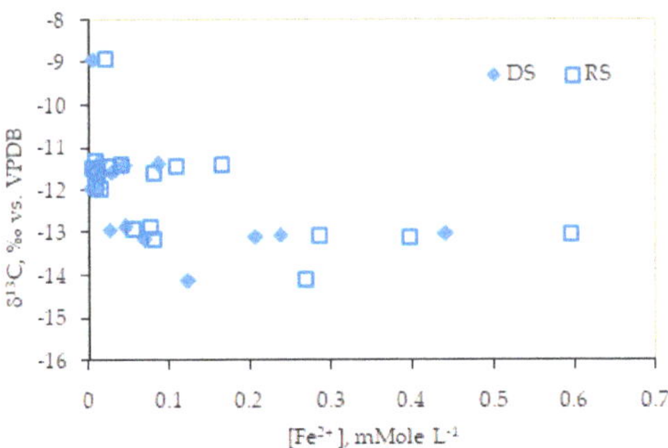

Figure 11. Scatter plot of $\delta^{13}C$ vs. $[Fe^{2+}]$ for water in Pleistocene, Neogene and Triassic aquifers in the study region.

Results in Table 4 showed that calcite and aragonite in most sampling points are dissolving as $SI_{cc} < 0$ and $SI_{arag} < 0$. Considering$\delta^{13}C$ in DIC originated from mineralization of organic matters of C3 plants be $-23‰$, $\delta^{13}C$ in DIC resulted from carbonates mineralization be $-1.5‰$ for the region, andaverage $\delta^{13}C$ in DIC of water samples be $-12.5‰$ (Figure 11), one could estimate the contribution ofinorganic carbonateto DICin water samples to be as high as around 50% based on binary mixing model.

5. Conclusions

Data of water isotopic compositions in groundwater in southern part of the Red River's Delta plain show that water in shallow Holocene aquifer is saline of Na-Cl type and currently affected by saltwater intrusion. Groundwater in deep Pleistocene, Neogene and Triassic aquifers of SW area is fresh, however that occurring across the NE region is brackish. Higher salinity of groundwater from deep aquifers of NE is due to migration of entrapped saline pore water from aquifer and/or aquitard deposits to aquifers. The hydro-geochemistry of the region is controlled by reduction of sulphate and iron-oxyhydroxide by organic matters and dissolution of inorganic carbonate minerals in aquifer sediments. Groundwater in deep aquifers of the study region is originated from meteoric water. The recharge area of deep aquifers is the northwest extension of the study region, at an altitude of 140-160m above sea level and water flows from northwesterly towards southeasterly to the sea coast at a rate of 2.5 m year^{-1}.

The extraction rate of fresh water from Pleistocene aquifer was over its recharge rate, thus leading to inter-aquifer leakage of water from Neogene and Triassic aquifers to Pleistocene aquifer through faults and fissures in the basement formation. At the same time, the over rate of freshwater mining from Pleistocene causes saline water from the seaside to flow backwards to the production well field that deteriorates the quality of drinking water supplied to local population. Therefore, it is highly advisable that a short-term appropriate strategy for groundwater management in the region is developed and implemented in order to avoid the threat of saltwater intrusion. The most effective measure for preventing seawater intrusion at present is control over groundwater mining at a rate not exceeding current production yield of 95,000 m^3 day^{-1}.

Supplementary Materials: The following are available online at http://www.mdpi.com/2073-4441/11/10/2120/s1, Table S1: Water isotopic compositions in water and carbon-13 in DIC and 14C-age of water samples under the study, Table S2: Chemistry of water samples under the study.

Author Contributions: Project administration, conceptualization, review: N.V.L.; software, validation, writing: H.V.H.; Methodology, analysis, writing-review and editing: D.D.N.

Funding: This research is funded by the Vietnam National Foundation for Science and Technology Development (NAFOSTED) under grant number 105.99-2014.19.

Acknowledgments: The authors would like to express their sincere thanks to anonymous reviewers for their helpful comments and review of the manuscript.

Conflicts of Interest: The authors declare no conflict of interest.

References

1. Frenken, K. (Ed.) *Irrigation in Southern and Eastern Asia in Figures*; Food and Agriculture Organization of the United Nations (FAO) Report No. 37; FAO: Rome, Italy, 2007.
2. Le, V.H.; Bui, H.; Chau, V.Q.; Dang, H.O.; Le, H.H.; Nguyen, T.T.; Tran, M. *Groundwater in the Bac Bo Plain (North Vietnam)*; A Technical Report to the General Department of Geology and Minerals of Viet Nam; Ministry of Natural Resources and Environment: Hanoi, Vietnam, 2000; 153p. (In Vietnamese)
3. Bui, D.D.; Kawamura, A.; Tong, N.T.; Amaguchi, H.; Nakagawa, N.; Iseri, Y. Identification of aquifer system in the whole Red River Delta. *Vietnam Geosci. J.* **2011**, *15*, 323–338. [CrossRef]
4. Hoang, H.V. Saltwater Intrusion in Quaternary Sediment in Nam Dinh Area. Ph.D. Thesis, Hanoi University of Mining and Geology, Hanoi, Vietnam, 5 September 2014. (In Vietnamese)
5. Tran, L.T.; Larsen, F.; Pham, N.Q.; Christiansen, A.V.; Tran, N.; Vu, H.V.; Tran, L.V., Hoang, H.V., Hinsby, K. Origin and extent of fresh groundwater, salty paleowaters and recent saltwater intrusions in the Red River flood plain aquifers, Vietnam. *Hydrogeol. J.* **2012**, *20*, 1295–1313. [CrossRef]

6. Nguyen, T.T.; Kawamura, A.; Tong, N.T.; Nakagawa, N.; Amaguchi, H.; Gilbuena, R., Jr. Hydrogeochemical characteristics of groundwater from the two aquifers in the Red River Delta, Vietnam. *J. Asian Earth Sci.* **2014**, *93*, 180–192. [CrossRef]

7. Nguyen, T.T.; Kawamura, A.; Tong, N.T.; Nakagawa, N.; Amaguchi, H.; Gilbuena, R., Jr. Clustering spatio-seasonal hydrogeochemical data using self-organizing maps for groundwater quality assessment in the Red River Delta, Vietnam. *J. Hydrol.* **2015**, *522*, 661–673. [CrossRef]

8. Nam Dinh Province Statistical Office (NDPSO) Area and Population of the Province. 2018. Available online: http://www.gso.gov.vn (accessed on 10 June 2019).

9. Nguyen, D.N.; Nguyen, T.H. *Climate and Climate Resources in Vietnam*; Agriculture Publishing House: Ha Noi, Vietnam, 2004; 210p. (In Vietnamese)

10. Tanabe, S.; Hori, K.; Saito, Y.; Haruyama, S.; Doanh, L.Q.; Sato, Y.; Hiraide, S. Sedimentary facies and radiocarbon dates of the Nam Dinh-1 core from the Song Hong (Red River) delta, Vietnam. *J. Asian Earth Sci.* **2003**, *21*, 503–513. [CrossRef]

11. Kasbohm, J.; Grothe, S.; Le, T.L. Province Nam Dinh an Analysis for a Future Integrated Water Resource Management. 2013. Available online: http://www.idm.gov.vn (accessed on 15 June 2019).

12. Luu, T.M.N.; Garnier, J.; Billen, G.; Orange, D.; Nemery, J.; Le, T.P.Q.; Tran, H.T.; Le, L.A. Hydrological regime and water budget of the Red River Delta (Northern Vietnam). *J. Asian Earth Sci.* **2010**, *37*, 219–228. [CrossRef]

13. National Center for Monitoring the Hydrology in the Marine Coast (NCMH). *The Hydrological Regime Along the Marine Coast, North Vietnam in the 2017 Year*; Annual Report to the General Directorate of Meteorology and Hydrology of Vietnam; NCMH: Hanoi, Vietnam, 2017; 53p. (In Vietnamese)

14. Erickson, E. *Stable Isotopes and Tritium in Precipitation. Guidebook on Nuclear Techniques in Hydrology*; IAEA Technical Report Series No. 91; IAEA: Vienna, Austria, 1983; pp. 19–33.

15. Clark, I.D.; Fritz, P. *Environmental Isotopes in Hydrology*; Lewis Publisher: New York, NY, USA, 1997; 328p.

16. Kalin, R.M. Radiocarbon dating of groundwater systems. In *Environmental Tracers in Subsurface Hydrology*; Cook, P.G., Herczeg, A.L., Eds.; Springer: Boston, MA, USA, 2000; pp. 57–68.

17. Zhu, C. Estimate of recharge from radiocarbon dating of groundwater and numerical flow and transport modeling. *Water Resour. Res.* **2000**, *36*, 2607–2620. [CrossRef]

18. MookW, G. (Ed.) *Environmental Isotopes in the Hydrological Cycle. Principles and Applications*; Volume II: Atmospheric Water; IAEA: Vienna, Austria, 2001; 288p.

19. Glynn, P.D.; Plummer, L.N. Geochemistry and the understanding of groundwater systems. *Hydrogeol. J.* **2005**, *13*, 263–287. [CrossRef]

20. Sánchez-Murillo, R.; Brooks, E.S.; Elliot, W.J.; Boll, J. Isotope hydrology and baseflow geochemistry in natural and human-altered watersheds in the Inland Pacific Northwest, USA. *Isot. Environ. Health Stud.* **2015**, *51*, 231–254. [CrossRef]

21. Craig, H. Isotopic variation in meteoric water. *Science* **1961**, *133*, 1702–1703. [CrossRef]

22. Stumm, W.; Morgan, J.J. *Aquatic Chemistry*, 2nd ed.; Wiley & Sons: New York, NY, USA, 1981; 780p.

23. Stookey, L.L. Ferrozine—A new spectrophotometric reagent for iron. *Anal. Chem.* **1970**, *42*, 779–781. [CrossRef]

24. International Atomic Energy Agency (IAEA). *Water and Environment Newsletter of the Isotope Hydrology Section*; International Atomic Energy Agency: Vienna, Austria, 2002; 8p.

25. *IsoPrime User's Guide*; Micromass UK Limited: Wemslow, UK, 2000; 18p.

26. Villa, M.; Manjon, G. Low-level measurements of tritium in water. *Appl. Radiat. Isot.* **2004**, *61*, 319–323. [CrossRef] [PubMed]

27. Plastino, W.; Chereji, I.; Cuna, S.; Kaihola, L.; de Felice, P.; Lupsa, N.; Balas, G.; Mirel, V.; Berdea, P.; Baciu, C. Tritium in water electrolytic enrichment and liquid scintillation counting. *Radiat. Meas.* **2007**, *42*, 68–73. [CrossRef]

28. Groening, M.; Dargier, M.; Tatzber, H. *Seventh IAEA Inter-Comparison of Low-Level Tritium Measurement in Water (TRIC-2004)*; International Atomic Energy Agency: Vienna, Austria, 2007. Available online: http://www-naweb.iaea.org/napc/ih/documents/IHL/TRIC/TRIC2004-Report.pdf (accessed on 15 June 2019).

29. Groening, M.; Tatzber, H.; Trinkl, A.; Klaus, B.; van Duren, M. *Eighth IAEA Inter-Comparison of Low-Level Tritium Measurement in Water (TRIC-2008)*; International Atomic Energy Agency: Vienna, Austria, 2009. Available online: http://www-naweb.iaea.org/napc/ih/documents/IHL/TRIC/TRIC2008-Report.pdf (accessed on 15 June 2019).

30. Tamers, M.A. Chemical yield optimization of the benzene synthesis for radiocarbon dating. *Int. J. Appl. Radiat. Isot.* **1975**, *26*, 676–682. [CrossRef]

31. Gupta, S.K.; Polach, H.A. *Radiocarbon Dating Practices at ANU*; Radiocarbon Laboratory, Research School of Pacific Studies, ANU: Canberra, Australia, 1985.

32. Mann, W.B. An international reference material for radiocarbon dating. *Radiocarbon* **1983**, *25*, 519–522. [CrossRef]

33. Salem, O.; Visser, J.M.; Deay, M.; Gonfiantini, R. Groundwater flow patterns in the western Lybian Arab Jamahitiya evaluated from isotope data. In *Arid Zone Hydrology: Investigation with Isotope Techniques*; IAEA: Vienna, Austria, 1980; pp. 165–179.

34. Bigeleisen, C.T.; Mayer, M.G. Calculation of equilibrium constants for isotopic exchange reactions. *J. Chem. Phys.* **1947**, *15*, 261–270. [CrossRef]

35. Mook, W.G.; Bommerson, J.C.; Staverman, W.H. Carbon isotope fractionation between dissolved and gaseous carbon dioxide. *Earth Planet Sci. Lett.* **1974**, *22*, 169–176. [CrossRef]

36. Appelo, C.A.J.; Postma, D. *Geochemistry, Groundwater and Pollution*, 2nd ed.; A.A.Balkema Publisher: Amsterdam, The Netherlands, 2007; pp. 175, 226.

37. Fontes, J.C.; Garnier, J.M. Determination of the initial activity of the total dissolved carbon: A review of the existing models and anew approach. *Water Resour. Res.* **1979**, *12*, 399–413. [CrossRef]

38. Fontes, J.C. *Dating of groundwater. Guidebook on Nuclear Techniques in Hydrology*; IAEA Technical Report Series No. 91; IAEA: Vienna, Austria, 1983; pp. 285–317.

39. Plummer, N.L.; Prestemon, E.C.; Parkhurst, D.L. *An Interactive Code (NETPATH) for Modeling Net Geochemical Reactions along a Flow Path, version 2.0*; US Geological Survey Water Resources Investigations Report 94–4169; USGS: Reston, VA, USA, 1994.

40. Nhan, D.D.; Lieu, D.B.; Minh, D.A.; Anh, V.T. Isotopic Compositions of Precipitation Over Red River's Delta Region (Vietnam): Data of the GNIP Hanoi. 2013. Available online: www.iaea/gnip (accessed on 15 March 2019).

41. Babu, M.M.; Viswanadh, G.K.; Rao, S.V. Assessment of saltwater intrusion along coastal areas of Nellore District, A.P. *Int. J. Sci. Eng. Res.* **2013**, *4*, 173–178.

42. Stiefel, J.M.; Melesse, A.M.; McClain, M.E.; René, M.P.; Anderson, E.P.; Chauhan, N.K. Effects of rainwater-harvesting induced artificial recharge on the groundwater of wells in Rajasthan, India. *Hydrogeol. J.* **2012**, *17*, 2061–2073. [CrossRef]

43. Nguyen, V.H. Investigation into the Rainwater Recharge to the Holocene Aquifer in Hanoi Area by Using Isotopicand Related Techniques. Master's Thesis, Hanoi University of Mining and Geology, Hanoi, Vietnam, 11 August 2009. (In Vietnamese)

44. Yurtsever, Y.; Payne, B.R. Application of environmental isotopes to groundwater investigations in Qatar. *Isot. Hydrol.* **1979**, *2*, 465–490.

45. Larsen, F.; Long, V.T.; Hoan, H.V.; Luu, T.T.; Christiansen, A.V.; Nhan, P.Q. Groundwater salinity influenced by Holocene seawater trapped in incised valleys in the Red River Delta plain. *Nat. Geosci.* **2017**, *10*, 376–381. [CrossRef]

46. Doan, V.C. *Investigation to Propose Criteria and Zones for Sustainable Exploitation and Protection of Groundwater Resources in the Red River's (Bac Bo) and Mekong River's (Nam Bo) Deltas*; Final Report to the Ministry of Science and Technology of Vietnam) for a Research Program; Ministry of Sci. & Technol. of Vietnam: Hanoi, Vietnam, 2015; 277p. (In Vietnamese)

47. Lindenmaier, F.; Bahls, R.; Wagner, F. *Assessment of Groundwater Resources in Nam Dinh Province*; Final Technical Report, Part B: Three Dimensional Structural and Numerical Modelling; Ministry of Natural Resources and Environment of Vietnam: Hanoi, Vietnam, 2011; 131p.

48. Hoang, H.T.; Bäumle, R. Complex hydrochemical characteristics of the Middle-Upper Pleistocene aquifer in Soc Trang province, Southern Vietnam. *Environ. Geochem. Health* **2019**, *41*, 325–341. [CrossRef] [PubMed]

Article

Moisture Sources for Precipitation and Hydrograph Components of the Sutri Dhaka Glacier Basin, Western Himalayas

Ajit T. Singh *, Waliur Rahaman, Parmanand Sharma, C. M. Laluraj, Lavkush K. Patel, Bhanu Pratap, Vinay Kumar Gaddam and Meloth Thamban

ESSO-National Centre for Polar and Ocean Research, Headland Sada, Vasco-da-Gama, Goa 403804, India; waliur@ncaor.gov.in (W.R.); pnsharma@ncaor.gov.in (P.S.); lalucm@ncaor.gov.in (C.M.L.); lavkushpatel@ncaor.gov.in (L.K.P.); bhanu@ncaor.gov.in (B.P.); gaddam_vinay@ymail.com (V.K.G.); meloth@ncaor.gov.in (M.T.)
* Correspondence: ajit.t.singh@gmail.com

Received: 26 August 2019; Accepted: 21 October 2019; Published: 26 October 2019

Abstract: Himalayan glaciers are the major source of fresh water supply to the Himalayan Rivers, which support the livelihoods of more than a billion people living in the downstream region. However, in the face of recent climate change, these glaciers might be vulnerable, and thereby become a serious threat to the future fresh water reserve. Therefore, special attention is required in terms of understanding moisture sources for precipitation over the Himalayan glaciers and the hydrograph components of streams and rivers flowing from the glacierized region. We have carried out a systematic study in one of the benchmark glaciers, "Sutri Dhaka" of the Chandra Basin, in the western Himalayas, to understand its hydrograph components, based on stable water isotopes ($\delta^{18}O$ and δ^2H) and field-based ablation measurements. Further, to decipher moisture sources for precipitation and its variability in the study region, we have studied stable water isotopes in precipitation samples (rain and snow), and performed a back-trajectory analysis of the air parcel that brings moisture to this region. Our results show that the moisture source for precipitation over the study region is mainly derived from the Mediterranean regions (>70%) by Western Disturbances (WDs) during winter (October–May) and a minor contribution (<20%) from the Indian Summer Monsoon (ISM) during summer season (June–September). A three-component hydrograph separation based on $\delta^{18}O$ and d-excess provides estimates of ice (65 ± 14%), snowpack (15 ± 9%) and fresh snow (20 ± 5%) contributions, respectively. Our field-based specific ablation measurements show that ice and snow melt contributions are 80 ± 16% and 20 ± 4%, respectively. The differences in hydrograph component estimates are apparently due to an unaccounted snow contribution 'missing component' from the valley slopes in field-based ablation measurements, whereas the isotope-based hydrograph separation method accounts for all the components, and provides a basin integrated estimate. Therefore, we suggest that for similar types of basins where contributions of rainfall and groundwater are minimal, and glaciers are often inaccessible for frequent field measurements/observations, the stable isotope-based method could significantly add to our ability to decipher moisture sources and estimate hydrograph components.

Keywords: Sutri Dhaka; Chandra Basin; Western Himalaya; hydrograph separation; stable water isotope; specific ablation

1. Introduction

The Himalayan-Karakorum mountain range has the largest concentration of glaciers outside the polar regions, out of which ~9600 glaciers lie in the Indian Himalayas, covering an area of

~40,000 km^2 [1]. These glaciers are the perennial source of runoff to major river systems, such as the Ganga, Brahmaputra and the Indus. These perennial rivers support more than a billion people living in the downstream region for their livelihood e.g., drinking, irrigation, industrial and sanitation [2–5]. Among all the major Himalayan river basins, the Indus Basin has the largest (~22,000 km^2) glacier extent [3]. The Ganga and the Brahmaputra River are primarily fed by monsoonal rain, whereas the Indus River receives the highest amount of water from snow and glacier melts [3,6]; the total glacier melt contribution to the Ganga, Brahmaputra and the Indus River estimated using snowmelt runoff model (SRM) are ~10%, ~12% and ~40%, respectively [3]. Previous studies have suggested that ~70% of Himalayan glaciers are receding at a faster rate, which has resulted in net loss of glacier volume [2,7,8]. As the Indus Basin receives its maximum runoff generated from the snow/ice-melt, it may face the most adverse effect of rising global temperatures [9], resulting in the initial rise in discharge, followed by the scarcity of freshwater supply leading to socio-economic instability in the downstream region [2,3]. Thus, considering a large number of Himalayan glaciers and their complex behavior and dynamics, it is imperative to have more studies and observations to improve our current knowledge and to address the pertinent questions related to moisture sources (rainfall/snowfall), snow/ice-melt contribution and their spatio-temporal variability.

Stable water isotope ratios of oxygen (δ^{18}O) and hydrogen (δ^2H), along with second order parameter, deuterium excess $\left(d - excess = \delta^2H - 8\delta^{18}O\right)$, have been widely used to trace moisture sources for precipitation, identify mixing water from various sources and to quantify their relative contributions [10,11]. Several isotope-based studies have been conducted in the Himalayan and polar regions to trace moisture sources and estimate the hydrograph components [12–23]. Moisture sources for precipitation over the central and eastern Himalayas are primarily derived from the Indian Summer Monsoon (ISM) during June-September, while moisture sources to the western Himalayas are predominantly derived from the Mediterranean region due to Western Disturbances (WDs) during winter (October-May) [6,24]. A previous study based on stable water isotopes shows that the WDs contribute the maximum (>70%) to the total annual precipitation in the Kashmir valley (western Himalayas), which is more than the Indian Summer Monsoon (ISM) (<30%) [15]. In contrast, a study in the Parbati Basin, western Himalayas, shows that WDs contribute a maximum up to 30% to the annual precipitation [17]. These reports clearly indicate large spatial variability in annual precipitation over the western Himalayas, particularly during the WDs.

Several studies have been conducted to estimate the contribution of snow and glacier melts in the Himalayan regions. A model-based water balance approach shows that the snow and glacier melt contribution to the Beas River at Pandoh Dam, western Himalayas, contributes ~35% to the annual flow [25]. On the contrary, another study based on a stable isotopes study suggests that the snow/glacier melt contribution to the Beas River, Western Himalaya, is up to ~50% [12]. A similar study in the Parbati River, a major tributary of the Beas River, has reported that glacier melt contributes up to ~44% (±15%) [17]. A model-based water balance approach for other Himalayan rivers like the Sutlej, Ganga and the Chenab, estimates the annual snow and glacier melt contributions up to ~60% (at Bhakra Dam), ~28% (at Devpryag), and ~49% (at Akhnoor), respectively [26,27]. Another isotope-based study reported up to ~32% contribution of snow and glacier melts to the Ganga River [28]. Such diverging results of glacier melt contributions for the Himalayan rivers could be due to the differences in the methods employed, sampling strategies, such as sampling frequency and locations (distance from glacier), uncertainty in constraining the glacier and snow melts end members, differences in defining terminologies such as glacier, snow and ice-melts, and local influences due to the diverse topography and variable climate regimes. Further, underlying assumptions involved in various methods have not been tested in these basins, and this therefore has resulted in large uncertainty and differences in the estimates [29,30].

It should be noted that the relative contribution of snow and ice-melts can be better constrained at the head-ward region of the glaciated catchment, since the contribution of precipitation and subsurface water increases substantially in the downstream region. A recent isotope-based study near the snout of the Gangotri Glacier (Upper Ganga Basin) have reported snow, glacier melts and direct runoff contributions of about 59.6, 36.8 and 3.6%, respectively [31]. Another study based on stable water isotope (δ^{18}O) and Electrical Conductivity (EC) has estimated contributions of supraglacial melt (~65%) and subglacial melt (~35%) from the Chhota Shigri Glacier [32]. Furthermore, few attempts have been made to validate the isotope-based precipitation source and hydrograph separation of snow/ice-melts. Therefore, to provide a baseline data for understanding moisture sources for precipitation, and to estimate the hydrograph components (snow and ice-melts) of stream flow during the peak ablation period, we have carried out a systematic study of a benchmark glacier (Sutri Dhaka) in the Chandra Basin, western Himalayas. We have employed two independent methods for the hydrograph component estimates, i.e. the stable water isotope method and field-based specific accumulation/ablation measurements. Similarly, we have also deciphered moisture sources for precipitation over the study region using stable water isotopes, which is further corroborated with the back-trajectory analysis of air parcels.

2. Study Area

The Sutri Dhaka Glacier catchment (32°22′49″ N and 77°33′05″ E) falls in the Chandra basin of the Western Himalayan region (Figure 1) [33,34]. This glacier is a clean type glacier (C-type) with less than 5% of total debris cover [34,35]. The total watershed area of the Sutri Dhaka Glacier is ~42 km^2, of which the glacier occupies an area of ~20 km^2, covering approximately 50% of the total watershed. The glacier elevation ranges from ~4500 m a.s.l. near the snout to an elevation of ~6000 m a.s.l at the bergschrund, with a mean length of about 11 km (Figure 1) [34,35]. A meltwater stream flows from the snout of the Sutri Dhaka Glacier in a south-east direction for ~3 km downstream and confluences with the Chandra River in the downstream region. A recent study based on remote sensing has reported that the Sutri Dhaka Glacier has shown a retreating trend from 1962 to 2013, with an annual retreat rate of 11.4 ± 0.7 m a^{-1} [35].

The climate in this region is dominated by long winters (November–March), followed by spring which lasts until the end of May [24]. The summer season starts at the end of May and lasts until the end of September, while October and early November mark a short autumn period [24]. The predominant precipitation occurs during winters (>70%) compared to summer months (<30%) [36]. The elevated Pir-Panjal range acts as an orographic barrier for the monsoonal clouds to reach the upper region of the Chandra Basin, resulting in limited rainfall during summer. However, few summer precipitation events occur in the form of drizzle. The aridity in the region is also shown by the lack of vegetation [36,37]. Since the study region comes under the rain shadow zone with high altitudinal variations due to steep topography, the local aquifer at higher altitude does not get recharged sufficiently to act as a potential groundwater reservoir for discharge at later stages [37]. A recent study by the Central Ground Water Board (CGWB) in the region has reported that the groundwater yield in the upstream regions of the Chandra Basin is less than 5 liters per second, which is meager compared to the total amount of discharge generated by the snow and ice-melts in the region [37]. Therefore, a major source of freshwater to the downstream settlement is mainly supplied by combined contribution of snow and ice-melts runoff. Thus, considering uniqueness of the study area in terms of the limited contribution of summer precipitation and groundwater contribution to the total discharge, the estimation of hydrograph components involves less complexity.

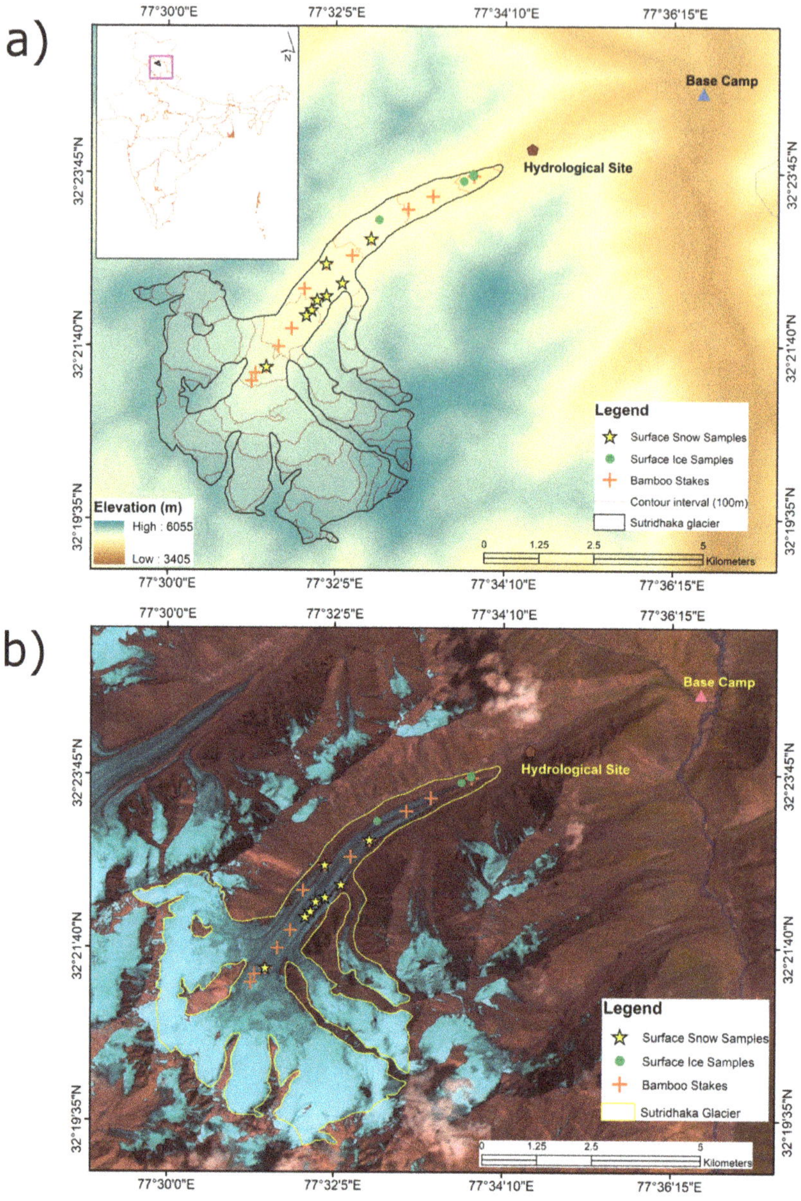

Figure 1. (**a**) Study area with Digital Elevation Model (DEM) derived from ASTER GDEM V2 along with sampling points; (**b**) Landsat 8 OLI image showing snow (light colored) and ice cover (dark colored) on the Sutri Dhaka glacier along with drainage (Image acquired on 20 August 2015).

3. Materials and Methods

An extensive field campaign was conducted in the Sutri Dhaka Glacier during summer–autumn (July–October) of the year 2015. Spatio temporal samples of glacier snowpack, glacier ice, fresh snow, rainwater and stream (a combination of snow and ice-melts from the glacier) were collected systematically during the field campaign (Table 1 and Figure 1).

An overview of the sample collection and in-situ field measurements of the glaciological and hydrological parameters on the Sutri Dhaka Glacier are shown in Figure 2, Table 1. The main tongue of the Sutri Dhaka Glacier is shown in Figure 2a. To measure the discharge of the Sutri Dhaka stream, a hydrological station was established nearly 200 m downstream of the present glacier snout (Figure 2b). Meltwater samples (n = 133) from the Sutri Dhaka stream were collected twice in a day at 10:00 hrs and 17:00 hrs (Figure 2c). Further, snowpack samples (n = 8) were collected from the glacier surface at various locations and certain intervals (Figure 2d), whereas fresh snowfall samples (n = 15) were collected from the base camp and discharge site (near snout) during a major snowfall event (20–24 September 2015). Glacial surface ice samples (n = 9) were strategically collected at an elevation ranging from 4550 m a.s.l to 4750 m a.s.l from the debris-covered as well as the debris-free part of the Sutri Dhaka Glacier. Precipitation samples were collected near the hydrological station (Figure 1b). However, we missed collecting samples of few rain events due to a lack of adequate logistic support. Therefore, for the present study, we have also used isotopic data of precipitation published during the same season (June to October 2015) for the Chhota Shigri Glacier, upper Chandra Basin [32]. Considering the proximity of these two glaciers, i.e., Chhota Shigri and Sutri Dhaka Glaciers, with a distance less than 15 km, we expect similar hydro-meteorological conditions. Additionally, we conducted sampling at a similar altitude for the present study; therefore, we assume minimum changes in the isotopic characteristics of precipitation in the Sutri Dhaka and the Chhota Shigri catchments. To avoid any evaporation and atmospheric exchange with collected samples, they were filled in 20 mL scintillation vials without any headspace and air bubbles, and sealed immediately.

Table 1. Detailed description of end member components and stream (mixed component) sampling at the Sutri Dhaka Glacier.

Sr. No	Sample Type	Sampling Time	No. of Samples (n)
1	Glacier snowpack	1 July 2015	5
		17 October 2015	3
2	Fresh Snow	21–24 September 2015	15
3	Glacier Ice	1 July 2015	9
4	Sutri Dhaka Stream	7 July 2015 to 9 October 2015	133
5	Rainwater at Sutri Dhaka	7 July 2015 to 9 October 2015	9

Discharge of the Sutri Dhaka stream was measured using the area-velocity method [38]. Wooden floats and Flow tracker (Son Tek Flowtracker, Son Tek, San Diego, CA, US) were used to determine the velocity of the stream. Excessive velocities, depth, boulder movement in the bed of stream and the floating drift of the instrument prohibited us from using the SonTek flow tracker during high flow. Therefore, our SonTek flow tracker measurements were only conducted during low flow conditions which showed a similar velocity reading to the float-based velocity measurement with an accuracy better than ± 10%. Depth profiles were measured using a metal gauge. To estimate daily discharge, daily gauge measurements were conducted at 10:00 Hrs (low flow) and 17:00 Hrs (high flow), and a level versus discharge relationship was established. The mean of high and low flow was considered as a daily mean discharge [33]. Since the bed topography of high mountainous streams are unstable, and the surface velocity of the stream is higher than bed velocity, the obtained discharge value was multiplied with a factor of 0.84 to estimate the discharge [38]. The meteorological parameters, i.e., temperature and relative humidity (RH%) were measured continuously using a temperature sensor (RHT-20, Extech, Waltham, MA, US) installed at the Sutri Dhaka Glacier (~5000 m a.s.l). Due to a technical problem with our rain gauge instrument installed at our study site, we could not measure precipitation during the field campaign.

Therefore, we have used precipitation data measured at the base camp of an adjacent glacier; the Chhota Shigri Glacier, located ~15 km away from the study region using an Automatic Weather Station (AWS) with accuracy better than 1%.

Figure 2. Fieldwork on the Sutri Dhaka Glacier and catchment: (**a**) Downstream synoptic view of the Sutri Dhaka Glacier showing the accumulation and ablation zone; (**b**) hydrological observation site; (**c**) meltwater sampling; (**d**) snowpack sampling; (**e**) stakes coordination and measurements; (**f**) snow pit excavation and density measurements.

All collected snow, ice, rain and meltwater samples were analyzed for ^{2}H/^{1}H, and ^{18}O/^{16}O ratio using OA-ICOS laser absorption spectroscopy (LGR, Triple Isotope Water Analyzer (TIWA-45EP), Los Gatos Research (LGR) Process Automation, Mountain View, CA, USA) at the National Centre for Polar and Ocean Research, Goa, India. The Analyzer uses LGR's Off-axis ICOS technology, a fourth-generation cavity ring down spectroscopy (CRDS) technique [39], which employs an optical cavity to greatly enhance spectral absorption and enable us to achieve the fastest and highest precision measurements of δ^{18}O and δ^2H. The isotopic ratios are reported in the standard δ-notation with relative to VSMOW-SLAP [40] and expressed as

$$\delta(\permil) = \frac{\left(R_{sample} - R_{standard}\right)}{R_{standard}} \times 1000 \tag{1}$$

where R represents either the ^{18}O/^{16}O or ^{2}H/^{1}H ratio. The overall accuracy of δ^2H and δ^{18}O measurements are better than ± 0.18‰ and ± 0.07‰, respectively, based on the known value of a laboratory standard with respect to V-SMOW-SLAP with six injections per samples. To understand the source of the precipitation in the study region and a better identification of end-member for hydrograph separation, d-excess (d − excess = δ^2H − 8δ^{18}O) was calculated for all samples using δ^2H and δ^{18}O [41]. The relationship between δ^2H and δ^{18}O in precipitation was defined using the least square regression method [10].

Meltwater of the Sutri Dhaka stream is predominantly sourced from two components (snow and ice-melts) and therefore a simple two component hydrograph separation using single tracer (δ^{18}O) can be used to determine the snow and ice-melt contributions to the meltwater stream [42].

However, it was difficult to constrain the end member value of δ^{18}O$_{snow}$ as a single component because they undergo several stages of post-depositional processes such as evaporation, sublimation and repetitive melting-freezing cycles, which could cause large isotope fractionation. In earlier studies, based on modeling and field evidences, it was suggested that $\delta^{18}O$ values in a snowpack could be heavier up to 3 to 5‰ than the fresh snow due to preferential removal of lighter isotopes in melts resulting in large uncertainty in hydrograph separation [42,43]. Therefore, in the present study, fresh snow (isotopically depleted) and snowpack (isotopically enriched) were considered as two separate components of snow, covering the entire spectrum of snow contribution, and a three-component hydrograph separation was performed to estimate the contribution of fresh snow, snowpack and ice-melt to the Sutri Dhaka stream.

In the case of three-component hydrograph separations based on a geochemical and isotope mass balance approach, at least two tracers are required. Using $\delta^{18}O$ along with electrical conductivity (EC), silica (SiO_2) and chloride (Cl^-) are among the most common tracers, and are widely used for three-component hydrograph separation (Klauss and McDonnell, 2013). However, they have limitations in separating snow and ice-melt contributions to glacier stream since the EC, SiO_2 and other dissolved solutes may get enriched due to water-rock interaction. In order to circumvent this problem, several studies have suggested that $\delta^{18}O$ and d-excess can be successfully used to trace the contribution of hydrological components [44,45].

Thus, we have used $\delta^{18}O$ and d-excess as tracers in constraining the end members for a three-component hydrograph separation of the Sutri Dhaka stream. Since the contribution of rainwater to total discharge is insignificant, we have not considered it as a major hydrograph component in our calculations [32,36,37,46]. The equation for three-component hydrograph separation can be written as follows [44].

$$Q_{St} = Q_i + Q_o + Q_n \tag{2}$$

where Q_i, Q_o and Q_n are the contribution of ice-melts, old snow (snowpack), fresh snow to the Sutri Dhaka stream discharge (Q_{st}).

$$1 = f_i + f_o + f_n \tag{3}$$

$$\delta_{st} = \delta_i \cdot f_i + \delta_o \cdot f_o + \delta_n \cdot f_n \tag{4}$$

δ_{st}, δ_i, δ_o and δ_n and are $\delta^{18}O$ for the stream, ice, snowpack and fresh snow respectively.

$$d_s = d_i \cdot f_i + d_o \cdot f_o + d_n \cdot f_n \tag{5}$$

d_{st}, d_i, d_o and d_n are d-excess tracer for the stream, ice, snowpack and fresh snow respectively.

$$Q_i(\%) = \frac{(d_{st} - d_n)(\delta_o - \delta_n) - (d_o - d_n)(\delta_{st} - \delta_n)}{(d_i - d_n)(\delta_o - \delta_n) - (d_o - d_n)(\delta_i - \delta_n)} \times 100 \tag{6}$$

$$Q_o(\%) = \frac{(d_{st} - d_n)}{(d_o - d_n)} \times 100 - \frac{(d_i - d_n)}{(d_o - d_n)} \times Q_i \tag{7}$$

$$Q_n(\%) = 100 - Q_i - Q_o \tag{8}$$

The ice-melt, snowpack and fresh snow contributions to the total discharge were calculated using Equations (6)–(8) respectively.

The total snow and ice-melt contributions were also estimated using a field-based ablation measurement of total snow and ice during the study period. In order to measure snow and ice ablation of the Sutri Dhaka Glacier, a network of 12–15 ablation stakes of ~6–10 m deep was installed along the center line of glacier surface at different altitudes, following the standard protocols published elsewhere [35,47]. To measure net ablation during the subsequent ablation period (July–October 2015), stakes were installed at the end of the ablation season i.e., September 2014. The lengths of the exposed stakes were measured on a monthly basis from 5 July to 5 October 2015 for the summer ablation measurements (Figure 2e). Net ablation was estimated based on the ice cover loss at each point multiplied with the density of ice. Ice density was measured at nine different locations in the ablation zone. The average density of 870 ± 10 kg m^{-3} based on nine measurements at different locations was used in ice ablation estimates in terms of water equivalent. For the snowmelt contribution, we have measured winter snow accumulation (4 July 2015) and annual/residual snow accumulation (September–October 2015) by excavating four snow pits followed by density measurements at different altitudes of the glacier surface (Figure 2f). The measured thickness was linearly extrapolated to the higher reaches (5350–6050 m a.s.l.) to accommodate the total snow accumulation in the glacierized zone. Total snowmelt during the study period (July to October 2015) was then calculated by subtracting winter accumulation from residual accumulation. A simple transient snow line (TSL)-snow pit method was also used to measure snow ablation. The snow line was measured before the study was conducted (27 June 2015), and at the maximum snowline elevation (30 August 2015), using Landsat 8 OLI satellite imagery [48]. Since the mean accumulated snow was known from snow pit estimates, the total snow cover area ablation estimated using the TSL method was multiplied with respective snow pit volume (m w.e.) and the total snow volume ablation for the study period were estimated [48]. Due to inaccessibility to the site for sampling during the spring season, our study was limited to the peak summer period (July–October 2015), when maximum melting occurs.

Further, to understand moisture sources for precipitation over the study region throughout the year, we performed monthly back trajectory analysis of air parcels reaching the sites using the NOAA HYSPLIT model together with a reanalysis model output from the Global Data Assimilation System (GDAS) dataset [49]. Four days back trajectory analysis was performed for all months of the year 2015. All trajectories were initialized at 1500 m above the surface because most of the water vapor in the atmosphere travel within 0–2 km above ground level [49]. Subsequently, trajectories obtained for each day for the respective months were clustered using Trajstat to obtain the mean monthly trajectories [50]. Similarly, four days back-trajectory analysis to capture major precipitation events was plotted using the Global Data Assimilation System (GDAS) dataset at different altitudes (100, 1500 and 2000 m AGL) and also using HYSPLIT online simulation developed by Air Resources Laboratory, NOAA [51,52].

4. Results and Discussion

4.1. Hydro-Meteorological Characteristics of the Sutri Dhaka Stream

Meteorological parameters play a significant role in controlling the glacier melt dynamics [3]. Temporal variations in air temperature, relative humidity, precipitation and discharge of the Sutri Dhaka Glacier are shown in Figure 3 and data provided in supplementary excel sheet. The discharge of the Sutri Dhaka stream during the study period varied between 0.2 m^3 s^{-1} to 20 m^3 s^{-1} with a mean of 8.9 m^3 s^{-1} (Figure 3a). The discharge during the study period increased from early July with rising temperatures and reached its peak by the end of July. The daily mean temperature during the study period ranged from -15.3 to 16.1 °C with a mean of 9.1 °C (Figure 3b). Similarly, daily mean RH varied from 34.8% to 99.9% with a mean of 59.8% (Figure 3c). The highest daily mean discharge (20 m^3 s^{-1}) and temperature (16.2 °C) were observed on 15 July 2015. A gradual declining trend in temperature and discharge were observed in the months of August, September and October. A significant correlation ($R^2 = 0.83$, n = 63; $p < 0.05$) was observed between the daily mean discharge and temperature (Figure 3), suggesting a dominant control of temperature on discharge. This relationship implies a ~9% increase in daily mean discharge per degree rise in daily mean air temperature. This finding confirms that as the temperature in the Himalayan region increases, it would lead to an initial rise in discharge due to an increase in glacier melt, followed by a drop in runoff and reduction in the glacierized area [53]. A total of ~110 mm of precipitation was recorded during the study period, of which two major precipitation events occurred on 12 July (20 mm) and on 23 September (52 mm) in the form of snow which account for ~66% of total precipitation during the study period (Figure 3d).

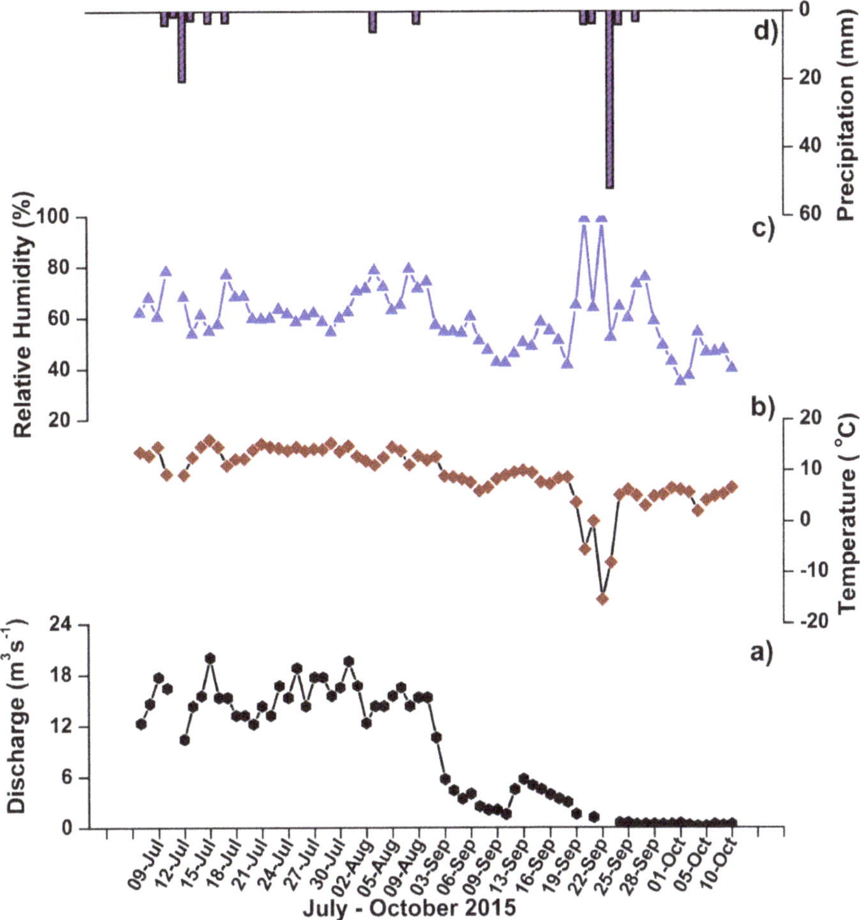

Figure 3. Daily distribution of the observed (**a**) Discharge; (**b**) Daily average temperature; (**c**) daily mean of Relative Humidity (RH%) (**d**) Daily mean precipitation measured during the study period.

4.2. Stable Isotope Characteristics and Its Relationship with Discharge

Details of samples and their isotopic characteristics are mentioned in Table 2 and data provided in supplementary excel sheet. $\delta^{18}O$ and δ^2H measured in rainwater samples varied from −13.9‰ to −5.4‰ and −107.2‰ to −32.5‰ with a mean of −11.2‰ and −81.6‰, respectively. Similarly, $\delta^{18}O$ in rainwater samples collected from the Chhota Shigri Glacier showed a median value of −11.2‰ [32]. The fresh snow samples collected at the base camp and hydrological station during the snowfall events, 20–24 September 2015, showed depleted $\delta^{18}O$ values compared to that of the snowpack. Enrichment of $\delta^{18}O$ in snowpack could be the result of isotopic fractionation between the melts and the snowpack. As isotopically-depleted snow starts melting, it leads to the removal of depleted meltwater that results in a heavier residual snowpack [30]. Glacier ice samples showed a narrower range compared to snow samples. $\delta^{18}O$ and δ^2H in ice samples ranged from −15.7‰ to −11.7‰ (mean −13.6 ± 1.2‰) and −106.3‰ to −73.2‰ (mean −91.2 ± 10.4‰) respectively. The $\delta^{18}O$ and δ^2H values in stream draining from the Sutri Dhaka Glacier varied from −15.7‰ to −13.3‰ (mean −14.4 ± 0.5‰) and −108.9‰ to −91.4‰ (mean −98.5 ± 3.8‰), respectively.

Table 2. δ^{18}O, δ^2H and d-excess values of rain, fresh snow, snowpack, glacier ice and meltwater of the Sutri Dhaka Glacier catchment.

Parameter	Rain (n = 9)		Fresh Snow (n = 15)		Snow Pack (n = 8)		Ice (n = 9)		Sutri Dhaka Stream (n = 133)	
	Mean	1 SD	Mean	1 SD	Mean	1 SD	Mean	1 SD	Mean	1 SD
δ^{18}O (‰)	−11.2	3.2	−20.3	0.2	−10.1	0.7	−13.6	1.25	−14.4	0.5
δ^2H (‰)	−81.6	26.5	−145.8	1.4	−67.7	5.5	−91.2	10.4	−98.5	3.9
d-excess (‰)	8.1	5.4	17.1	0.7	13.2	2.5	18	1.4	17.1	0.9

The mean d-excess values of snow, ice, meltwater and rainwater are provided in Table 2. The deuterium excess (d-excess) values for all samples range from 0.47‰ to 20.2‰ with a mean of 16.5 ± 2.7‰. Lower d-excess values were observed in rain events, whereas higher values were observed in snowfall events and the snowpack. A similar trend was also reported in precipitation events collected from the Chhota Shigri Glacier during the same period [32]. Glacier ice samples showed the highest d-excess values among all followed by the Sutri Dhaka stream. The higher d-excess (>12‰) of fresh snow, old snow, ice and Sutri Dhaka stream suggest Western Disturbance (WD) as their common moisture source, derived from the Mediterranean regions [19].

4.3. Local Meteoric Water Line (LMWL), d-Excess and Moisture Sources for Precipitation

In an earlier study, a Local Meteoric Water Line (LMWL) plotted based on precipitation samples collected from the Chhota Shigri shows a slope of 7.9 and an intercept of 21.4 [32] (Figure 4). This slope is similar to the Global Meteoric Water Line (GMWL) within their uncertainty (δ^2H $= 8 * \delta^{18}$O $+ 10$), suggesting their marine origin, while a higher intercept indicates mixing of air moisture masses derived from different sources. The overall least square regression line or best fit line constructed based on precipitation event samples collected from the Sutri Dhaka Glacier shows that the slope (7.4 ± 0.4) is similar to the Chhota Shigri Glacier with lower intercept (2.0 ± 5.4) (Figure 4). The lower intercept of the Sutri Dhaka Glacier could be because of a smaller number of representative samples for the entire season. Few precipitation events at the Sutri Dhaka Glacier fall close to GMWL, suggesting a contribution of ISM and the effect of sub-cloud evaporation during the precipitation events (Figure 4).

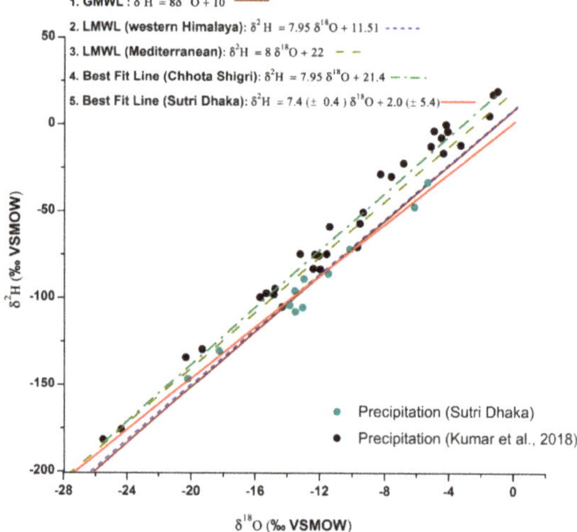

Figure 4. The regression lines in the cross plot (δ^2H vs. δ^{18}O) for precipitation samples from the Sutri Dhaka Glacier and Chhota Shigri Glacier are compared with Local Meteoric Water Line (LMWL) of the Western Himalaya, Mediterranean region and Global Meteoric Water Line (GMWL) (Kumar et al., 2018).

δ^2H vs. δ^{18}O plot for fresh snow, snowpack, glacier ice, rain is shown in Figure 5, and the slope and intercept of each regression line are also provided in Table 3. The ice samples collected from the Sutri Dhaka Glacier show a similar slope (8.1 ± 0.4) and intercept (20.3 ± 5.6) similar to the Chhota Shigri and the Mediterranean region. Lower slope (6.5 ± 1.2) and intercept (−1.2 ± 12.3) were reported for the snowpack, indicating the effect of non-equilibrium fractionation leading to isotopic enrichment of the snowpack due to a preferential removal of lighter isotopes in melts during sublimation processes [13,19]. Unlike snowpack, fresh snow samples showed slightly higher slope (7.3 ± 0.2) and intercept of (4.1 ± 5.1), while rainwater showed a slope of (7.7 ± 0.6) and intercept of (6.1 ± 7.1). The Sutri Dhaka stream showed a slope of (7.1 ± 0.2) with an intercept of (5 ± 2.7). The slopes of the regression lines of all components (snow, ice, rain, stream) are similar within their uncertainty, which confirms that they have common moisture sources [20]. However, the intercepts of all components are highly variable, suggesting the effect of secondary isotope fractionation processes during precipitation events and melting processes. The slopes and intercepts of LMWL in the Himalayan region varies significantly, indicating variable sources of precipitation and environmental conditions, such as temperature of condensation, local moisture recycling and amount of sub-cloud evaporation during precipitation (Table 3) [10,54–56].

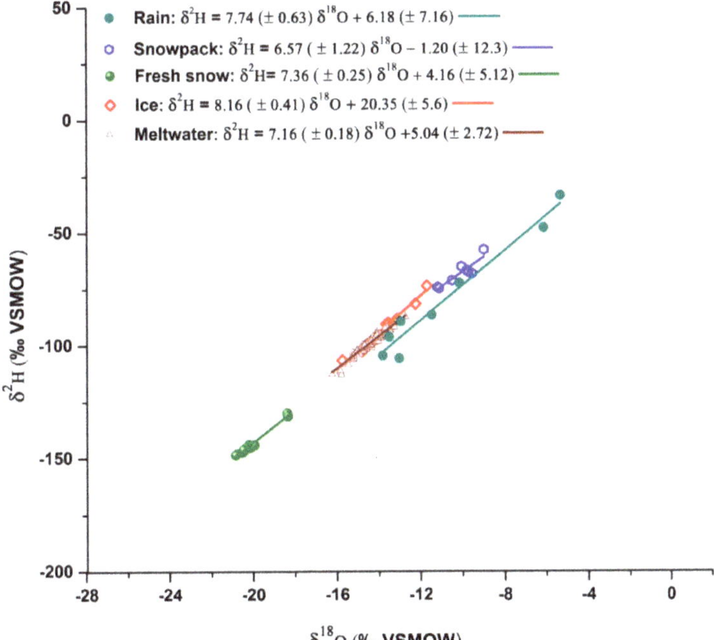

Figure 5. The stable isotope regression plot (δ^2H vs. δ^{18}O) for Rain, Fresh snow, Snowpack, Ice and daily Sutri Dhaka Glacier stream.

Table 3. Compilation of local meteoric water lines (LMWL) of various studies carried out in the Himalayan regions for Rain (R), Fresh Snow (SF), Snowpack/firn (SP), Glacier ice (GI) and meltwater stream or River (SR).

Glacier/Region	Latitude	Longitude	Altitude (m)	LMWL	R^2	n	Reference
Sutri Dhaka Glacier	32°22′49″ N	77°33′05″ E	4500–6200	$\delta^2H = 6.5\ (\pm1.2) * \delta^{18}O - 1.2\ (\pm12.3)$ (SP)	0.99	8	Present Study
				$\delta^2H = 7.3\ (\pm0.2) * \delta^{18}O + 4.1\ (\pm5.1)$ (SF)	0.98	17	
				$\delta^2H = 8.1\ (\pm0.4) * \delta^{18}O + 20.3\ (\pm5.6)$ (GI)	0.97	11	
				$\delta^2H = 7.1\ (\pm0.2) * \delta^{18}O + 5.04\ (\pm2.7)$ (SR)	0.94	65	
				$\delta^2H = 7.7\ (\pm0.6) * \delta^{18}O + 6.18\ (\pm7.1)$ (R)	0.95	9	
Chota Shigri Glacier	32°16′48″ N	77°34′ 48″ E	4050–6263	$\delta^2H = 7.8 * \delta^{18}O + 25$ (SP)	0.99	10	[32]
				$\delta^2H = 6.3 * \delta^{18}O + 3.6$ (GI)	0.76	15	
				$\delta^2H = 7.9 * \delta^{18}O + 21.4$ (R)			
Chorabari Glacier	30°46′20.58″ N	79°02′59.381″ E	4400–6200	$\delta^2H = 8.1 * \delta^{18}O + 24.1$(SF)	0.9	45	[16]
				$\delta^2H = 7.7 * \delta^{18}O + 21.2$ (GI)	1	13	
				$\delta^2H = 6.51 * \delta^{18}O - 0.0$ (SR)	0.8	116	
				$\delta^2H = 7.98 * \delta^{18}O + 16.8$ (R)	0.98	35	
QS and Glacier no.12, T.P	39°26.4″ N	96°32.5″ E	4260–5481	$\delta^2H = 8.2* \delta^{18}O + 21.68$ (SP)	0.95		[57]
				$\delta^2H = 7.7 * \delta^{18}O + 15.7$ (GI)	0.83		
				$\delta^2H = 7.8 * \delta^{18}O + 16.8$ (R)	0.95		
Kashmir Drass and Ladakh Zanskar	32°50′–34°18′ N	74°45′–78°20′ E	3250–4345	$\delta^2H = 8.2 * \delta^{18}O + 23.8$ (SR)			[19]
				$\delta^2H = 6.6 * \delta^{18}O - 1$ (SR)			
				$\delta^2H = 9.5 * \delta^{18}O + 38.7$ (GI)			
Jammu and Kashmir	33°20′–34°15′ N	74°30′–75°35′ E	1592–3248	$\delta^2H = 7.6* \delta^{18}O + 11.8$ (R)			[13,14,58]
				$\delta^2H = 7.6 * \delta^{18}O + 15$ (SP)	0.95	39	
				$\delta^2H = 6.7 * \delta^{18}O + 8.1$ (SR)	0.87	155	
Nam Co Basin, T.P	30°39″ N	90°38″ E	4730	$\delta^2H = 8.3 * \delta^{18}O + 7.8$ (SP/GI)	0.98		[59]
				$\delta^2H = 7.6* \delta^{18}O - 2.30$ (SR)	0.99		

The $\delta^{18}O$ values of the precipitation samples collected during the major precipitation events which occurred on 12 July and 20–24 September 2015 show depleted values and higher d-excess (Figure 6a–c).

A significant reduction in discharge after the major precipitation events coincides with temperature drop (Figure 6a). As the fresh snow after major precipitation events melts it also results in depleted $\delta^{18}O$ in stream water with higher d-excess (Figure 6a–c).

Several studies have shown that the higher d-excess (>12‰) in precipitation was generally related to the precipitation sourced from the high evaporation and low humidity regions, such as the Mediterranean, Caspian and Black seas through the western disturbances [19,58]. Monthly back-trajectory analysis of air parcels also reveals that the moisture parcel is primarily derived from the Mediterranean Sea and the Persian Gulf during the winter months due to WDs. Similarly, the moisture sources during the summer months are supplied by WDs with a minor contribution by ISM (Figure 7a). Few studies have also been carried out to decipher moisture sources for precipitation over other parts of the Himalayan regions using d-excess and back trajectory analysis [12,15,16,20,28,59,60]. However, other studies in the lower region of the western Himalayas have reported a stronger influence of ISM with high spatio-temporal variability in the sources for precipitation over the western Himalayan region [12,17,28]. Our event-based back trajectory analysis of the major summer precipitation events shows that the air parcel comprises both ISM and WDs origins (Figure 7b). An extensive study conducted in the western Himalayas found that the precipitation derived from WDs shows a much higher d-excess value (mean 18.9‰) compared to that of ISM (mean 9.4‰) [58]. Considering the higher d-excess (> 17‰) in our major precipitation event clearly indicates that they are mainly derived from the WDs. A two-component mixing model using ISM and WDs d-excess values as end members shows that more than 80% of the total precipitation is derived from WDs while the ISM contributes more than 20%. Since the d-excess values remain constant during phase change at the time of the rainout events, we can confidently deduce that the precipitation source to the major precipitation event is predominantly derived from the WDs. Similarly, a study conducted at a similar duration of our study at the Chhota Shigri Glacier in the Chandra Basin, western Himalayas, also reported higher d-excess values during heavy summer precipitation events, attributing to WDs being the precipitation source [32].

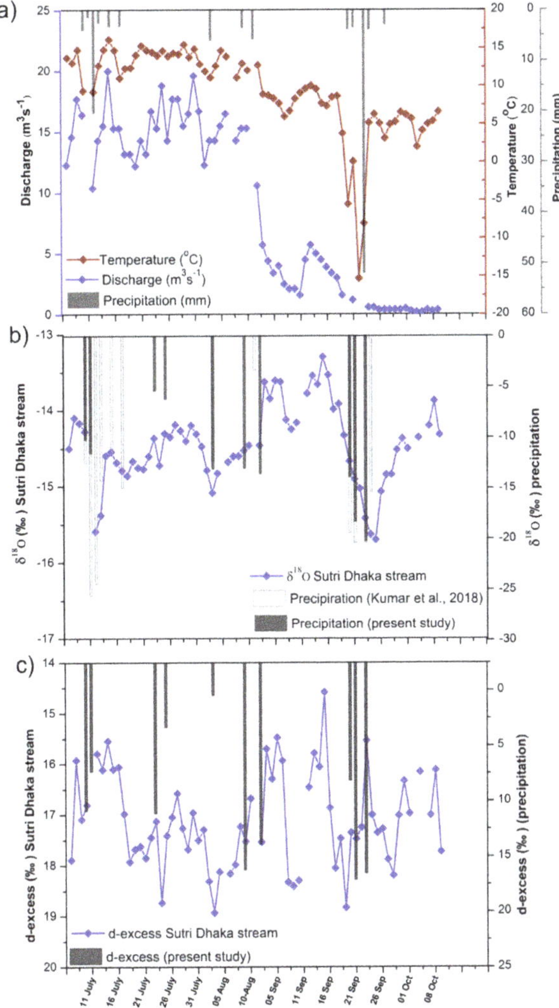

Figure 6. Time series of the measured parameters; (**a**) discharge and precipitation at the study region during the study period along with their associated (**b**) $\delta^{18}O$ ‰ and (**c**) d-excess of the stream water and precipitation at the Sutri Dhaka Glacier (present study), and the Chhota Shigri Glacier (Kumar et al., 2018) Upper Chandra Basin.

A positive correlation ($R^2 = 0.3$, n = 33; $p < 0.05$) was observed between discharge and $\delta^{18}O$ values of the stream water samples during July to mid-August, which was further improved during September and early October ($R^2 = 0.57$, n = 31; $p < 0.05$) (Figure 8). Variable correlations between discharge and $\delta^{18}O$ values suggest that the initial stream water was sourced from both snow as well as ice-melt waters with variable contributions and as summer progressed, snow contribution got reduced with an increasing contribution from ice-melts to stream discharge. However, it is noteworthy to observe a significant declining trend in discharge and $\delta^{18}O$ values after intense precipitation events on 12 July 2015 and 20–24 September 2015 (Figure 8). A significant drop in temperature, glacier melt and discharge indicates that the precipitation on the glacier occurred in the form of snow. Therefore, a major supply of meltwater contributed to the downstream region was supplied by the melting of isotopically-depleted fresh snow, which resulted in depleted meltwater.

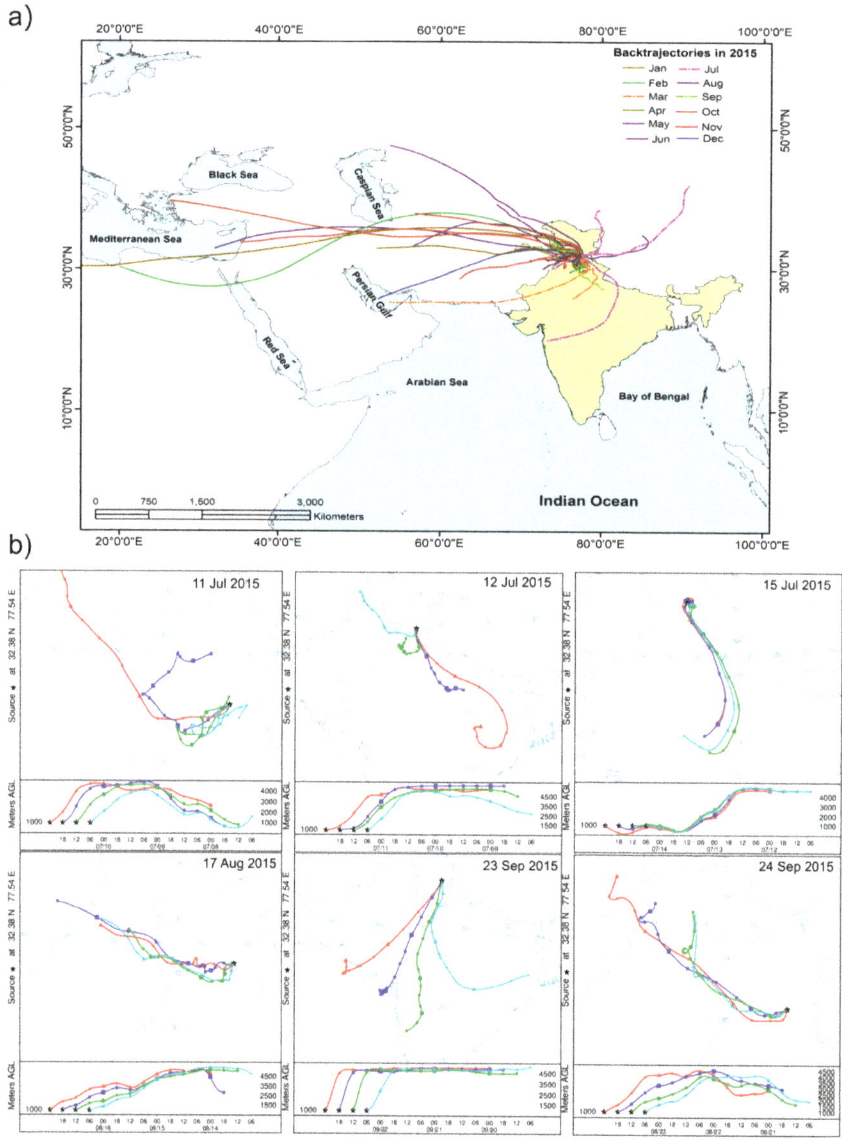

Figure 7 Back trajectory analysis at the study region (**a**) for entire year (January–December 2015) (**b**) for major precipitation events initiated at 1000, 1500 and 2000 m AGL for four days.

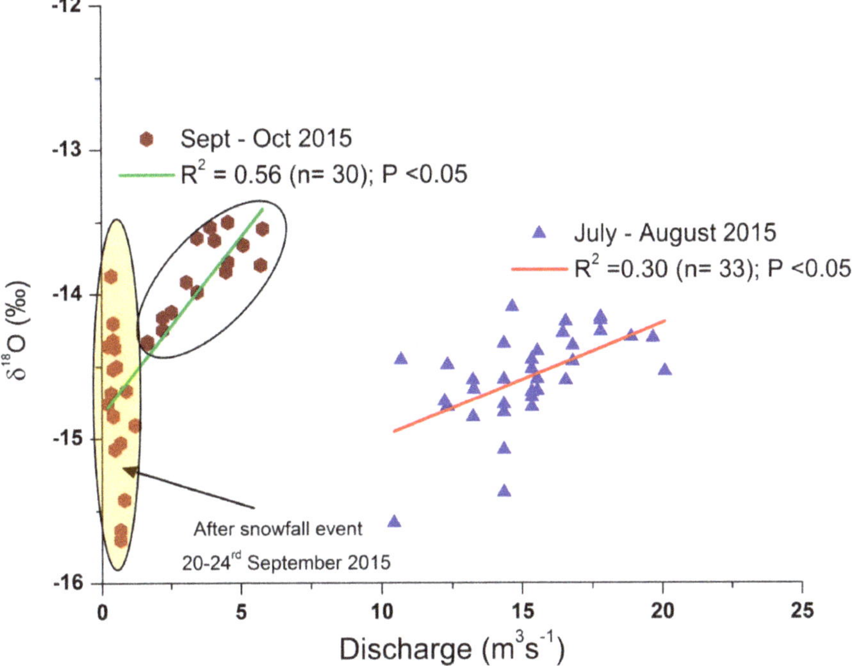

Figure 8. Relationships between $\delta^{18}O$ and stream discharge during the peak ablation period (July and August) and at the end of ablation season (September and October).

4.4. Snow and Ice-Melt Contribution to the Sutri Dhaka Stream

4.4.1. Hydrograph Separation

The hydrological process can be well understood using a mixing plot (d-excess vs. $\delta^{18}O$) [44,45]. Mixing plots in the present study clearly shows that the stable isotope signatures of the Sutri Dhaka stream are more close to the glacier ice-melt end member compared to snowmelt and rainwater (Figure 9). This indicates that the fresh snow on the glacier surface might undergo several stages of post-depositional processes, leading to large isotopic fractionation.

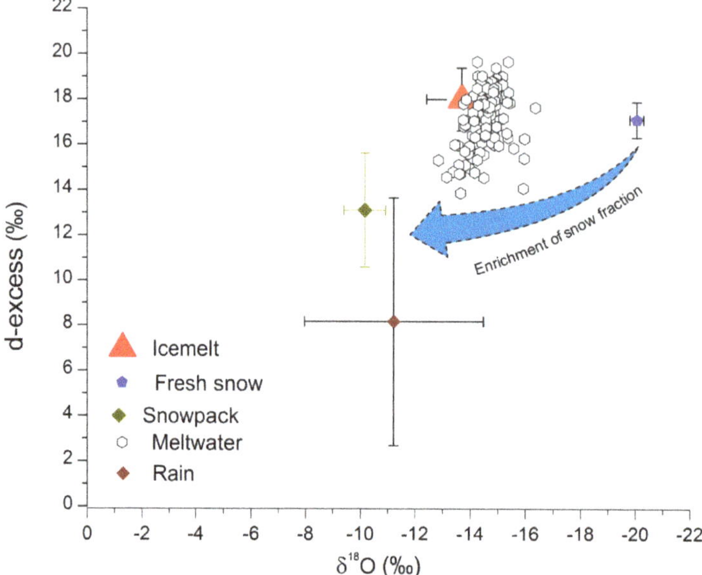

Figure 9. A mixing diagram showing d-excess vs. δ^{18}O of the mean value of snow, ice and rainwater and daily stream meltwater samples. The arrow indicates the evolution of the fresh snow due to the effect of secondary processes after the precipitation.

A three-component hydrograph separation of the Sutri Dhaka stream revealed that the contribution of ice-melt, snowpack and fresh snow are 65 ± 14, 15 ± 9 and 20 ± 5%, respectively (Figure 10). Results of hydrograph separation suggest that the ice-melt is the dominant contributor to the Sutri Dhaka stream water during July–October. However, we observed a significant declining trend in ice-melt and an increase in fresh snow, followed by a snowpack contribution to the Sutri Dhaka stream after the major snowfall events on 12 July and 20–24 September 2015. The overall contribution from the snowmelt suggests that they are mainly derived from the isotopically-depleted snow as well as the enriched snowpack. Despite a spike in fresh snow contribution followed by a snowpack, the ice-melt remained a dominant contributor throughout the study period. A significant decline in temperature has been observed after the major snowfall event (20–24 September 2015), which significantly reduces the surface melting. Therefore, the main source of ice-melt during September and October could be from the subglacial ice-melt due to pressure melting [32]. It has been observed that by the end of September the supraglacial melting reduced substantially, and a major contribution to the stream melt water were supplied by the melting of subglacial ice [61].

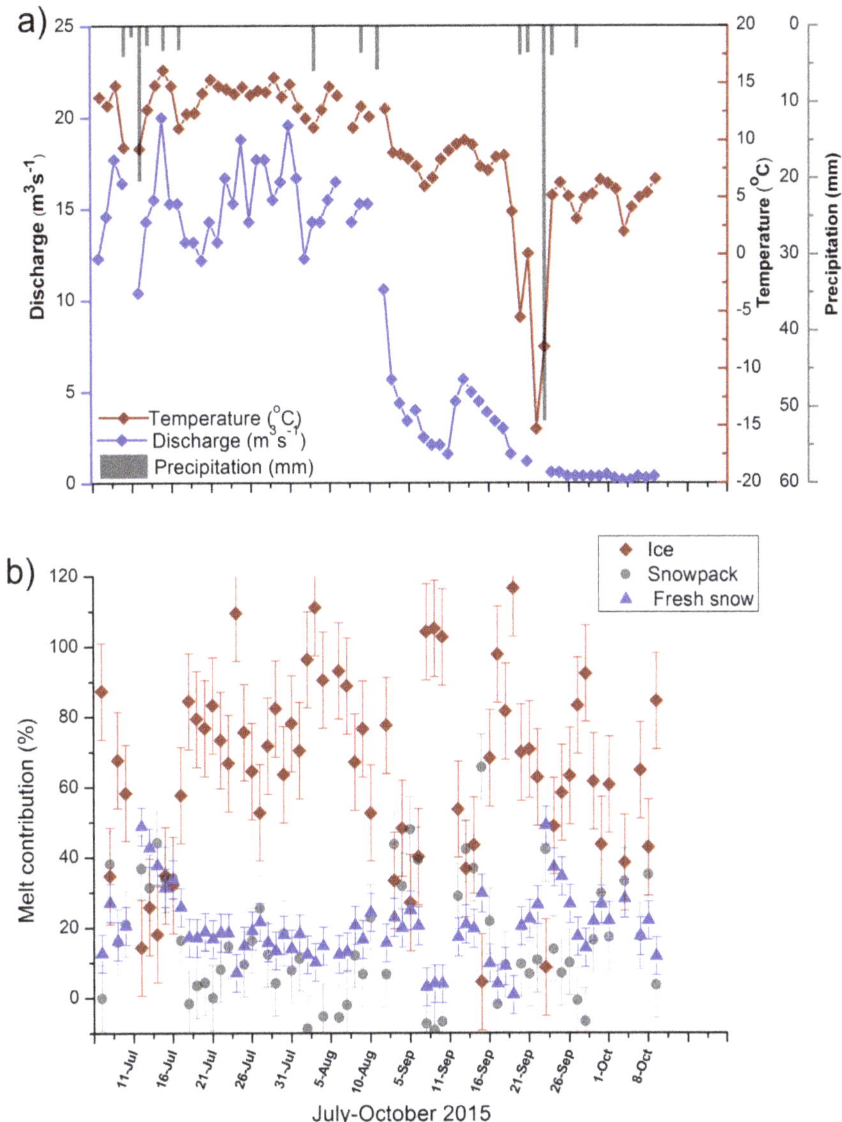

Figure 10. Hydro-meteorological characteristics of the Sutri Dhaka Glacier (**a**) Discharge is compared with temperature and precipitation; (**b**) contribution of ice-melt, snowpack and fresh snow to the Sutri Dhaka stream estimated based on three-component separation.

4.4.2. Specific Ablation of Snow and Ice

To validate our estimates of hydrograph separation based on stable isotope method, we have compared with the field estimates obtained from stake-based ablation and snow accumulation measurements. The estimated winter snow accumulation based on several snow-pit measurements on the glacierized area on 4 July 2015 varied from 0.24 to 1.17 m w.e (meter water equivalent) between the elevation ranges 4500 to 5300 m a.s.l. The total snow accumulation estimated for the glacierized area is $14.0 \pm 2.28 \times 10^6$ m^3 w.e.

As the summer progresses, the accumulated snow starts melting, and by the end of the ablation period most of the area (<5320 m a.s.l) becomes snow-free. At the end of the study period (28 September), estimation of the residual (annual) snow accumulation at ~5300 m a.s.l was $9.04 \pm 1.8 \times 10^6$ m³ w.e (Figure 11). The depletion of snow line over the glacier surface defines the zero balance area, known as the Equilibrium Line Altitude (ELA). Based on the measurements of stakes installed on the glacier, the ELA observed at the end of the ablation season was at 5320 m a.s.l. To estimate the total snowmelt from 4 July to 28 September (nearly three months), the annual snow accumulation was subtracted from the winter snow accumulation. The difference between the winter accumulation and residual accumulation (*Cw-Ca*) is $-5.0 \pm 1.0 \times 10^6$ m³ w.e., which accounts for snowmelt contribution to the stream runoff (Figure 11).

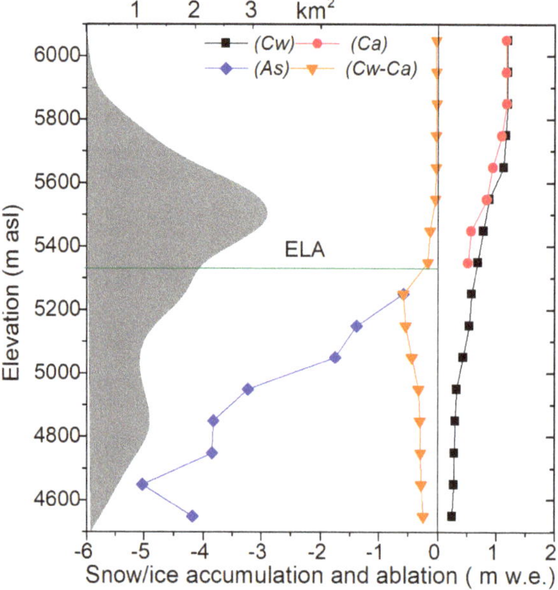

Figure 11. The observed winter snow accumulation (*Cw*), annual/residual snow accumulation (*Ca*), summer ice ablation (*As*) and summer snowmelt (*Cw-Ca*), and the area altitude distribution of Sutri Dhaka Glacier for the period of 5 July–28 September 2015.

A transient snowline on the glacier was estimated using Landsat 8 OLI images and demarcated using ASTER GDEM which was further validated with the ground control points. A transient snowline-snow pit method shows that the snowline position before the study commenced was at an altitude of 4511 (±46) m a.s.l (27 June 2015) (Supplementary Figure S1). As the melting progressed, the snowline reached up to 5601 (±46) m a.s.l (30 August 2015) at peak of ablation season with total snow area loss of 10.41 km² and snow volume loss of -5.8×10^6 m³ w.e (Figure 12). It is noteworthy to observe that the snow ablation estimate using the transient snowline-snow pit method agrees with our field-based snow ablation measurement within the range of uncertainty. As the snow cover melts, the ice becomes exposed, and starts melting with the increase in air temperature. The cumulative summer ice ablation (*As*) for the study period (5 July to 28 September 2015) between the elevation of 4500 to 5300 m a.s.l, yield a total glacier ice-melt of $-20 \pm 4 \times 10^6$ m³ w.e. Maximum ice ablation was observed at an altitude of 4600–4700 m a.s.l, which was reduced progressively towards higher altitude due to lower air temperature. The total cumulative melt contribution from snow and ice during this period provides an estimate of $-25 \pm 5 \times 10^6$ m³ w.e (Figure 11). Contributions of snow and ice-melt towards the production of meltwater discharge are approximately 20 ± 4% and 80 ± 16%, respectively.

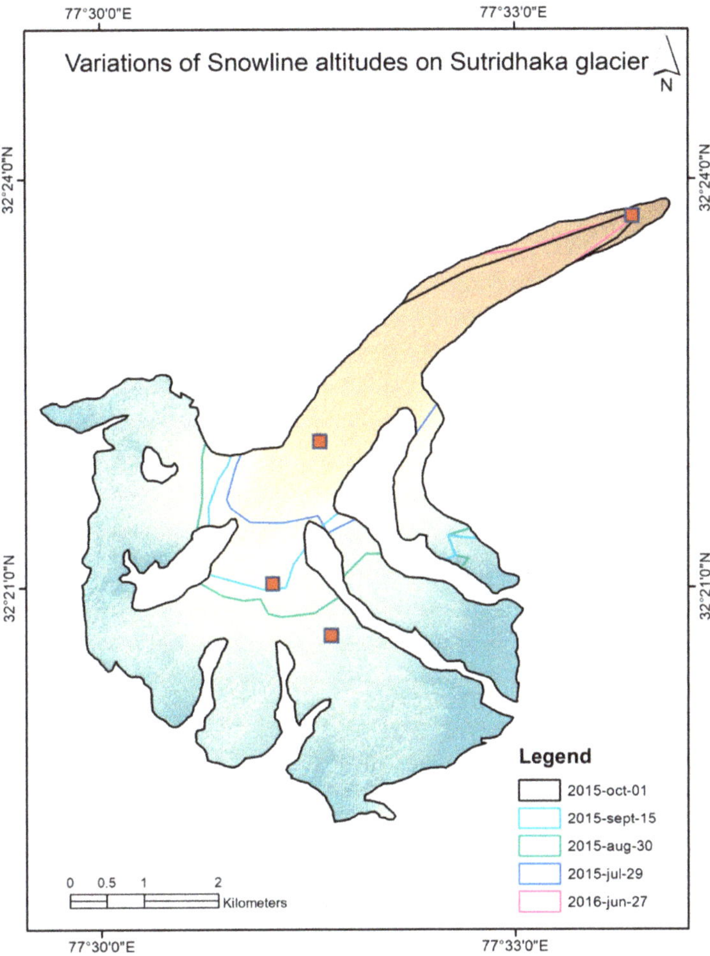

Figure 12. Transient snowline estimated using Landsat 8 OLI images of Sutri Dhaka Glacier and excavated snow pits (brown square) during the study period.

4.4.3. Uncertainty in Hydrograph Separation Estimates

Hydrograph separations and their quantitative assessments using the stable isotope method are often a challenging task due to a large spread on the $\delta^{18}O$ values of the hydrograph components. Stable water isotope compositions of snow and ice vary spatially and temporally, which contributes to uncertainties associated with the estimates of hydrograph components [62,63]. Three-component hydrograph estimates are critically dependent on the end-member values of $\delta^{18}O$ and d-excess and uncertainties associated with them. To better constrain the uncertainty associated with the estimates of hydrograph components based on $\delta^{18}O$ and d-excess, we used a Monte Carlo error propagation method [64,65]. This simulation method was used to solve the Equations (7)–(9) with 50,000 iterations for each stream samples and uncertainty, with a 68% confidence interval (CI) acquired for each mixing fraction. Based on Monte Carlo simulation, the uncertainty in ice-melt, snowpack and fresh snow estimates are 14, 9 and 5%, respectively. The overall uncertainty in each component is shown with error bars (Figure 10).

The accuracy of specific ablation using a network of bamboo stakes and snow pits cannot be evaluated strictly, as some of the random errors in this method are unknown. Uncertainty in this method is mainly derived from glacier area estimates, stake measurements and snow depth and density measurements [66,67]. Based on the field data, uncertainty in glaciological field measurements has been reported to be maximum up to ± 20% [35,65]. Therefore, in the present study, we have considered an error of ± 20% for specific ablation measurements.

Comparison of our hydrograph components estimates-based two independent methods i.e., the field-based ablation measurement and the stable isotope-based hydrograph method, the latter method provides ~15% higher estimates of the snowmelt component. This difference arises due to an unaccounted contribution from the deposited snow over the valley slope of the glacierized area in the field-based method. However, the isotope-based method provides an estimate of an integrated average of snow and ice-melt contributions from the entire basin to the downstream. Therefore, we suggest that for basins in the upper Indus Basin, where the contribution of rainfall and groundwater are minimal, a stable isotope method can be complimentary along with the field-based ablation measurements. Further, the glaciers which are inaccessible for the field measurements, the stable isotope method could add to our ability to evaluate snow /ice-melt contribution from high altitude Himalayan glaciers.

5. Conclusions

The present study provides insights on the moisture sources for the precipitation and hydrograph components of the Sutri Dhaka Glacier basin in the western Himalayas during the peak ablation period in the year 2015. Stable isotope fingerprinting of moisture sources together with back trajectory analysis indicate that the moisture to the study area is predominantly derived from the Mediterranean Sea regions through the Western Disturbances (WDs). However, the major sources for the precipitation during summer are supplied by the ISM as well as WDs. A combination of stable water isotope data and field ablation measurements provide information about the dominant sources of water contributing to the stream runoff of the Sutri Dhaka as well as the major moisture sources for precipitation over the study region. Three-component hydrograph separation of the Sutri Dhaka Glacier based on stable isotope methods shows a dominant contribution of ice-melt (65% ± 14) to the stream discharge, followed by snowpack (15% ± 9) and fresh snow (20 ± 5%). Despite the uncertainties associated with these estimates, the results of isotope hydrograph separation are overall consistent with that of stake-based field measurements; the contribution of ice-melt and snowmelt are 80 ± 16% and 20 ± 4%, respectively. However, the stable isotope method provides relatively more accurate estimates of hydrograph components compared to field-based ablation measurements, as it integrates over the entire catchments in the upstream of the sampling site, whereas the field-based method does not account for part of the snow component in the valley slope. Considering the limited information on the hydrograph components of the Himalayan glaciers, this study will enhance our current knowledge and understanding of hydrological processes in high altitude western Himalayan regions.

Supplementary Materials: A excel sheet containing all numerical data used for present study is available online at http://www.mdpi.com/2073-4441/11/11/2242/s1. Excel 1, Figure S1: Transient snowline estimated using Landsat 8 OLI images of Sutri Dhaka glacier during the study period.

Author Contributions: A.T.S., W.R., C.M.L., P.S. and M.T. defined the objectives of the study and the writing of the manuscript. A.T.S. collected field samples and analyzed them for the stable water isotope at Ice core laboratory. A.T.S., L.K.P. collected data in the field, and B.P. interpreted the infield ablation results. V.K.G. generated the study area map and its attribute table using satellite images in ArcGIS version 10.4. All authors contributed to the data interpretation and discussion of the manuscript.

Funding: This research was funded by the Ministry of Earth Sciences (MoES), Government of India, through the project 'Cryosphere and Climate' under the 'Polar Sciences and Cryosphere Research (PACER)' scheme.

Acknowledgments: We thank the Director ESSO—National Centre for Polar and Ocean Research for encouragement and Ministry of Earth Sciences for financial support through the project PACER Cryosphere and Climate. The authors also wish to thank BL Redkar and Ashish Painginkar for assistance with laboratory analysis. The first author would like to thank Rasik Ravindra for encouragement and Runa Antony, K. Mahalinganathan, Rahul Dey, and Sunil Oulkar for their valuable comments. This is NCPOR contribution J-40/2019-20.

Conflicts of Interest: The authors declare that they have no conflict of interest.

References

1. AGRIS. Available online: http://agris.fao.org (accessed on 16 July 2018).
2. Bolch, T.; Kulkarni, A.; Kaab, A.; Huggel, C.; Paul, F.; Cogley, J.G.; Frey, H.; Kargel, J.S.; Fujita, K.; Scheel, M.; et al. The State and Fate of Himalayan Glaciers. *Science* **2012**, *336*, 310–314. [CrossRef] [PubMed]
3. Immerzeel, W.W.; van Beek, L.P.H.; Bierkens, M.F.P. Climate Change Will Affect the Asian Water Towers. *Science* **2010**, *328*, 1382–1385. [CrossRef] [PubMed]
4. Kaser, G.; Grosshauser, M.; Marzeion, B. Contribution potential of glaciers to water availability in different climate regimes. *Proc. Natl. Acad. Sci. USA* **2010**, *107*, 20223–20227. [CrossRef] [PubMed]
5. Milner, A.M.; Khamis, K.; Battin, T.J.; Brittain, J.E.; Barrand, N.E.; Fureder, L.; Cauvy-Fraunie, S.; Gislason, G.M.; Jacobsen, D.; Hannah, D.M.; et al. Glacier shrinkage driving global changes in downstream systems. *Proc. Natl. Acad. Sci. USA* **2017**, *114*, 9770–9778. [CrossRef]
6. Bookhagen, B.; Burbank, D.W. Topography, relief, and TRMM-derived rainfall variations along the Himalaya (vol 33, art no L08405, 2006). *Geophys. Res. Lett.* **2006**, *33*. [CrossRef]
7. Bhambri, R.; Bolch, T. Glacier mapping: A review with special reference to the Indian Himalayas. *Prog. Phys. Geogr.* **2009**, *33*, 672–704. [CrossRef]
8. Casper, J.K. *Global Warming Cycles: Ice Ages and Glacial Retreat*; Infobase Publishing: New York City, NY, USA, 2010.
9. Parry, M.; Parry, M.L.; Canziani, O.; Palutikof, J.; Van der Linden, P.; Hanson, C. *Climate Change 2007-Impacts, Adaptation and Vulnerability: Working Group II Contribution to the Fourth Assessment Report of the IPCC*; Cambridge University Press: Cambridge, MA, USA, 2007; Volume 4.
10. Craig, H. Isotopic Variations in Meteoric Waters. *Science* **1961**, *133*, 1702–1703. [CrossRef]
11. Rahaman, W.; Thamban, M.; Laluraj, C.J.T.H. Twentieth-century sea ice variability in the Weddell Sea and its effect on moisture transport: Evidence from a coastal East Antarctic ice core record. *Holocene* **2016**, *26*, 338–349. [CrossRef]
12. Ahluwalia, R.S.; Rai, S.P.; Jain, S.K.; Kumar, B.; Dobhal, D.P. Assessment of snowmelt runoff modelling and isotope analysis: A case study from the western Himalaya, India. *Ann. Glaciol.* **2013**, *54*, 299–304. [CrossRef]
13. Jeelani, G.; Kumar, U.S.; Kumar, B. Variation of delta O-18 and delta D in precipitation and stream waters across the Kashmir Himalaya (India) to distinguish and estimate the seasonal sources of stream flow. *J. Hydrol.* **2013**, *481*, 157–165. [CrossRef]
14. Jeelani, G.; Shah, R.A.; Fryar, A.E.; Deshpande, R.D.; Mukherjee, A.; Perrin, J. Hydrological processes in glacierized high-altitude basins of the western Himalayas. *Hydrogeol. J.* **2018**, *26*, 615–628. [CrossRef]
15. Jeelani, G.; Shah, R.A.; Jacob, N.; Deshpande, R.D. Estimation of snow and glacier melt contribution to Liddar stream in a mountainous catchment, western Himalaya: An isotopic approach. *Isot. Environ. Health Stud.* **2017**, *53*, 18–35. [CrossRef] [PubMed]
16. Kumar, A.; Tiwari, S.K.; Verma, A.; Gupta, A.K. Tracing isotopic signatures (δD and δ18O) in precipitation and glacier melt over Chorabari Glacier–Hydroclimatic inferences for the Upper Ganga Basin (UGB), Garhwal Himalaya. *J. Hydrol. Reg. Stud.* **2018**, *15*, 68–89. [CrossRef]
17. Laskar, A.H.; Bhattacharya, S.K.; Rao, D.K.; Jani, R.A.; Gandhi, N. Seasonal variation in stable isotope compositions of waters from a Himalayan river: Estimation of glacier melt contribution. *Hydrol. Process.* **2018**, *32*, 3866–3880. [CrossRef]
18. Nijampurkar, V.N.; Rao, D.K. Accumulation and Flow-Rates of Ice On Chhota Shigri Glacier, Central Himalaya, Using Radio Active and Stable Isotopes. *J. Glaciol.* **1992**, *38*, 43–50. [CrossRef]
19. Rai, S.P.; Thayyen, R.J.; Purushothaman, P.; Kumar, B. Isotopic characteristics of cryospheric waters in parts of Western Himalayas, India. *Environ. Earth Sci.* **2016**, *75*, 9. [CrossRef]
20. Verma, A.; Kumar, A.; Gupta, A.K.; Tiwari, S.K.; Bhambri, R.; Naithani, S. Hydroclimatic significance of stable isotopes in precipitation from glaciers of Garhwal Himalaya, Upper Ganga Basin (UGB), India. *Hydrol. Process.* **2018**, *32*, 1874–1893. [CrossRef]
21. Shi, M.; Wang, S.; Argiriou, A.A.; Zhang, M.; Guo, R.; Jiao, R.; Kong, J.; Zhang, Y.; Qiu, X.; Zhou, S. Stable Isotope Composition in Surface Water in the Upper Yellow River in Northwest China. *Water* **2019**, *11*, 967. [CrossRef]

22. Kurita, N.; Sugimoto, A.; Fujii, Y.; Fukazawa, T.; Makarov, V.N.; Watanabe, O.; Ichiyanagi, K.; Numaguti, A.; Yoshida, N. Isotopic composition and origin of snow over Siberia. *J. Geophys. Res. Atmos.* **2005**, *110*. [CrossRef]

23. Suzuki, K.; Konohira, E.; Yamazaki, Y.; Kubota, J.; Ohata, T.; Vuglinsky, V. Transport of organic carbon from the Mogot Experimental Watershed in the southern mountainous taiga of eastern Siberia. *Hydrol. Res.* **2006**, *37*, 303–312. [CrossRef]

24. Owen, L.A.; Bailey, R.M.; Rhodes, E.J.; Mitchell, W.A.; Coxon, P. Style and timing of glaciation in the Lahul Himalaya, northern India: A framework for reconstructing late Quaternary palaeoclimatic change in the western Himalayas. *J. Quat. Sci.* **1997**, *12*, 83–109. [CrossRef]

25. Kumar, V.; Singh, P.; Singh, V. Snow and glacier melt contribution in the Beas River at Pandoh Dam, Himachal Pradesh, India. *Hydrol. Sci. J.* **2007**, *52*, 376–388. [CrossRef]

26. Singh, P.; Jain, S.K. Snow and glacier melt in the Satluj River at Bhakra Dam in the western Himalayan region. *Hydrol. Sci. J.* **2002**, *47*, 93–106. [CrossRef]

27. Singh, P.; Jain, S.K.; Kumar, N. Estimation of snow and glacier-melt contribution to the Chenab River, Western Himalaya. *Mt. Res. Dev.* **1997**, *17*, 49–56. [CrossRef]

28. Maurya, A.S.; Shah, M.; Deshpande, R.D.; Bhardwaj, R.M.; Prasad, A.; Gupta, S.K. Hydrograph separation and precipitation source identification using stable water isotopes and conductivity: River Ganga at Himalayan foothills. *Hydrol. Process.* **2011**, *25*, 1521–1530. [CrossRef]

29. Burns, D.A. Stormflow-hydrograph separation based on isotopes: The thrill is gone—What's next? *Hydrol. Process.* **2002**, *16*, 1515–1517. [CrossRef]

30. Unnikrishna, P.V.; McDonnell, J.J.; Kendall, C. Isotope variations in a Sierra Nevada snowpack and their relation to meltwater. *J. Hydrol.* **2002**, *260*, 38–57. [CrossRef]

31. Rai, S.P.; Singh, D.; Jacob, N.; Rawat, Y.S.; Arora, M.; Kumar, B. Identifying contribution of snowmelt and glacier melt to the Bhagirathi River (Upper Ganga) near snout of the Gangotri Glacier using environmental isotopes. *Catena* **2019**, *173*, 339–351. [CrossRef]

32. Kumar, N.; Ramanathan, A.; Keesari, T.; Chidambaram, S.; Ranjan, S.; Soheb, M.; Tranter, M. Tracer-based estimation of temporal variation of water sources: An insight from supra- and subglacial environments. *Hydrol. Sci. J.* **2018**, *63*, 1717–1732. [CrossRef]

33. Singh, A.T.; Laluraj, C.M.; Sharma, P.; Patel, L.K.; Thamban, M. Export fluxes of geochemical solutes in the meltwater stream of Sutri Dhaka Glacier, Chandra basin, Western Himalaya. *Environ. Monit. Assess.* **2017**, *189*. [CrossRef]

34. Pratap, B.; Sharma, P.; Patel, L.; Singh, A.T.; Gaddam, V.K.; Oulkar, S.; Thamban, M. Reconciling High Glacier Surface Melting in Summer with Air Temperature in the Semi-Arid Zone of Western Himalaya. *Water* **2019**, *11*, 1561. [CrossRef]

35. Sharma, P.; Patel, L.K.; Ravindra, R.; Singh, A.; Mahalinganathan, K.; Thamban, M. Role of debris cover to control specific ablation of adjoining Batal and Sutri Dhaka glaciers in Chandra Basin (Himachal Pradesh) during peak ablation season. *J. Earth Syst. Sci.* **2016**, *125*, 459–473. [CrossRef]

36. Azam, M.F.; Wagnon, P.; Vincent, C.; Ramanathan, A.; Favier, V.; Mandal, A.; Pottakkal, J.G. Processes governing the mass balance of Chhota Shigri Glacier (western Himalaya, India) assessed by point-scale surface energy balance measurements. *Cryosphere* **2014**, *8*, 2195–2217. [CrossRef]

37. Pradesh, I. Central Ground Water Board. Available online: http://cgwb.gov.in/District_Profile/HP/Lahul%20Spiti.pdf (accessed on 25 October 2019).

38. Srivastava, P.K.; Han, D.; Rico-Ramirez, M.A.; Islam, T. Sensitivity and uncertainty analysis of mesoscale model downscaled hydro-meteorological variables for discharge prediction. *Hydrol. Process.* **2014**, *28*, 4419–4432. [CrossRef]

39. Chakraborty, S.; Sinha, N.; Chattopadhyay, R.; Sengupta, S.; Mohan, P.M.; Datye, A. Atmospheric controls on the precipitation isotopes over the Andaman Islands, Bay of Bengal. *Sci. Rep.* **2016**, *6*. [CrossRef]

40. Coplen, T.; De Bièvre, P.; Krouse, H.; Vocke, R., Jr.; Gröning, M.; Rozanski, K. Ratios for light-element isotopes standardized for better interlaboratory comparison. *Eos Trans. Am. Geophys. Union* **1996**, *77*, 255. [CrossRef]

41. Dansgaard, W. Stable isotopes in precipitation. *Tellus* **1964**, *16*, 436–468. [CrossRef]

42. Taylor, S.; Feng, X.H.; Williams, M.; McNamara, J. How isotopic fractionation of snowmelt affects hydrograph separation. *Hydrol. Process.* **2002**, *16*, 3683–3690. [CrossRef]

43. He, Y.; Pang, H.; Theakstone, W.; Zhang, D.; Lu, A.; Song, B.; Yuan, L.; Ning, B. Spatial and temporal variations of oxygen isotopes in snowpacks and glacial runoff in different types of glacial area in western China. *Ann. Glaciol.* **2006**, *43*, 269–274. [CrossRef]

44. Gibson, J.J.; Price, J.S.; Aravena, R.; Fitzgerald, D.F.; Maloney, D. Runoff generation in a hypermaritime bog-forest upland. *Hydrol. Process.* **2000**, *14*, 2711–2730. [CrossRef]

45. Throckmorton, H.M.; Newman, B.D.; Heikoop, J.M.; Perkins, G.B.; Feng, X.; Graham, D.E.; O'Malley, D.; Vesselinov, V.V.; Young, J.; Wullschleger, S.D.; et al. Active layer hydrology in an arctic tundra ecosystem: Quantifying water sources and cycling using water stable isotopes. *Hydrol. Process.* **2016**, *30*, 4972–4986. [CrossRef]

46. Rao, S.; Rao, M.; Ramasastri, K.; Singh, R. A study of sedimentation in Chenab basin in western Himalayas. *Hydrol. Res.* **1997**, *28*, 201–216. [CrossRef]

47. Østrem, G.; Brugman, M. Glacier Mass Balance Measurement Techniques: A Manual for Field and Office Work. Available online: https://wgms.ch/downloads/Oestrem_Brugman_GlacierMassBalanceMeasurements_1991.pdf (accessed on 25 October 2019).

48. Mernild, S.H.; Pelto, M.; Malmros, J.K.; Yde, J.C.; Knudsen, N.T.; Hanna, E. Identification of snow ablation rate, ELA, AAR and net mass balance using transient snowline variations on two Arctic glaciers. *J. Glaciol.* **2013**, *59*, 649–659. [CrossRef]

49. Draxler, R.R.; Rolph, G.D. Hysplit (hybrid single-particle lagrangian integrated trajectory) model access via NOAA ARL ready website. *Silver Spring MD* **2011**, *8*, 26.

50. Stein, A.F.; Draxler, R.R.; Rolph, G.D.; Stunder, B.J.B.; Cohen, M.D.; Ngan, F. Noaa's Hysplit Atmospheric Transport And Dispersion Modeling System. *Bull. Am. Meteorol. Soc.* **2015**, *96*, 2059–2077. [CrossRef]

51. Wallace, J.M.; Hobbs, P.V. *Atmospheric Science: An Introductory Survey*; Elsevier: Amsterdam, The Netherlands, 2006; Volume 92.

52. Wang, Y.; Zhang, X.; Draxler, R.R. TrajStat: GIS-based software that uses various trajectory statistical analysis methods to identify potential sources from long-term air pollution measurement data. *Environ. Model. Softw.* **2009**, *24*, 938–939. [CrossRef]

53. Kraaijenbrink, P.D.A.; Bierkens, M.F.P.; Lutz, A.F.; Immerzeel, W.W. Impact of a global temperature rise of 1.5 degrees Celsius on Asia's glaciers. *Nature* **2017**, *549*, 257–260. [CrossRef]

54. Zhou, S.; Nakawo, M.; Hashimoto, S.; Sakai, A. The effect of refreezing on the isotopic composition of melting snowpack. *Hydrol. Process.* **2008**, *22*, 873–882. [CrossRef]

55. Stichler, W.; Schotterer, U.; Frohlich, K.; Ginot, P.; Kull, C.; Gaggeler, H.; Pouyaud, B. Influence of sublimation on stable isotope records recovered from high-altitude glaciers in the tropical Andes. *J. Geophys. Res. Atmos.* **2001**, *106*, 22613–22620. [CrossRef]

56. Rozanski, K.; Araguás-Araguás, L.; Gonfiantini, R. Isotopic patterns in modern global precipitation. *Clim. Chang. Cont. Isot. Rec.* **1993**, *78*, 1–36.

57. Wu, J.-K.; Ding, Y.-J.; Yang, J.-H.; Liu, S.-W.; Zhou, J.-X.; Qin, X. Spatial variation of stable isotopes in different waters during melt season in the Laohugou Glacial Catchment, Shule River basin. *J. Mt. Sci.* **2016**, *13*, 1453–1463. [CrossRef]

58. Jeelani, G.; Deshpande, R.D.; Shah, R.A.; Hassan, W. Influence of southwest monsoons in the Kashmir Valley, western Himalayas. *Isot. Environ. Health Stud.* **2017**, *53*, 400–412. [CrossRef] [PubMed]

59. Zhou, S.; Wang, Z.; Joswiak, D.R. From precipitation to runoff: Stable isotopic fractionation effect of glacier melting on a catchment scale. *Hydrol. Process.* **2014**, *28*, 3341–3349. [CrossRef]

60. Lone, S.A.; Jeelani, G.; Deshpande, R.D.; Mukherjee, A. Stable isotope (delta O-18 and delta D) dynamics of precipitation in a high altitude Himalayan cold desert and its surroundings in Indus river basin, Ladakh. *Atmos. Res.* **2019**, *221*, 46–57. [CrossRef]

61. Kumar, A.; Gokhale, A.A.; Shukla, T.; Dobhal, D.P. Hydroclimatic influence on particle size distribution of suspended sediments evacuated from debris-covered Chorabari Glacier, upper Mandakini catchment, central Himalaya. *Geomorphology* **2016**, *265*, 45–67. [CrossRef]

62. Khan, A.A.; Pant, N.C.; Sarkar, A.; Tandon, S.K.; Thamban, M.; Mahalinganathan, K. The Himalayan cryosphere: A critical assessment and evaluation of glacial melt fraction in the Bhagirathi basin. *Geosci. Front.* **2017**, *8*, 107–115. [CrossRef]

63. Pfahl, S.; Sodemann, H. What controls deuterium excess in global precipitation? *Clim. Past* **2014**, *10*, 771–781. [CrossRef]

64. Bazemore, D.E.; Eshleman, K.N.; Hollenbeck, K.J. The role of soil water in stormflow generation in a forested headwater catchment: Synthesis of natural tracer and hydrometric evidence. *J. Hydrol.* **1994**, *162*, 47–75. [CrossRef]

65. Genereux, D. Quantifying uncertainty in tracer-based hydrograph separations. *Water Resour. Res.* **1998**, *34*, 915–919. [CrossRef]

66. Funk, M.; Morelli, R.; Stahel, W. Mass balance of Griesgletscher 1961-1994: Different methods of determination. With 10 figures. *Z. Gletsch. Glazialgeol.* **1997**, *33*, 41–56.

67. Jansson, P. Effect of uncertainties in measured variables on the calculated mass balance of Storglaciaren. *Geogr. Ann. Ser. A Phys. Geogr.* **1999**, *81A*, 633–642. [CrossRef]

Article

Developing Meteoric Water Lines for Iran Based on Air Masses and Moisture Sources

Mojtaba Heydarizad [1], Ezzat Raeisi [1], Rogert Sorí [2] and Luis Gimeno [2,*

[1] Department of Geology, Faculty of Sciences, Shiraz University, Shiraz 71946-84695, Iran; mojtabaheydarizad@shirazu.ac.ir (M.H.); e_raeisi@yahoo.com (E.R.)

[2] Environmental Physics Laboratory (EphysLab), Facultad de Ciencias, Universidade de Vigo, 32004 Ourense, Spain; rogert.sori@uvigo.es

* Correspondence: l.gimeno@uvigo.es

Received: 28 August 2019; Accepted: 6 November 2019; Published: 10 November 2019

Abstract: Iran is a semi-arid to arid country that faces a water shortage crisis. Its weather is also influenced by various air masses and moisture sources. Therefore, applying accurate stable isotope techniques to investigate Iran's precipitation characteristics and developing Iran meteoric water lines (MWLs) as an initial step for future isotope hydrology studies is vitally important. The aim of this study was to determine the MWLs for Iran by considering air masses and dominant moisture sources. The Hybrid Single-Particle Lagrangian Integrated Trajectory (HYSPLIT) model backward analysis was used to determine the trajectories of various air masses in 19 weather stations in Iran and the areas covered by them. $\delta^{18}O$ and $\delta^{2}H$ contents were obtained for precipitation events from 32 stations in Iran and four in Iraq. Stable isotope samples were gathered from different sources and analyzed in various laboratories across the world. Three MWLs for north of Iran, south Zagros, and west Zagros, were determined based on the locations of dominant air masses and moisture sources. The proposed MWLs were validated by comparison with fresh karstic spring isotope data across Iran. In addition, Iran main moisture sources MWLs were used to determine dominant moisture sources role in karstic springs and surface water resources recharge.

Keywords: stable isotopes; HYSPLIT model; MWL validation; karstic springs

1. Introduction

Isotope composition of hydrogen ($\delta^{2}H$) and oxygen ($\delta^{18}O$) of precipitation provides important fingerprint information and allows for the identification of moisture sources for precipitation, evaporated atmospheric moisture conditions, and air mass trajectory patterns [1–7]. The source of the moisture is the most important predictor of the isotope content of precipitation [1,8]. There are numerous numerical models to study moisture sources, but a Hybrid Single-Particle Lagrangian Integrated Trajectory (HYSPLIT) [9] model has been used in numerous stable isotope studies for tracking moisture sources of precipitation [5,10,11]. Some of these studies, such as [11], consider *d*-excess as reliable fingerprints to study moisture sources responsible for precipitation. Moisture released from water bodies with a high sea surface temperature (SST) and a low relative humidity normally show a high *d*-excess (*d*-excess = $\delta^{2}H - 8 \times \delta^{18}O$ [12]), whereas precipitation originating from water bodies with a low SST and a high relative humidity will normally have a moderate to low *d*-excess [12–14].

Stable isotope technique is a precious method to study water resources characteristics, mainly in semi-arid and arid regions like Iran. To apply the stable isotope technique, it is important to develop MWLs (as an indicator of the local precipitation). For accurate interpretation in water resource studies, stable isotope data in water supply (surface and groundwater resources) should be compared with local precipitation via MWLs. In most stable isotope studies in Iran, the global meteoric water line

(GMWL) and the Eastern Mediterranean MWL (EMMWL) were used. However, some authors have developed local MWLs for certain regions. References [15–24] developed MWLs for Mashhad, Sirjan, Tehran, Zarivar and Marivan, Khersan, western Zagros, Shahrood, Lar National Park, northeast Iran, and North Khorasan Province, respectively. Each of these MWLs is applicable to the region for which it was developed and cannot be used in other parts of Iran because of differences in altitude, latitude, precipitation, and temperature [25,26]. Therefore, there is a need to develop regional Iran MWLs that can be applied to larger areas of the country. These MWLs will solve crucial obstacles regarding stable isotope studies and will be applied in the future isotope hydrology studies in Iran. In addition, to develop an MWL, the statistical approach should also be applied on precipitation isotopes data. In some studies, including [27,28], a statistical approach has been used in developing MWLs.

Developing unique MWLs for Iran is not practical or reliable due to various weather conditions across Iran. Although Shamsi and Kazemi [29] tried to present a unique MWL with a limited number of samples and stations for Iran, their MWL was not reliable due to various climate conditions that govern in different parts of Iran. Therefore, the authors tried to find a way and presented a method to consider moisture sources in developing MWLs. In previous studies [30,31], it has been determined that Iran is under the influence of various air masses with different isotope characteristics. The aim of this study was to determine the sources of precipitation moisture for different regions of Iran using the HYSPLIT model backward trajectories, and also to determine various air masses' dominance zones. Iran MWLs were developed based on various air masses' dominance zones (presented by HYSPLIT) and validated by comparison with fresh karstic springs. Finally, the MWLs for the main Iran moisture sources were developed. These MWLs were used to study the role and contribution of various air masses in karstic springs and surface water recharges across Iran. The application of moisture sources MWLs to study the role and contribution of various moisture sources in karstic springs and surface water resources recharge is a new method.

2. Iran Climatology and Weather Conditions

Iran is known for its diverse topography and climatology. A number of large water bodies border Iran (the Caspian Sea to the north, and the Persian Gulf and Oman Sea to the south). There are also high mountain ranges in Zagros (west and southwest Iran) and Alborz (north of Iran), which surround two large deserts (Dasht-e Lut and Dasht-e Kavir in central Iran). These features influence the climate of Iran, particularly the distribution of precipitation across the country. Average precipitation in Iran is 250 mm/year which varies from less than 100 mm/year in central Iran to higher than 1000 mm/year in the Caspian Sea coastal area [31]. There is a dry period (May to October) and a wet period (November to April) [31,32]. Four air masses including maritime polar (mP), Mediterranean (MedT), continental polar (also called the Siberian high-pressure system, cP), and continental tropical (also known as Sudan, cT) normally influence Iran in the wet period (November to April). However, the maritime tropical (mT) air mass only influences Iran in the dry period (May to October) [25,26]. The cP air mass predominantly supplies moisture from the Caspian Sea, and to a lesser extent from the Black Sea. It enters from the north and influences the north part of Iran. The mP air mass enters Iran from the northwest and mainly affects northwest Iran. The mP air mass predominantly supplies moisture from the Atlantic Ocean and the Black Sea to Iran. MedT is one of the most active air masses and affects almost all parts of Iran. The MedT air mass supplies moisture from the Mediterranean Sea and the Atlantic Ocean, and to a lesser extent the Black Sea. The cP and mP air masses dominantly influence Iran during December, January and February, while the MedT air mass dominantly influences Iran during March and April. The cT air mass affects most parts of Iran (like the MedT air mass) but affects the south most strongly. The cT air mass enters from the south and rarely affects areas outside of Iran. The cT air mass transports a considerable amount of moisture from the Persian Gulf, the Red Sea, and the Arabian Sea. The cT air mass has a crucial role in Iran precipitation during all of the wet period [31]. Precipitation southeast of Iran is predominantly influenced by the mT air mass [25]. The mT air mass supplies moisture from the Arabian Sea and the Indian Ocean to southeast and south of Iran. Karimi and Farajzadeh [32] calculated air mass trajectories for Iran using 40-year reanalysis datasets provided by the European Center for Medium Range Weather Forecasts, the US National Centers for Environmental Prediction, and the National Center for Atmospheric Research. Numerical models such as HYSPLIT and FLEXible_PARTicle dispersion model (FLEXPART) have been used widely to identify the most important moisture sources and to determine the roles of moisture sources in supplying precipitation to remote regions in several parts of the world including Iran [1,5,10,33–36].

3. Materials and Methods

The dominant air masses and moisture sources for 900 precipitation events (for the period of 2010 to 2016) at 19 meteorological stations were determined using 120-hour backward trajectories. A precipitation event was considered to have occurred when precipitation was > 5 mm/day. These trajectories were obtained using the online version of the HYSPLIT model called READY (Real Time Environmental Application and Display System) [9,37]. HYSPLIT was initially developed by the US National Oceanic and Atmospheric Administration Air Resources Laboratory in 1982 and has been improved markedly since then. The HYSPLIT model can compute simple air parcel trajectories (backwards and forwards) and complex simulations involving dispersion and deposition [37]. The position of an air parcel at a particular time is computed after wind speed, temperature, pressure, and solar radiation data have been input into the HYSPLIT model from the US National Oceanic and Atmospheric Administration "FNL" meteorological database [37].

$\delta^{18}O$ and $\delta^{2}H$ data were obtained from 36 precipitation sampling stations. In addition to precipitation, the isotope composition of hydrogen ($\delta^{2}H$) and oxygen ($\delta^{18}O$) of fresh karstic springs and surface water resources was also studied. All the isotope data were collected from previous publications in academic journals, PhD theses, MSc dissertations, and reports/data from the Global Network of Isotopes in Precipitation (GNIP) stations, the Karst Research Center of Iran, the Iran Regional Water Authorities, and the Iran Water Resources Institute. For event-based determination of stable isotope composition of precipitation, samples were collected in 25 mL polyethylene bottles after each precipitation event and they were sent to the laboratory for analyses. However, for monthly based samples, the procedure presented by the GNIP was used. To avoid and minimize evaporation in monthly samples, these samples were taken in an event-based approach according to the rainwater collected in the rain gauges after each event, and were transferred into a monthly accumulation bottle [38]. These accumulation bottles were sent to the laboratory for further stable isotope analyses. All the samples were analyzed for $\delta^{2}H$ and $\delta^{18}O$ (δ (‰) = (R sample /R standard − 1) × 10^{3}), where R is $^{2}H/H$ or $^{18}O/^{16}O$ ratio. The isotopic composition is expressed in δ per mil, and ‰ shows $^{2}H/H$ and $^{18}O/^{16}O$ deviation from the reference VSMOW (Vienna Standard Mean Ocean Water). Analysis of the water samples was performed in several laboratories including G.G. HATCH Stable Isotope Laboratory at the University of Ottawa, Canada; Stable Isotope Laboratory at the University of Waterloo, Canada; IAEA laboratories; Federal Institute for Geosciences and Natural Resources in Hannover, Germany; Isotope Science Laboratory at the University of Calgary, Canada; National Research Center for Environment and Health (GSF), Neuherberg, Germany; the Isotope Hydrology Laboratory at Kumamoto University, Kumamoto, Japan, and several other laboratories across the world. Samples were analyzed for isotope composition of hydrogen ($\delta^{2}H$) and oxygen ($\delta^{18}O$) using isotope ratio mass spectrometer (IRMS) (Thermo Finigan, Bremen, Germany) and Los Gatos Research, Inc. (LGR) (ABB/LGR group, San Jose, CA, USA) instruments. The analytical standard uncertainties for most of the samples were ± 0.1 ‰ and ± 1‰ for $\delta^{18}O$ and $\delta^{2}H$, respectively.

The MWLs for Iran were developed using available isotope data and moisture sources obtained by HYSPLIT model backward trajectories. A linear regression model was used to determine the trend line between $\delta^{18}O$ and δ^2H data and develop MWLs. Linear regression was done in Microsoft Excel with Analysis ToolPack of Microsoft Office (2016) Professional Plus (Microsoft Corporation, Redmond, WA, USA) with License from the University of Vigo [39]. To understand how well the linear regression model fits the data, R-squared (R^2) value was used. The higher R^2 values demonstrate smaller differences between variable data and also show how strongly the variables are correlated with each other. The developed MWLs of Iran were validated using the $\delta^{18}O$ and δ^2H contents of fresh karstic springs across the country. The application of karstic spring as a natural pluviometer has been done previously in several studies such as [40]. In addition to regression models, the Analysis of Covariance test (ANCOVA) was also applied to the precipitation and karstic springs isotope data. An ANCOVA test is just like an ANOVA test, but ANCOVA takes into account the influence of the covariate (a covariate is a variable which has influence on the dependent variable/variables). The ANCOVA test checks the effect of the covariate on the dependent variable/variables. Authors used R programming (R Core team, Vienna, Austria) to calculate the ANCOVA test [41].

4. Results and Discussion

The contributions of different air masses which caused precipitation at 19 weather stations in Iran were determined using the HYSPLIT model backward trajectories. The results are presented in Figure 1. In addition, the main air mass trajectories toward Iran and areas covered by the different air masses are also shown in this figure. Precipitation in large parts of Iran is influenced by several air masses, but in some parts of the country just one or two air masses predominate.

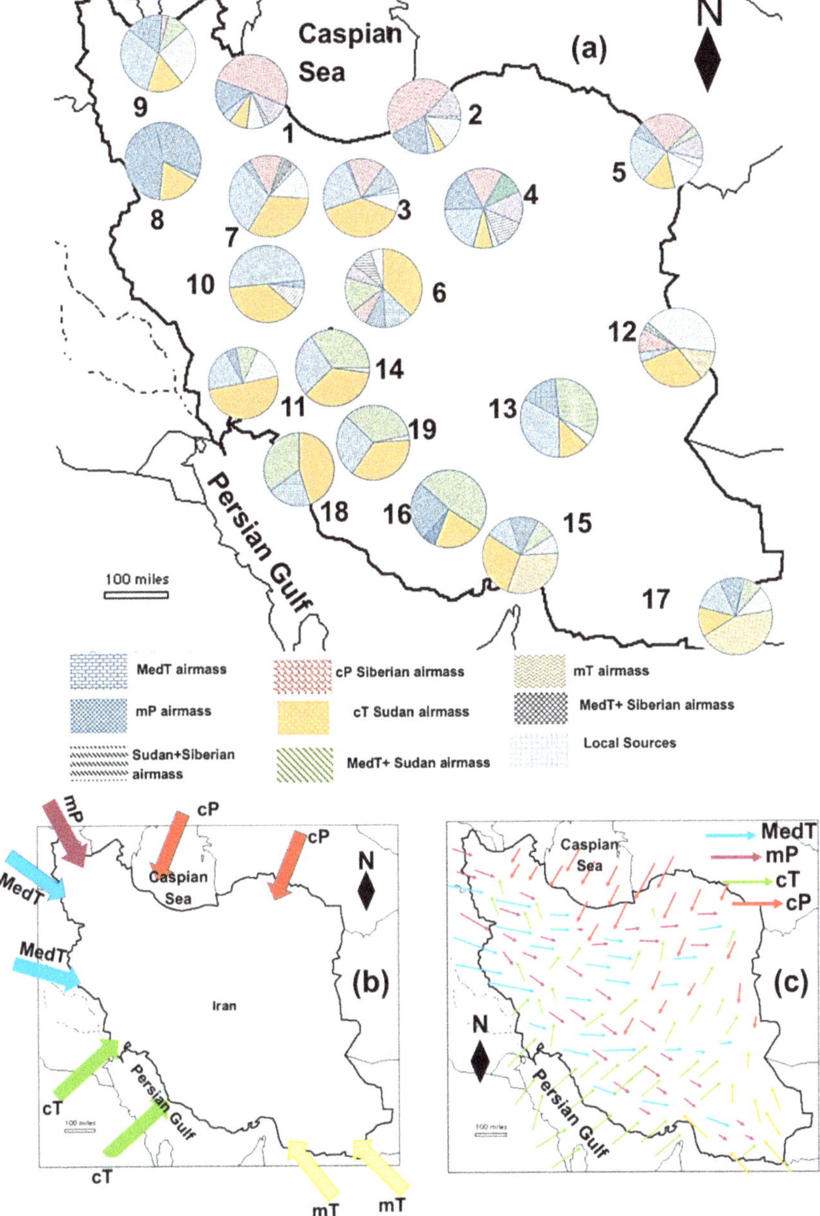

Figure 1. Contributions of different air masses causing precipitation at 19 weather stations in Iran for the period of 2010–2016. (**a**) The dominant air mass trajectories toward Iran. (**b**) The spatial distributions of these air masses across Iran. (**c**) The length of each arrow indicates approximate intensity of the air mass. Station names are as follows: 1 Bandar Anzali, 2 Gorgan, 3 Tehran, 4 Shahrood, 5 Mashhad, 6 Isfahan, 7 Arak, 8 Marivan, 9 Tabriz, 10 Shahrekord, 11 Ahvaz, 12 Zahedan, 13 Sirjan, 14 Samyrom, 15 Bandar Abas, 16 Darab, 17 Chabahar, 18 Bushehr, and 19 Shiraz).

4.1. Developing MWLs for Iran

The mean isotope composition of hydrogen (δ^2H) and oxygen (δ^{18}O) of precipitation as well as *d*-excess, station elevation, precipitation, and air temperature for each station in Iran and Iraq are presented in Supplementary Table S1 [15,16,18,19,21,42–57]. The spatial distribution of the precipitation sampling stations is shown in Figure 2a and karstic springs sampling stations in Figure 2b. These stations were separated into three groups according to air mass dominance zones presented by the HYSPLIT model backward trajectories presented in Figure 1.

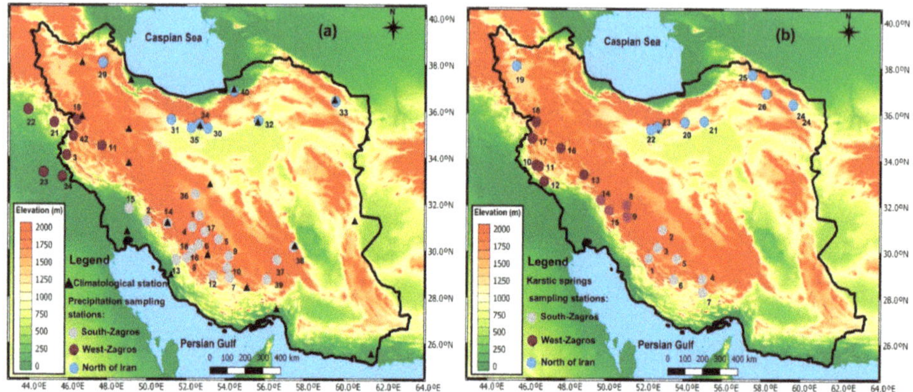

Figure 2. Climatological stations (black triangles) and precipitation sampling sites for isotope composition of hydrogen (δ^2H) and oxygen (δ^{18}O) analysis in North Iran (sky blue circles), West Zagros (brown circles) and South Zagros (gray circles) (**a**). Karstic springs sampling sites for isotope composition of hydrogen (δ^2H) and oxygen (δ^{18}O) analysis in Iran (**b**). The geographic location, boundaries of Iran (black line), and elevation map are derived from the Hydrosheds project [58].

As Iran weather is influenced by various air masses and moisture sources, it has not been possible to develop a single MWL for Iran because the MWL and stable isotopes in precipitation mainly depend on air masses and moisture sources [1,2]. Thus, three MWLs were developed for Iran (one each for Zagros-west, Zagros-south, and north of Iran) according to the various air mass dominance zones across Iran presented in Figure 1. To understand the effect of air mass dominance zones on the developed Iran MWLs, the unique Iran MWL was developed by linear regression model and its R^2 was compared with the R^2 of Iran MWLs developed based on air masses and moisture sources. Results showed that the R^2 values for Iran's unique MWL was lower than the R^2 for Iran MWLs developed based on air masses and moisture sources (Supplementary Table S2). This confirmed that developing MWLs based on air mass and moisture source dominance zones was much more accurate compared to developing a unique MWL for the whole of Iran. The linear regression model was used to develop a trend line between δ^{18}O and δ^2H data and develop MWLs. The proposed Iran MWLs and the relevant equations are shown in Figure 3.

The intercept for the North of Iran MWL (+3.86‰) was markedly lower than the intercept for Zagros-south (14.82‰) and Zagros-west (16.99‰) MWLs, while the North of Iran MWL slope (7.11) was between the slope of Zagros-south (7.06) and Zagros-west (8.12), respectively. The much lower intercept and lower slope of the North of Iran MWL compared to the Zagros region's MWLs was due to various isotope characteristics of moisture sources that influence these regions. The higher slope of the Zagros-west MWL compared to other Iran MWLs and also GMWL δ^2H = (8.17 ± 0.06) δ^{18}O + (10.35 ± 0.65) [59] was due to high relative humidity in this region. This region is covered by high mountains and local jungles with high average annual precipitation and relative humidity (>85%) which is higher than the global average [31]. Furthermore, the average *d*-excess for the north of Iran stations (9.45‰) was considerably lower than the averaged *d*-excess values for the south of

Zagros (18.57‰) and west of Zagros (16.31‰), respectively. This is because the SSTs for the water bodies providing moisture to the Zagros regions (mainly the Mediterranean Sea, the Arabian Sea and the Persian Gulf) are higher during the primary evaporation stage than the SSTs for the water bodies providing moisture to the north of Iran (mainly the Caspian Sea, the Black Sea, and the Atlantic Ocean) [60].

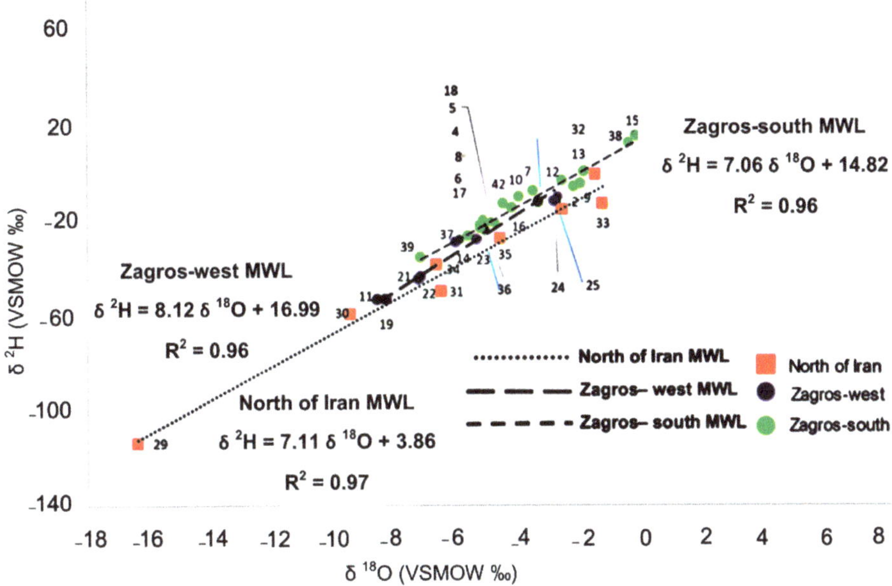

Figure 3. Average δ^2H and δ^{18}O in the sampling station's precipitation, and Iran developed meteoric water lines (MWLs).

In the ANCOVA test on the precipitation samples, air temperature, precipitation amount, and sampling station elevation were considered as covariate variables which influenced the stable isotope content of precipitation. Air mass dominance zones were considered as fixed parameters, while δ^{18}O, δ^2H and d-excess were considered depended variables. After checking the first assumption of the ANCOVA test, only elevation showed $p > 0.05$ and met covariate requirements (Supplementary spreadsheet S1). For the next assumption of the ANCOVA test, the homogeneity of the regression was checked. To check this assumption, air mass dominance zones were considered the fixed factor, elevation was a covariate, and δ^{18}O, δ^2H and d-excess were dependent variables. The p values were >0.05 for all the variables and homogeneity of regression was met (Supplementary spreadsheet S1). Elevation is variable which directly influences the stable isotope content of precipitation in Iran.

To validate the developed Iran MWLs, they were compared with fresh karstic springs across Iran. The δ^{18}O, δ^2H, electrical conductivity (EC), and d-excess values for over 200 karstic freshwater springs are shown in Supplementary Table S3 [17,19,20,43,45,54,61–71]. Evaporation did not markedly affect the selected karstic springs, and the EC of the karstic springs water was < 1000 μS/cm. The isotopic ratio of the karstic springs could therefore be used to check the validity of the MWLs. The proposed MWLs were validated by matching the isotopic ratio of the karstic springs with the MWLs, and the results indicated that the proposed MWLs were appropriate (Figure 4). The linear regression model was applied to trend the karstic springs water line (Supplementary Table S2). Trending lines on karstic springs showed very mild deviation from MWLs which confirmed the very low and negligible evaporation in most of the samples and the reliability of karstic springs for validation of the developed Iran MWLs.

Studying the isotope data in karstic springs using an ANCOVA test demonstrated important results. Authors considered EC as a covariate which influenced the stable isotope content of karstic springs. Authors checked the validity of EC as a covariate and confirmed that this variable met the covariate requirements. Therefore, the first assumption of the ANCOVA test was achieved. Furthermore, the homogeneity of the regression was also tested. Karstic regions of Iran were considered as the fixed parameter, EC as the covariate variable, and δ^{18}O, δ^2H and *d*-excess as dependent variables (Supplementary spreadsheet S1). The homogeneity of regression was met for all the dependent variables and thus two ANCOVA test assumptions were achieved. EC directly influenced the stable isotope content of karstic springs, and thus choosing the karstic springs with low EC values for validation of MWLs was a wise decision as high EC values (>1000 µS/cm) can influence the stable isotope content of karstic springs.

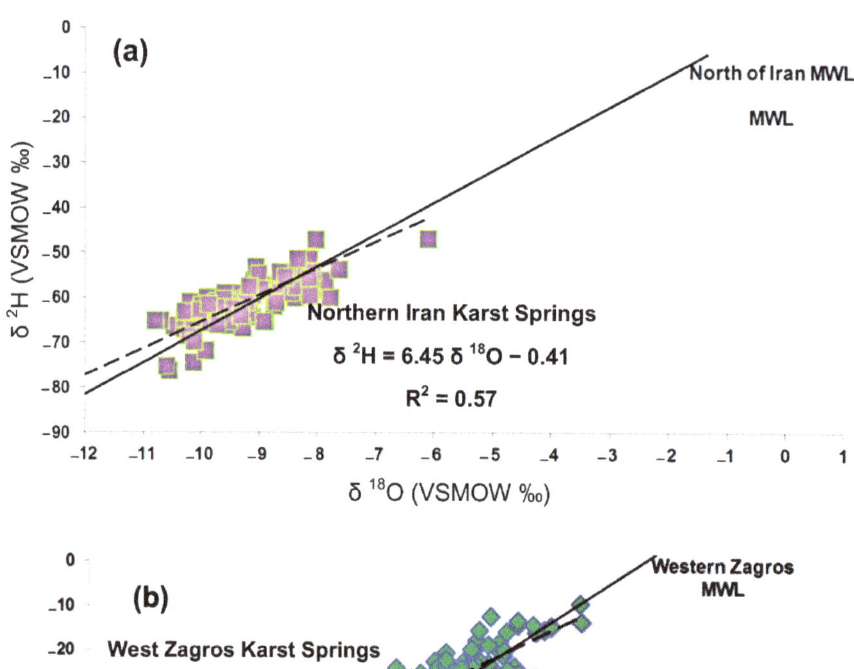

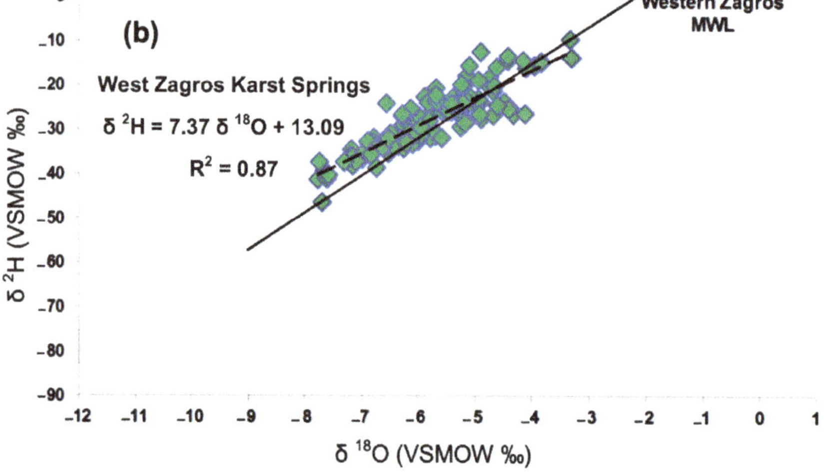

Figure 4. *Cont.*

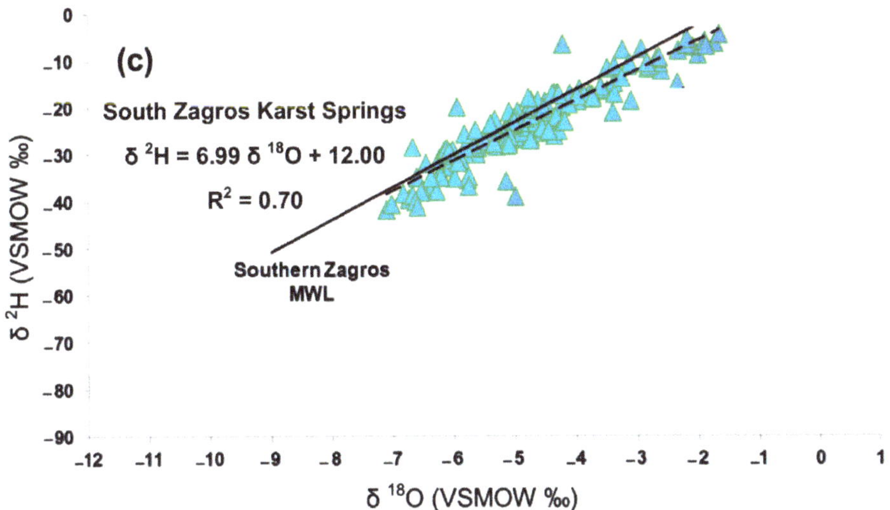

Figure 4. Karstic springs isotope values $\delta^{18}O$ and δ^2H plot on (**a**) North of Iran, (**b**) Zagros-west, and (**c**) Zagros-south MWLs.

The areas covered by the Iran MWLs are shown in Figure 5. No isotope data were available for coastal lowland areas along the Persian Gulf, central Iran, east and southeast Iran. Therefore, MWLs were not developed for these areas. The HYSPLIT model output indicates that the monsoon is the dominant source of precipitation for the southeast of Iran. Karachi (Pakistan) is also influenced by monsoon moisture sources [72], so the Karachi station MWL (KMWL) $\delta^2H = 7.56\ \delta^{18}O + 0.34$ [73] can be used as an alternative MWL for the southeast of Iran. No alternative MWL is available for coastal areas along the Persian Gulf and central Iran.

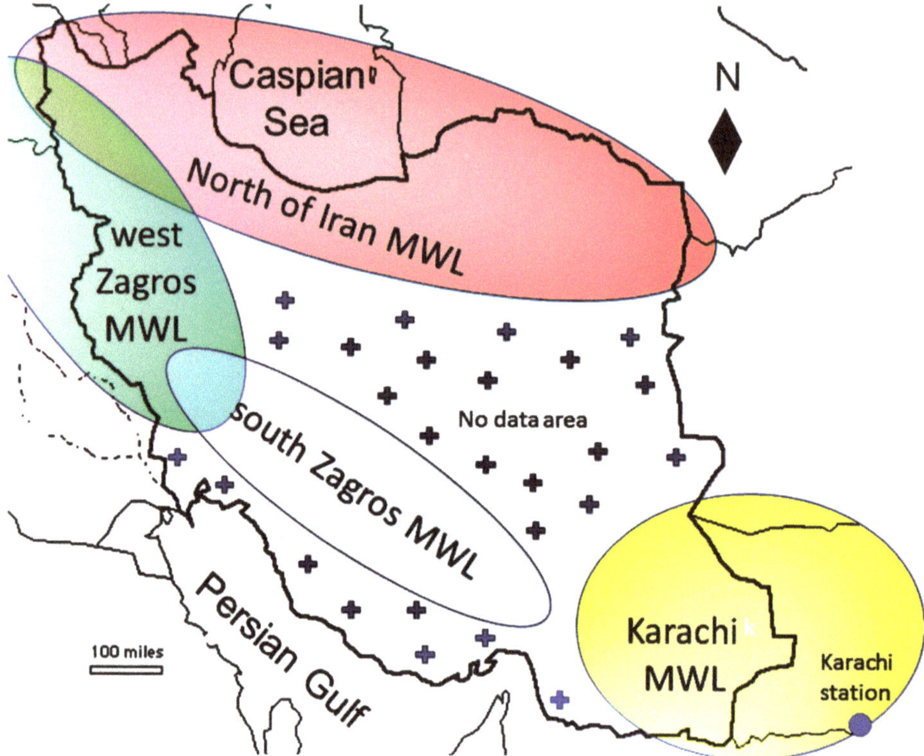

Figure 5. Map of areas covered by Iran MWLs and areas with no suggested MWLs. (The plus signs cover the areas with no suggested MWLs in Iran).

4.2. MWLs for Specific Moisture Sources

The HYSPLIT model output for 161 precipitation events for which isotope data were available (12 of the 32 studied stations in Iran) indicated that moisture predominantly originated from various moisture sources (mainly the Caspian Sea, the Mediterranean Sea, and the Persian Gulf). In previous studies, only the stable isotope characteristics of different moisture sources were determined [1–5,30], but herein an attempt was also made to determine MWLs for the precipitation events originated from the main moisture sources (Figure 6). These MWLs could be used to study the roles and contributions of various moisture sources in surface water and karstic springs recharge in Iran.

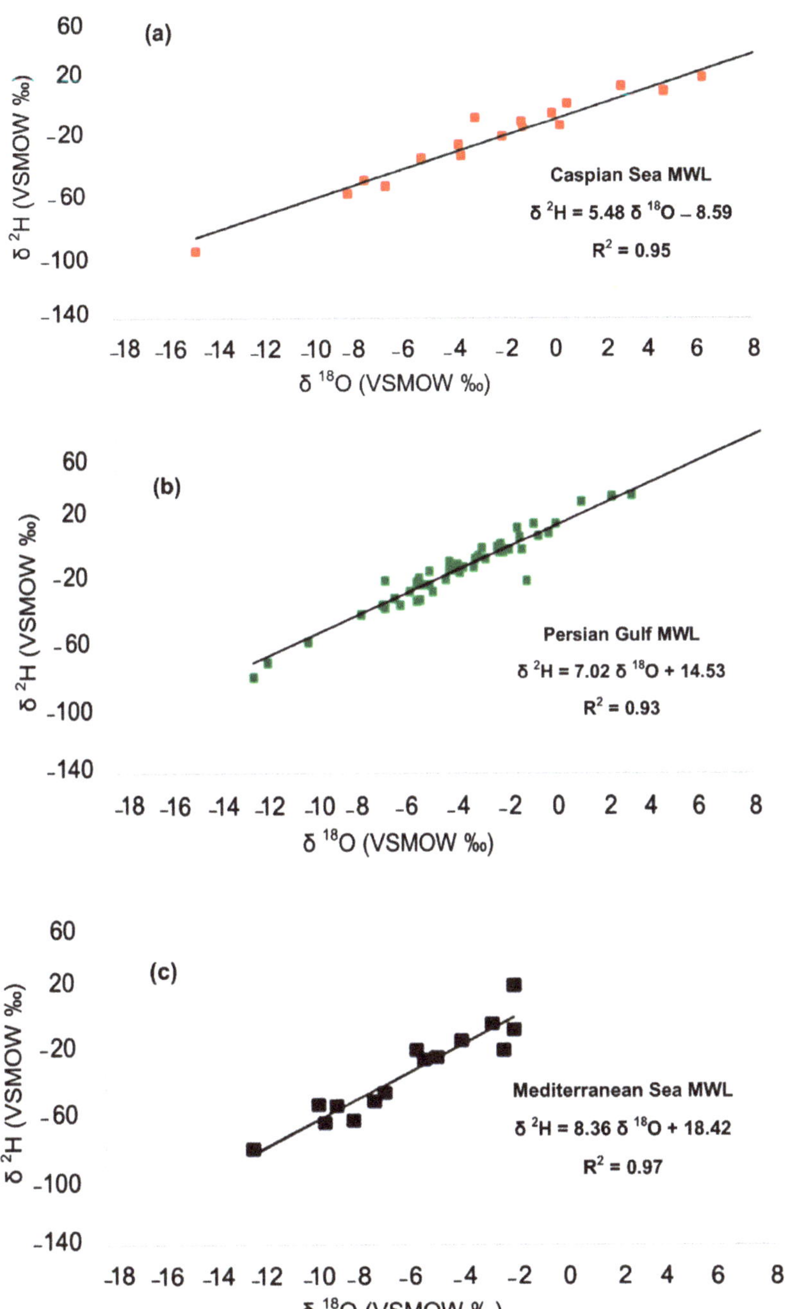

Figure 6. MWLs for specific moisture sources from the Caspian Sea (**a**), the Persian Gulf (**b**), and the Mediterranean Sea (**c**), all of which supply precipitation to Iran.

The developed Iran MWLs and dominant moisture sources MWLs are shown in (Table 1). Comparing the slope and intercept of the MWLs demonstrated very valuable results. In the west Zagros region where precipitation is dominantly provided by MedT air mass, the Zagros-west MWL is similar to the Mediterranean Sea MWL in both slope and intercept. However, in the south Zagros region where precipitation is provided dominantly by MedT and cT air masses, Zagros-south MWL slope and intercept values were between the Persian Gulf and the Mediterranean Sea MWLs slope and intercept. In the north of Iran where precipitation is provided mainly by the simultaneous influence of MedT and cP air masses, the North of Iran MWL slope and intercept values were between the Mediterranean and the Caspian Sea MWL slope and intercept values.

Table 1. Regional MWLs for Iran and MWLs for specific moisture sources.

MWL	Slope	Intercept	Dominant Air Masses	Dominant Moisture Sources
Zagros-south	7.06	14.82	MedT & cT	Arabian Sea, Mediterranean Sea, Persian Gulf, and continental sources
Zagros-west	8.12	16.99	MedT, cT, & mP	Mediterranean Sea, Black Sea, Persian Gulf, and continental sources
North of Iran	7.11	3.86	MedT, cT, mP, & cP	Caspian Sea, Mediterranean Sea, Black Sea, Persian Gulf, and continental sources
Karachi MWL (KMWL) for Southeast Iran	7.56	0.34	mT	Indian Ocean and Arabian Sea
Caspian Sea	5.48	−8.59	cP	Caspian Sea
Mediterranean Sea	8.36	18.42	MedT	Mediterranean Sea and Black Sea
Persian Gulf	7.02	14.53	cT	Persian Gulf

4.3. The Role of Various Moisture Sources in Surface Water and Karstic Springs Recharge in Iran

As mentioned earlier, MWLs developed for the dominant Iran moisture sources can be used in the study of moisture source contribution rate in karstic springs and surface water resources recharge. Surface water resources isotope data were collected from three dominant zones in the north of Iran, west Zagros and south Zagros [17,18,20,43,54,57,62,67,68,74–85] and presented in Supplementary Table S4. To study the role of the dominant moisture sources in surface water resources recharge, surface water samples were plotted on the Caspian Sea, the Persian Gulf and the Mediterranean Sea MWLs (Figure 7). Surface water resources in the north of Iran were closely plotted on the Caspian Sea MWL. This demonstrated the considerable role of the Caspian Sea moisture in surface water resources recharge in this region. However, the surface water resources in the south Zagros region were closely plotted on the Persian Gulf MWL. This confirmed that surface water resources in this region were dominantly under the influence of the moisture originating from the Persian Gulf. Some of the surface water resources in the south Zagros region deviated considerably from the Persian Gulf MWL due to huge evaporation effect. These surface water samples (mainly Parishan and Dasht-Arjan lakes) faced huge evaporation from their surfaces (Supplementary Table S4). Finally, plotting surface water resources on the Mediterranean Sea MWL showed that most of the surface water resources in all three dominate regions were plotted closely on this MWL. This confirmed the dominant role of the Mediterranean Sea moisture source in precipitation and surface water resources recharge in all parts of Iran. Some of the surface water resources in the north of Iran (Caspian Sea and Bazangan Lake) and also south Zagros (Parishan and Dasht-Arjan) dominantly deviated from the Mediterranean Sea MWL which was due to an intense evaporation effect on these resources (Supplementary Table S4). A linear regression model was used to develop a water line for surface water resources. The developed surface water isotope lines showed huge decline in both slope and intercept compared to both Iran MWLs and also to karstic springs isotope lines. This was due to the huge evaporation effect on surface water resources in the studied regions (Supplementary Table S4).

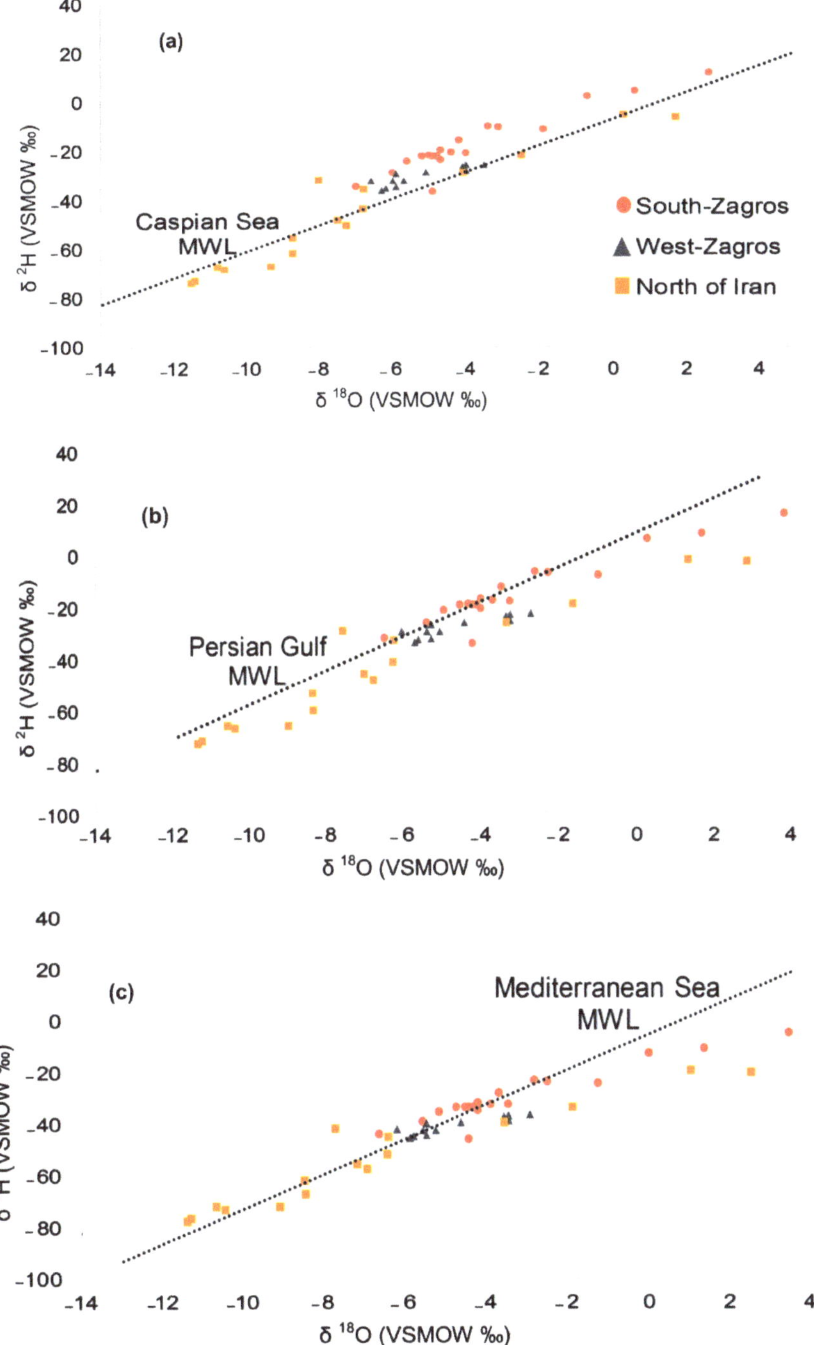

Figure 7. Plotting surface water resources $\delta^{18}O$ and δ^2H on the Caspian Sea (**a**), the Persian Gulf (**b**), and the Mediterranean Sea (**c**) MWLs. (The Mediterranean Sea, the Persian Gulf, and the Caspian Sea MWLs are taken from Figure 6.).

Plotting karstic springs on the dominant moisture sources MWL also demonstrated very valuable results (Figure 8). Karstic springs in the north of Iran were plotted closely on the Caspian Sea and the Mediterranean Sea MWLs. This is due to the fact that the Caspian and Mediterranean seas moisture have a dominant role in karstic springs recharge in the north of Iran. However, karstic springs in the Zagros regions (west and south) were mainly plotted on the Persian Gulf and the Mediterranean Sea MWLs. Karstic springs in the Zagros regions are mainly recharged by the precipitation events originating from these water bodies.

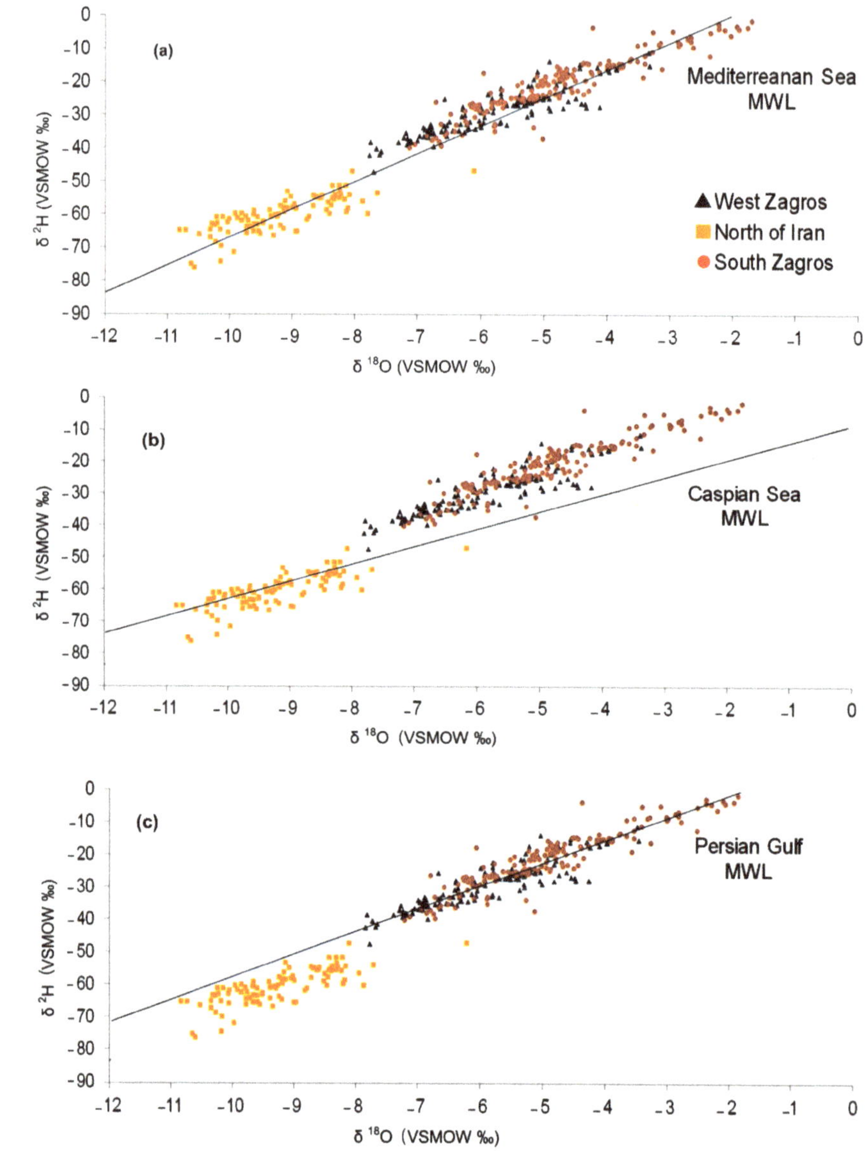

Figure 8. Plotting karstic springs $\delta^{18}O$ and $\delta^{2}H$ across Iran on the Mediterranean Sea (**a**), the Caspian Sea (**b**), and the Persian Gulf (**c**) MWLs. (The Mediterranean Sea, the Persian Gulf, and the Caspian Sea MWLs are taken from Figure 6).

Studying the role of various moisture sources on karstic springs and surface water resources across Iran confirmed that various moisture sources dominantly recharged surface and karstic springs resources across Iran. In the north of Iran, the Caspian and Mediterranean seas influence surface and karstic springs resources, while in the Zagros regions in the west and southwest of Iran, the Persian Gulf and the Mediterranean Sea moisture have a dominant role in surface and karstic springs recharge.

5. Conclusions

Most parts of Iran are influenced by several air masses, but specific air masses are dominant in each part of the country. Thus, a single MWL for Iran is not appropriate. Three MWLs were developed for Iran based on the main moisture sources and air masses which influence this country (Zagros-west, Zagros-south, and North of Iran). The proposed MWLs for Iran were validated by matching karstic spring $\delta^{18}O$ and δ^2H values to the proposed MWLs. The *d*-excess values were higher for the west and south Zagros regions compared to the north of Iran, because the water bodies supplying moisture to west and south Zagros' precipitation have higher SSTs and lower humidity than those supplying moisture to the north of Iran. Furthermore, MWLs were also developed for the main Iran moisture sources (the Caspian Sea, the Mediterranean Sea, and the Persian Gulf). Plotting karstic springs and surface water resources on the main moisture sources MWLs showed that both karstic springs and surface water samples in the north of Iran were mainly plotted on the Caspian Sea and the Mediterranean Sea MWLs. However, most of the karstic springs and surface water samples were plotted on the Persian Gulf and the Mediterranean Sea MWLs in the south and west Zagros regions. The methods proposed here can be applied in other regions influenced by various air masses and moisture sources.

Supplementary Materials: The following are available online at http://www.mdpi.com/2073-4441/11/11/2359/s1, Table S1. Station elevation, precipitation depth (P), air temperature (T), mean isotope composition of hydrogen (δ^2H) and oxygen ($\delta^{18}O$) of precipitation, and mean *d*-excess for each studied station in Iran and Iraq; Table S2. The linear regression models for precipitation, karstic springs and surface water resources data; Table S3. Mean isotope composition of hydrogen (δ^2H) and oxygen ($\delta^{18}O$), electrical conductivity (EC), and *d*-excess values in karstic springs sampling stations for the regions in Iran covered by the MWLs, and the number of karst springs in each region; Table S4. Mean isotope composition of hydrogen (δ^2H) and oxygen ($\delta^{18}O$), and *d*-excess values in surface water resources for the regions in Iran covered by the MWLs, and the number of surface water resources in each region; Spreadsheet S1: ANCOVA test results for the precipitation and karstic springs stable isotope data.

Author Contributions: Conceptualization, M.H. and E.R.; investigation, M.H. and R.S.; methodology, M.H. and L.G.; project administration, L.G. and M.H.; software, M.H., L.G. and R.S.; supervision, E.R., L.G. and R.S.; writing—original draft, M.H. and L.G.

Funding: This research received no external funding.

Acknowledgments: The support of researchers at the Iran Regional Water Authorities and researchers at the Iran Water Research Institute is greatly appreciated. Special thanks to our colleagues and students in the EPhysLab Group at the Ourense campus of Vigo University in Spain and at Shiraz University in Iran for their help and support during this study.

Conflicts of Interest: The authors declare no conflicts of interest.

References

1. Sjostrom, D.J.; Welker, J.M. The influence of air mass source on the seasonal isotopic composition of precipitation, eastern USA. *J. Geochem. Explor.* **2009**, *102*, 103–112. [CrossRef]
2. Birks, J.; EDWARDS, T.W. Atmospheric circulation controls on precipitation isotope—Climate relations in western Canada. *Tellus B Chem. Phys. Meteorol.* **2009**, *61*, 566–576. [CrossRef]
3. Lykoudis, S.P.; Kostopoulou, E.; Argiriou, A.A. Stable isotopic signature of precipitation under various synoptic classifications. *Phys. Chem. Earth Parts A/B/C* **2010**, *35*, 530–535. [CrossRef]
4. Liu, Z.; Bowen, G.J.; Welker, J.M. Atmospheric circulation is reflected in precipitation isotope gradients over the conterminous United States. *J. Geophys. Res. Space Phys.* **2010**, *115*. [CrossRef]

5. Crawford, J.; Hughes, C.E.; Parkes, S.D. Is the isotopic composition of event based precipitation driven by moisture source or synoptic scale weather in the Sydney Basin, Australia? *J. Hydrol.* **2013**, *507*, 213–226. [CrossRef]

6. Kong, Y.; Wang, K.; Li, J.; Pang, Z. Stable Isotopes of Precipitation in China: A Consideration of Moisture Sources. *Water* **2019**, *11*, 1239. [CrossRef]

7. Fan, Y.; Chen, Y.; He, Q.; Li, W.; Wang, Y. Isotopic Characterization of River Waters and Water Source Identification in an Inland River, Central Asia. *Water* **2016**, *8*, 286. [CrossRef]

8. Burnett, A.W.; Mullins, H.T.; Patterson, W.P. Relationship between atmospheric circulation and winter precipitation $\delta18$ O in central New York State. *Geophys. Res. Lett.* **2004**, *31*, 209. [CrossRef]

9. Stein, A.F.; Draxler, R.R.; Rolph, G.D.; Stunder, B.J.B.; Cohen, M.D.; Ngan, F.; Stein, A. NOAA's HYSPLIT Atmospheric Transport and Dispersion Modeling System. *Bull. Am. Meteorol. Soc.* **2015**, *96*, 2059–2077. [CrossRef]

10. Soderberg, K.; Good, S.P.; O'Connor, M.; Wang, L.; Ryan, K.; Caylor, K.K. Using atmospheric trajectories to model the isotopic composition of rainfall in central Kenya. *Ecosphere* **2013**, *4*, 1–18. [CrossRef]

11. Juhlke, T.R.; Meier, C.; Van Geldern, R.; Vanselow, K.A.; Wernicke, J.; Baidulloeva, J.; Barth, J.A.; Weise, S.M. Assessing moisture sources of precipitation in the Western Pamir Mountains (Tajikistan, Central Asia) using deuterium excess. *Tellus B Chem. Phys. Meteorol.* **2019**, *71*, 1–16. [CrossRef]

12. Dansgard, W. Stable isotopes in precipitation. *Tellus* **1964**, *16*, 436–468. [CrossRef]

13. Gat, J.R.; Carmi, I. Evolution of the isotopic composition of atmospheric waters in the Mediterranean Sea area. *J. Geophys. Res. Space Phys.* **1970**, *75*, 3039–3048. [CrossRef]

14. Clark, I.D. *Environmental Isotopes in Hydrogeology*; CRC Press/Lewis Publishers: Boca Raton, FL, USA, 1997; ISBN 1566702496.

15. Mohammadzadeh, H. The Meteoric Relationship for ^{18}O and ^{2}H in Precipitations and Isotopic Compositions of water resources in Mashhad Area (NE Iran). In *The 1st International Applied Geological Congress*; Islamic Azad University—Mashad Branch: Mashhad, Iran, 2010; pp. 555–559.

16. Jahanshahi, R. *Environmental Effects of GoleGohar Iron Ore Mine on Groundwater of the Area*; Shiraz University: Shiraz, Iran, 2013.

17. Heydarizad, M. *Investigation of Hydrochemistry of Karde Dam and its Hydraulically Connection with Downstream Water Resources*; Ferdowsi University of Mashhad: Mashhad, Iran, 2011.

18. Mohammadzadeh, H.; Ebrahimpour, S. Application of stable isotopes and hydrochemistry to investigate sources and quality exchange Zarivar catchment area. *J. Water Soil* **2012**, *26*, 1018–1031.

19. Mohammadi, Z. *Method of Leakage Study at Karst Dam Site, the Zagros Region*; Shiraz University: Shiraz, Iran, 2006.

20. Osati, K.; Koeniger, P.; Salajegheh, A.; Mahdavi, M.; Chapi, K.; Malekian, A. Spatiotemporal patterns of stable isotopes and hydrochemistry in springs and river flow of the upper Karkheh River Basin, Iran. *Isot. Environ. Health Stud.* **2014**, *50*, 169–183. [CrossRef]

21. Kazemi, G.A.; Ichiyanagi, K.; Shimada, J. Isotopic characteristics, chemical composition and salinization of atmospheric precipitation in Shahrood, northeastern Iran. *Environ. Earth Sci.* **2015**, *73*, 361–374. [CrossRef]

22. Shamsi, A.; Karami, G.H.; Hunkeler, D.; Taheri, A. Isotopic and hydrogeochemical evaluation of springs discharging from high-elevation karst aquifers in Lar National Park, northern Iran. *Hydrogeology* **2019**, *27*, 655. [CrossRef]

23. Mohammadzadeh, H.; Heydarizad, M. δ^{18}O and δ^{2}H characteristics of moisture sources and their role in surface water recharge in the north-east of Iran. *Isot. Environ. Health Stud.* **2019**, 1–16. [CrossRef]

24. Bagheri, R.; Bagheri, F.; Karami, G.H.; Jafari, H. Chemo-isotopes (^{18}O & ^{2}H) signatures and HYSPLIT model application: Clues to the atmospheric moisture and air mass origins. *Atmos. Environ.* **2019**, *215*, 116892.

25. Alijani, B. *Iran Climatology*, 5th ed.; Payam Nour Publication: Tehran, Iran, 2000; ISBN 978-964-455-621-0.

26. Heydarizad, M.; Raeisi, E.; Sori, R.; Gimeno, L.; Nieto, R. The Role of Moisture Sources and Climatic Teleconnections in Northeastern and South-Central Iran's Hydro-Climatology. *Water* **2018**, *10*, 1550. [CrossRef]

27. Crawford, J.; Hughes, C.E.; Lykoudis, S. Alternative least squares methods for determining the meteoric water line, demonstrated using GNIP data. *J. Hydrol.* **2014**, *519*, 2331–2340. [CrossRef]

28. Boschetti, T.; Cifuentes, J.; Iacumin, P.; Selmo, E. Local Meteoric Water Line of Northern Chile (18° S–30° S): An Application of Error-in-Variables Regression to the Oxygen and Hydrogen Stable Isotope Ratio of Precipitation. *Water* **2019**, *11*, 791. [CrossRef]

29. Shamsi, A.; Kazemi, G.A. A review of research dealing with isotope hydrology in Iran and the first Iranian meteoric water line. *JGeope* **2014**, *4*, 73–86.

30. Heydarizad, M.; Raeisi, E.; Sori, R.; Gimeno, L. An overview of the atmospheric moisture transport effect on stable isotopes ($\delta 18O$, $\delta 2H$) and D excess contents of precipitation in Iran. *Theor. Appl. Clim.* **2019**, *138*, 47–63. [CrossRef]

31. Heydarizad, M. *Meteoric Water Lines of Iran for Various Precipitation Sources*; Shiraz University: Shiraz, Iran, 2018.

32. Karimi, M.; Farajzadeh, M. Spatial and Temporal distribution of Iran's precipitation moisture. *J. Geogr. Sci. Stud.* **2011**, *19*, 109–127.

33. Ciric, D.; Stojanovic, M.; Drumond, A.; Nieto, R.; Gimeno, L. Tracking the Origin of Moisture over the Danube River Basin Using a Lagrangian Approach. *Atmosphere* **2016**, *7*, 162. [CrossRef]

34. Drumond, A.; Taboada, E.; Nieto, R.; Gimeno, L.; Vicente-Serrano, S.M.; López-Moreno, J.I. A Lagrangian analysis of the present-day sources of moisture for major ice-core sites. *Earth Syst. Dyn.* **2016**, *7*, 549–558. [CrossRef]

35. Heydarizad, M.; Raeisi, E.; Sori, R.; Gimeno, L. The Identification of Iran's Moisture Sources Using a Lagrangian Particle Dispersion Model. *Atmosphere* **2018**, *9*, 408. [CrossRef]

36. Nieto, R.; Gimeno, L.; Gallego, D.; Trigo, R. Contributions to the moisture budget of airmasses over Iceland. *Meteorol. Z.* **2007**, *16*, 37–44. [CrossRef]

37. NOAA. Available online: https://www.esrl.noaa.gov (accessed on 20 October 2019).

38. IAEA/GNIP. *Precipitation Sampling Guide*; International Atomic Energy Agency (IAEA): Vienna, Austria, 2014.

39. *Microsoft Office (2016) Professional PLus Microsoft Excel 2016*; Microsoft corporation: Redmond, WA, USA; Available online: https://www.office.com/ (accessed on 10 November 2019).

40. Minissale, A.; Vaselli, O. Karst springs as "natural" pluviometers: Constraints on the isotopic composition of rainfall in the Apennines of central Italy. *Appl. Geochem.* **2011**, *26*, 838–852. [CrossRef]

41. R core Team. *R: A Language and Environment for Statistical Computing*; R Foundation for Statistical Computing: Vienna, Austria, 2018; Available online: https://www.R-project.org/ (accessed on 10 November 2019).

42. Faroughi, A. *Characterizing Isotopic Signature of Precipitation in Fars Province, Iran*; Shiraz University: Shiraz, Iran, 2008.

43. Zarei, H.; Akhondali, A.M.; Mohammadzadeh, H.; Radmanesh, F.; Laudon, H. Runoff generation processes during the wet-up phase in a semi-arid basin in Iran. *Hydrol. Earth Syst. Sci. Discuss.* **2014**, *11*, 3787–3810. [CrossRef]

44. Karimi, H. Investigating the Oxygen-18 and Deuterium Isotope Composition of Precipitations in Western Zagros. In Proceedings of the 1st National Conference on Application of Stable Isotopes, Mashhad, Iran, 8 May 2013.

45. Rezaei, M.; Karimi, H.; Jokar, B. Studies of Kazeroon-Persian Gulf karstic basins. *Karst Res. Cent. Iran* **1998**, *2*, 75–120.

46. Ghazban, F. Geological and geochemical investigation of Damavand geothermal prospect, central Alborz mountain, northern Iran. *Geotherm. Resour. Counc. Trans.* **2000**, *24*, 229–234.

47. Nickghoujag, Y.; Mohammadzadeh, H.; Naseri, H.R. The application of stable isotopes (^{18}O and ^2H) to determine the role of surface water in shallow groundwater resources in Gorgan rood basin. In Proceedings of the Second National Conference on Application of Stable Isotopes, Mashhad, Iran, 11 May 2016.

48. Feyzi, D. *Traceing Studies in Roudball Dam Site*; Iran Regional Water Authorities: Shiraz, Iran, 1998.

49. Saadati, H.; Sharifi, F.; Mahdavi, M.; Ahmadi, H.; Mohseni, M. Determining origin of groundwater recharge resources, drought and wet periods by isotopic tracers in Hashtgerd plain. *Manag. J. Range Water Shed* **2009**, *62*, 49–63.

50. Khademi, H.; Mermut, A.; Krouse, H. Isotopic composition of gypsum hydration water in selected landforms from central Iran. *Chem. Geol.* **1997**, *138*, 245–255. [CrossRef]

51. Mohammadzadeh, H.; Eskandari, E.; Najafi, M. Studying the stable isotope content of precipitation in Paveh region. In Proceedings of the Second National Conference on Application of Stable Isotopes, Mashhad, Iran, 11 May 2016.

52. Farpoor, M.; Khademi, H.; Eghbal, M.; Krouse, H. Mode of gypsum deposition in southeastern Iranian soils as revealed by isotopic composition of crystallization water. *Geoderma* **2004**, *121*, 233–242. [CrossRef]

53. Parizi, H.S.; Samani, N. Environmental Isotope Investigation of Groundwater in the Sarcheshmeh Copper Mine Area, Iran. *Mine Water Environ.* **2014**, *33*, 97–109. [CrossRef]

54. Kalantari, N.; Mohamadi Behzad, H. Investigation source of recharge to Sabzab and Bibi Talkhon karstic springes by application of ^{18}O and ^{2}H stable isotopes. In Proceedings of the 1st National Conference on Application of Stable Isotopes, Mashhad, Iran, 8 May 2013.

55. Ali, K.K.; Al-Kubaisi, Q.Y.; Al-Paruany, K.B. Isotopic study of water resources in a semi-arid region, western Iraq. *Environ. Earth Sci.* **2015**, *74*, 1671–1686. [CrossRef]

56. IAEA/GNIP Global Network of Isotopes in Precipitation (GNIP). Available online: https://www.iaea.org/servises/networks/gnip. (accessed on 8 November 2019).

57. Porkhial, S.; Ghomshei, M.M.; Yousefi, P. Stable Isotope and Elemental Chemistry of Mt. Sabalan Geothermal Field, Ardebil Province of North West Iran. In Proceedings of the World Geothermal Congress, Bali, Indonesia, 25–30 April 2010.

58. Lehner, B.; Verdin, K.; Jarvis, A. New Global Hydrography Derived From Spaceborne Elevation Data. *Eos* **2008**, *89*, 93. [CrossRef]

59. Rozanski, K.; Araguas-Araguas, L.; Gonfiantini, R. Isotopic patterns in modem global precipitation. In *Climate Change in Continental Isotopic Records*; Swart, P.K., Lohmann, K.C., Mckenzie, J., Savin, S., Eds.; Geophysical Monograph Series; American Geophysical Union: Washington, DC, USA, 1993; pp. 1–36. ISBN 9781118664025.

60. Kalnay, E.; Kanamitsu, M.; Kistler, R.; Collins, W.; Deaven, D.; Gandin, L.; Iredell, M.; Saha, S.; White, G.; Woollen, J.; et al. The NCEP/NCAR 40-Year Reanalysis Project. *Bull. Am. Meteorol. Soc.* **1996**, *77*, 437–471. [CrossRef]

61. Asari, A. *Source of Salinity at the Plunge of Gar and Barm Firooz Karstic Anticlines*; Shiraz University: Shiraz, Iran, 2011.

62. Bagheri Sheshdeh, R. *Leakage Potential in Seymareh Dam Site*; Shiraz University: Shiraz, Iran, 2007.

63. Masjedi, M.; Jahani, H.; Khalaj Amir hosseini, Y.; Hatami, F.; Kooh Pour, E. *Tracer Studies in Seymareh Dam Site in Ilam Province*; Iran Water Resources Institute: Tehran, Iran, 1995.

64. Rafighdoust, Y.; Eckstein, Y.; Harami, R.M.; Gharaie, M.H.M.; Griffith, E.M.; Mahboubi, A. Isotopic analysis, hydrogeochemistry and geothermometry of Tang-Bijar oilfield springs, Zagros region, Iran. *Geothermics* **2015**, *55*, 24–30. [CrossRef]

65. Ahmadipour, M. reza Karst springs of Alashtar, Iran. *Acta Carsolog.* **2003**, *32*, 244–254.

66. Chitsazan, M.; Vardanjani, H.K.; Karimi, H.; Charchi, A. A comparison between karst development in two main zones of Iran: Case study—Keyno anticline (Zagros Range) and Shotori anticline (Central Iran). *Arab. J. Geosci.* **2015**, *8*, 10833–10844. [CrossRef]

67. Khalaj Amirhosseini, Y. Application of isotopes and dye tracing methods in hydrology (case study: Havasan dam construction). In Proceedings of the An International Symposium of Isotopes in Hydrology, Marine Ecosystems and Climate Change Studies, Monaco-Ville, Monaco, 27 March–1 April 2011.

68. Mohammadzadeh, H.; Heydarizad, M. Investigating geochemistry and the stable isotope (δ^{18}O & δ^{2}H) composition of Karde Carbonate Lake water (NE Iran). In Proceedings of the 22th Goldschmidt Conference, Monteral, QC, Canada, 24–29 June 2012; p. 2123.

69. Faezi, N. *Hydrogeology and Vulenerability Assessment of Ghaleh Dokhtar Spring Catchment Area*; Shiraz University: Shiraz, Iran, 2015.

70. Mohammadzadeh, H.; Kazemi, M. Geofluids Assessment of the Ayub and Shafa Hot Springs in Kopet-Dagh Zone (NE Iran): An Isotopic Geochemistry Approach. *Geofluids* **2017**, *2017*, 1–11. [CrossRef]

71. Karimi, H.; Raeisi, E.; Bakalowicz, M. Characterising the main karst aquifers of the Alvand basin, northwest of Zagros, Iran, by a hydrological approach. *Hydrogeol. J.* **2005**, *13*, 787–799. [CrossRef]

72. Araguás-Araguás, L.; Froehlich, K.; Rozanski, K. Stable isotope composition of precipitation over southeast Asia. *J. Geophys. Res. Space Phys.* **1998**, *103*, 28721–28742. [CrossRef]

73. Mashiatullah, A.; Qureshi, R.; Tasneem, M.; Javed, T.; Gaye, C.; Ahmad, E.; Ahmad, N. Isotope hydrochemical investigation of saline intrusion in the coastal aquifer of Karachi, Pakistan. *Radioact. Environ.* **2006**, *8*, 382–393.

74. Niroomand, M.H.; Pakzad, M. *Studies of Bakhtegan Karstic Water Resources*; Karst Research Center of Iran: Shiraz, Iran, 1997.

75. Zarei, H.; Damough, N.A. Determination the recharge zones of Karstic springs downstram of Karoun3 dam reservoir. In *The Fifth Engineering Geology and Enviromental Geology Meeings*; Engineering Geology Assossiation: Tehran, Iran, 2007.

76. Mirnejad, H.; Sisakht, V.; Mohammadzadeh, H.; Amini, A.H.; Rostron, B.J.; Haghparast, G. Major, minor element chemistry and oxygen and hydrogen isotopic compositions of Marun oil-field brines, SW Iran: Source history and economic potential. *Geol. J.* **2011**, *46*, 1–9. [CrossRef]

77. Ehghahi, M. *Traceing Studies of Doupulan Tapeh Karstic Region*; Iran Regional Water Authorities: Tehran, Iran, 1990.

78. Pakzad, M. *Tracer Studies in Mirza Shirazi (Kovar) Dam Site in Fars Province*; Iran Water Research Institute: Tehran, Iran, 1993.

79. Rezaei, M.; Karami, H.; Jokar, B. *Studies of Hajiabad Karstic Water Resources*; Karst Research Center of Iran: Shiraz, Iran, 2000.

80. Rezaei, M.; Krimi, H.; Jokar, B. *Semi Detail Studies of Kazeroun and Persian Gulf Karstic Mega Basin*; Karst Research Center of Iran: Shiraz, Iran, 1998.

81. Mohammadzadeh, H. ^{18}O, ^{2}H, and ^{13}C isotopic compositions of water in Torogh and Kardeh dams, Mashhad. In Proceedings of the Joint European Stable Isotope User Meeting—JESIUM, Presqu'ile de Giens, France, 31 August–5 September 2008.

82. Pakzad, M. *Isotopic Studies of Karstic Resources in Kashaf Rood Basin in Khorasan Province*; Iran Water Resources Institute: Tehran, Iran, 2000.

83. Mohammadzadeh, H.; Robin, M.; Khanehbad, M. An Investigating of the origin and the groundwater discharge to Bazangan Lake, Eastern Kopet-Dagh Basin-Iran, using geochemistry and stable isotopes approaches. *Geochmica Cosmochim. Acta* **2008**, *72*, A641.

84. Lar dam consulting bureau. *Tracing Studies in Lar Dam Site*; Iran Water Resources Institute: Tehran, Iran, 1983.

85. Mousavi Shalmani, M.A.; Lakzian, A.; Khorasani, A.; Feiziasl, V.; Mahmoudi, A.; Pourmohammad, N. Use of Multiple Linear Regression Method for Modelling Seasonal Changes in Stable Isotopes of ^{18}O and ^{2}H in 30 Pouns in Gilan Province. *J. Water Soil* **2014**, *28*, 239–252.

 *water* 

Article

Characteristics of Water Isotopes and Water Source Identification During the Wet Season in Naqu River Basin, Qinghai–Tibet Plateau

Xi Chen [1], Guoli Wang [1], Fuqiang Wang [2,3,*], Denghua Yan [4] and Heng Zhao [2,3]

[1] School of Hydraulic Engineering, Dalian University of Technology, Dalian 116024, China; chenxi218@mail.dlut.edu.cn (X.C.); wanggl@dlut.edu.cn (G.W.)

[2] Department of Water Conservancy Engineering, North China University of Water Resources and Electric Power, Zhengzhou 450046, China; zhaoheng@ncwu.edu.cn

[3] Collaborative Innovation Center of Water Resources Efficient Utilization and Support Engineering, Zhengzhou 450046, China

[4] Water Resources Department, China Institute of Water Resources and Hydropower Research, Beijing 100038, China; yandh@iwhr.com

* Correspondence: wangfuqiang@ncwu.edu.cn; Tel.: +86-137-0371-4661

Received: 8 October 2019; Accepted: 15 November 2019; Published: 18 November 2019

Abstract: Climate change is affecting the discharge of headstreams from mountainous areas on the Qinghai–Tibet Plateau. To constrain future changes in discharge, it is important to understand the present-day formation mechanism and components of runoff in the basin. Here we explore the sources of runoff and spatial variations in discharge through measurements of δ^2H and $\delta^{18}O$ in the Naqu River, at the source of the Nu River, on the Qinghai–Tibet plateau, during the month of August from 2016 to 2018. We established thirteen sampling sites on the main stream and tributaries, and collected 39 samples from the river. We examined all the water samples and analyzed them for isotopes. We find a significant spatial variation trend based on one-way analysis of variance (ANOVA) ($p < 0.05$) between Main stream-2 and tributaries. The local meteoric water-line (LMWL) can be described as: $\delta^2H = 7.9\delta^{18}O + 6.29$. Isotopic evaporative fractionation in water and mixing of different water sources are responsible for the spatial difference in isotopic values between Main stream-2 and tributaries. Based on isotopic hydrograph separation, the proportion of snowmelt in runoff components ranges from 15% to 47%, and the proportion of rainwater ranges from 3% to 35%. Thus, the main components of runoff in the Naqu River are snowmelt and groundwater.

Keywords: stable isotopes; spatial variations; hydrograph separation; Naqu River basin; Qinghai–Tibet Plateau

---

1. Introduction

The gradual trend of global warming will affect the discharge of headstreams to plateau rivers, including on the Qinghai–Tibet Plateau [1–4]. Therefore, it is important to explore the formation mechanism and identify the components of runoff on the Qinghai–Tibet Plateau [5,6].

The Naqu River basin is sensitive to environmental change due to its high altitude. Studying its water cycle is not straightforward due to the lack of hydrological data and harsh natural conditions. Meanwhile, little is known about the water source contribution and the mechanism of the runoff. To constrain future changes in discharge, it is important to understand the present-day formation mechanism and components of runoff in the basin.

There are many methods to identify runoff components [7–13]. Recent studies have shown that hydrograph separation based on stable isotopes is an effective way to study the runoff mechanism [14–16]. In general, river components can be divided into precipitation, groundwater, soil water, and snowmelt based on isotope hydrological separation [17–25]. For example, based on the isotopic values of river waters, significant spatial and temporal variations of the Xijiang River were investigated [26]; Kong et al. found that the snowmelt water accounted for more than 57% of runoff of the Kumalak River [1], and more than 53% during the wet season [6]. Based on isotopes and geochemical tracers, streams in plateau regions are mainly replenished by snowmelt and groundwater [27–38]. There has been relatively little research on the composition and mechanism of water sources on the Qinghai–Tibet Plateau. The advantages of isotope techniques in the hydrologic cycle are obvious and water samples can be obtained easily in the plateau region due to the lack of hydrological and meteorological data [18].

In this study, we analyze the spatial variation of isotopes in the runoff and compute the proportions of runoff components in the month of August based on hydrograph separation. We hope that the research results of this paper will provide a relevant theoretical basis for the formation mechanism of runoff on the Qinghai–Tibet Plateau.

2. Study Areas

The Naqu River basin is the source of the Nu River in southwest China (Figure 1). The Naqu River has several main tributaries, such as the Najinqu, Sangqu, Bazongqu, Mumuqu, Chengqu, Zongqungqu, Mugequ, and Gongqu Rivers. There are many seasonal streams and mountain streams flowing into the Naqu River. The average annual temperature in this area is −0.6 °C. The drainage area of the Naqu River basin is 16,350 km^2, at a high altitude of 4600 m above sea level [39].

Annual precipitation is 531 mm. From May to October, precipitation accounts for about 82% of the total annual rainfall, with less precipitation from November to April. Although there is not a significant amount of snowfall throughout the whole year, snowmelt has a strong replenishment effect on the runoff in the flood season. The climate is affected by Indian Ocean southwest monsoon in summer. The water vapor of precipitation comes from water vapor that evaporated under wetter conditions. This is consistent with summer southwest monsoon precipitation in the region coming directly from the Bay of Bengal. The d-excess value in the precipitation directly from the Bay of Bengal is lower due to the high relative humidity of the sea surface [37–42]. For this reason, the precipitation is coincident with the annual peak of snowmelt during the wet season. And they become the main components of runoff in the Naqu River basin.

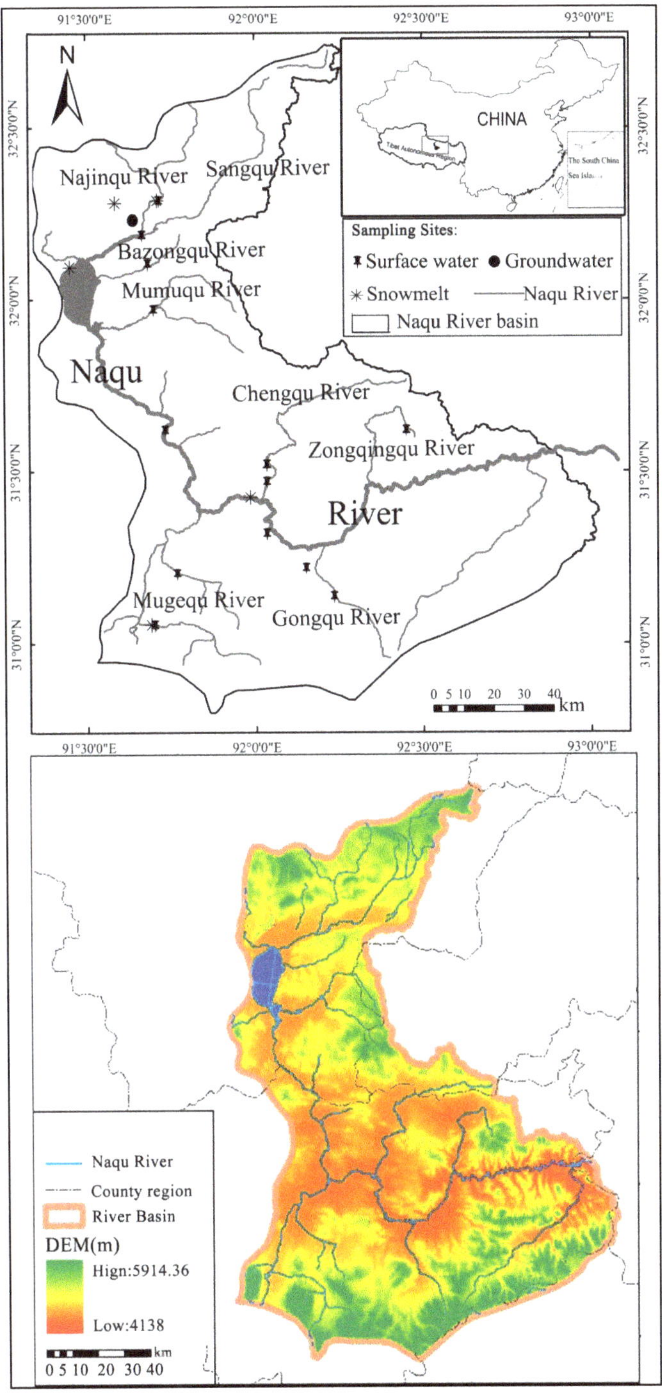

Figure 1. Research area and geographic location.

3. Materials and Methods

3.1. Field Sampling

We installed thirteen sampling sites on the main stream and tributaries of the Naqu River, with the sampling sites of Main stream-1 and Main stream-2 along the main channel, and Najinqu, Sangqu, Bazongqu, Mumuqu, Chengqu, Zongqungqu, Mugequ, and Gongqu on eight tributaries (Figure 1). We collected a total of 39 samples from the river in the month of August from 2016 to 2018.

In general, water samples included 39 runoff samples, two groundwater samples, two rain samples, and five snowmelt samples during the wet season from 2016 to 2018. We collected two rain samples on 13 August 2018.

3.2. Measurement

δ^{18}O and δ^{2}H analysis: Wavelength-scanned cavity ring down spectroscopy (WS-CRDS) (Picarro L1115-I, Picarro, Santa Clara, CA, USA) was used to measure water isotope composition, which were corrected using the Vienna Standard Mean Ocean Water (VSMOW, δ^{2}H = 0‰, δ^{18}O = 0‰) and Standard Light Antarctic Precipitation (δ^{2}H = −428‰, δ^{18}O = −55.5‰). The analytical precision was generally 0.5‰ for δ^{2}H and 0.1‰ for δ^{18}O [39]. The δ^{18}O and δ^{2}H values are expressed as follows:

$$\delta^{2}\mathrm{H}_{\text{V-SMOW}} = \left(\frac{{}^{2}H/{}^{1}H_{sample}}{{}^{2}H/{}^{1}H_{standard}} - 1 \right) \times 1000(‰) \tag{1}$$

$$\delta^{18}\mathrm{O}_{\text{V-SMOW}} = \left(\frac{{}^{18}O/{}^{16}O_{sample}}{{}^{18}O/{}^{16}O_{standard}} - 1 \right) \times 1000(‰) \tag{2}$$

EC analysis: Electrical conductivity (EC) was measured in situ with a conductivity meter. EC was measured concurrently with stream sampling using a standard conductivity cell (WTW Cond 340iTM). The standard conductivity cell was calibrated to correct for water temperature to 25 °C.

D-excess calculation: The deuterium excess (d-excess) was used to measure the isotopic variability [19] and is defined as:

$$\text{d-excess} = \delta^{2}\mathrm{H} - 8 \times \delta^{18}\mathrm{O} \tag{3}$$

3.3. Data Analysis

We explored the spatial characteristics of the isotopes based on one-way analysis of variance (ANOVA) by using SPSS 17.0. Isotopic contents in water bodies of different main stream tributaries are expressed by box plot. ArcGIS of ESRI is applied to display spatial features of the Naqu River basin based on inverse distance weighting (IDW).

We analyzed samples for the two tracers collected from snowmelt, rain, stream water, and groundwater. Respecting the water and tracer mass conservation, electrical conductivity was measured in situ with a conductivity meter. The percentage of different components in the total runoff was determined using isotope hydrologic separation. If we suppose the objective percentage of n components are evaluated based on n parts and $n-1$ measuring factors t_1, $t_2, \cdots, t_{n-1}$, there are n linear mixing equations. These are defined as follows:

$$Q_T = Q_1 + Q_2 + \cdots + Q_n \tag{4}$$

$$C_T^{t_i} Q_T = C_1^{t_i} Q_1 + C_2^{t_i} Q_2 + \cdots + C_n^{t_i} Q_n \tag{5}$$

where Q_T is the total flow of the river; $Q_1, Q_2, \cdots, Q_n$ represent the flows of different water sources of runoff, and $C_1^{t_1}, C_2^{t_2}, \cdots, C_i^{t_i}$ represent the concentrations of relevant observed tracer t_i.

When the runoff contains only two sources of water, the percentages of different components of the runoff can be expressed as follows:

$$Q_1/Q_s = (C_2 - C_s)/(C_2 - C_1) \tag{6}$$

$$Q_2/Q_s = (C_s - C_1)/(C_2 - C_1) \tag{7}$$

where Q is the runoff of each component, C is the concentrations of relevant observed tracer, and s is the total flow.

When the runoff contains only three sources of water, groundwater, snowmelt, and precipitation are the main components of runoff in August in the Naqu River basin. Hydrograph separation is used to calculate the various composition of the runoff based on two tracers (δ^2H and EC). Supposing the river flow is a function of snowmelt, groundwater, and precipitation, then the three-component sources model can be defined as follows:

$$f_p + f_g + f_m = 1 \tag{8}$$

$$f_pQ_p + f_gQ_g + f_mQ_m = Q_r \tag{9}$$

$$f_pE_p + f_gE_g + f_mE_m = E_r \tag{10}$$

where f_p, f_g, f_m represent the shares of the individual components in the total runoff, and Q and E represent the concentrations of tracers.

4. Results

4.1. Spatial Characteristics of $\delta^{18}O$ and δ^2H

In August 2017, δ^{18}O values of runoff vary from −15.6‰ to −10.5‰ with a mean of −15.49‰. The δ^{18}O values of snowmelt water range from −15.0‰ and −7.6‰ with an average of −11.4‰. The δ^{18}O values of runoff vary from −15.49‰ to −14.27‰ (Table 1). For groundwater, the δ^{18}O values are relatively stable, ranging from −19.03‰ to −17.66‰, which indicated that the surrounding environment had little influence on groundwater and the recharge source of groundwater was relatively stable (Table 1).

Table 1. Oxygen isotope composition of different types of water in the Naqu River basin.

Water type		2016.8	2017.8	2018.8
Stream water	Sample number	13	13	13
	Mean of δ^{18}O (‰)	−14.27	−15.49	−14.83

Water type		2016–2018
Snowmelt water	Sample number	5
	Mean of δ^{18}O (‰)	−11.37
Groundwater	Sample number	2
	Mean of δ^{18}O (‰)	−18.51
Rain	Sample number	2
	Mean of δ^{18}O (‰)	−18.35

The results of elevation effect analysis on the collected rivers (Table 2) show that the isotopes in runoff do not change with elevation (Figure 2). All values are plotted against altitude. We hypothesize that the water body experienced intense evaporative fractionation due to the slow river flow rate in the Naqu River basin.

Table 2. Average values of δ^2H and δ^{18}O of main streams and tributaries.

Location	Sampling Sites	Sample Number	δ^{18}O (‰)	δ^2H (‰)	Longtitude (E)	Latitude (N)	Altitude (m a.s.l.)
Main stream−1	4	3	−15.68	−118.58	92°02′38.7″	31°19′52″	4451
Main stream−2	8	3	−15.48	−116.32	91°44′17.7″	31°37′15.2″	4551
Bazongqu	9	3	−16.88	−124.93	91°42′51.9″	31°58′40.2″	4622
Chengqu	3	3	−16.37	−124.20	92°03′34.9″	31°29′44.7″	4503
	13	3	−17.03	−127.50	92°02′19.3″	31°31′39.5″	4519
Gongqu	5	3	−15.83	−119.46	92°09′30.5″	31°13′32.5″	4498
	6	3	−16.21	−121.16	92°14′23.2″	31°08′21.8″	4578
Mugequ	2	3	−15.17	−113.81	91°41′23.3″	31°3′17.9″	4681
	14	3	−15.26	−115.65	91°46′33.3″	31°11′56″	4591
Mumuqu	10	3	−14.97	−113.63	91°41′23.4″	32°06′5.1″	4712
Najinqu	12	3	−14.81	−111.99	91°42′38.5″	32°22′30.3″	4771
Sangqu	11	3	−14.57	−111.32	91°40′44.3″	32°11′22″	4626
Zongqingqu	7	3	−13.18	−103.02	92°25′42.2″	31°41′12.6″	4567

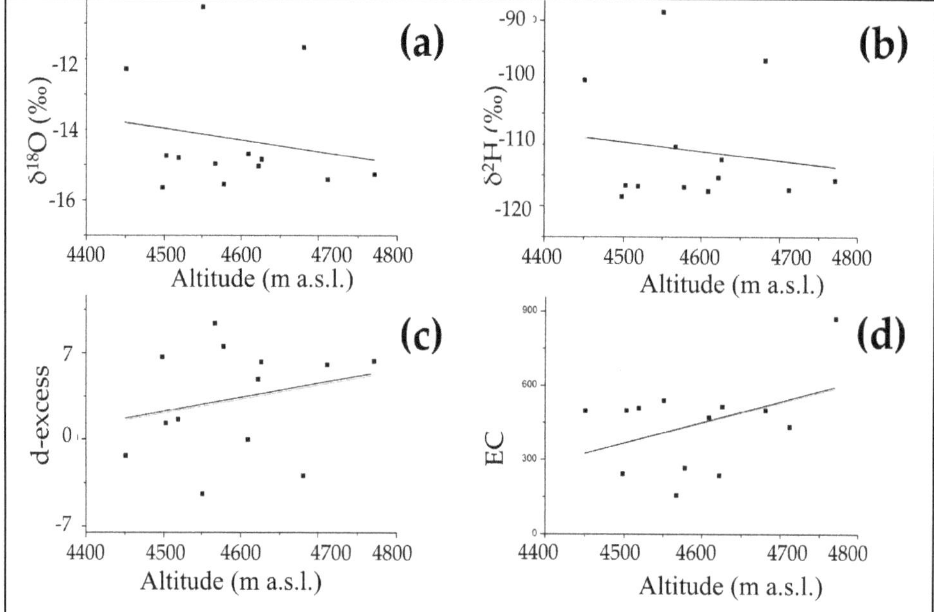

Figure 2. (a) δ^{18}O altitude, (b) δ^2H altitude, (c) d-excess altitude, and (d) electrical conductivity (EC) altitude relationship.

δ^{18}O and δ^2H values are shown in box plots for all the sampling sites (Figure 3).

Our analysis showed a significant spatial trend based on one-way ANOVA ($p < 0.05$) at 13 sampling sites between Main stream-2 and tributaries (Najinqu, Sangqu, Bazongqu, Mumuqu, Chengqu, Zongqungqu, Mugequ, and Gongqu). We speculate that isotopic evaporative fractionation in water and mixing of different water sources are the reasons for the spatial difference in isotopic values between Main stream-2 and tributaries.

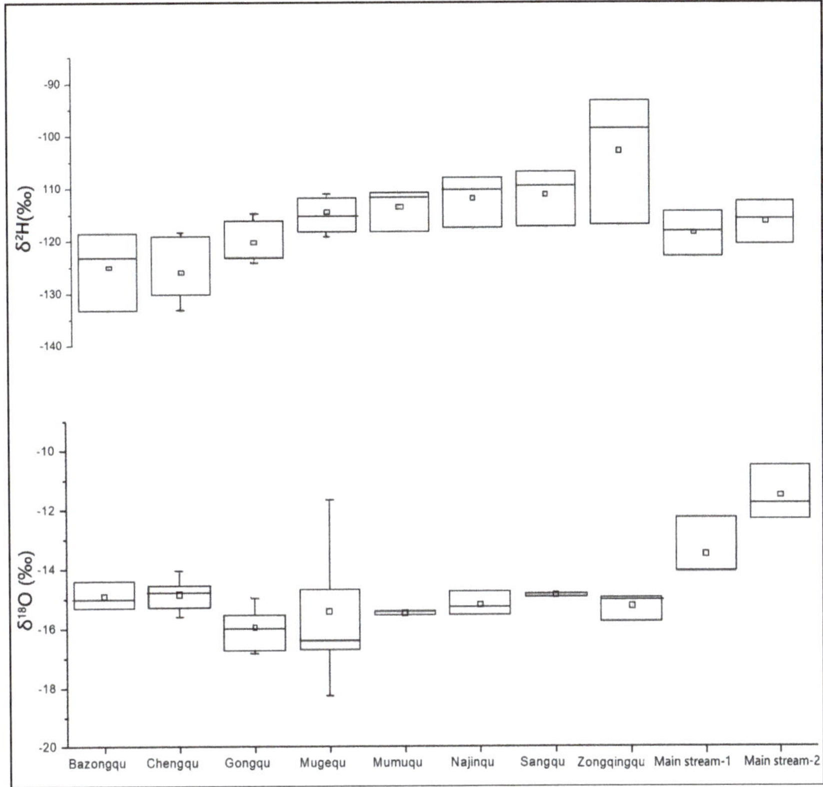

Figure 3. Box plots for δ^2H and δ^{18}O of runoff.

4.2. Isotopic Characterization of River

Craig [12] found that stable isotope ratios of δ^{18}O and δ^2H in precipitation correlate at a global scale in a linear relationship known as the global meteoric water line (GMWL). A linear relationship between δ^{18}O and δ^2H was established for average local meteoric waters as the local meteoric water line (LMWL). Important information about the water sources of precipitation can be revealed based on the deviation between LMWL and GMWL. By the location characteristics of different water samples, the water sources of rivers and the isotopic evaporative fractionation can be analyzed. In this paper, the LMWL of Lhasa region is adopted to replace the LMWL of the Naqu River basin. The LMWL can be described as: δ^2H = 7.9δ^{18}O + 6.29 [43]. Compared to the LMWL, some sets of isotopic data with high δ^{18}O values are below the LMWL, which signifies the effect of intensive evaporation processes.

By comparing different water samples with the LMWL, the water sources of the river and the isotopic evaporative fractionation can be analyzed. Most of the river sampling sites are close to LMWL (δ^2H = 7.9δ^{18}O + 6.29) (Figure 4). At the same time, many samples are close to each other, indicating that the water sources of these tributaries are relatively similar. The river water line is δ^2H = 5.75δ^{18}O − 27.98. The groundwater and snowmelt samples are distributed around the river samples, indicating that the water is originated from local rainfall and runoff is recharged by groundwater, snowmelt, and precipitation.

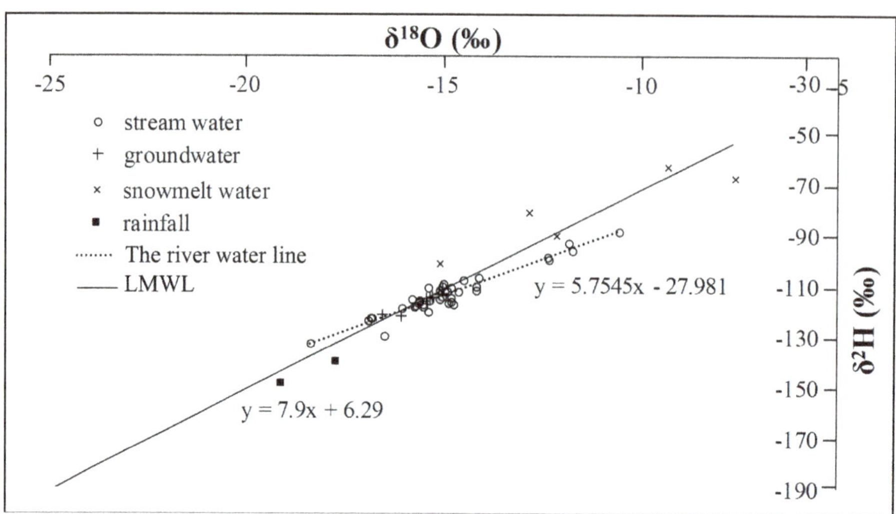

Figure 4. Plot of δ^2H versus δ^{18}O for different water sources.

The isotopic values of river water samples are closer to groundwater than those of snowmelt, indicating the frequent interaction between groundwater and runoff (Figure 4). Meanwhile, the slope and intercept are both smaller than that of LMWL, indicating that the water body in the Naqu River basin have experienced an obvious evaporation process.

The isotopes of snowmelt in winter appear to be the most enriched compared with other water sources, which is due to evaporation. When the snow begins to melt, the influence of evaporative fractionation increases, and the content of heavy isotopes in the meltwater increases.

4.3. Hydrograph Separation

Based on the formulas provided above, we calculated the contributions of rain, groundwater, and snowmelt by isotopic hydrograph separation in 2018 (Table 3, Figure 5). The proportion of snowmelt in runoff components ranges from 15% to 47%, and the proportion of rainwater ranges from 3% to 35%. The main components of runoff in the Naqu River are snowmelt and groundwater.

Table 3. Contribution of different water sources (^{2}H, ‰; EC, ms/cm).

Tributary	Mean Elevation	River Water		Snowmelt		Groundwater		Rainfall		Contribution (%)		
	(m a.s.l.)	D	EC	D	EC	D	EC	D	EC	Snowmelt	Groundwater	Rainfall
Bazongqu	4622	−112	0.21	−87	0.12	−122	0.50	−144	0.01	45%	30%	24%
Chengqu	4519	−115	0.31	−87	0.12	−122	0.50	−144	0.01	29%	56%	16%
Gongqu	4578	−119	0.22	−87	0.12	−122	0.50	−144	0.01	29%	36%	35%
Mugequ	4609	−121	0.31	−87	0.12	−122	0.50	−144	0.01	18%	58%	24%
Mumuqu	4712	−117	0.43	−87	0.12	−122	0.50	−144	0.01	15%	82%	3%
Najinqu	4771	−114	0.21	−87	0.12	−122	0.50	−144	0.01	40%	32%	29%
Sangqu	4626	−112	0.21	−87	0.12	−122	0.50	−144	0.01	45%	30%	25%
Zongqingqu	4567	−113	0.17	−87	0.12	−122	0.50	−144	0.01	46%	23%	31%

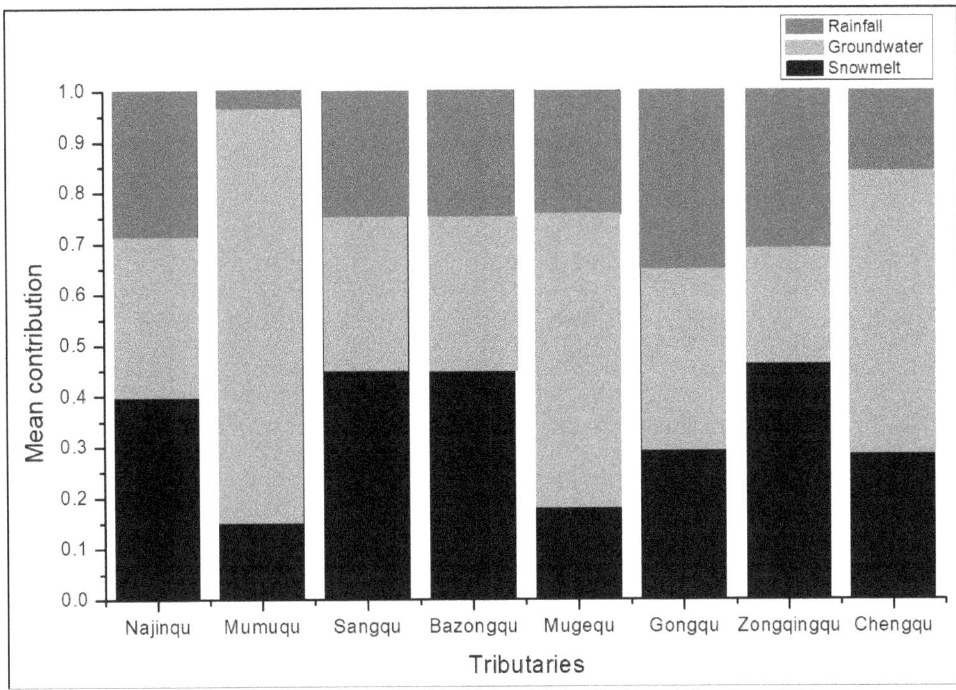

Figure 5. Contribution of different water sources.

5. Discussion

5.1. Analysis of Spatial Variations of δ^2H and $\delta^{18}O$ Values of the River

Our analysis showed an insignificant spatial trend of either δ^2H or $\delta^{18}O$ among tributaries Najinqu, Sangqu, Bazongqu, Mumuqu, Chengqu, Zongqungqu, Mugequ, and Gongqu in August. However, there is a significant spatial variation trend based on one-way ANOVA between Main stream-2 and tributaries (Najinqu, Sangqu, Bazongqu, Mumuqu, Chengqu, Zongqungqu, Mugequ, and Gongqu) (Figure 3). Although elevation effects play an important role in isotopic variation in large topographic area, there was no obvious elevation effects between Main stream-2 and tributaries (Figure 2). Surface and groundwater samples are often below the LMWL and GMWL under intense evaporative fractionation and low humidity. In the Naqu River, some water samples deviate from the LMWL, and the waters experience intense evaporative fractionation due to the slow river flow rate.

For groundwater, the $\delta^{18}O$ values were relatively stable, ranging from −19.03‰ to −17.66‰, indicating that the surrounding environment has little influence on groundwater and the recharge source of groundwater is relatively stable (Table 1)., The groundwater was recharged by old water stored in the basin previously.

5.2. Estimation of Different Water Sources Contribution to the River Flow

Based on the analysis of runoff components, runoff of the Naqu River can be divided into three water sources by isotopic hydrograph separation: Groundwater, rain, and snowmelt. The calculation results show that snowmelt of most tributaries contributed more than 30% to the runoff, while the proportion of rain ranges from 3% to 35% in the Naqu River basin. The results of hydrologic separation show that during the wet season, the river sources are mainly meltwater, and groundwater, with groundwater accounting for the largest proportion (more than 50%). Groundwater and snowmelt account for a greater proportion of runoff composition in the Naqu River basin. Such results are different from the runoff data collected at the hydrological station (Figure 6). In the past, we believed that the changing trend of runoff was completely controlled by precipitation. We speculate that the main components of runoff are snowmelt water and groundwater, while rain affects the change of runoff in the Naqu River.

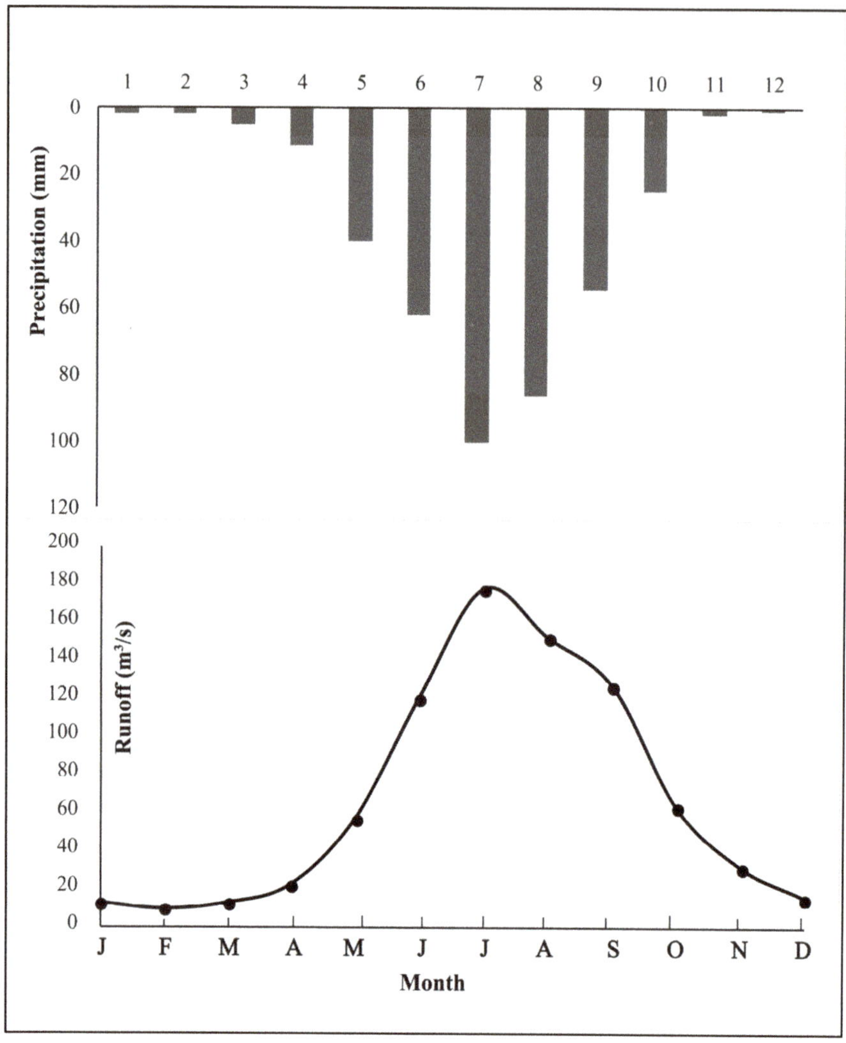

Figure 6. Monthly variations of precipitation and runoff in the Naqu River basin.

As shown in Figure 7, the contributions of groundwater of Mumuqu, Mugequ, and Chengqu are bigger than those of other tributaries. The contributions of snowmelt of Najinqu, Sangqu, Bazongqu, and Zongqingqu Rivers are bigger than those of other tributaries. We speculate that this phenomenon is related to the elevation characteristics of the Naqu River basin (Figure 1). At lower altitudes to the south, the recharge of groundwater in the river is stronger. The loose structure of the rocks, large areas of grassland, and abundant melt-water make the area relatively permeable. At higher altitudes in the north and east, the recharge of snowmelt to the river is stronger. The results of elevation effect analysis on the collected samples of the main stream and tributaries showed that isotopes in runoff do not change with elevation. There is a certain correlation between runoff composition and elevation, particularly related to the proportion of groundwater and snowmelt. Groundwater contributes more to the river in the central and western regions. Spatially, in the Naqu River, meltwater contributes more than 30% to runoff in the north, east, and south. Our results show that the groundwater and snowmelt water have different dominant effects on runoff composition from the upper to the lower reaches in the Naqu River. And the results can be referred to for near-future assessments of changes in discharge in the basin.

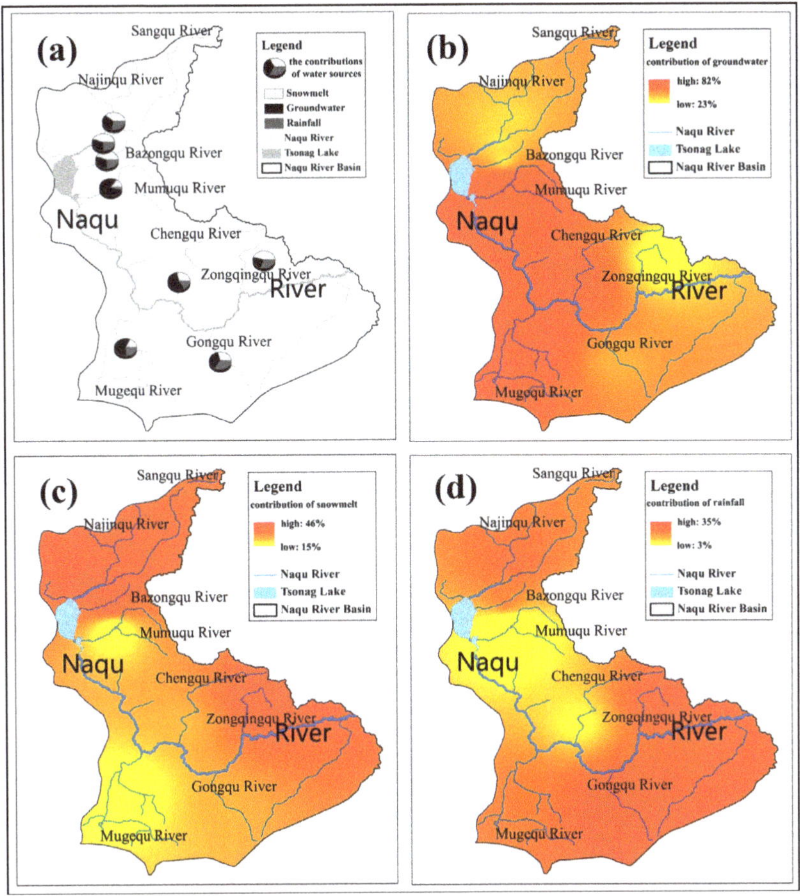

Figure 7. Spatial variation of contributions of numerous water sources in the Naqu River basin: (**a**) numerous water sources (**b**) groundwater, (**c**) snowmelt, and (**d**) rainfall.

6. Conclusions

We analyze the spatial variations of $\delta^2 H$ and $\delta^{18}O$ with influencing factors and the sources of runoff in August, 2016–2018, for the Naqu River at the source of the Nu River on the Qinghai–Tibet. Our analysis showed an insignificant spatial trend of either $\delta^2 H$ or $\delta^{18}O$ values among the tributaries Najinqu, Sangqu, Bazongqu, Mumuqu, Chengqu, Zongqungqu, Mugequ, and Gongqu in August. However, there is a significant spatial variation trend based on one-way ANOVA at 13 sampling sites between Main stream-2 and tributaries (Najinqu, Sangqu, Bazongqu, Mumuqu, Chengqu, Zongqungqu,

Mugequ, and Gongqu). Isotopic evaporative fractionation in water and mixing of different water sources are the reasons for the spatial difference of isotopic values between Main stream-2 and tributaries. Runoff of the Naqu River can be divided into three water sources: Groundwater, rainwater, and snowmelt. The proportion of snowmelt in runoff components ranges from 15% to 47%, and the proportion of rainwater ranges from 3% to 35%. Thus, the main components of runoff are snowmelt and groundwater, while rain affects the change of runoff.

Author Contributions: X.C. and F.W. conceived, designed, and drafted the manuscript; X.C., G.W., D.Y. and H.Z. planned and designed the methodology; X.C. revised the manuscript; F.W. guided and supervised the whole process; and all authors read and approved the final manuscript.

Funding: This work was supported by the National Natural Science Foundation of People's Republic of China (51879106, 51709111), the Major Research Plan of the National Natural Science Foundation of China (91547209), National Key Research and Development Program of China (2016YFC0401401), the Distinguished Young Scholar of Science and Technology Innovation (184100510014), the Science and Technology Innovation Team in Universities of Henan Province (20IRTSTHN010).

Acknowledgments: The authors would like to express their sincere gratitude to the anonymous reviewers for their constructive comments and useful suggestions that helped us improve our paper.

Conflicts of Interest: No conflict of interest exists in the submission of this manuscript, and the manuscript is approved by all authors for publication.

References

1. Kong, Y.; Pang, Z. Evaluating the sensitivity of glacier rivers to climate change based on hydrograph separation of discharge. *J. Hydrol.* **2012**, *434*, 121–129. [CrossRef]

2. Wang, X.; Li, Z.; Ross, E.; Tayier, R.; Zhou, P. Characteristics of water isotopes and hydrograph separation during the spring flood period in Yushugou River basin, Eastern Tianshans, China. *J. Earth Syst. Sci.* **2015**, *124*, 115–124. [CrossRef]

3. Wang, Z.; Zhou, C.; Guan, B.; Deng, Z.; Zhi, Y.; Liu, Y.; Xu, C.; Fang, S.; Xu, Z.; Yang, H. The Headwater Loss of the Western Plateau Exacerbates China's Long Thirst. *AMBIO* **2006**, *35*, 271. [PubMed]

4. Zhou, J.; Wu, J.; Liu, S.; Zeng, G.; Qin, J.; Wang, X.; Zhao, Q. Hydrograph Separation in the Headwaters of the Shule River Basin: Combining Water Chemistry and Stable Isotopes. *Adv. Meteorol.* **2015**, *2015*, 1–10. [CrossRef]

5. Barthold, F.K.; Wu, J.K.; Vaché, K.B.; Schneider, K.; Frede, H.G.; Breuer, L.; Tetzlaff, D.; Carey, S.K.; Laudon, H.; Mcguire, K. Identification of geographic runoff sources in a data sparse region: Hydrological processes and the limitations of tracer-based approaches. *Hydrol. Process.* **2010**, *24*, 2313–2327. [CrossRef]

6. Pu, T.; He, Y.; Zhu, G.; Zhang, N.; Du, J.; Wang, C. Characteristics of water stable isotopes and hydrograph separation in Baishui catchment during the wet season in Mt.Yulong region, south western China. *Hydrol. Process.* **2013**, *27*, 3641–3648. [CrossRef]

7. Buttle, J.M. Isotope hydrograph separations and rapid delivery of pre-event water from drainage basins. *Progr. Phys. Geogr.* **1994**, *18*, 16–41. [CrossRef]

8. Liu, Y.; Fan, N.; An, S.; Bai, X.; Liu, F.; Xu, Z.; Wang, Z.; Liu, S. Characteristics of water isotopes and hydrograph separation during the wet season in the Heishui River, China. *J. Hydrol.* **2008**, *353*, 314–321. [CrossRef]

9. Sun, C.; Chen, Y.; Li, W.; Li, X.; Yang, Y. Isotopic time series partitioning of streamflow components under regional climate change in the Urumqi River, northwest China. *Hydrol. Sci. J.* **2016**, *61*, 1443–1459.

10. Wu, H.; Zhang, X.; Li, X.; Li, G.; Huang, Y. Seasonal variations of deuterium and oxygen-18 isotopes and their response to moisture source for precipitation events in the subtropical monsoon region. *Hydrol. Process.* **2015**, *29*, 90–102. [CrossRef]

11. James, C. Hydrogeological Conceptual Model of Groundwater from Carbonate Aquifers Using Environmental Isotopes (^{18}O, ^{2}H) and Chemical Tracers: A Case Study in Southern Latium Region, Central Italy. *J. Water Resour. Prot.* **2012**, *4*, 695–716.

12. Craig, H. Isotopic variations in meteoric waters. *Science* **1961**, *133*, 1702–1703. [CrossRef] [PubMed]

13. Longinelli, A.; Selmo, E. Isotopic composition of precipitation in Italy: A first overall map. *J. Hydrol.* **2003**, *270*, 75–88.

14. Gibson, J.J.; Edwards, T.W.D.; Birks, S.J.; St Amour, N.A.; Buhay, W.M.; McEachern, P.; Wolfe, B.B.; Peters, D.L. Progress in isotope tracer hydrology in Canada. *Hydrol. Process.* **2005**, *19*, 303–327. [CrossRef]

15. Rank, D.; Wyhlidal, S.; Schott, K.; Weigand, S.; Oblin, A. Temporal and spatial distribution of isotopes in river water in Central Europe: 50 years experience with the Austrian network of isotopes in rivers. *Isot. Environ. Health Stud.* **2018**, *54*, 115–136. [CrossRef]

16. Zhang, W.; Cheng, B.; Hu, Z.; An, S.; Xu, Z.; Zhao, Y.; Cui, J.; Xu, Q. Using stable isotopes to determine the water sources in alpine ecosystems on the east Qinghai-Tibet plateau, China. *Hydrol. Process.* **2010**, *24*, 3270–3280. [CrossRef]

17. Kendall, C.; Coplen, T.B. Distribution of oxygen-18 and deuterium in river waters across the United States. *Hydrol. Process.* **2001**, *15*, 1363–1393. [CrossRef]

18. Liu, W.; Chen, S.; Xiang, Q.; Baumann, F.; Scholten, T.; Zhou, Z.; Sun, W.; Zhang, T.; Ren, J.; Qin, D. Storage, patterns, and control of soil organic carbon and nitrogen in the northeastern margin of the Qinghai–Tibetan Plateau. *Environ. Res. Lett.* **2012**, *7*, 35401–35412. [CrossRef]

19. Dansgaard, W. Stable isotopes in precipitation. *Tellus* **1964**, *16*, 436–468. [CrossRef]

20. Price, R.M.; Swart, P.K.; Willoughby, H.E. Seasonal and spatial variation in the stable isotopic composition (δ^{18}O and δ^2H) of precipitation in south Florida. *J. Hydrol.* **2008**, *358*, 193–205. [CrossRef]

21. Song, X.; Kayane, I.; Tanaka, T.; Shimada, J. A study of the ground cycle in Sri Lanka using stable isotopes. *Hydrol. Process.* **2015**, *13*, 1479–1496. [CrossRef]

22. Harrington, G.A.; Cook, P.G.; Herczeg, A.L. Spatial and temporal variability of ground water recharge in central Australia: A tracer approach. *Groundwater* **2010**, *40*, 518–527. [CrossRef] [PubMed]

23. He, Y.; Theakstone, W.H.; Yao, T.; Shi, Y. The isotopic record at an alpine glacier and its implications for local climatic changes and isotopic homogenization processes. *J. Glaciol.* **2017**, *47*, 147–151.

24. Fan, Z.; Ma, Y. Some Issues about the Exploitation and the Rational Utilization of Water Resources in the Arid Areas. *Arid Zone Res.* **2000**, *17*, 6–11.

25. Xu, Q. Application of Environmental Isotopes in Water Cycle of Forest Ecosystem. *World For. Res.* **2008**, *21*, 11–15.

26. Han, G.; Lv, P.; Tang, Y.; Song, Z. Spatial and temporal variation of H and O isotopic compositions of the Xijiang River system, Southwest China. *Isot. Environ. Health Stud.* **2018**, *54*, 137–146. [CrossRef] [PubMed]

27. Qu, S.; Wang, Y.; Zhou, M.; Liu, H.; Shi, P.; Yu, Z.; Xiang, L. Temporal ^{18}O and deuterium variations in hydrologic components of a small watershed during a typhoon event. *Isot. Environ. Health Stud.* **2017**, *53*, 172–183. [CrossRef]

28. Yang, Y.; Xiao, H.; Qin, Z.; Zou, S. Hydrogen and oxygen isotopic records in monthly scales variations of hydrological characteristics in the different landscape zones of alpine cold regions. *J. Hydrol.* **2013**, *499*, 124–131. [CrossRef]

29. Meredith, K.T.; Hollins, S.E.; Hughes, C.E.; Cendón, D.I.; Chisari, R.; Griffiths, A.; Crawford, J. Evaporation and concentration gradients created by episodic river recharge in a semi-arid zone aquifer: Insights from Cl$^-$, δ^{18}O, δ^2H, and ^{3}H. *J. Hydrol.* **2015**, *529*, 1070–1078. [CrossRef]

30. Biggs, T.W.; Lai, C.T.; Chandan, P.; Lee, R.M.; Messina, A.; Lesher, R.S.; Khatoon, N. Evaporative fractions and elevation effects on stable isotopes of high elevation lakes and streams in arid western Himalaya. *J. Hydrol.* **2015**, *522*, 239–249. [CrossRef]

31. Edirisinghe, E.A.; Pitawala, H.M.; Dharmagunawardhane, H.A.; Wijayawardane, R.L. Spatial and temporal variation in the stable isotope composition (δ^{18}O and δ^2H) of rain across the tropical island of Sri Lanka. *Isot. Environ. Health Stud.* **2017**, *53*, 628–645. [CrossRef] [PubMed]

32. Wassenaar, L.I.; Athanasopoulos, P.; Hendry, M.J. Isotope hydrology of precipitation, surface and ground waters in the Okanagan Valley, British Columbia, Canada. *J. Hydrol.* **2011**, *411*, 37–48. [CrossRef]

33. Sánchez-Murillo, R.; Esquivel-Hernández, G.; Sáenz-Rosales, O.; Piedra-Marín, G.; Fonseca-Sánchez, A.; Madrigal-Solís, H.; Ulloa-Chaverri, F.; Rojas-Jiménez, L.D.; Vargas-Víquez, J.A. Isotopic composition in precipitation and groundwater in the northern mountainous region of the Central Valley of Costa Rica. *Isot. Environ. Health Stud.* **2016**, *53*, 1–17. [CrossRef] [PubMed]

34. Jin, L.; Siegel, D.I.; Lautz, L.K.; Lu, Z. Identifying streamflow sources during spring snowmelt using water chemistry and isotopic composition in semi-arid mountain streams. *J. Hydrol.* **2012**, *470*, 289–301. [CrossRef]

35. Rock, L.; Mayer, B. Isotope hydrology of the Oldman River basin, southern Alberta, Canada. *Hydrol. Process.* **2010**, *21*, 3301–3315. [CrossRef]

36. Pinder, G.F.; Jones, J.F. Determination of the Ground-Water Component of Peak Discharge from the Chemistry of Total Runoff. *Water Resour. Res.* **1969**, *5*, 438–445. [CrossRef]

37. Aggarwal, P.K. Isotope hydrology at the International Atomic Energy Agency. *Hydrol. Process.* **2002**, *16*, 2257–2259. [CrossRef]

38. Christophersen, N.; Hooper, R.P. Multivariate Analysis of Stream Water Chemical Data: The Use of Principal Component Analysis for the End-Member Mixing Problem. *Water Resour. Res.* **1992**, *28*, 99–107. [CrossRef]

39. Chen, X.; Wang, G.; Wang, F. Classification of Stable Isotopes and Identification of Water Replenishment in the Naqu River Basin, Qinghai-Tibet Plateau. *Water* **2019**, *11*, 46. [CrossRef]

40. Vespasiano, G.; Apollaro, C.; De Rosa, R.; Muto, F.; Larosa, S.; Fiebig, J.; Mulch, A.; Marini, L. The Small Spring Method (SSM) for the definition of stable isotope-elevation relationships in Northern Calabria (Southern Italy). *Appl. Geochem.* **2015**, *63*, 333–346. [CrossRef]

41. Vespasiano, G.; Apollaro, C.; Muto, F.; Dotsika, E.; De Rosa, R.; Marini, L. Chemical and isotopic characteristics of the warm and cold waters of the Luigiane Spa near Guardia Piemontese (Calabria, Italy) in a complex faulted geological framework. *Appl. Geochem.* **2014**, *41*, 73–88. [CrossRef]

42. Apollaro, C.; Dotsika, E.; Marini, L.; Barca, D.; Bloise, A.; De Rosa, R.; Doveri, M.; Lelli, M.; Muto, F. Chemical and isotopic characterization of the thermo mineral water of Terme Sibarite springs (Northern Calabria, Italy). *Geochem. J.* **2012**, *46*, 117–129. [CrossRef]

43. Li-De, T.; Shan-Dong, Y.; Wei-Zhen, S.; Stievenard, M.; Jouzel, J. Relationship between δ^2H and $\delta^{18}O$ and water vapour circulation in the precipitation of the qinghai-tibet plateau. *Sci. Sin. Terrae* **2001**, *31*, 214–220.

Article

What Do Plants Leave after Summer on the Ground?—The Effect of Afforested Plants in Arid Environments

César Dionisio Jiménez-Rodríguez [1,2,*], **Miriam Coenders-Gerrits** [1], **Stefan Uhlenbrook** [3] and **Jochen Wenninger** [4]

[1] Department of Water Management, Water Resources Section, Delft University of Technology, 2600 GA Delft, The Netherlands; A.M.J.Coenders@tudelft.nl

[2] Escuela de Ingeniería Forestal, Tecnológico de Costa Rica, P.O. Box 159-7050, Cartago 30101, Costa Rica

[3] UNESCO World Water Assessment Programme, Via dei Ceraioli, 45, 06134 Perugia, PG, Italy; s.uhlenbrook@unesco.org

[4] Water Science and Engineering Department, IHE Delft, 2611 AX Delft, The Netherlands; j.wenninger@un-ihe.org

* Correspondence: cdjimenezcr@gmail.com

Received: 15 October 2019; Accepted: 30 November 2019; Published: 4 December 2019

Abstract: The implementation of afforestation programs in arid environments in northern China had modified the natural vegetation patterns. This increases the evaporation flux; however, the influence of these new covers on the soil water conditions is poorly understood. This work aims to describe the effect of Willow bushes (*Salix psammophila* C. Wang and Chang Y. Yang) and Willow trees (*Salix matsudana* Koidz.) on the soil water conditions after the summer. Two experimental plots located in the Hailiutu catchment (Shaanxi province, northwest China), and covered with plants of each species, were monitored during Autumn in 2010. The monitoring included the soil moisture, fine root distribution and transpiration fluxes that provided information about water availability, access and use by the plants. Meanwhile, the monitoring of stable water isotopes collected from precipitation, soil water, groundwater and xylem water linked the water paths. The presence of Willow trees and Willow bushes reduce the effect of soil evaporation after summer, increasing the soil moisture respect to bare soil conditions. Also, the presence of soil water with stable water isotope signatures close to groundwater reflect the hydraulic lift process. This is an indication of soil water redistribution carried out by both plant species.

Keywords: stable water isotopes; hydrogen; oxygen; soil water; fine root system

1. Introduction

Continental arid environments are characterized by excessive heat and variable precipitation distributed all over the year, with a tendency to peak during summer months [1–3]. These conditions favoured the presence of a discontinuous vegetation cover characterized by banded and spotted shapes, large size variability and specialized plant species [3–5]. The northern arid lands in China are an example of this type of environment, where the landscape is shaped by eolic erosion due to the high erodability of this soil type and the scarce ground cover protection [6–10]. Consequently, desertification in this region registered a strong growth of barren areas before 1999 [11]. However, after 2005 the plant cover experienced a positive change, reducing the areas affected by desertification thanks to the rehabilitation and afforestation programs established in the region [11,12]. The current implementation of afforestation and agricultural programs modified the landscape cover with additional crop areas. These afforestation practices trigger a series of impacts to the environment due to the inadequate selection of plant species [13,14]. This increment in vegetation cover reduces the local surface temperature [15] and affects the local evaporation flux due to the increment of plant transpiration which depends mostly on groundwater [16–18].

The evaporation (E) of arid environments is mainly composed of soil evaporation (E_s) and a small proportion of intercepted water by plant surfaces (E_i) and transpiration (E_t) [19–21]. The low precipitation rates underline the importance of soil water and groundwater availability for the plants. Rainfall interception decreases the water infiltration rates of vegetated areas in respect to bare soil conditions in arid and semi-arid regions [22,23]. This is the result of the quick evaporation of the intercepted water on the leaves, branches and stem of the plants [19,21]. The relevance of interception increases considering the precipitation characteristics of the arid and semi–arid regions where the low volume, high intensity, lower and irregular frequency hinder the plant water acquisition [5]. Due to the scarce water resources in these regions the plants are adapted to quickly respond to environmental triggers such as the irregular rains [24]. Thus increases the soil water acquisition by the plants and consequently its transpiration momentarily [25,26].

The plant root system provides anchorage for the plant and an effective water extraction system [27] which is powered by the plant transpiration [28]. This system absorbs the water close to the meristematic region of the root, transporting it through the xylem towards the leaves and using it during photosynthesis [28–30]. However, the presence of young roots in soil layers does not mean effective absorption of water from those zones [31]. Instead, some species are able to absorb water through suberized roots under soft drought or winter conditions [28,30]. As a consequence, the identification of plant water sources is a difficult task that requires the use of tracers.

Determination of water sources for the plants has been successfully done with the stable water isotopes oxygen ($\delta^{18}O$) and hydrogen (δ^2H) [18,27,32–40]. The specific isotopic signatures of soil water is the result of a fractionation process that modifies the isotope composition [41,42], allowing to trace the water paths within the ecosystem [43]. The isotope signature of the absorbed water is not modified by plant uptake until the water reaches the photosynthetic tissues [27,31]. Here, the leaf tissues will become enriched by the escape of lighter isotopes [44]. Although the roots do not modify the soil water during uptake, the isotope signature of xylem water is affected by mixing processes when different water sources are used by the same plant. Barbeta et al. [41] briefly describe a series of analysis tools used for the determination of water sources used by plants. Some of these methods are the Bayesian isotope mixing models such as SIAR [45,46] and MixSIAR [47] or standard linear mixing models such as IsoSource [48,49]. The IsoSource model provides all the feasible combination of water source contributions keeping the mass balance principle. It uses only the isotope signature of the water sources and the xylem water as the final mixture. SIAR and MixSIR models require more complex data sets. These models require the isotope signatures of the sources and mixtures as well as their standard deviations and an enrichment factor. As a result, the models provide the statistical uncertainties and the optimal solution for the analyzed mixture. The IsoSource tool has been used to study sand dunes bushes, corn and cotton plantations, woody species and estuarine vegetation to determine the water sources of those covers [33,34,36,37,39,40]. Thus can provide information of the origin of water within the plant and if this water can be redistributed on the soil profile.

The implementation of afforestation programs in arid environments modify the distribution patterns of local vegetation, influencing the ratio between transpiration and evaporation ($\frac{E_t}{E}$) [50]. These changes together with the usual omission of interception of precipitation [19,21,51], the irregular rains [5] and the large capacity to transpire soil water by arid plants [52]; exert a lot of pressure on the scarce water resources of arid environments. This has been the case with the introduction of Willow trees (*Salix matsudana*) and Willow bushes (*Salix psammophila*) in afforestation programs in the Hailiutu catchment [16–18]. The transpiration of these species increased the demand on the groundwater resource, however its influence on the soil water conditions are poorly understood. This work aims to describe the effect of Willow trees and Willow bushes on the soil water conditions after the summer. The monitoring included the soil moisture, fine root distribution and transpiration fluxes that provided information about water availability, access and use by the plants. Meanwhile the monitoring of stable water isotopes collected from precipitation, soil water, groundwater and xylem water linked the water fluxes. This information provided an indication of the vegetation influence on the soil water conditions beneath the covers.

2. Materials and Methods

2.1. Study Site

The study site is located within the Hailiutu catchment (area: 2645 km^2) in Yulin County; Shaanxi province; Northwest China (Figure 1). This catchment is part of the Maowusu semi-desert, which is characterized by undulating sand dunes over and dominated by a xeric scrubland. The nearest meteorological stations (Dong Shen: 39.833° N–109.983° E; Yanchi: 37.800° N–107.383° E; and Yulin: 38.233° N–109.700° E) described a semi-arid continental climate with a mean annual precipitation of 386.1 mm year^{-1} and a mean annual temperature of 8.6 °C (seasonal range: -17.4 °C to 27.1 °C) based on 12 years of meteorological records (period: 2000–2011). The soil type is classified as Calcaric Arenosols (ARc) with a high concentration of basic cations (Ca^{2+}, Mg^{2+}, K$^+$ and Na$^+$) and a pH value over 8.0; with an excessive drainage due to its sandy texture [53]. The study site is composed of two experimental plots (see Appendix A Figure A1) located at 300 m from each other. The first plot is dominated by Willow bushes (*Salix psammophila* C. Wang & Chang Y. Yang) and has an area of 625 m^2 (25 m × 25 m). The second plot covers 81 m^2 (9 m × 9 m) and contains mainly individuals of Willow trees (*Salix matsudana* Koidz.) and Poplar trees (*Populus simonii* Carr.). In both plots soil water, groundwater, plant parameters and soil variables were measured between September and October 2010.

2.2. Hydrologic Data

Meteorological data was retrieved from the stations Dong Shen (1459 m a.s.l.), Yanchi (1356 m a.s.l.) and Yulin (1058 m a.s.l.). The climatic data was downloaded from the National Oceanic and Atmospheric Administration (NOAA) [54]. This data set contains daily values of total precipitation (mm day^{-1}) and daily means for temperature (°C), dewpoint (°C), wind speed (m s^{-1}) and atmospheric pressure (mbar). Due the lack of solar radiation measurements in the selected study period, this variable was estimated according to Allen et al. [55] for missing data. Once all data were determined, the reference evaporation (E_o) in mm day^{-1} was calculated with the FAO Penman–Monteith equation:

$$E_o = \frac{\Delta(R_n - G) + \rho_a C_p \frac{(e_s - e_a)}{r_a}}{\Delta + \gamma\left(1 + \frac{r_s}{r_a}\right)}, \tag{1}$$

where net radiation (R_n) and soil heat flux (G) are expressed in MJ m^{-2} day^{-1}. The vapour pressure deficit of the air ($e_s - e_a$) is based on the saturation vapor pressure (e_s) and actual vapor pressure (e_a) both measured in KPa. Δ is the slope of the vapour-pressure relationship (kPa °C^{-1}), γ is the psychrometric constant (0.054 kPa °C), ρ_a is the air density (1.225 kg m^{-3}) and c_p is the specific heat of the air (1.013 ×10^{-3} MJ kg^{-1} °C^{-1}). The wind speed (m s^{-1}) at 2 m height (u_2) was used to determine the aerodynamic resistance (r_a) and surface resistance (r_s). For daily time steps the soil heat flux is considered to equal 0 MJ m^{-2}day^{-1} due the small daily differences [55].

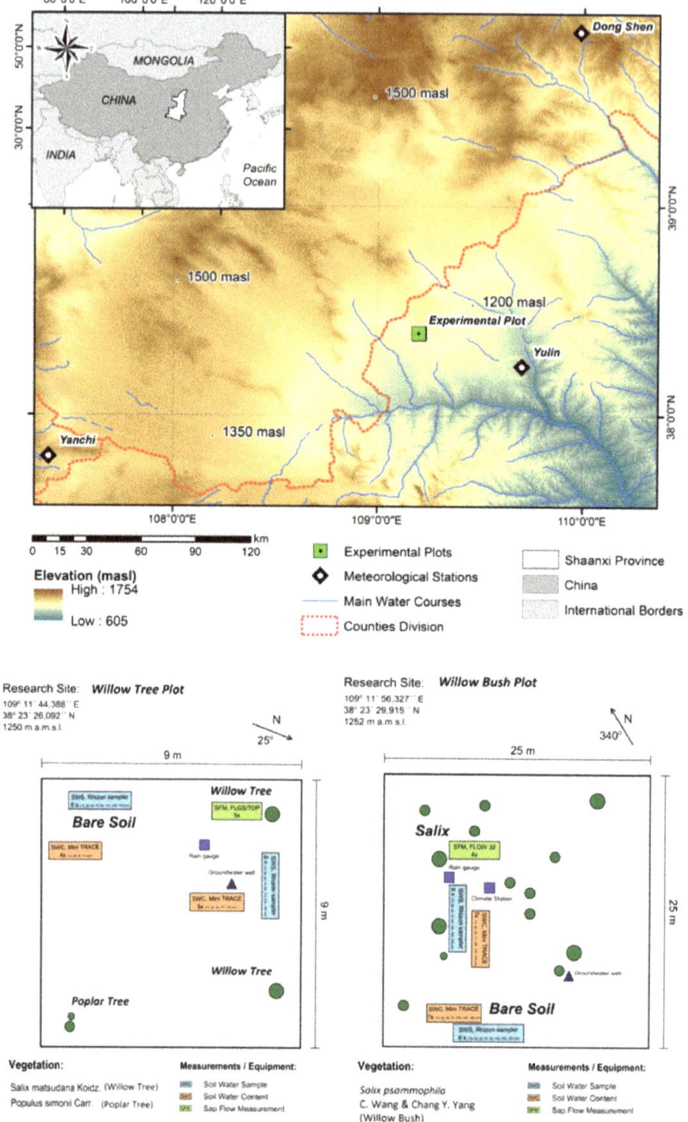

Figure 1. Geographical location of the experimental site and the meteorological stations Dong Shen, Yanchi and Yulin used during the study period in the Shaanxi province, China. The experimental design of both plots is shown on the bottom of the map.

Soil moisture (θ, m^3 m^{-3}) and groundwater level (h, m) measurements were carried out to describe the soil water dynamics in both sites. Soil moisture measurements were carried out sporadically along the study period. The reference values of soil moisture in sandy soils for permanent wilting point (θ_{WP}), field capacity (θ_{FC}) and saturation point (θ_{SP}) were 0.05 m^3 m^{-3}, 0.1 m^3 m^{-3} and 0.46 m^3 m^{-3}, respectively [56]. Soil moisture was monitored with a Mini-TRASE sensor (type: 6050X3K1B) and the probes were located at 10 cm, 20 cm, 40 cm, 70 cm and 100 cm depth beneath each species. On the Willow bush plot two more depths were monitored: 120 cm and 140 cm. Considering the presence of bare soil areas within the plots, the soil moisture was also monitored at the same depths as Willow bush. The groundwater monitoring wells were constructed with a manual soil auger thanks to the shallow groundwater level and sandy texture of the soil. The groundwater level was measured on a daily basis from the ground surface as the reference point with a Mini-Diver (type: DI 501) in each plot. Groundwater depth from the surface in both plots oscillates between 136 cm to 164 cm beneath the Willow bush plot and between 150 cm to 172 cm beneath Willow tree plot period between 21 August 2010 to 20 April 2011).

2.3. Water Sampling

Water samples were collected after each rainy day to determine the isotopic signature of the precipitation, groundwater, soil water and xylem water throughout the monitoring period. Soil water samples were taken with a Macro Rhizon SMS Eijkelkamp (length: 9 cm, diameter: 4.5 mm, porous diameter: 0.15 μm, part number: 19.21.SA) soil moisture sampler in both plots. The samples were collected at nine depths (10 cm, 20 cm, 40 cm, 70 cm, 90 cm, 110 cm, 140 cm, 150 cm and 160 cm), while the groundwater sampling depended on the water head elevation during the samplings. Xylem water was collected from an incision done at the twig of each tree; removing the bark, phloem and cambium to prevent the collection of fractionated sap water. The incision location was far from the meristematic region, avoiding the fractionation linked to photosynthesis. Rain water was collected during the events to prevent fractionation by evaporation on an event basis. Each sample was sealed hermetically in 1.5 mL vials and transported to The Netherlands for their analysis. The isotopic composition was determined with a LGR Liquid Water Isotope Analyzer (type: DLT-100) with a precision of $<0.3‰$ for $^{18}O/^{16}O$ and $<1.0‰$ for $^{2}H/^{1}H$ and expressed in respect to the Vienna Standard Mean Ocean Water (VSMOW). The isotopic signature of each sample was expressed in respect to the VSMOW through the following equation [57]:

$$\delta = \left(\frac{R_{sample}}{R_{standard}} - 1 \right) \tag{2}$$

where δ (‰) is the relative isotope composition of ^{18}O and ^{2}H, R_{sample} and $R_{standard}$ are the ratios of heavy to light isotopes ($^{18}O/^{16}O$ or $^{2}H/^{1}H$) of the sample and standard water, respectively.

2.4. Plant Parameters

For each plot the plant densities (plants ha^{-1}), canopy heights (m) and leaf area index (LAI, m^2 m^{-2}) were measured to describe the stand conditions. Transpired water (E_t) was monitored in the Willow shrubs establishing four ring gauges (type: Dynagage Energy Balance sensor, model: SGA3-WS and SGA5-WS) in an individual of Willow bush at 35 cm height; while five probes (type: Thermal Dissipation Probes, model: TDP-50) were installed in an individual of Willow at 1.3 m height. Each probe recorded the data at 10 minute intervals and those were summarized in an hourly and daily time step. Total mobilized water as transpiration was calculated with the product between the sapwood area and flow velocity. Considering the physiognomic differences between both plant species, the sapwood area was estimated accordingly with the plant type. Willow bush is a bush up to 4 m tall with numerous branchlets per plant [58], where most of the xylem within the branchlets is able to transport water. As a consequence, the sapwood area was measured through the average diameter of the measured branchlets. Willow tree is able to grow up to 10 m height with a symmetrical crown with a sole stem [59]. It has a clear differentiation between sapwood and hardwood, allowing to measure directly the sapwood from a tree wood ring obtained from the measured tree. The wood ring area was measured from inked water transported by capillary rise within the active sapwood sections. Sapwood area (A) for the Willow tree was 274.6 cm^2 and the average area for Willow bush was 5.1 cm^2. Transpiration flow for each plant was obtained though the empirical equation developed by Granier [60]:

$$E_t = 3600 \times 0.0119 \times \left(\frac{\partial T_m - \partial T}{\partial T} \right) \times A \times \rho, \qquad (3)$$

where E_t is the transpiration ($g\,h^{-1}$), ∂T is the vertical temperature difference (°C) measured within the plants, ∂T_m is the maximum temperature difference with zero E_t (°C), A is the cross section area (cm^2) and ρ is the water density ($g\,cm^{-3}$).

The fine root system was described through the total root biomass (TRB, $kg\,m^{-3}$) and the root length density (RLD, $cm\,cm^{-3}$). The survey involved the collection of 80 samples of soil per species with an auger of $300\,cm^3$ within a radius of 4.0 m. The sampling procedure was based on eight equidistant points from the stem towards the canopy edge, extracting 10 samples per point until a depth of 150 cm was reached. The samples were sieved to separate the soil from the roots, photographed on a scaled paper and dried up following the procedure proposed by Cornelissen et al. [61] to determine the root length density (RLD, $cm\,cm^{-3}$). The total root biomass was determined by weighing the dry cleaned roots with a digital balance. The total root length (cm) was determined by processing the root images with the use of the GIS free source software (www.gvsig.org). The total root length density was obtained dividing the total root length (cm) by the core volumes (cm^3) [62].

2.5. Data Analysis

Plant differences were determined using an Analysis of Covariance (ANCOVA) with a p_{value} of 0.05. Statistical differences were determined with a Tukey HSD analysis. A Pearson correlation analysis was applied to evaluate the influence of meteorological conditions on plant transpiration. All the statistical analyses are based on normal distributions, so the normality, variance homogeneity and presence of outliers were tested. The plant water source of transpiration was determined using the software IsoSource [48]. This model provides the relative contributions of soil water sources to sap flow in both species, based on the isotopic mass balance principle. Consequently, the isotopic soil water contribution analysis followed the "a posteriori aggregation" method proposed by Phillips et al. [48]. This method allows the aggregation of sources with similar isotopic signatures based on specific characteristics showed by the sources, reducing the number of contributing factors.

3. Results

Total precipitation in 2010 was $401.0\,mm\,year^{-1}$ at the experimental site, registering a slightly wet condition in respect to the regional average of $386.1\,mm\,year^{-1}$. However, this amount of precipitation does not supply the reference evaporation (E_o) of $1339.1\,mm\,year^{-1}$ at this site as a consequence of the irregular rain events (Figure 2). The $938.1\,mm\,year^{-1}$ difference between precipitation and reference evaporation support the Arid Steppe classification due to its annual water deficit [3,5,63,64]. September and October 2010 registered $48.2\,mm\,month^{-1}$ and $40.5\,mm\,month^{-1}$ of precipitation accounting for 12.0% and 10.1% of the annual amount, respectively (see Appendix B Figure A2). The water availability experienced during the study period allowed the presence of soil moisture above the permanent wilting point (θ_{WP}) for sandy soils ($0.05\,m^3\,m^{-3}$) while the field capacity (θ_{FC}) was exceeded only in the deepest layers in both plots (Figure 3). Additionally, soil moisture increases with depth in Willow bush and Willow tree stands, keeping higher values than under bare soil conditions. Soil moisture under both plant species has larger values in respect to bare soil condition until a depth of 100 cm (ANCOVA, F = 37.91, p = 0.0000). Average soil moisture shows the following order: Willow bush (θ: $0.11\,m^3\,m^{-3}$) > Willow tree (θ: $0.10\,m^3\,m^{-3}$) > Bare Soil (θ: $0.08\,m^3\,m^{-3}$).

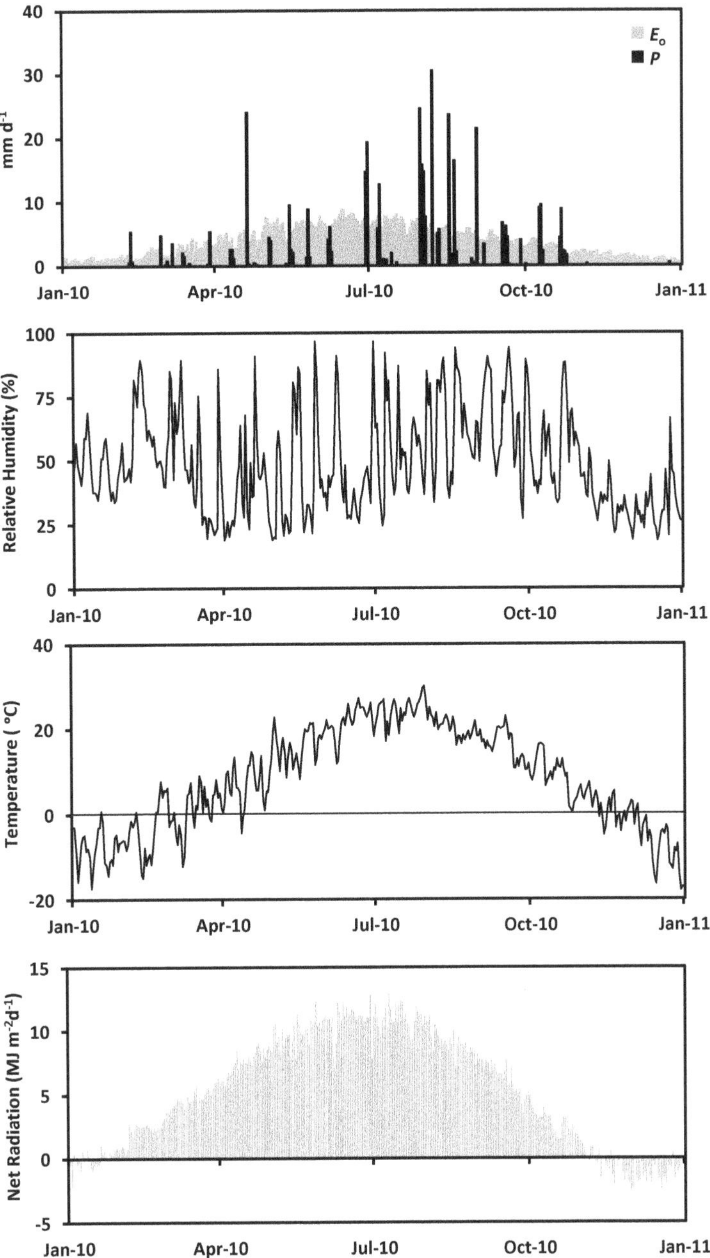

Figure 2. Meteorological conditions registered during 2010 at the research site based on the data of Dong Shen, Yanchi and Yulin meteorological stations.

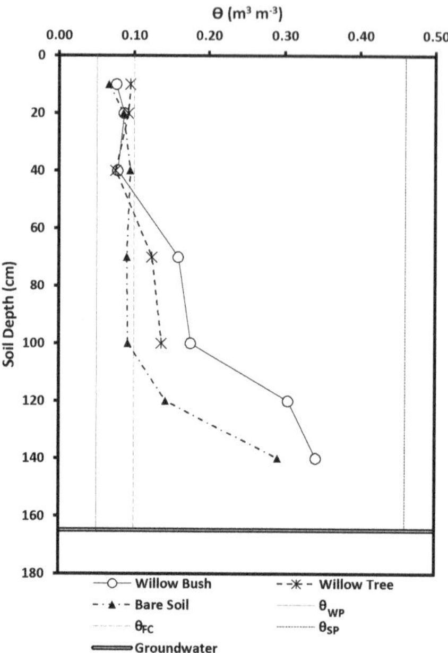

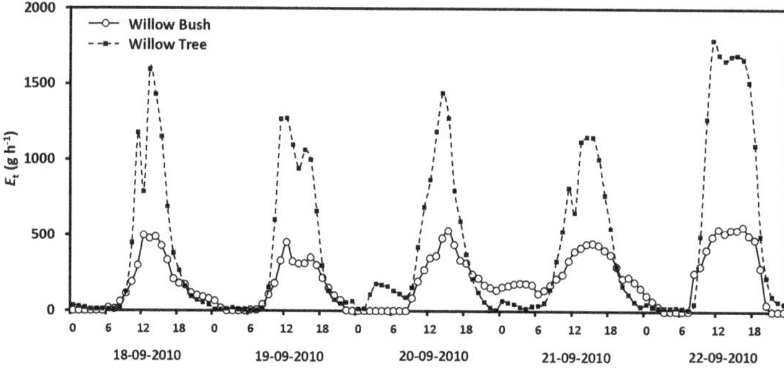

Figure 3. Soil moisture ($m^3\,m^{-3}$) measured in both plots for Willow bush, Willow tree and bare soil conditions during the study period.

Hourly transpiration differs in amount and timing between species. Figure 4 shows the differences along five days where the sap flux for Willow tree is remarkably higher than Willow bush. Willow tree shows a larger capacity to transpire water with peak fluxes averaging $1549.1\,g\,h^{-1}$; whereas Willow bush peaks do not exceed $500\,g\,h^{-1}$ on average. Daily transpiration rates in both species depict a significant decreasing trend (ANCOVA, F = 36.09; n = 87, p = 0.0000) and a statistical difference between total daily rates (ANCOVA, F = 63.05, n = 87, p = 0.0000), where Willow bush transport an average of $4.57\,kg\,day^{-1}$ being three times smaller than Willow tree fluxes ($12.82\,kg\,day^{-1}$). In addition, as transpiration is a physiological response to environmental climatic parameters the Pearson correlation analysis (p < 0.001) shows a significant positive correlation with temperature (r = 0.47) and net radiation (r = 0.35); while wind speed (r = 0.05) and relative humidity (r = −0.27) are not significant.

Figure 4. Hourly transpiration flow measured in Willow tree and Willow bush plants during the study period.

Rain during the study period has a wide range of isotope signatures (see Appendix C Figure A3). The evaporation front is identifiable at 40 cm depth for Willow and at 20 cm for Willow bush in both isotopes (Figure 5). The isotope signature of groundwater samples (Willow Bush: δ^{18}O: -9.2 ‰, δ^2H: -66.1 ‰ and Willow Tree: δ^{18}O: -8.59 ‰, δ^2H: -60.66 ‰) lie close to the rain water signature, depicting the effect of local groundwater recharge having a similar signature to local rains. Sap water signature in both species seems to contain fractionated and non-fractionated water. However, both stable isotopes do not show statistical differences between species ($p > 0.05$) as a consequence of the wide variation in isotope signatures. After a preliminary run of the IsoSource the soil water contribution to xylem water from deeper soil layers show a similar proportion in both species. It showed that only the 40 cm and 10 cm soil layers provide a strong contribution in Willow and Willow bush, respectively. Therefore "a posteriori aggregation" [48] was performed, grouping the soil layers according to their similarities between isotopic signatures, evaporation front presence and proximity within the soil profile. The grouping was settled as: 0–30 cm, 30–60 cm, >60 cm; including in the last soil layer the groundwater due its isotopic similarity with the deeper soil waters. The IsoSource output shows all the possible solutions to match the sap water mixture of δ^2H and δ^{18}O (Figure 6). The Willow tree stand shows a well-defined proportion of soil water contributions among the three water sources. The deep water source (>60 cm) contributes, with a proportion lower than 0.08, to the sap water mixture, while the upper soil layers (<30 cm) provides between 0.28 and 0.48 of the mixed water and the intermediate soil layers (30–60 cm) own the higher contribution values from 0.50 to 0.64. The clear differentiation between soil water sources in Willow is not visible for Willow bush. This species shows overlapping contributions of the water sources mainly for the superficial soil layers (0–60 cm), showing the deepest water source a contribution ranging from 0.21 to 0.54 (Figure 6).

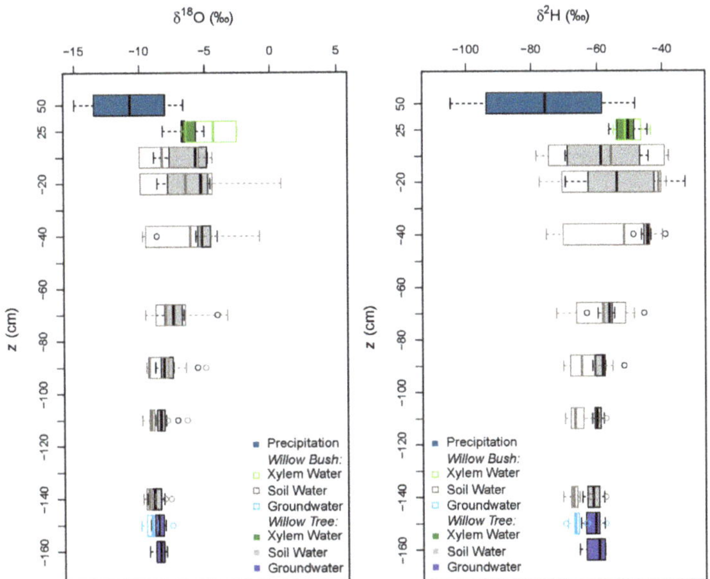

Figure 5. Isotopic profile of the stable water isotopes sampled in both stands during autumn 2010. Each boxplot describes the data set with the median (thick vertical line within the box), the first and third quartiles (edges of the box) and the minimum and maximum values (whiskers).

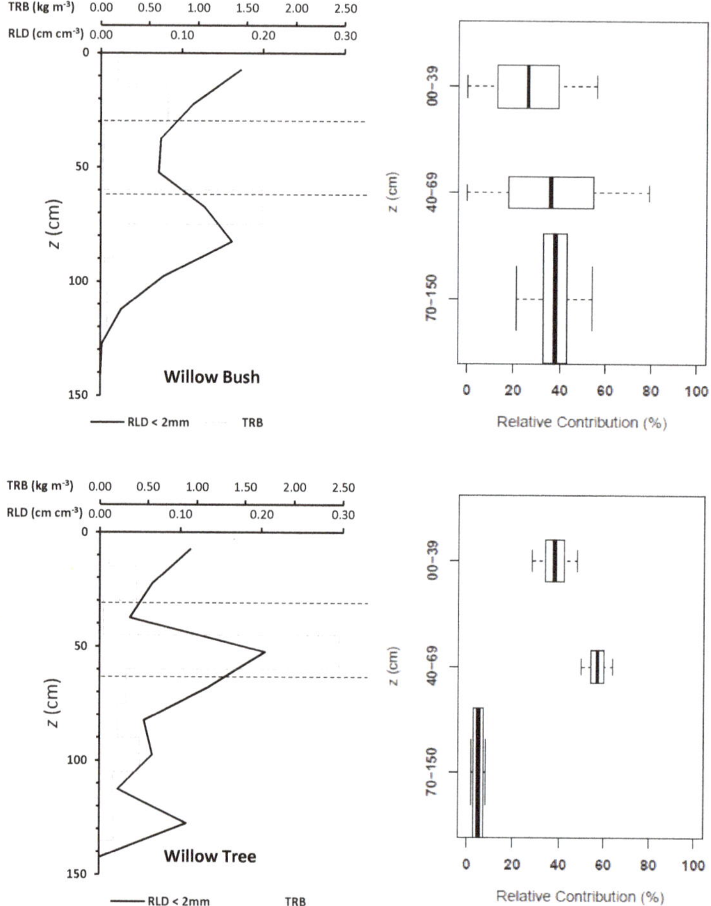

Figure 6. Root length density (RLD) and total root biomass (TRB) distribution along the soil profile and its relation with the relative contribution to sap water mixture of Willow tree and Willow bush based on δ^2H and δ^{18}O isotope signatures per group of soil depth. Each boxplot describes the the median (thick vertical line within the box), the first and third quartiles (edges of the box) and the minimum and maximum values (whiskers). The boxplot height is proportional to the soil depth range.

Plant densities differ between stands, where the Willow bush stand has the higher plant density (900 trees ha^{-1}) with an average height of 2.6 ± 0.6 m. In contrast, the Willow tree stand has a plant density three times smaller (300 trees ha^{-1}) but with higher trees (3.5 ± 0.5 m). However, the LAI is affected by the leaf size and canopy diameter of the individual plants, where Willow bush register a leaf area index of 0.39 m^2 m^{-2} which is twice smaller than Willow tree (0.68 m^2 m^{-2}). Underground stand characteristics also differ between species. Willow trees fix a larger root biomass beneath the 45 cm depth than Willow bush shrubs. Moreover, the root length density distribution shows a bimodal accumulation in Willow bush: at the soil surface (0–30 cm) and at mid depth (55–70 cm). Oppositely, Willow tree has three sections with high RLD values. The first two sections follow the Willow bush pattern, with an additional accumulation bellow 105 cm. The fine root distribution in both species expressed as RLD, provide them a good system for soil water acquisition for the superficial soil layers (Figure 6).

4. Discussion

The main differences in plant size, fine root distribution and water uptake capacity between Willow tree and Willow bush underline the importance of selecting plant species with low water requirements in respect to their biomass for afforestation programs. Willow tree is capable to withdraw up to 12.8 kg day^{-1} of water, extracting more than 90% from soil layers above 60 cm depth. This species is capable to make use of the superficial soil water during the autumn period, even if the groundwater level is shallow. Conversely, Willow bush show lower transpiration rates not higher than 5.0 kg day^{-1} extracted uniformly from the whole soil profile including the groundwater. This extraction pattern shown by Willow bush depicts a more efficient root system acquiring water from different soil water sources due their fine root distribution. During this period, both species extract more than 50% of the water from the upper soil layers, taking advantage of the sporadic autumn rains and residual soil moisture. These results are congruent with the behavior of Willow bush during the growing season (May–July), where Willow bush uses water from both sources—soil and groundwater [18]. On the other hand, the soil water dependency during autumn of Willow trees differ in their summer behavior as documented by Yin et al. [17]. During summer, Willow trees have access to soil and groundwater to maintain their water consumption.

Shallow groundwater levels prevent desiccation processes in scarce rainfall environments, providing a vast water source for adapted plants that use the water economically [65]. Even if both species do not differ in the root amount, their vertical distribution shows different root spots. Willow bush root distribution displays two zones, supporting the hierarchy theory proposed by Schwinning and Sala [66]. The Willow bush can withdraw water from rains as stemflow, while the deeper roots can obtained from a constant source (groundwater in this case). The fine root distribution beneath the Willow tree exemplifies woody patches capacity to use rain water in a short time response [67], as well as the hierarchy theory of Schwinning and Sala [66]. The fine root distribution of Willow tree with three dense regions with RLD higher than 0.1 cm cm^{-3} allow them to use different soil water source depending on soil water availability.

The isotopic values of groundwater are similar to local rain water (see Appendix C Figure A3), depicting a local groundwater recharge documented for the Hailiutu catchment [18]. This is the consequence of the high capacity to infiltrate water by the sandy soils [21]. Consequently, infiltrated water will be available for longer periods because soil water evaporation at soil depths between 10–30 cm can take several weeks in arid environments [24]. The shallow groundwater recharge occurred during the previous growing season due to the high rainfall intensities ($>$5 mm day^{-1}) between July and September. This phenomenon has also been documented by Li et al. [68] in Taihang (China), reporting a daily groundwater recharge with rains ranging from 3.2 mm day^{-1} to 3.8 mm day^{-1}. This recharge capacity has been registered in the provinces of Shangxi and Inner Mongolia, gattering from 9% to 12% of the long term annual precipitation [22].

Conversely to groundwater, the isotopic composition of the soil water in the unsaturated zone is affected by the interaction between vegetation cover and soil evaporation (see Appendix C Figure A3). Soil evaporation affects the isotopic signature of soil water in the unsaturated zone providing particular signatures at different soil layers [69–72]. Meanwhile the plant cover type reduces the soil evaporation, where lower θ in the top soil layer (0–10 cm) were registered for Willow bushes in comparison to Willow trees. Conversely, the high θ under Willow bush in respect to Willow tree reflects capacity to fix more root biomass below 40 cm depth. This enhanced the infiltration capacity by the presence of a low plant cover with a large alive root system [73,74].

However, the stable isotope signatures of soil water beneath the plant cover differs considerably. Beneath Willow trees, both isotopes depict the theoretical evaporation front. This as a consequence of the evaporation process in the superficial soil layers, enabling the generation of heavy isotope enrichment [67–69,75,76] (see Figure 5). On the other hand, beneath Willow bush only the δ^{18}O profile shows the theoretical evaporation front. The homogeneity of δ^2H beneath Willow bush indicates a recent redistribution of groundwater along the soil profile, which can be linked to hydraulic lift processes carried out by this bush.

Lower evaporation rates during the study period depict a lower water need for both species, that is visible in the diminution of sap flow rates. Solar radiation and air temperature are the limiting factors for transpiration as it was showed by the p_{values}. The diminution of solar radiation and air temperature in the region are the clear indication of the arrival of autumn [3], which reduces the available energy for the plants to carry out the photosynthesis. Also, the access to the groundwater reservoir allowed the plants to prevent dehydration, reducing the effect of wind speed and relative humidity as triggers of the transpiration process as it happens during summer with both species [17,62]. This reduction in water needs affects the water uptake of Willow tree, which registered a lower contribution of deep soil water sources while the water uptake by superficial roots is more constant. On the other hand, Willow bush shows a high dynamic root system which extracts water from all the available sources indifferently from the upper soil layers and a strong contribution of the deep sources. This contribution is linked to the root distribution, keeping a high root length density in comparison to the Willow tree. The groundwater dependency of Willow bush [18] implies a permanent deep water extraction during summer and autumn, extracting more deep water than Willow trees during the autumn season.

Despite the few rains, water used of both plant species does not reduce the soil water storage on the soil layers above 100 cm. This can be linked to the presence of hydraulic lift, where the root system prevents the soil water depletion on upper soil layers thanks to the redistribution of deeper soil water (in this case, groundwater). The hydraulic lift allows the formation of water pools along the soil profile in water scarce environments [77,78]. This process requires the movement of soil water by the potential difference between roots and the soil [77–80], allowing the diffusion of water through the roots cell membranes. The hydraulic lift had been identified in different plant species such as *Prosopis tamarugo*, *Artemisia tridentata*, *Acer saccharum* and *Madicago sativa* [77].

The hydraulic lifted water has an isotope signature close to the groundwater. It is relocated during night periods [81] and once it is on the superficial soil layers evaporation will happen affecting the isotope signature of soil water [82]. This water relocation is maintained by Willow trees, which despite the larger transpiration rates the soil water is not shortened. Liste and White [78] mention a Willow as a tree with the water redistribution capacity, providing evidence related to the potential of Willow to use groundwater through this process. Other tree species such as *Eucalyptus kochii* has the capacity to redistribute groundwater [79] or use it as an strategy of competition in saline conditions like *Juniperus phoenicea* and *Pistacia lentiscus* [83].

The replacement of bare soil areas with different plant covers none adapted to arid environments, speed up the water use in those regions. Water needs of plants such as the Willow tree (*Salix matsudana*) are high and require a constant water supply [17,59]. On the other hand, the use of plants adapted to arid environments such as the Willow bush (*Salix psammophila*) [58] ensure the success of the afforestation programs without risking the scarce water resources. The plant water use during summer months is the largest of the year, as a consequence of the long light hours in temperate regions [3]. During this time of the year the newly afforested zones extract more water from the soil and groundwater reservoirs. However, the diminution of solar radiation and temperature during autumn reduces the water demand by all the plants. These plants can redistribute part of the groundwater to the upper soil layers, making it available for the periods with no rains. Also, these plants have the capacity to reduce soil evaporation thanks to the shadow effect of their canopy.

5. Conclusions

The presence of Willow trees (*Salix matsudana*) and Willow bushes (*Salix psammophila*) reduced the effect of soil evaporation after summer, allowing a larger soil moisture beneath both species than bare soil conditions. Also, the plant cover allowed the soil moisture below 60 cm depth to be larger than the field capacity for sandy soils. This augment in soil water can be linked to water redistribution thanks to the presence of fine roots along the soil profile and the hydraulic lift carried out by the plants.

This process redistributes groundwater on the spots with larger fine root allocation, enabling the plants to allocate it at night and using it later during day time. Willow trees uses more water for transpiration than willow bushes, this difference in water consumption allowed the Willow bushes to kept a higher soil moisture after summer (θ: $0.11\,\mathrm{m}^3\,\mathrm{m}^{-3}$) followed by Willow trees (θ: $0.10\,\mathrm{m}^3\,\mathrm{m}^{-3}$) and bare soil ($\theta$: $0.08\,\mathrm{m}^3\,\mathrm{m}^{-3}$). The larger transpiration rates of Willow trees respect to Willow bushes do not match with the water source of the xylem water as it is showed by the IsoSource model. This is linked with the hydraulic lift capacity of Willow tree, redistributing groundwater that is quickly affected by evaporation processes. Fine root distribution along the soil profile allowed the water redistribution and later absorption by both plants. This is supported by both species preferences to withdraw water from the upper soil layers. The water use by Willow bush does not show a strong differentiation among water sources. This species is capable of extract soil and ground water with different proportions according to water availability. On the other hand, Willow tree is able to extract soil water and groundwater with specific proportions. The species selection for afforestation programs has to be carried out carefully to not endanger the scarce water resources in arid regions. Thus considering that species such as Willow trees use more water than Willow bushes, despite the diminution in solar radiation and air temperature during autumn.

Author Contributions: Project administration, J.W.; conceptualization, J.W. and S.U.; methodology, J.W., S.U. and C.D.J.-R.; formal analysis, C.D.J.-R., M.C.-G. and J.W.; isotope measurements, J.W. and C.D.J.-R.; data curation, C.D.J.-R.; writing—original draft preparation, C.D.J.-R. with inputs from all co-authors; writing—review and editing, C.D.J.-R. with inputs from all co-authors; funding acquisition, J.W., S.U., M.C.-G. and C.D.J.-R.

Funding: This study was supported by the Dutch government's Asia Facility for China project Partnership for education and research in water and ecosystem interactions, the Groundwater Circulation and Rational Development in the Ordos Plateau project (1212010634204), Groundwater monitoring in the Ordos Basin, the National Natural Sciences Foundation of China (4103752), Shaanxi Science and Technology Research and Development Program (2011KJXX56), Honor Power Foundation. We also received support from Ministerio de Ciencia, Tecnología y Telecomunicaciones (PINN-MICITT, contract: PED-032-2015-1) and The Netherlands Organization for Scientific Research (NWO, grant: 863.15.022).

Acknowledgments: Special thanks to NUFFIC program through a NFP-Fellowship. Also, to L. Yin and J. Huang by their collaboration in the field. The authors thank the reviewers for the comments and suggestions that improved the manuscript.

Conflicts of Interest: The authors declare no conflict of interest.

Appendix A

Figure A1. Photographs of the experimental sites and different sampling procedures carried out in Yulin County; Shaanxi province–China. Picture (**A**): panoramic view of the bush lands dominated by Willow bushes. Picture (**B**): bare soil conditions close to the experimental plots. Picture (**C**): experimental plot with Willow trees. Picture (**D**): Thermal dissipation probe installed in a Willow tree. Picture (**E**): Root sampling within Willow Tree Plot. Picture (**F**): print screen of the fine root measuring procedure.

Appendix B

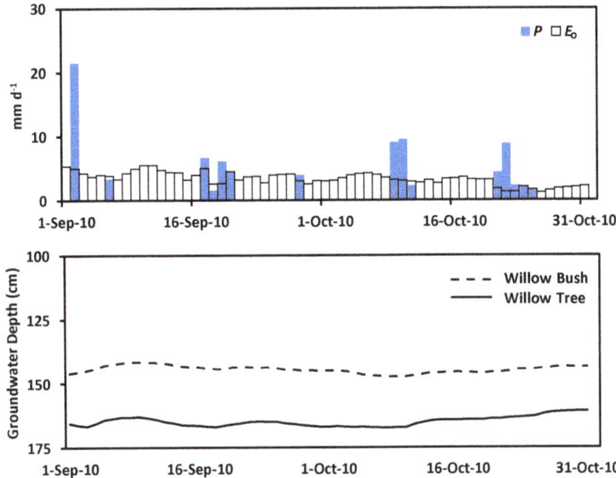

Figure A2. Daily measurements of precipitation, evaporation and groundwater depth during the monitoring period.

Appendix C

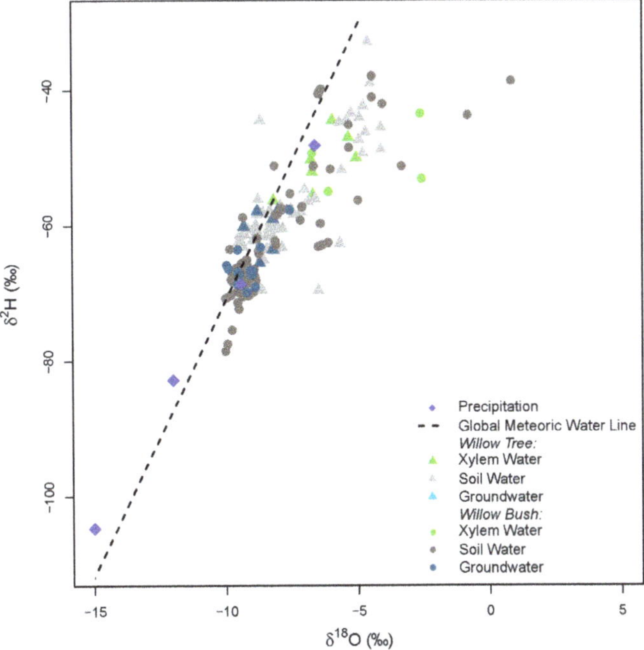

Figure A3. Dual isotope plot of δ^2H and δ^{18}O for the water samples analyzed in the study.

References

1. Abd El-Ghani, M.M.; Huerta-Martínez, F.M.; Hongyan, L.; Qureshi, R. Arid Deserts of the World: Origin, Distribution, and Features. In *Plant Responses to Hyperarid Desert Environments*; Springer International Publishing: Cham, Switzerland, 2017; pp. 1–7. [CrossRef]
2. Salem, B. *Arid Zone Forestry: A Guide for Field Technicians*; Number 20; Food and Agriculture Organization (FAO): Rome, Italy, 1989.
3. Bonan, G. *Ecological Climatology: Concepts and Applications*; Cambridge University Press: Cambridge, UK, 2002.
4. Aguiar, M.R.; Sala, O.E. Patch structure, dynamics and implications for the functioning of arid ecosystems. *Trends Ecol. Evol.* **1999**, *14*, 273–277. [CrossRef]
5. Wainwright, J.; Mulligan, M.; Thornes, J. Plants and water in drylands. In *Eco-Hydrology: Plants and Water in Terrestrial and Aquatic Environments*; Baird, A., Wilby, R., Eds.; Routledge: Abingdon, UK, 1999; pp. 78–126.
6. FAO. *Guidelines for Soil Description*; Food and Agriculture Organization (FAO): Rome, Italy, 2006.
7. Huggett, R. *Fundamentals of Geomorphology*, 2nd ed.; Routledge Fundamentals of Physical Geography: Abingdon, UK, 2007.
8. Summerfield, M. *Global Geomorphology*; Longman: Harlow, UK, 1991.
9. Yang, X.; Zhang, K.; Jia, B.; Ci, L. Desertification assessment in China: An overview. *J. Arid. Environ.* **2005**, *63*, 517–531. [CrossRef]
10. Young, A. *Agroforestry for Soil Conservation*; CAB International: Wallingford, UK, 1989.
11. Han, K.S.; Park, Y.Y.; Yeom, J.M. Detection of change in vegetation in the surrounding Desert areas of Northwest China and Mongolia with multi-temporal satellite images. *Asia-Pac. J. Atmos. Sci.* **2015**, *51*, 173–181. [CrossRef]
12. Song, X.; Wang, T.; Xue, X.; Yan, C.; Li, S. Monitoring and analysis of aeolian desertification dynamics from 1975 to 2010 in the Heihe River Basin, northwestern China. *Environ. Earth Sci.* **2015**, *74*, 3123–3133. [CrossRef]
13. Cao, S.; Tian, T.; Chen, L.; Dong, X.; Yu, X.; Wang, G. Damage Caused to the Environment by Reforestation Policies in Arid and Semi-Arid Areas of China. *AMBIO* **2010**, *39*, 279–283. [CrossRef]
14. Cao, S.; Chen, L.; Shankman, D.; Wang, C.; Wang, X.; Zhang, H. Excessive reliance on afforestation in China's arid and semi-arid regions: Lessons in ecological restoration. *Earth-Sci. Rev.* **2011**, *104*, 240–245. [CrossRef]
15. Peng, S.S.; Piao, S.; Zeng, Z.; Ciais, P.; Zhou, L.; Li, L.Z.X.; Myneni, R.B.; Yin, Y.; Zeng, H. Afforestation in China cools local land surface temperature. *Proc. Natl. Acad. Sci. USA* **2014**, *111*, 2915–2919. [CrossRef]
16. Yang, Z.; Zhou, Y.; Wenninger, J.; Uhlenbrook, S. The causes of flow regime shifts in the semi-arid Hailiutu River, Northwest China. *Hydrol. Earth Syst. Sci.* **2012**, *16*, 87–103. [CrossRef]
17. Yin, L.; Zhou, Y.; Huang, J.; Wenninger, J.; Hou, G.; Zhang, E.; Wang, X.; Dong, J.; Zhang, J.; Uhlenbrook, S. Dynamics of willow tree (Salix matsudana) water use and its response to environmental factors in the semi-arid Hailiutu River catchment, Northwest China. *Environ. Earth Sci.* **2014**, *71*, 4997–5006. [CrossRef]
18. Zhou, Y.; Wenninger, J.; Yang, Z.; Yin, L.; Huang, J.; Hou, L.; Wang, X.; Zhang, D.; Uhlenbrook, S. Groundwater–surface water interactions, vegetation dependencies and implications for water resources management in the semi-arid Hailiutu River catchment, China—A synthesis. *Hydrol. Earth Syst. Sci.* **2013**, *17*, 2435–2447. [CrossRef]
19. Roberts, J. Plants and water in forests and woodlands. In *Eco-Hydrology: Plants and Water in Terrestrial and Aquatic Environments*; Routledge: Abingdon, UK, 1999; pp. 181–236.
20. Savenije, H.H.G. The importance of interception and why we should delete the term evapotranspiration from our vocabulary. *Hydrol. Process.* **2004**, *18*, 1507–1511. [CrossRef]
21. Yaseef, N.R.; Yakir, D.; Rotenberg, E.; Schiller, G.; Cohen, S. Ecohydrology of a semi-arid forest: Partitioning among water balance components and its implications for predicted precipitation changes. *Ecohydrology* **2009**, *3*, 143–154. [CrossRef]
22. Scanlon, B.R.; Keese, K.E.; Flint, A.L.; Flint, L.E.; Gaye, C.B.; Edmunds, W.M.; Simmers, I. Global synthesis of groundwater recharge in semiarid and arid regions. *Hydrol. Process.* **2006**, *20*, 3335–3370. [CrossRef]
23. Zhang, Y.F.; Wang, X.P.; Hu, R.; Pan, Y.X.; Paradeloc, M. Rainfall partitioning into throughfall, stemflow and interception loss by two xerophytic shrubs within a rain-fed re-vegetated desert ecosystem, northwestern China. *J. Hydrol.* **2015**, *527*, 1084–1095. [CrossRef]

24. Noy-Meir, I. Desert Ecosystems: Environment and Producers. *Annu. Rev. Ecol. Syst.* **1973**, *4*, 25–51. [CrossRef]

25. Chesson, P.; Gebauer, R.L.E.; Schwinning, S.; Huntly, N.; Wiegand, K.; Ernest, M.S.K.; Sher, A.; Novoplansky, A.; Weltzin, J.F. Resource pulses, species interactions, and diversity maintenance in arid and semi-arid environments. *Oecologia* **2004**, *141*, 236–253. [CrossRef]

26. Ivans, S.; Hipps, L.; Leffler, A.J.; Ivans, C.Y. Response of Water Vapor and CO_2 Fluxes in Semiarid Lands to Seasonal and Intermittent Precipitation Pulses. *J. Hydrometeorol.* **2006**, *7*, 995–1010. [CrossRef]

27. Ogle, K.; Wolpert, R.L.; Reynolds, J.F. Reconstructing plant root area and water uptake profiles. *Ecology* **2004**, *85*, 1967–1978. [CrossRef]

28. Hopkins, W.G.; Hüner, N.P. *Introduction to Plant Physiology*; Wiley: Hoboken, NJ, USA, 2008.

29. Cardon, Z.G.; Whitbeck, J.L. *The Rhizosphere. An Ecological Perspective*; Elsevier Academic Press: Amsterdam, The Nertherlands, 2007.

30. Curl, E.; Truelove, B. *The Rhizosphere*; Springer: New York, NY, USA, 1986.

31. Dawson, T.E.; Mambelli, S.; Plamboeck, A.H.; Templer, P.H.; Tu, K.P. Stable Isotopes in Plant Ecology. *Annu. Rev. Ecol. Syst.* **2002**, *33*, 507–559. [CrossRef]

32. Evaristo, J.; McDonnell, J.J.; Clemens, J. Plant source water apportionment using stable isotopes: A comparison of simple linear, two-compartment mixing model approaches. *Hydrol. Process.* **2017**, *31*, 3750–3758. [CrossRef]

33. Jia, Z.; Zhu, Y.; Liu, L. Different Water Use Strategies of Juvenile and Adult *Caragana intermedia* Plantations in the Gonghe Basin, Tibet Plateau. *PLoS ONE* **2012**, *7*, e45902. [CrossRef] [PubMed]

34. Nie, Y.P.; Chen, H.S.; Wang, K.l.; Tan, W.; Deng, P.Y.; Yang, J. Seasonal water use patterns of woody species growing on the continuous dolostone outcrops and nearby thin soils in subtropical China. *Plant Soil* **2011**, *341*, 399–412. [CrossRef]

35. Palacio, S.; Montserrat-Martí, G.; Ferrio, J.P. Water use segregation among plants with contrasting root depth and distribution along gypsum hills. *J. Veg. Sci.* **2017**, *28*, 1107–1117. [CrossRef]

36. Rossatto, D.R.; da Silveira Lobo Sternberg, L.; Franco, A.C. The partitioning of water uptake between growth forms in a Neotropical savanna: Do herbs exploit a third water source niche? *Plant Biol.* **2012**, *15*, 84–92. [CrossRef] [PubMed]

37. Swaffer, B.A.; Holland, K.L.; Doody, T.M.; Li, C.; Hutson, J. Water use strategies of two co-occurring tree species in a semi–arid karst environment. *Hydrol. Process.* **2013**, *28*, 2003–2017. [CrossRef]

38. Voltas, J.; Lucabaugh, D.; Chambel, M.R.; Ferrio, J.P. Intraspecific variation in the use of water sources by the circum-Mediterranean conifer Pinus halepensis. *New Phytol.* **2015**, *208*, 1031–1041. [CrossRef] [PubMed]

39. Wang, P.; Song, X.; Han, D.; Zhang, Y.; Liu, X. A study of root water uptake of crops indicated by hydrogen and oxygen stable isotopes: A case in Shanxi Province, China. *Agric. Water Manag.* **2010**, *97*, 475–482. [CrossRef]

40. Wei, L.; Lockington, D.A.; Poh, S.C.; Gasparon, M.; Lovelock, C.E. Water use patterns of estuarine vegetation in a tidal creek system. *Oecologia* **2013**, *172*, 485–494. [CrossRef]

41. Barbeta, A.; Ogée, J.; Peñuelas, J. Stable-Isotope Techniques to Investigate Sources of Plant Water. In *Advances in Plant Ecophysiology Techniques*; Sánchez-Moreiras, A.M., Reigosa, M.J., Eds.; Springer International Publishing: Cham, Switzerland, 2018; pp. 439–456. [CrossRef]

42. Geyh, M. Groundwater. Saturated and unsaturated zone. In *Environmental Isotopes in the Hydrological Cycle: Principles and Applications*; Mook, W., Ed.; UNESCO: Paris, France, 2000; Volume IV.

43. Leibundgut, C.; Seibert, J. 2.09. Tracer Hydrology. In *Treatise on Water Science*; Wilderer, P., Ed.; Elsevier: Amsterdam, The Netherlands, 2011; pp. 215–236.

44. Butt, S.; Ali, M.; Fazil, M.; Latif, Z. Seasonal variations in the isotopic composition of leaf and stem water from an arid region of Southeast Asia. *Hydrol. Sci. J.* **2010**, *55*, 844–848. [CrossRef]

45. Parnell, A.C.; Inger, R.; Bearhop, S.; Jackson, A.L. Source partitioning using stable isotopes: Coping with too much variation. *PLoS ONE* **2010**, *5*, e9672. [CrossRef]

46. Parnell, A.C.; Phillips, D.L.; Bearhop, S.; Semmens, B.X.; Ward, E.J.; Moore, J.W.; Jackson, A.L.; Grey, J.; Kelly, D.J.; Inger, R. Bayesian stable isotope mixing models. *Environmetrics* **2013**, *24*, 387–399. [CrossRef]

47. Moore, J.W.; Semmens, B.X. Incorporating uncertainty and prior information into stable isotope mixing models. *Ecol. Lett.* **2008**, *11*, 470–480. [CrossRef] [PubMed]

48. Phillips, D.L.; Newsome, S.D.; Gregg, J.W. Combining sources in stable isotope mixing models: Alternative methods. *Oecologia* **2005**, *144*, 520–527. [CrossRef] [PubMed]

49. Phillips, D.L.; Gregg, J.W. Source partitioning using stable isotopes: Coping with too many sources. *Oecologia* **2003**, *136*, 261–269. [CrossRef] [PubMed]

50. Zhu, J.; Sun, D.; Young, M.H.; Caldwell, T.G.; Pan, F. Shrub spatial organization and partitioning of evaporation and transpiration in arid environments. *Ecohydrology* **2015**, *8*, 1218–1228. [CrossRef]

51. Walker, B.H.; Langridge, J.L. Modelling plant and soil water dynamics in semi-arid ecosystems with limited site data. *Ecol. Model.* **1996**, *87*, 153–167. [CrossRef]

52. Schlesinger, W.H.; Fonteyn, P.J.; Marion, G.M. Soil moisture content and plant transpiration in the Chihuahuan Desert of New Mexico. *J. Arid. Environ.* **1987**, *12*, 119–126. [CrossRef]

53. IIASA/FAO. *Global Agro-Ecological Zones (GAEZ v3.0)*; IIASA: Laxenburg, Austria; FAO: Rome, Italy, 2012.

54. NCDC. NNDC Climate Data Online. 2012. Available online: http://www7.ncdc.noaa.gov/CDO/cdoselect.cmd (accessed on 23 January 2012).

55. Allen, R.G.; Pereira, L.S.; Raes, D.; Smith, M. *Crop Evapotranspiration-Guidelines for Computing Crop Water Requirements–FAO Irrigation and Drainage Paper 56*; FAO: Rome, Italy, 1998.

56. Saxton, K.E.; Rawls, W.J. Soil water characteristic estimates by texture and organic matter for hydrologic solutions. *Soil Sci. Soc. Am. J.* **2006**, *70*, 1569–1578. [CrossRef]

57. Craig, H. Standard for Reporting Concentrations of Deuterium and Oxygen-18 in Natural Waters. *Science* **1961**, *133*, 1833–1834. [CrossRef]

58. Wang, Z.; Chang, Y. *Salix Psammophila*; Bulletin of Botanical Laboratory of North-Eastern Forestry Institute: Harbin, China, 1980; Volume 9.

59. Gilman, E.; Watson, G. *Salix Matsudana 'Tortuosa' Corkscrew Willow*; Fact Sheet ST-577 Environmental Horticulture Department, Florida Cooperative Extension Service, Institute of Food and Agricultural Sciences, University of Florida: Gainesville, FL, USA, 1994.

60. Granier, A. Une nouvelle méthode pour la mesure du flux de sève brute dans le tronc des arbres. *Ann. Des Sci. For.* **1985**, *42*, 193–200. [CrossRef]

61. Cornelissen, J.; Lavorel, S.; Garnier, E.; Diaz, S.; Buchmann, N.; Gurvich, D.; Reich, P.; Ter Steege, H.; Morgan, H.; Van Der Heijden, M.; et al. A handbook of protocols for standardised and easy measurement of plant functional traits worldwide. *Aust. J. Bot.* **2003**, *51*, 335–380. [CrossRef]

62. Huang, J.; Zhou, Y.; Yin, L.; Wenninger, J.; Zhang, J.; Hou, G.; Zhang, E.; Uhlenbrook, S. Climatic controls on sap flow dynamics and used water sources of Salix psammophila in a semi-arid environment in northwest China. *Environ. Earth Sci.* **2015**, *73*, 289–301. [CrossRef]

63. Kottek, M.; Grieser, J.; Beck, C.; Rudolf, B.; Rubel, F. World map of the Köppen-Geiger climate classification updated. *Meteorol. Z.* **2006**, *15*, 259–263. [CrossRef]

64. Peel, M.C.; Finlayson, B.L.; Mcmahon, T.A. Updated world map of the Köppen-Geiger climate classification. *Hydrol. Earth Syst. Sci. Discuss.* **2007**, *4*, 439–473. [CrossRef]

65. Jiang, G.; He, W. Species- and habitat-variability of photosynthesis, transpiration and water use efficiency of different plant species in Maowusu Sand Area. *Acta Bot. Sin.* **1999**, *41*, 1114–1124.

66. Schwinning, S.; Sala, O.E. Hierarchy of responses to resource pulses in arid and semi-arid ecosystems. *Oecologia* **2004**, *141*, 211–220. [CrossRef]

67. Midwood, A.; Boutton, T.; Archer, S.; Watts, S. Water use by woody plants on contrasting soils in a savanna parkland: assessment with δ^2H and δ^{18}O. *Plant Soil* **1998**, *205*, 13–24. [CrossRef]

68. Li, F.; Song, X.; Tang, C.; Liu, C.; Yu, J.; Zhang, W. Tracing infiltration and recharge using stable isotope in Taihang Mt., North China. *Environ. Geol.* **2007**, *53*, 687–696. [CrossRef]

69. Barnes, C.J.; Allison, G. Tracing of water movement in the unsaturated zone using stable isotopes of hydrogen and oxygen. *J. Hydrol.* **1988**, *100*, 143–176. [CrossRef]

70. Brunel, J.P.; Walker, G.R.; Kennett-Smith, A.K. Field validation of isotopic procedures for determining sources of water used by plants in a semi-arid environment. *J. Hydrol.* **1995**, *167*, 351–368. [CrossRef]

71. Rothfuss, Y.; Biron, P.; Braud, I.; Canale, L.; Durand, J.L.; Gaudet, J.P.; Richard, P.; Vauclin, M.; Bariac, T. Partitioning evapotranspiration fluxes into soil evaporation and plant transpiration using water stable isotopes under controlled conditions. *Hydrol. Process.* **2010**, *24*, 3177–3194. [CrossRef]

72. Schwinning, S.; Ehleringer, J.R. Water use trade-offs and optimal adaptations to pulse-driven arid ecosystems. *J. Ecol.* **2001**, *89*, 464–480. [CrossRef]

73. Basche, A.D.; DeLonge, M.S. Comparing infiltration rates in soils managed with conventional and alternative farming methods: A meta-analysis. *PLoS ONE* **2019**, *14*, e0215702. [CrossRef] [PubMed]

74. Fischer, C.; Tischer, J.; Roscher, C.; Eisenhauer, N.; Ravenek, J.; Gleixner, G.; Attinger, S.; Jensen, B.; de Kroon, H.; Mommer, L.; et al. Plant species diversity affects infiltration capacity in an experimental grassland through changes in soil properties. *Plant Soil* **2015**, *397*, 1–16. [CrossRef]

75. Sutanto, S.J.; Wenninger, J.; Coenders-Gerrits, A.M.J.; Uhlenbrook, S. Partitioning of evaporation into transpiration, soil evaporation and interception: A comparison between isotope measurements and a HYDRUS-1D model. *Hydrol. Earth Syst. Sci.* **2012**, *16*, 2605–2616. [CrossRef]

76. Wenninger, J.; Beza, D.T.; Uhlenbrook, S. Experimental investigations of water fluxes within the soil–vegetation–atmosphere system: Stable isotope mass-balance approach to partition evaporation and transpiration. *Phys. Chem. Earth Parts A/B/C* **2010**, *35*, 565–570. [CrossRef]

77. Horton, J.L.; Hart, S.C. Hydraulic lift: A potentially important ecosystem process. *Trends Ecol. Evol.* **1998**, *13*, 232–235. [CrossRef]

78. Liste, H.H.; White, J.C. Plant hydraulic lift of soil water—Implications for crop production and land restoration. *Plant Soil* **2008**, *313*, 1–17. [CrossRef]

79. Brooksbank, K.; White, D.A.; Veneklaas, E.J.; Carter, J.L. Hydraulic redistribution in Eucalyptus kochii subsp. borealis with variable access to fresh groundwater. *Trees* **2011**, *25*, 735–744. [CrossRef]

80. Niinemets, U. Responses of forest trees to single and multiple environmental stresses from seedlings to mature plants: Past stress history, stress interactions, tolerance and acclimation. *For. Ecol. Manag.* **2010**, *260*, 1623–1639. [CrossRef]

81. Caldwell, M.M.; Dawson, T.E.; Richards, J.H. Hydraulic lift: Consequences of water efflux from the roots of plants. *Oecologia* **1998**, *113*, 151–161. [CrossRef] [PubMed]

82. Dawson, T.E.; Pate, J.S. Seasonal water uptake and movement in root systems of Australian phraeatophytic plants of dimorphic root morphology: A stable isotope investigation. *Oecologia* **1996**, *107*, 13–20. [CrossRef] [PubMed]

83. Armas, C.; Padilla, F.M.; Pugnaire, F.I.; Jackson, R.B. Hydraulic lift and tolerance to salinity of semiarid species: Consequences for species interactions. *Oecologia* **2010**, *162*, 11–21. [CrossRef] [PubMed]

Article

Long-Term Isotope Records of Precipitation in Zagreb, Croatia

Ines Krajcar Bronić [1],*, Jadranka Barešić [1], Damir Borković [1], Andreja Sironić [1], Ivanka Lovrenčić Mikelić [1] and Polona Vreča [2]

[1] Laboratory for Low-level Radioactivities, Division of Experimental Physics, Ruđer Bošković Institute, 10000 Zagreb, Croatia; jbaresic@irb.hr (J.B.); damir.borkovic@irb.hr (D.B.); asironic@irb.hr (A.S.); ivanka.lovrencic@irb.hr (I.L.M.)

[2] Department of Environmental Sciences, Jožef Stefan Institute, 1000 Ljubljana, Slovenia; polona.vreca@ijs.si

* Correspondence: krajcar@irb.hr

Received: 29 November 2019; Accepted: 10 January 2020; Published: 14 January 2020

Abstract: The isotope composition of precipitation has been monitored in monthly precipitation at Zagreb, Croatia, since 1976. Here, we present a statistical analysis of available long-term isotope data (^{3}H activity concentration, δ^2H, δ^{18}O, and deuterium excess) and compare them to basic meteorological data. The aim was to see whether isotope composition reflected observed climate changes in Zagreb: a significant increase in the annual air temperature and larger variations in the precipitation amount. Annual mean δ^{18}O and δ^2H values showed an increase of 0.017‰ and 0.14‰ per year, respectively, with larger differences in monthly mean values in the first half of the year than in the second half. Mean annual d-excess remained constant over the whole long-term period, with a tendency for monthly mean d-excess values to decrease in the first half of the year and increase in the second half due to the influence of air masses originating from the eastern Mediterranean. Changes in the stable isotope composition of precipitation thus resembled changes in the temperature, the circulation pattern of air masses, and the precipitation regime. A local meteoric water line was obtained using different regression methods, which did not result in significant differences between nonweighted and precipitation-weighted slope and intercept values. Deviations from the Global Meteoric Water Line GMWL (lower slopes and intercepts) were observed in two recent periods and could be explained by changes in climate parameters. The temperature gradient of δ^{18}O was 0.33‰/°C. The tritium activity concentrations in precipitation showed slight decreases during the last two decades, and the mean A in the most recent period, 2012–2018, was 7.6 ± 0.8 Tritium Units (TU).

Keywords: precipitation; Zagreb; Croatia; stable isotope ratios; ^{2}H/^{1}H and ^{18}O/^{16}O; deuterium excess; local meteoric water line; δ^{18}O–temperature relation; tritium

1. Introduction

Water, especially groundwater, has become an invaluable natural resource, and the availability of freshwater is one of the greatest issues facing mankind today [1]. A consistent and careful assessment and management of water resources is crucial for their sustainable development. This can be performed by various methodologies, among which isotope methods using environmental (stable and radioactive) and artificial radioactive isotopes have proven to be effective tools for solving many critical hydrological problems and processes [2–9]. In many cases, isotope techniques have provided information that could not be obtained through any other conventional means [10–12].

Isotopes that are constituent elements of a water molecule are of special interest as perfect candidates for water tracers: hydrogen (^{1}H, ^{2}H, ^{3}H) and oxygen (^{16}O, ^{17}O, ^{18}O) isotopes. Among these, only ^{3}H is a radioactive isotope, while the others are stable isotopes.

Precipitation presents an input to groundwater, and therefore knowledge on the isotope composition of precipitation is a prerequisite for groundwater studies. Temporal and spatial patterns of isotopes in precipitation (expressed as δ^2H and δ^{18}O values and the tritium activity concentration, A) have been observed since the 1950s and have contributed to hydrological and hydrogeological research [13–16], climate and paleoclimate studies [4,15,17–22], and ecological research [23–25]; further, precipitation isotope mapping has been widely implemented during recent decades [25,26].

The International Atomic Energy Agency (IAEA) and the World Meteorological Organization (WMO) have recognized the importance of the isotopic composition of precipitation on a global scale. A program involving the worldwide monitoring of the isotopic composition of monthly precipitation (called the Global Network of Isotopes in Precipitation (GNIP)) was therefore established in 1961. The objective of the network was a systematic collection of data on the isotopic composition of precipitation across the globe to determine temporal and spatial variations of isotopes in precipitation. Isotopic data include the tritium activity concentration, A (expressed in tritium units, TU), and stable isotopes of hydrogen and oxygen isotopes (δ^2H and δ^{18}O values), as well as climatological data (mean monthly temperature, monthly precipitation amount, and atmospheric water vapor pressure) [27]. The collected records have enabled the establishment of seasonal variations and various correlations among the data. Seasonal variations in δ^{18}O and δ^2H values of precipitation and their weighted mean annual values have remained fairly constant from year to year at a given location as long as the annual range and sequence of climatic conditions did not change significantly from year to year.

However, in recent years, we have become aware of climatic changes that have caused an increase in global temperature and changes in the precipitation pattern, as well as severe and extreme weather events (droughts, heavy storms, and temperature records). The recent increase in global temperature has exceeded the natural variabilities during the Holocene [28]. The planet's average surface temperature has risen about 1 °C (between 0.8 °C and 1.2 °C [29]) since the late 19th century, a change driven largely by increased carbon dioxide and other human-made emissions into the atmosphere. Most of the warming has occurred in the past 35 years, with 16 of the 17 warmest years on record occurring since 2001. The 10 hottest years ever recorded have all occurred since 1998 [30]. The hottest years on record globally have been the last five (2014–2018), with 2016 being the hottest year [31], and eight months in 2016 (from January through September, except for June) were the warmest on record for those respective months. October, November, and December of 2016 were the second warmest of those months on record: in all three cases, behind records set in 2015. Correlations between the precipitation isotope ratios recorded in the GNIP and meteorological quantities may provide additional evidence of recent climate change that appears to have manifested globally as well as evidence of the local weather situation. To find such evidence, one should have sufficiently long records of both climate data and the isotopic composition of precipitation.

Monitoring of the isotope composition of monthly precipitation at a station in Zagreb (Croatia) has been performed since 1976 (tritium activity concentration, A) and since 1980 (stable isotope ratios of hydrogen (^{2}H/^{1}H) and oxygen (^{18}O/^{16}O)). Isotope data up to 2003 are available in the GNIP database [27]. This work presents details on the history of monitoring the isotope composition of precipitation in Zagreb, Croatia, for the period 1976–2018. Such series of isotope data are rather scarce in Europe [27], and the present time series analysis can be a first step toward a more detailed comparison of data reported for sites with similar long-term records, such as Vienna (Austria), Krakow (Poland), and Ljubljana (Slovenia), in order to obtain a wider spatial and temporal pattern of the isotope composition of precipitation. Statistical analyses of isotope (δ^2H, δ^{18}O, deuterium excess, and tritium activity concentration) and basic meteorological data (temperature and precipitation amount) were performed. The complete 43-year-long record was divided into subperiods in order to better investigate climatological and isotope-in-precipitation changes. Subperiods should be neither too long nor too short, so we arbitrarily chose four almost equally long subperiods: 1976–1985, 1986–1995, 1996–2006, and 2007–2018.

In the following section, we introduce the notation (δ^2H, δ^{18}O) used for stable isotope composition as well as the concepts of meteoric water lines and deuterium excess. We describe the behavior of tritium activity concentration in the atmosphere and give a brief description of measurement techniques that have changed during the studied period. An overview of sampling locations of monthly precipitation in Croatia and climate characteristics of the area are presented, which will help in discussing the data from Zagreb. In Section 3, we show the results of the monthly data, while in Section 4, we present a discussion of the statistical analyses of the annual data and average values in the subperiods, the observed temporal trends, and various correlations among the data. We discuss how the observed temperature and precipitation amount changes were recorded in the isotopic composition of precipitation in Zagreb.

2. Materials and Methods

2.1. Stable Isotopes of Hydrogen and Oxygen

The isotopic composition of water constituents depends on isotope fractionation caused by phase transfers of water masses (evaporation/precipitation), which depend on the area of water origin (latitude, altitude, continent or maritime, climate region) and the precipitation amount [2,4,32–34]. Therefore, the isotopic composition of precipitation of different origins and seasons provides for the application of stable water isotopes as tracers of the hydrological cycle.

The results are reported as δ-values per mill (‰) relative to the standard [4,35–38]:

$$\delta_{S/R} = \frac{R_{Sample}}{R_{Reference}} - 1. \tag{1}$$

Here, R_{Sample} and $R_{Reference}$ stand for the isotope ratio ($R = {}^2$H/^{1}H and $R = {}^{18}$O/^{16}O) in the sample and the reference material (standard), respectively. Standard mean ocean water, SMOW, has been proposed as a (virtual) standard for reporting measured values [39]. SMOW is an arbitrary mean value based on the Epstein–Mayeda oxygen scale obtained from deep ocean water, since it does not interact with the atmosphere and has a stable isotopic composition [40], and it is defined in terms of an actual water reference standard, the NBS-1 (National Bureau of Standards, USA). In 1968, the IAEA established an international standard, the Vienna SMOW (VSMOW), which has been replaced by the VSMOW2 [38,41].

The δ^2H and δ^{18}O isotopic compositions of meteoric waters (precipitation and atmospheric water vapor) are strongly correlated. If δ^2H is plotted versus δ^{18}O, the data cluster along a straight line is

$$\delta^2\text{H} = 8.0 \cdot \delta^{18}\text{O} + 10. \tag{2}$$

The relation in Equation (2) is referred to as the global meteoric water line (GMWL) [2,39,42]. It describes the general relation between δ^2H and δ^{18}O on a global scale reasonably well. However, the intercept is higher for precipitation originating from the Mediterranean area [2,4,43–45], while the slope does not change, as in the case of the eastern Mediterranean meteoric water line, δ^2H = 8 δ^{18}O + 22 [43]. This example shows that for applications in hydrogeological studies, regional local meteoric water lines (LMWLs), either long-term or for certain shorter periods, can be more appropriate. Generally, an LMWL has the form δ^2H = a δ^{18}O + b, where a is the slope and b is the intercept. LMWLs can differ from the GMWL in terms of both the slope and intercept values, depending on the conditions for forming a local water source [1,4,46,47]. There are different ways of calculating a- and b-values. Traditionally, the ordinary least squares regression (OLSR) is used, and more recently, the reduced major axis regression (RMA) and the major axis least squares regression (MA, sometimes also called the orthogonal regression) [48,49] have been applied. The precipitation-weighted OLSR approach (PWLSR) was introduced to reduce the impact of small precipitation events that are hydrogeologically not significant [49]. Similarly, precipitation-weighted RMA and MA, i.e., PWRMA and PWMA, regressions have been applied to data from the GNIP database, with at least 36 monthly datapoints available [48].

Deuterium excess (*d*-excess, or *d*) is defined as [2]

$$d = \delta^2 H - 8\, \delta^{18}O, \tag{3}$$

which can be related to the meteorological conditions in the source region from which the water vapor is obtained [1,4,35,43,46]; therefore, it can be used to identify vapor source regions, and it is often considered to be the most useful parameter in characterizing vapor origin [43]. Winter precipitation originating from the Mediterranean Sea is characterized by distinctly higher *d*-excess values ($d > 18‰$) than is precipitation coming from the Atlantic ($d \sim 10‰$), reflecting the specific source conditions during water vapor formation. Increased deuterium excess in precipitation can also arise from a significant addition of re-evaporated moisture from continental basins to water vapor traveling inland [1,4,27,35,44–46,50,51].

2.2. Tritium

Tritium (^{3}H) is a natural cosmogenic isotope of hydrogen that is formed in the upper atmosphere through reactions of thermal neutrons with ^{14}N. It oxidizes to tritiated water, H^{3}HO, and thus enters the natural water cycle. The half-life of ^{3}H is 12.32 years [52], and it decays to ^{3}He by emitting beta particles with a maximal energy of 18.6 keV.

Tritium is also an anthropogenically produced isotope, and it can be differentiated as "bomb-produced" tritium and technogenic tritium. Massive injections of ^{3}H from weapons tests in the 1950s and 1960s, mostly in the Northern Hemisphere, caused an almost 100-fold increase in the tritium activity concentration in precipitation [1], known as the bomb peak. The highest concentration of tritium, about 6000 TU (1 TU = 0.118 Bq/l), was observed in 1963 in precipitation at continental stations in the Northern Hemisphere [1,27,53], while at maritime stations, the maximal values were lower (about 2000 TU). The data at the marine stations were systematically lower than at the continental stations because moisture evaporated from the ocean has a low ^{3}H activity concentration due to the long residence time of water in the ocean. After the cessation of atmospheric nuclear weapons tests, a gradual decrease in ^{3}H activity concentration at all stations in both hemispheres was observed due to natural decay and the washout of tritium into the oceans and groundwater. The levels of tritium have declined globally and regionally, approaching the natural pre-bomb level. The pre-bomb natural tritium activity concentration is assumed to be about 1 TU in oceanic regions, about 10 TU in inland areas, about 5 TU in central Europe [1], and about 5 TU on average globally [54]. Monitoring of the tritium level in precipitation at several short-distance stations showed that there was no significant systematic discrepancy between them [1]. The "anthropogenically modified natural distributions" present now are "new natural global" environmental levels.

Technogenic tritium is produced in various industries, such as nuclear power plants, nuclear reactors, future fusion reactors, fuel reprocessing plants, heavy water production facilities, medical diagnostics, radiopharmaceuticals, luminous paints, sign illumination, self-luminous aircraft, airport runway lights, luminous dials, and gauges and wrist watches [55–58]. Technogenic tritium causes deviations from the "anthropogenically modified natural tritium distribution" at a local or regional level.

The seasonal and spatial distribution of tritium activity concentration in precipitation around the globe has been found to be dominated by the annual stratosphere–troposphere exchange at high latitudes in early spring, in combination with latitudinal and continental effects [1]. The latitude effect is described as the highest ^{3}H activity concentrations observed between the 30th and 60th parallel, with values lower by a factor of approximately five at low-latitude and tropical stations.

It should be noted that the seasonal variations in ^{3}H activity concentration in precipitation do not have the same origin as the seasonal variations in δ^2H and δ^{18}O: variations in ^{3}H are caused by the exchange between the stratosphere and the troposphere and are not caused or influenced by the temperature, while the local seasonal variations in δ^2H and δ^{18}O show a close relation with the local temperature [1,2,4,47].

2.3. Sampling Sites and Climate

There are three main climate types prevailing in Croatia: continental, maritime, and mountain. Such a climate distribution is determined by the geographical position of Croatia in the northern midlatitudes and the corresponding weather processes. Croatia is a relatively small country (56,594 km^2) positioned between the Pannonian Plain and the Adriatic Sea and has a large orographic variety. Therefore, the most important climate modifiers are the Adriatic Sea in the southwest, the mountain chain Dinarides in the central part, and openness to the Pannonian plain in the northeast [59]. Accordingly, most of Croatia has a temperate rainy climate (Köppen code Cxx [60,61]). For example, the Zagreb climate zone is described as Cfb: a temperate climate without a dry season and with a warm summer. The highest mountain areas have a cold snow and forest climate (Köppen code Dxx). The complete designation of climates for particular sampling sites is presented in Table S1 (Supplementary Materials). A comparison of the 30-year period 1981–2010 to the standard climatological period, 1961–1990, showed a significant increase in the mean annual temperature at all 20 studied stations in Croatia, with stronger warming at the continental stations than along the coast and with the largest changes in the summer [59]. It was also noted that both minimal and maximal temperatures increased by larger amplitudes inland compared to along the coast. The precipitation pattern has also changed, but differently in different parts of Croatia: an increase in the precipitation amount has been observed inland, with a statistically significant increase in autumn [59]. The observed climate changes (temperature increase and the precipitation regime) have resulted in changes in climate classes for some stations. The stations with isotope-in-precipitation data for which the climate class changed are Dubrovnik and Zadar (from Cfa to Csa) and Puntijarka on Mt. Medvednica, near Zagreb (from Dfb to Cfb) (Table S1).

A long-term isotopes-in-precipitation record (1976–2018) exists in Croatia only for the Zagreb station [27,47,62] (although microlocations have changed, as will be explained later). Data for some other stations with shorter monitoring periods were obtained during various individual projects (location numbers 1 to 12, Figure 1) [18,47,62–68]. Details on sampling sites are presented in Table S1. Some projects have also included monitoring at stations in Slovenia (location numbers 13 to 15, Figure 1), e.g., in Ljubljana [47,62,69–71], Portorož, and Kozina [47,72,73].

Precipitation monthly composite samples for the Zagreb station were collected at the Ruđer Bošković Institute (RBI, 45.817° N, 15.967°, 165 m a.s.l.) from 1976 until 1995. In 1994 and 1995, higher ^{3}H activity concentrations were measured due to experimental research in the nearby department in which technogenic tritium was used [74,75], while there was no influence on stable isotope data. As a consequence of local tritium contamination, a new location for precipitation sampling had to be found. During 1995 and 1996, precipitation samples were additionally collected at Puntijarka, on Mt. Medvednica (45.917° N, 15.95°, 988 m a.s.l.), 15 km north of the city of Zagreb [62], and at the Zagreb–Grič site at the Croatian Meteorological and Hydrological Service (45.814° N, 15.972° E, 157 m a.s.l.) in the center of Zagreb (in 1996). ^{3}H activity concentrations in precipitation at the stations Zagreb–Grič and Puntijarka in 1996 were almost identical, 10.7 ± 4.0 TU and 10.4 ± 6.0 TU, respectively [62]. For the analyses in this paper, the Zagreb–RBI data for tritium activity concentration were used for up to 1993, data from Puntijarka were taken as representative for Zagreb in 1995, and for 1996 and on, data from Zagreb–Grič were used. Tritium data exist for 1994, but they will not be discussed here because they present technogenic (local) tritium. Data for the stable isotope composition of Zagreb precipitation are not available for the 2007–2009 period and for 2011.

Daily rain events were collected in Zagreb for the period from October 2002 to March 2003, and the obtained stable isotope composition was compared to the monthly δ^{18}O and δ^2H data [76].

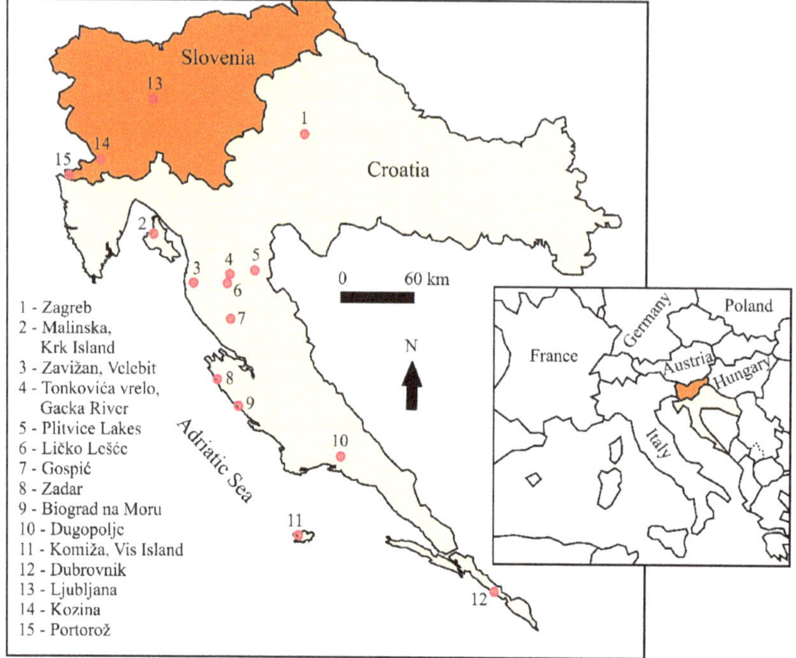

Figure 1. Map of stations with isotope-in-precipitation data in Croatia (locations 1–12) and in Slovenia (locations 13–15). Details on stations are in Table S1.

2.4. Meteorological Data

The meteorological data consisted of monthly precipitation amount and average monthly air temperature. Data were obtained on request from the Croatian Meteorological and Hydrological Service (CMHS). Meteorological records for Zagreb, Croatia, exist for the period since 1862 and have been analyzed by CMHS for the usual climatological periods [77]. Here, we used data from only the 1976–2018 period, for which we had records of the isotope composition of precipitation. Minimal and maximal monthly values within a year were identified, and the mean annual values of temperature and the total annual precipitation amount were determined.

2.5. Measurement of δ^2H and $\delta^{18}O$

The stable isotope composition of the precipitation samples for the period 1980–2003 was measured on a Varian MAT 250 dual inlet isotope ratio mass spectrometer (IRMS) at the Jožef Stefan Institute in Ljubljana [49,69,70,73]. The isotopic composition of hydrogen (δ^2H) was determined by means of the H_2 generated by the reduction of water over hot zinc (up to 1998) [69], and later over hot chromium [78]. The oxygen isotopic composition ($\delta^{18}O$) was measured by means of the water–CO_2 equilibration technique [40]. All measurements were carried out together with laboratory standards that were calibrated periodically against international standards, as recommended by the IAEA. The measurement precision of duplicates was better than ±0.1‰ for $\delta^{18}O$ and ±1‰ for δ^2H. The stable isotope composition of the precipitation sampled between 2004 and 2006 was determined at SILab (Stable Isotope Laboratory at the Physics Department, School of Medicine, University of Rijeka, Rijeka, Croatia). An HDO Equilibration Unit (ISO Cal) attached to the dual inlet port of a DeltaPlusXP (Thermo Finnigan) IRMS was used [68,79]. The $\delta^{18}O$ and the δ^2H were obtained from CO_2 and H_2 gas, respectively, after equilibration with a 4 ml water sample. The measurement reproducibility of duplicates was better than ±0.1‰ for $\delta^{18}O$ and ±1‰ for δ^2H. The stable isotope composition of precipitation from 2010 and on was analyzed with a Liquid Water Isotope Analyzer (LWIA-24d, Los Gatos Research) at the Institute for Geochemical Research, the Hungarian Academy of Sciences, Budapest, Hungary. The uncertainty of the measurements was reported to be ±0.2‰ for $\delta^{18}O$ and ±0.6‰ for δ^2H [66]. The stable isotope composition of Zagreb's precipitation for the 2012–2018 period was determined at the Laboratory for Spectroscopy of the Faculty of Mining, Geology, and Petroleum Engineering, University of Zagreb, with a Liquid Water Isotope Analyzer (LWIA-45-EP, Los Gatos Research). Data were analyzed by the Laboratory Information Management System (LIMS) [80]. The measurement precision of duplicates was ±0.1‰ for $\delta^{18}O$ and ±0.3‰ for δ^2H.

2.6. Measurement of Tritium Activity Concentration

Tritium activity concentration (A) in all monthly samples was determined at the Ruđer Bošković Institute in Zagreb. The results are expressed in tritium units (1 TU = 0.118 Bq l^{-1}) [1] since the same units are used by the IAEA–WMO/GNIP database [27]. A tritium unit represents one ^3H atom in 10^{18} atoms of hydrogen. The gas proportional counting technique (GPC) was used up to 2009 [81–83]. Methane (CH_4) was used as a counting gas in a multiwire proportional counter. It was obtained through the reaction of water (50 ml) with aluminum carbide at 150 °C [81]. The counting energy window was set to energies between 1 keV and 10 keV to obtain the best figure of merit. Gas quality control was performed by simultaneously monitoring the count rate above the tritium channel, i.e., above 20 keV [82]. The detection limit was 2.5 TU, and the measurement uncertainty was between 2 and 5 TU, depending on the activity concentration. In 2008, a technique of liquid scintillation counting of electrolytically enriched samples (LSC-EE) was introduced, and between 2008 and 2009, GPC and LSC-EE techniques were used [83,84]. Since 2010, samples have been measured using the LSC-EE technique only [71,83–86].

The electrolytic enrichment system at the RBI was produced by the AGH University of Science and Technology, Krakow, Poland [87,88]. It consists of 20 cells 500 ml in volume (stainless steel anodes and mild steel cathodes). Each sample was distilled before electrolysis, as the required conductivity is <50 µS/cm. An enrichment run contained 15 unknown samples, 3 spike waters (water of known tritium activity concentration, 500–600 TU) used for monitoring the electrolysis performances and the calculation of enrichment factor, and 2 tritium-free samples used for system control. The enrichment procedure at RBI took 1420 Ah distributed over 8 days. Enriched samples were distilled again after electrolysis, with 6–8 g of $PbCl_2$ added to each sample. The scintillation cocktails for measurement in LSC were prepared by 8 ml of sample and 12 ml of scintillator Ultima Gold LLT in high-density low-diffusion polyethylene vials. Measurements were performed by an ultra-low-level liquid scintillation counter, the Quantulus 1220, in 10 cycles of 50 min each. Each measurement run consisted of 24 scintillation cocktails: 20 enriched samples, 1 nonenriched spike sample, 1 international standard, and 2 nonenriched background samples. On the basis of the initial and final mass of water in cells and the individual count rates of the spike water before and after enrichment, the enrichment factor E was calculated [86,87]. The average 10-year E value of the system was 26 ± 2. The detection limit obtained by the LSC-EE technique was around 0.5 TU, and the measurement uncertainty was between 0.5 and 3 TU [71,83–86].

2.7. Data Evaluation

Mean $\delta^{18}O$, δ^2H, and *d*-excess values, weighted by precipitation amount, were calculated from all monthly data and then summed over all collected samples per month and per year. Thus, the monthly mean values for a specific month over a certain period were obtained, as were the annual mean values. The number of datapoints per year for all years (Table 1) fulfilled the requirement for the calculation of annual mean values, i.e., the lowest number of monthly isotope datapoints per year was 9, while at least 7 monthly samples were required [89]. All years with an incomplete number of monthly samples, however, met the requirement that the available isotope data comprise more than 70% of the total precipitation amount collected per year (Table 1) [89].

Table 1. Mean annual temperature T, ranges of monthly temperatures within a year, annual precipitation amount P, weighted mean (w.m.) annual $\delta^{18}O$, δ^2H, d-excess (d), and the mean annual tritium activity concentration A in precipitation at Zagreb, 1976–2018. Here, n: number of monthly datapoints; %P: percentage of precipitation if different from 100%, comprised by $\delta^{18}O$, δ^2H, and d-excess data; TU: tritium unit. Bold font: the lowest and highest values in a series.

Year	T (°C) Mean	T (°C) Range	P (mm)	$\delta^{18}O$ (‰) w.m.	n, %P	δ^2H (‰) w.m.	n, %P	d (‰) w.m.	n, %P	A (TU) Mean
1976	10.6	1.3–21.4	908	–	–	–	–	–	–	**100.5**
1977	11.4	0.2–20.1	1014	–	–	–	–	–	–	79.4
1978	9.8	0.8–**18.9**	758	–	–	–	–	–	–	73.8
1979	11.0	−1.1–20.9	792	–	–	–	–	–	–	36.8
1980	**9.6**	−1.5–20.0	931	−8.92	12	−62.65	12	8.86	12	39.5
1981	11.6	−0.8–21.2	871	−9.44	12	−67.32	12	8.20	12	38.3
1982	11.7	−0.9–22.2	805	−8.31	10, 97	−59.21	10, 97	7.27	10, 97	25.1
1983	12.1	0.7–23.8	755	−9.07	12	−65.22	12	7.35	12	25.8
1984	10.9	1.3–19.9	897	−9.25	12	−64.24	12	9.75	12	20.6
1985	11.0	−3.4–22.0	800	−9.12	12	−66.85	12	8.84	12	18.4
1986	11.1	−2.1–21.8	786	−8.96	12	−64.45	12	7.20	12	19.3
1987	11.4	−1.8–23.2	816	−9.21	12	−64.64	12	9.04	12	23.2
1988	11.9	2.6–23.3	749	−7.29	10, 81	−55.17	11, 86	6.90	10, 81	17.8
1989	12.0	0.1–21.8	957	−6.54	11, 95	−43.23	11, 89	8.46	10, 84	23.2
1990	12.5	1.1–22.2	694	−7.81	12	−56.12	12	6.44	12	16.1
1991	11.4	0.2–23.0	787	−8.37	12	−61.94	12	5.06	12	14.5
1992	13.0	2.3–**25.8**	808	−8.96	11, 99	−63.54	11, 99	8.11	11, 99	11.1
1993	12.1	2.3–22.3	928	−8.48	10, 97	−58.22	10, 97	9.65	10, 97	17.3
1994	13.2	3.3–23.9	962	−7.20	12	−48.02	12	9.59	12	–
1995	12.0	1.8–23.8	962	−9.15	11, 99	−62.49	11, 99	**10.74**	11, 99	11.9
1996	11.0	−0.3–21.1	959	−8.30	12	−52.46	10, 90	6.79	10, 90	10.5
1997	12.2	−0.3–21.7	723	−8.02	11, 92	−56.33	11, 92	7.85	11, 92	9.7
1998	12.4	−1.5–22.5	1000	−6.87	10, 93	−47.38	10, 81	7.21	10, 81	9.1
1999	12.5	1.7–22.4	997	−8.55	12	−64.53	12	3.89	12	8.8
2000	13.8	−0.2–24.4	725	**−5.54**	10, 85	**−39.68**	9, 81	**2.28**	9, 81	9.2
2001	12.7	−0.7–23.6	813	−7.97	12	−56.68	12	7.10	12	9.5
2002	13.2	2.2–22.5	1064	−8.21	12	−56.03	12	9.65	12	8.6
2003	12.9	−0.1–**25.8**	623	−7.74	10, 95	−55.29	12	9.56	10, 95	7.3
2004	12.0	0.8–21.7	993	−8.12	12	−60.99	12	8.41	12	**5.4**
2005	11.7	−0.1–22.1	988	−9.22	12	−64.38	12	9.40	12	9.7
2006	12.7	−0.1–24.6	754	−8.23	12	−58.10	12	7.76	12	8.5
2007	13.6	1.4–23.8	896	–	–	–	–	–	–	9.5
2008	13.4	3.8–22.8	769	–	–	–	–	–	–	9.1
2009	13.4	0.0–23.6	795	–	–	–	–	–	–	9.5
2010	12.2	0.3–24.1	1155	**−9.67**	12	**−68.6**	12	8.76	12	7.9
2011	13.2	2.9–24.3	**521**	–	–	–	–	–	–	9.4
2012	13.7	−0.2–25.4	813	−6.33	9, 77	−45.22	9, 77	5.41	9, 77	6.7
2013	12.9	2.4–24.5	1092	−8.77	9, 94	−61.82	9, 94	8.37	9, 94	8.5
2014	13.8	**5.4**–22.4	**1234**	−7.69	12	−53.64	12	7.90	12	7.4
2015	13.7	3.6–25.4	824	−7.81	10, 92	−55.35	10, 92	7.16	10, 92	7.9
2016	13.1	1.0–24.2	854	−8.54	11, 99	−61.00	11, 99	7.30	11, 99	7.3
2017	13.6	−2.3–25.0	889	−7.44	11, 95	−50.40	11, 95	9.14	11, 95	6.8
2018	14.1	0.9–25.0	827	−9.27	12	−63.85	12	10.33	12	8.7

Correlations between various datapoints were obtained as ordinary least squares regressions, and the Pearson's coefficient r is given, as are the number of data pairs n and the p-value describing the statistical significance of the correlations. Data taken from the literature usually have the adjacent r^2 value reported.

In the special case of correlations between δ^2H and $\delta^{18}O$ (i.e., for LMWLs), different methods were applied: an ordinary least squares regression (OLSR), a reduced major axis regression (RMA), and a major axis least squares regression (MA) [48,49,89]. In addition, we calculated precipitation weighted regressions (PWLSR, PWRMA, and PWMA) [48,49], which took into account the precipitation amount in a particular month. The local meteoric water lines are defined as LMWL$_{OLSF}$, LMWL$_{RMA}$, LMWL$_{MA}$, LMWL$_{PWLSR}$, LMWL$_{PWRMA}$, and LMWL$_{PWMA}$. While OLSR regressions were found by commercially available software (MS Excel), for other regressions we used Local Meteoric Water Line Freeware [90]. The software also calculated an average of the root mean square sum of squared errors (*rmSSE$_{av}$*), which is a relative error that allows for a comparison of different methods: the closer the value of *rmSSE$_{av}$* is to 1.0, the better the regression method for that set of data [48].

Deuterium excess *d* was calculated from paired monthly data according to Equation (3). Some precipitation samples showed very low (highly negative; in our case, the lowest value was −13‰) *d*-excess values. This can be caused by improper sampling (e.g., the precipitation stayed for the whole month in a sample collector), by evaporation due to a low amount of precipitation and/or high temperatures, or by the evaporation/sublimation of raindrops falling in a dry atmosphere [50,51,71].

3. Results

Data on monthly temperatures, monthly precipitation amount, and the tritium activity concentration of precipitation at Zagreb for the 1976–2018 period, as well as the stable isotope composition of monthly precipitation (δ^{18}O, δ^2H, *d*-excess) for the 1980–2018 period, are shown in Table S2 (Supplementary Materials).

3.1. Meteorological Data

Minimal monthly (mean air) temperatures at Zagreb for the 1976–2018 period ranged from −3.4 °C (in 1985) to 5.4 °C (in 2014) (Figure 2a). They were measured in January (in 21 cases, or 49%), February (12 cases, or 27.9%), and December (9 cases, or 20.9%), and only once, in 1988, was the coldest month November. The highest monthly temperatures, in a range from 18.9 °C (1978) to 25.8 °C (1992, 2003) were measured in 26 out of 43 years (60.5%) in July, 15 times (35%) in August, and only in 2 cases in June (1979 and 1996).

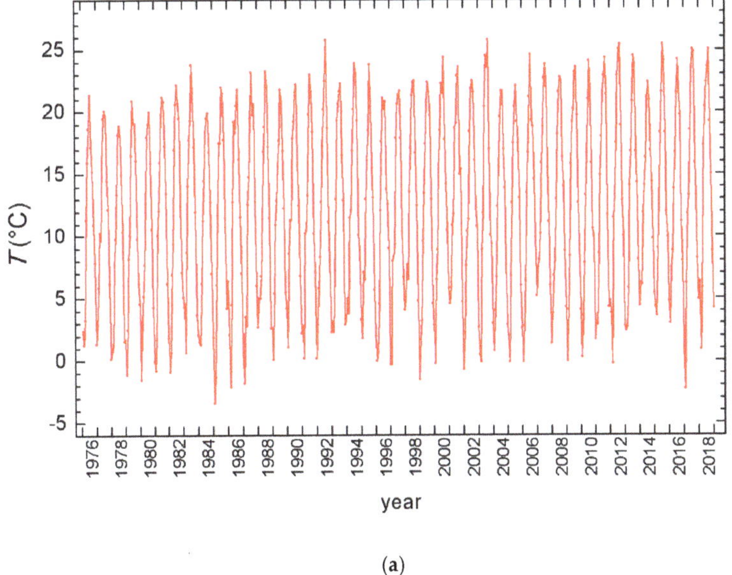

(a)

Figure 2. *Cont.*

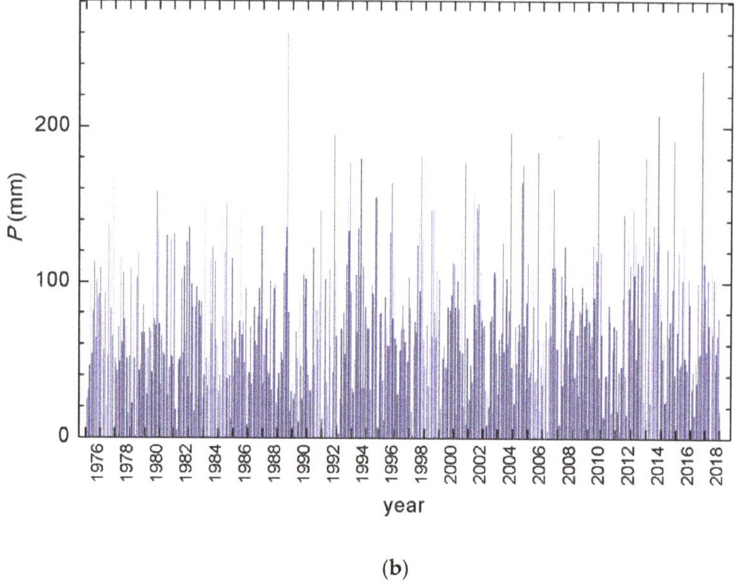

(b)

Figure 2. (**a**) Mean monthly air temperatures; (**b**) monthly precipitation amount. Data from station Zagreb–Grič for the 1976–2018 period. Data obtained from the Croatian Meteorological and Hydrological Service (CMHS).

Mean annual temperatures, together with yearly minimal and maximal monthly temperatures and the total yearly amount of precipitation at Zagreb for the 1976–2018 period, are shown in Table 1. Mean annual temperatures ranged from 9.6 °C in 1980 to 14.1 °C in 2018.

The annual precipitation amount ranged from 521 mm (2011) to 1234 mm (2014) (Table 1), while the maximal monthly precipitation amount occurred in August 1989 (260 mm), followed by 236 mm in September 2017 and 208 mm in September 2014. In all other months, the monthly amount of precipitation was below 200 mm (Figure 2b).

3.2. Stable Isotopes

Monthly δ^{18}O values in precipitation at Zagreb ranged from -17.6‰ (January 2005) to -0.5‰ in June 1998 (Figure 3). Similarly, the lowest δ^{2}H = -133.1‰ (January 2005) and the highest δ^{2}H = -11.4‰ (June 1998) were determined in the same months (Figure S1, Supplementary Materials).

The lowest weighted mean annual values were observed in 2010 (δ^{18}O = -9.7‰, δ^{2}H = -68.0‰) and the highest in 2000 (δ^{18}O = -5.54‰, δ^{2}H = -39.68‰) (Table 1). The number of monthly isotope datapoints (n) available in each year as well as the percentage of annual precipitation amount (%P) comprised by the isotope data during the n months (Table 1) satisfied the requirements for the calculation of mean annual values [89]. For 15 years out of 39 (with the available stable isotope data), the number of monthly samples was less than 12, but it was never less than 9, and the available isotope data comprised at least 77% of annual precipitation in these years.

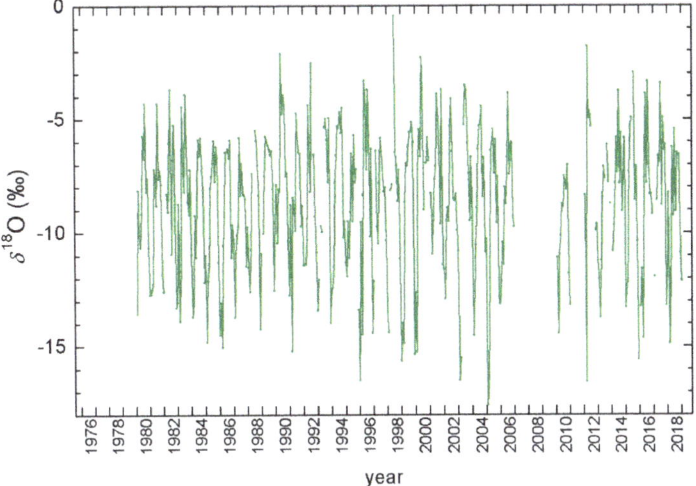

Figure 3. Monthly $\delta^{18}O$ in precipitation at Zagreb, 1980–2018 period.

Monthly values of deuterium excess (Figure 4, Table S2) ranged from −6.9‰ in May 2000 to 22.5‰ in October 1994, with 110 mm rain, probably from the Mediterranean. It should be noted here that for six months of the entire studied period, the monthly d-excess values were lower than −7‰, i.e., they were more than three standard deviations lower than the overall mean d-excess value and were therefore deleted from the record and further analysis. Three out of six cases were caused by the evaporation of small monthly precipitation amounts (11 mm in March 1996, 22 mm in February 2000, and 2 mm in December 2016), and the other three (January 1996, June 1998, and July 2012) were probably from the evaporation/sublimation of raindrops. Mean annual d-excess values ranged between 2.28‰ and 10.74‰ in 2000 and 1995, respectively (Table 1).

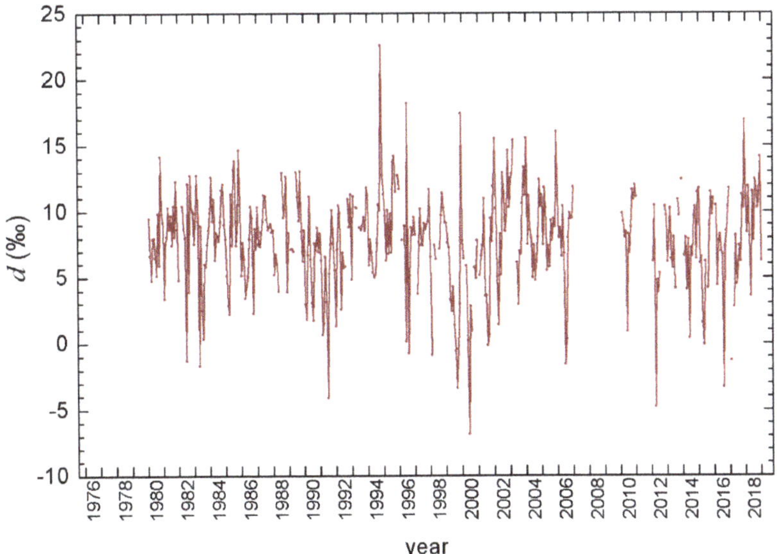

Figure 4. Monthly deuterium excess values for precipitation at Zagreb, 1980–2018.

3.3. Tritium Activity Concentration in Precipitation at Zagreb

The complete record of tritium activity concentration (*A*) in precipitation for Zagreb, 1976–2018 (Figure 5), exhibited a pattern typical of continental stations of the Northern Hemisphere. Seasonal variations were superposed on the basic decreasing trend of mean annual values until approximately 1996. The maximal monthly ^{3}H activity concentration at the Zagreb station was observed between May and July, mostly in June. A secondary maximum was also observed three times in January and February. The lowest ^{3}H activity concentrations were almost uniformly distributed from October to February, with a slightly more frequent occurrence in December.

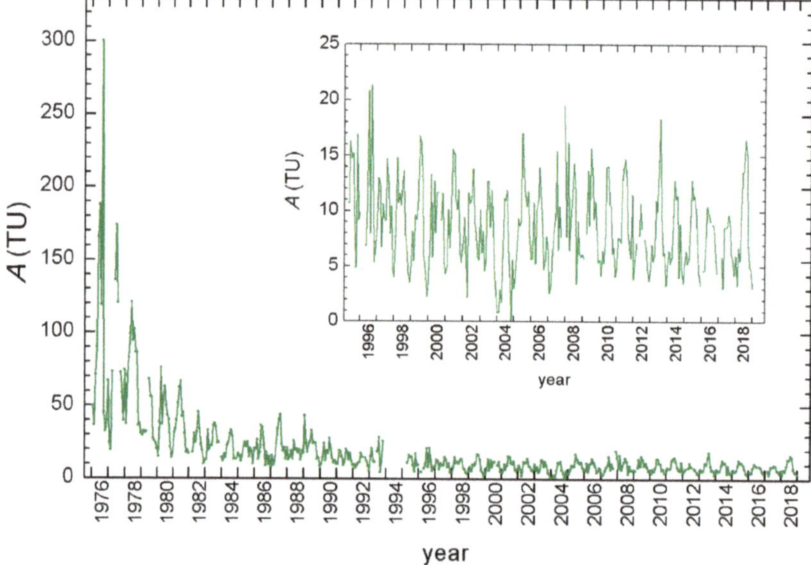

Figure 5. Complete record of monthly tritium activity concentration (*A*) in precipitation at Zagreb, 1976–2018. Insert: for the 1995–2018 period.

Data recorded from 1995 to 2018 (insert in Figure 5) showed an almost constant mean annual ^{3}H activity concentration ranging between 5.4 TU (in 2004) and 13.5 TU (in 1995), with a mean value of 8.5 ± 1.2 TU. Seasonal variations remained observable, with winter activities close to the natural pre-bomb ^{3}H activity concentrations ($\leq$5 TU) and summer values up to 21 TU [91,92].

4. Discussion

4.1. Trends in Meteorological Parameters

The annual precipitation amount *P* at Zagreb for the 1976–2018 period (Figure 6) showed a slight increase (1.4 ± 1.7 mm/y, *r* = 0.13, *p* = 0.4), as did the maximal monthly values within a year (0.7 ± 0.5 mm/y, *r* = 0.23, *p* = 0.14), while the minimal monthly values within a year, including no-rain months, showed a slight decrease (−0.1 ± 0.1 mm/y, *r* = −0.13, *p* = 0.41). However, the trends (statistically not significant) were not the most prominent characteristics of the data. Higher dispersion/fluctuations from the mean value for the whole period, 1976–2018 (867 ± 138 mm), were obvious (Figure 6). In the period 1976–2000, practically all values lay within ±1 standard deviation (±1 σ), while later on there were some years with deviations from the mean over ±2 σ. The mean values of the precipitation amount in the subperiods (1980–1985, 1986–1995, 1996–2006, and 2012–2018) for which isotope data were available showed larger fluctuations in the precipitation amount in more recent periods (Table 2).

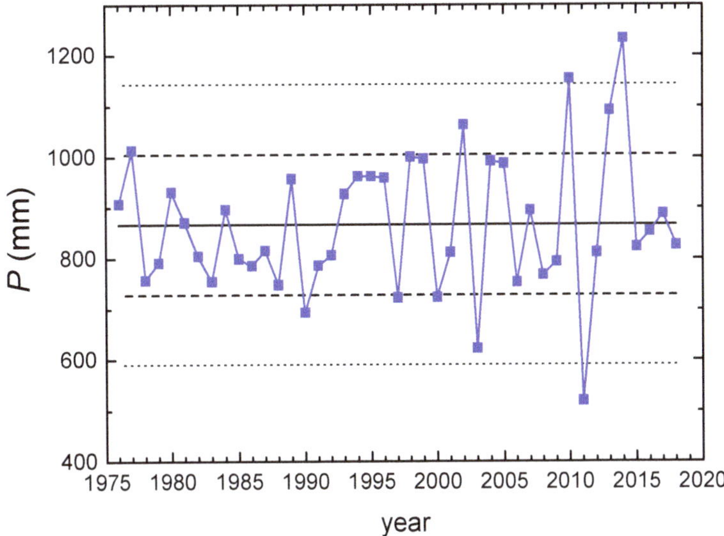

Figure 6. Annual precipitation amount *P* in Zagreb, 1976–2018 period. Solid line: the mean value for the whole period; dashed lines: ±1σ; dotted lines: ±2σ.

Table 2. Comparison of mean values of meteorological parameters (*T*, *P*) and isotopic data ($\delta^{18}O$, δ^2H, and *d*-excess) for different periods.

Period	*T* (°C)	*P* (mm)	$\delta^{18}O_{w.m.}$ (‰)	$\delta^2H_{w.m.}$ (‰)	$d_{w.m.}$ (‰)
1980–1985	11.2 ± 0.9	843 ± 67	−9.0 ± 0.4	−64.2 ± 3.0	8.4 ± 0.9
1986–1995	12.1 ± 0.7	845 ± 98	−8.2 ± 0.9	−57.8 ± 7.3	8.1 ± 1.7
1996–2006	12.5 ± 0.8	876 ± 150	−7.9 ± 1.0	−55.6 ± 7.2	7.3 ± 2.3
2012–2018	13.5 ± 0.4	933 ± 164	−8.0 ± 1.0	−55.9 ± 6.8	7.9 ± 1.6

Mean annual temperature and the minimal and maximal monthly temperature within a year (Figure 7) showed a significant increase at a 95% confidence level ($p < 0.05$) (a mean value of 0.071 ± 0.008 °C per year ($r = 0.82$), a minimal value of 0.05 ± 0.08 °C/year ($r = 0.33$), and a maximal value of 0.09 ± 0.02 °C per year ($r = 0.69$)). Since both the minimal and maximal monthly temperatures increased, but with different gradients, the result was that the increase in the amplitudes of the air temperatures (0.04 ± 0.03 °C per year, $r = 0.26$, $p = 0.09$) was significant at a 90% significance level (Figure 7).

It is also interesting to look at the monthly mean *P* and *T* values at the Zagreb station averaged for each month over a certain period (Figure 8) and calculated for four subperiods. The monthly amount of precipitation was relatively uniformly distributed throughout the year. However, the average values in January–April and in December were lower (<61 mm) than in May–November (>70 mm). The month with the highest average precipitation was September, with 97 mm. September was also the only month showing a constant increase in the amount of precipitation over the studied periods. Monthly precipitation for the period 1976–1996 (i.e., the first two periods from Figure 8a) also showed lower precipitation (<60 mm) for January to April. However, maximum precipitation amounts were observed in June (94 mm) and August (91 mm) [62]. Such a shift in the precipitation regime is in accordance with observations of climate changes in Croatia [59].

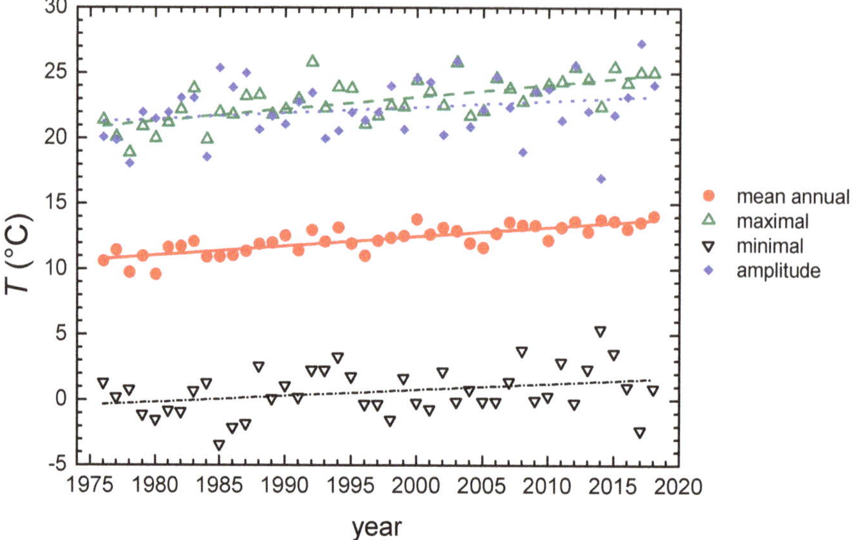

Figure 7. Mean annual temperature, minimaland maximal monthly temperatures within a year, and the temperature amplitudes in Zagreb, 1976–2018 period. Trend lines for each dataset are shown in the same color.

Mean monthly temperatures (Figure 8b) for all months in the 1976–1985 period were lower than in the most recent period, 2007–2018. The difference ranged from 1.1 °C in October and December to 3.5 °C in August and 3.6 °C in April. This observation once more corroborates observations of constant recent temperature increases that are more pronounced in the spring–summer periods [59].

4.2. Trends in Stable Isotope Data

Both the arithmetic mean annual values ($\delta^{18}O_a$, δ^2H_a, d_a) and the weighted mean annual values by amount of precipitation ($\delta^{18}O_{w.m.}$, $\delta^2H_{w.m.}$, $d_{w.m.}$) were calculated (Table S3). Due to the relatively homogeneous distribution of the annual precipitation amount (Figure 8a), there was no significant difference between the two types of annual means; in fact, a very good correlation was obtained, and here we present an example of the $\delta^{18}O$ values:

$$\delta^{18}O_{w.m.} = (1.006 \pm 0.16)\,\delta^{18}O_a + (0.48 \pm 1.4),\ n = 34,\ r = 0.73. \tag{4}$$

Changes in the weighted mean annual values of $\delta^{18}O$, δ^2H, and d-excess in the studied 1980–2018 period are shown in Figure 9 together with their respective trends. The $\delta^{18}O$ and δ^2H values exhibited increases, with an increase rate of 0.017‰ ± 0.014‰ per year ($r = 0.21$, $p = 0.23$) for $\delta^{18}O$ and 0.14‰ ± 0.11‰ per year ($r = 0.23$, $p = 0.19$) for δ^2H. The annual mean d-excess remained constant (slope ≈ 0). The corresponding weighted mean values for the periods 1980–2006 (as well as for the shorter subperiods) and 2012–2018 are shown in Table 2. Both the $\delta^{18}O$ and δ^2H values were higher in the more recent period, 2012–2018, than from 1980 to 2006. We observed a similar trend earlier: the mean $\delta^{18}O$ in the 2001–2003 period (−8.3‰) was more positive than the long-term mean $\delta^{18}O$ (−8.8‰) [76]. However, it was noticed (Table 2) that the values in the 1996–2006 and 2012–2018 periods were practically the same, although the temperature differed in these periods.

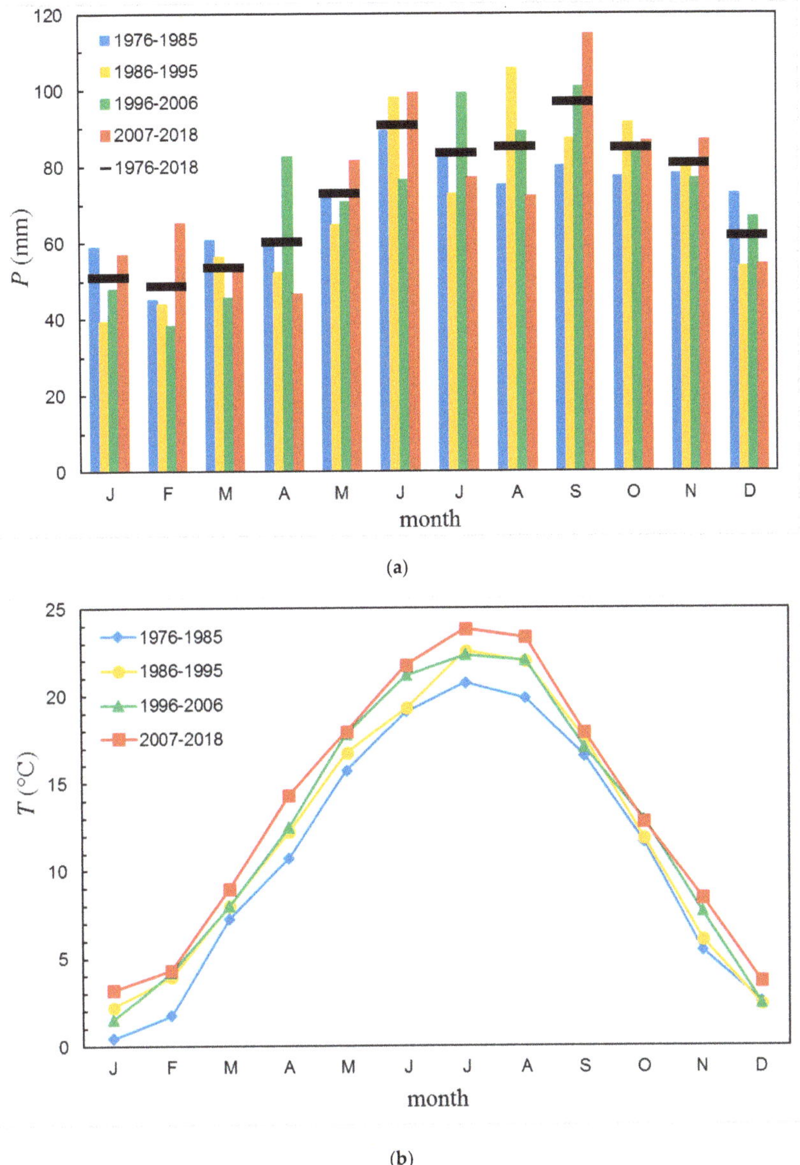

(**a**)

(**b**)

Figure 8. (**a**) Monthly mean precipitation amounts for the four subperiods and the average value for the whole period; (**b**) monthly mean air temperature for the subperiods.

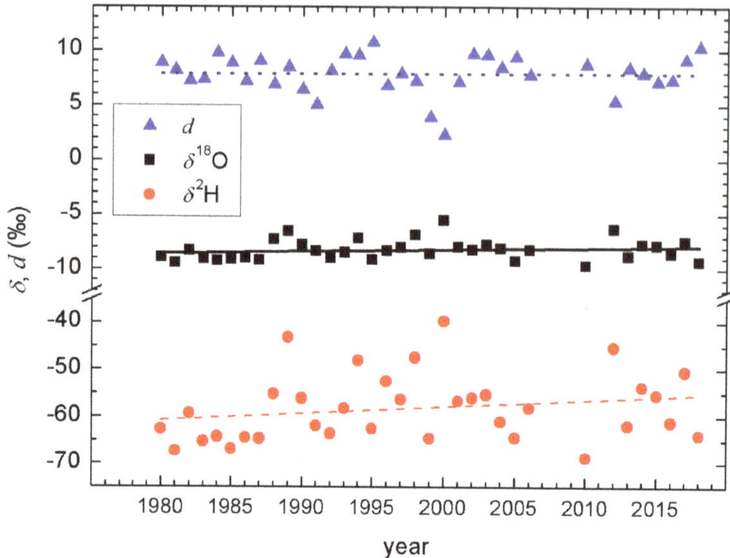

Figure 9. Temporal changes of weighted mean annual $\delta^{18}O$, δ^2H, and *d*-excess values.

To investigate which months contributed most to the changes in the mean annual values of $\delta^{18}O$ and δ^2H, we calculated the monthly mean $\delta^{18}O$ values in the four subperiods (Figure 10). The most pronounced differences were observed in the first half of the year, with higher values in the 2012–2018 period than in earlier years in January, March, and April, while in February, the most recent $\delta^{18}O$ and δ^2H were the lowest. The differences in the second half of the year were not large. This is behavior similar to that described earlier for the monthly mean temperature (Figure 8b). However, while the summer temperatures (June, July, and August) were the highest in the most recent period, the $\delta^{18}O$ and δ^2H were not.

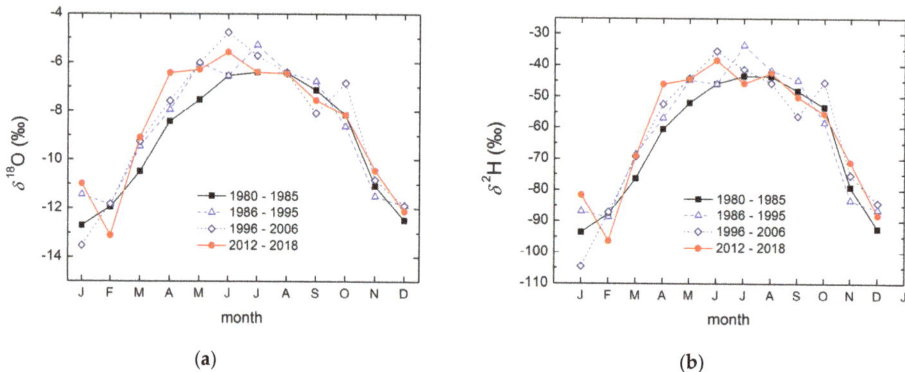

Figure 10. (**a**) Monthly mean values of $\delta^{18}O$ in the four subperiods; (**b**) monthly mean values of δ^2H in the four subperiods.

No significant correlation was observed between $\delta^{18}O$ and *P*. This was expected for the midlatitude continental station of Zagreb, which had an expressed seasonality in temperature (Figure 8b) and $\delta^{18}O$ values (Figure 10a) and a lack of clear seasonality in precipitation amount (Figure 8a) [4]. No amount effect was observed or reported for stations in Croatia and Slovenia [17,47,62,70–73].

4.3. Deuterium Excess

In previous analyses [47,76], it was demonstrated that the deuterium excess values for Zagreb's precipitation were higher in autumn than in the spring. The same occurred in this study using monthly mean values of *d*-excess for the four periods, 1980–1985, 1986–1995, 1995–2006, and 2012–2018 (Figure 11). In all periods, values from January to June (<8‰) were lower than those in the second half of the year (>8‰). While in the first half, there was a slight decrease in the newer periods (7.6 ± 0.6‰ in 1980–1985 to 6.0 ± 1.6‰ in 2012–2018), the *d*-excess values in the second half of the year showed an increasing trend. Such an interplay of decreases and increases of the monthly mean values eventually resulted in no change in the mean annual *d*-excess values over the whole studied period, as was shown earlier (Figure 9, Table 2). The higher *d*-excess in autumn (Figure 11) indicates a higher influence of the Mediterranean air masses in these months. It is interesting to note the shift of the autumn peak in the *d*-excess value from October (in 1980–1985) to November in the 2012–2018 period. Such behavior resembled the changes in the monthly precipitation amount in autumn (Figure 8a). November was the only month in which a significant increase in *d*-excess in the whole period was observed, with a rate/slope of 0.14‰ ± 0.04‰ per year ($n = 35$, $r = 0.54$, $p < 0.05$).

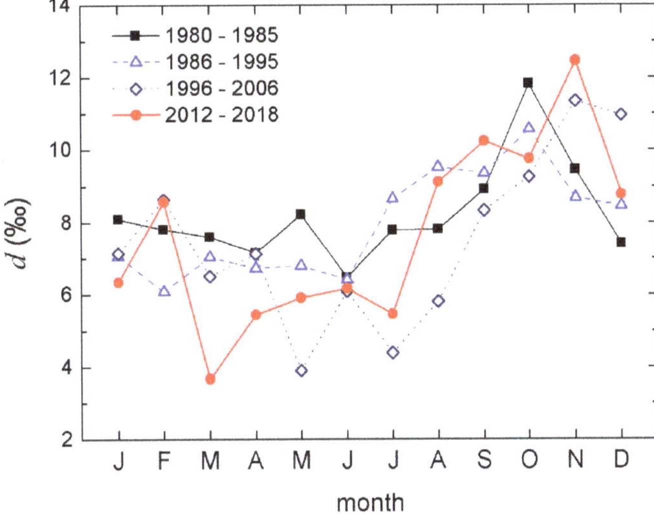

Figure 11. Monthly mean values of *d*-excess in the four subperiods.

A similar pattern of monthly *d*-excess distribution was also observed for all other Croatian stations (Figure 1, Table S1) [27,47]. The mean *d*-excess value depended on the location and altitude of the station, but in all cases, higher *d*-excess monthly values were observed in autumn–winter precipitation, usually from September to December [47]. The lowest mean monthly *d*-excess values were observed in summer months. At the South Adriatic stations (Komiža, Dubrovnik), the distribution of the precipitation amount had higher seasonality (a low amount in summer and high in winter) than it did in Zagreb, and the *d*-excess values in summer months (from May to August) were lower than 8‰, which is indicative of secondary evaporation of raindrops falling in a warm and dry atmosphere [47,50,51].

The very striking behavior of *d*-excess in March of the 2012–2018 period (Figure 11) can be explained as follows. During this period, the precipitation amount in March (57%) in four out of seven months was below 23 mm, resulting in *d*-excess values between −4.8‰ (2012, $P = 4$ mm) and 4.1‰ (2014, $P = 22$ mm), which was fully in accordance with observations from the station in Ljubljana [71] d < 5‰) and corresponded with months with low precipitation. These *d*-excess values were low, but were still within 3σ of the mean value, and therefore they were not excluded from the analyses. When these lower *d*-excess values were excluded from the mean *d*-values for March from 2012 to 2018, the mean *d*-excess for 2012–2018 became 7.4‰, which was not different from other periods. In contrast, in the 1980–2006 period, the total number of months with precipitation below 23 mm was only two (i.e., the occurrence of low precipitation in March was only 8%), and both *d*-excess values were excluded from the analyses. This observation also corroborated observed changes in the seasonal distribution of precipitation at continental stations in Croatia [59].

4.4. Trends in Tritium Activity Concentration

The seasonal variations in the tritium activity concentration in precipitation at Zagreb for the period from 1976 to 1993 were superposed onto a generally decreasing trend (Figure 5) that could be approximated by an exponential decay curve with a half-life of about 6 years, in accordance with the estimated residence time of tritiated water vapor in the lower stratosphere (of the order of a few years) [1]. The same pattern for the tritium-in-precipitation regime (maxima from June to July and minima in winter) was observed in other continental stations (Figure 1): Ljubljana [47,70,71], Plitvice Lakes [65,67], and Gacka [68]. The ratio of the maximum to minimum value for Zagreb precipitation up to 1996 ranged from 2.2 to 5.7 without a significant trend [62], similarly to other Northern Hemisphere stations (between 2.5 and 6 [1]). The ratio was similar in the 1996–2018 period, between 2.3 and 5.3 in most cases, and was higher only in years with tritium activity concentrations close to the GPC detection limit.

The decrease in mean annual tritium activity concentration values continued after 1996 (Figure 5 insert, Figure 12), but to a much lesser extent of 0.08 TU per year, resulting in mean values of 8.8 ± 1.4 TU in 1996–2006 and 7.6 ± 0.8 TU in 2012–2018. The mean values for the station Ljubljana were 9.1 TU and 8.3 TU during 1998–2010 and 2007–2010, respectively [71], showing the same trend.

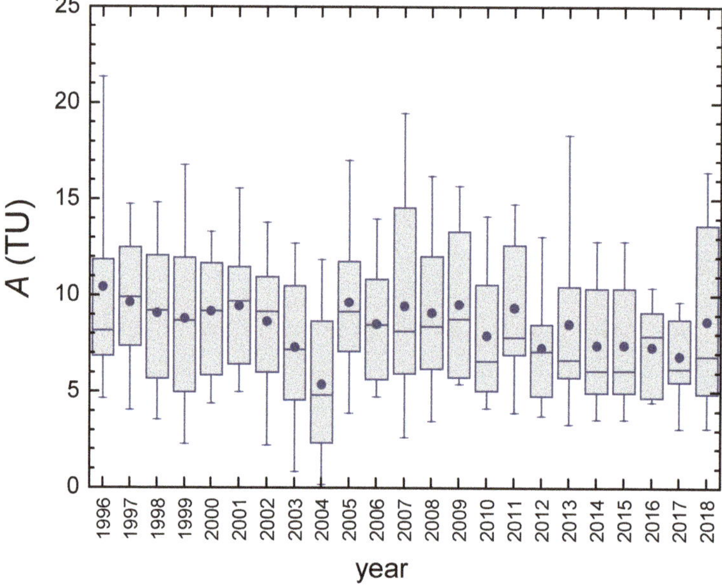

Figure 12. Tritium activity concentration in precipitation at Zagreb during the period 1996–2018. The boxplot shows the median value and the percentile range 25%–75% in a shadowed box, — shows the percentile range 5%–95%, and the average value is shown by the symbol ●.

Bomb-produced tritium in precipitation until about 1995 prevented studies on whether the natural production of tritium was influenced by variations in solar activities. The modulation of cosmogenic tritium production by an 11-year solar cycle has been recently shown in precipitation at several stations worldwide [93]. Local maxima in the tritium activity concentration in precipitation were observed simultaneously with maxima in neutron flux (minima in sunspot numbers). Our data (Figure 12) also showed local maxima in mean annual values and larger variability in 1996, 2007, and 2018 in accordance with the observations presented in Reference [93].

4.5. Local Meteoric Water Line

The slopes (*a*) and intercepts (*b*) of LMWLs were obtained using different regression methods (Table 3). The whole 1980–2018 period was taken, as were individual subperiods, and the present values were compared to the available data on LMWLs for Zagreb precipitation from different earlier periods (obtained using different regression methods). The slopes and intercepts determined using different regression methods increased from the OLSR to the RMA and MA (in the whole period, as well as in the subperiods (Table 3)), as was also observed for most continental stations [48]. The same was also valid for precipitation-weighted regressions. No difference was observed between the corresponding nonweighted and weighted types of regression, as was expected from the rather homogeneous distribution of the monthly precipitation amount. If the $rmSSE_{av}$ value was taken into account, all values were close to 1 (closer to 1 for RMA and PWRMA over the whole long-term period). In subperiod 1980–1985, PWMA represented the LMWL equally as well as PWRMA and RMA did; in subperiods 1986–1995 and 1996–2006, the best fits were obtained by RMA and PWMA; and in the most recent subperiods, the best fits were obtained by RMA and PWLSR (Table 3). It may be concluded that the local meteoric water line for Zagreb was best described by the nonweighted RMA regression method (Figure 13):

$$\delta^2 H_{RMA,all} = (7.74 \pm 0.06)\,\delta^{18}O + (5.6 \pm 0.6),\ n = 389. \tag{5}$$

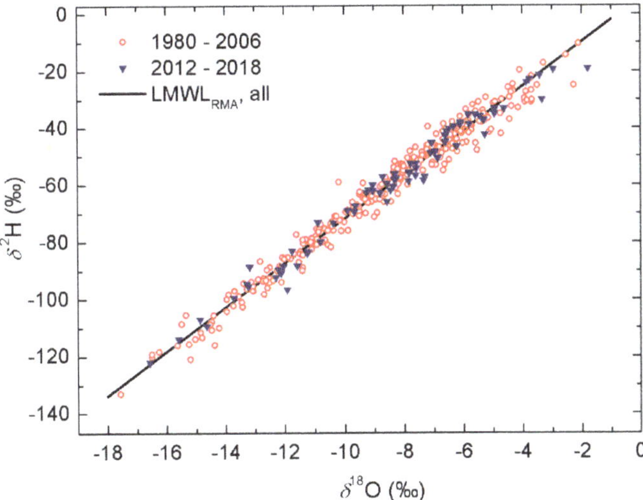

Figure 13. Local meteoric water line (LMWL) for Zagreb precipitation. Data from two periods are shown by different symbols. The fitted line is expressed by Equation (5).

A closer look at previously published LMWLs points to the difference in data up to 1996, with slope and intercept values close to 8 [27,62], while in 2001–2003, the slope was lower (7.3 + 0.2) and the intercept much lower (2.8 + 1.8) [47]. A similar conclusion was obtained from the present calculations (Table 3): all regression methods resulted in a slope close to 8 and an intercept in the range 7.3 to 9.3 in the subperiods 1980–1985 and 1986–1995. However, the slope values ranged from 7.4 to 7.8 and the intercept values from 2.6 to 6.3 in the subperiods 1996–2006 and 2012–2018. This difference can be explained by increases in temperature, especially in the summer months, and higher variability in the precipitation amount, which both led to more precipitation with lower d-excess, i.e., below the LMWL, as can be seen from Figure 13.

Table 3. Slopes (a) and intercepts (b) of the local meteoric water line (LMWL) for Zagreb in different periods and obtained using different regression methods. OLSR: ordinary least squares regression; RMA: reduced major axis regression; MA: major axis least squares regression; PWLSR, PWRMA, and PWMA: precipitation-weighted respective regressions; n: number of datapoints included; r and r^2: regression coefficients; $rmSSE_{av}$: average of the root mean square sum of squared errors [48].

Period	Method	a	b	n	r or r^2	Ref.
1980–1996	OLSR	7.9 ± 0.1	7.9 ± 0.1	194	$r = 0.985$	[62]
1980–1995 Zagreb	OLSR	7.91 ± 0.09	7.33 ± 0.83	182	$r^2 = 0.98$	[27]
	RMA	8.00 ± 0.09	8.13 ± 0.83	182	$r^2 = 0.98$	
	PWLSR	7.88 ± 0.09	7.52 ± 0.82	182	$r^2 = 0.95$	
1996–2003 Zagreb–Grič	OLSR	7.32 ± 0.17	0.68 ± 1.57	89	$r^2 = 0.95$	[27]
	RMA	7.50 ± 0.17	2.16 ± 1.55	89	$r^2 = 0.95$	
	PWLSR	7.22 ± 0.16	0.50 ± 1.39	89	$r^2 = 0.96$	
1980–2003	OLSR	7.8 ± 0.1	5.7 ± 0.8	271	$r = 0.98$	[47]
2001–2003	OLSR	7.3 ± 0.2	$2.8 + 1.8$	37	$r = 0.99$	[47]
					$rmSSE_{av}$	
1980–2018	OLSR	7.65 ± 0.06	4.79 ± 0.55	389	1.0047	This work
	RMA	7.74 ± 0.06	5.57 ± 0.55	389	1.0019	
	MA	7.83 ± 0.06	6.36 ± 0.56	389	1.0047	
	PWLSR	7.64 ± 0.06	5.24 ± 0.54	389	1.0060	
	PWRMA	7.73 ± 0.06	6.00 ± 0.54	389	1.0019	
	PWMA	7.82 ± 0.06	6.76 ± 0.55	389	1.0035	
1980–1985	OLSR	7.92 ± 0.14	7.45 ± 1.35	70	1.0044	This work
	RMA	8.00 ± 0.14	8.23 ± 1.33	70	1.0018	
	MA	8.09 ± 0.14	9.00 ± 1.36	70	1.0044	
	PWLSR	7.87 ± 0.14	7.26 ± 1.36	70	1.0075	
	PWRMA	7.96 ± 0.14	8.07 ± 1.36	70	1.0018	
	PWMA	8.05 ± 0.15	8.86 ± 1.38	70	1.0018	
1986–1995	OLSR	7.94 ± 0.11	7.46 ± 1.03	112	1.0045	This work
	RMA	8.03 ± 0.11	8.22 ± 1.02	112	1.0018	
	MA	8.11 ± 0.11	8.96 ± 1.04	112	1.0044	
	PWLSR	7.90 ± 0.12	7.73 ± 1.03	112	1.0093	
	PWRMA	7.99 ± 0.12	8.51 ± 1.03	112	1.0024	
	PWMA	8.08 ± 0.12	9.28 ± 1.04	112	1.0016	
1996–2006	OLSR	7.43 ± 0.11	2.59 ± 0.97	121	1.0048	This work
	RMA	7.52 ± 0.10	3.35 ± 0.96	121	1.0019	
	MA	7.60 ± 0.11	4.08 ± 0.98	121	1.0047	
	PWLSR	7.38 ± 0.11	2.68 ± 0.95	121	1.0095	
	PWRMA	7.47 ± 0.11	3.44 ± 0.95	121	1.0024	
	PWMA	7.56 ± 0.11	4.19 ± 0.96	121	1.0018	
2012–2018	OLSR	7.52 ± 0.13	3.56 ± 1.17	74	1.0043	This work
	RMA	7.60 ± 0.13	4.24 ± 1.16	74	1.0017	
	MA	7.68 ± 0.13	4.90 ± 1.19	74	1.0043	
	PWLSR	7.61 ± 0.12	5.16 ± 1.09	74	1.0008	
	PWRMA	7.68 ± 0.12	5.74 ± 1.09	74	1.0029	
	PWMA	7.75 ± 0.12	6.32 ± 1.10	74	1.0084	

4.6. Temperature Dependence of $\delta^{18}O$

The relation between all values of the mean monthly air temperature T and the monthly $\delta^{18}O$ values (Figure 14) can be described as

$$\delta^{18}O = (0.331 \pm 0.013)\, T - (12.8 \pm 0.2),\ n = 394,\ r = 0.795. \tag{6}$$

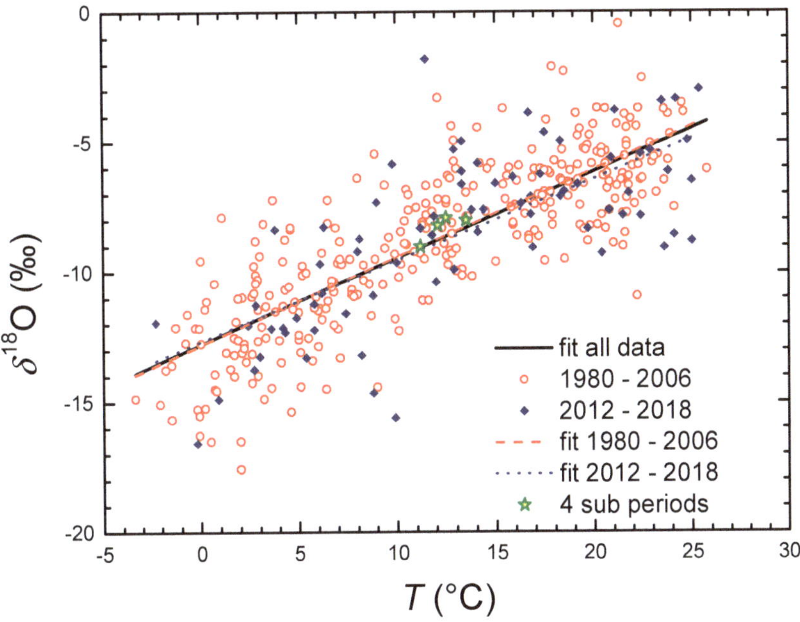

Figure 14. Temperature dependence of $\delta^{18}O$ in monthly precipitation at the Zagreb station during different periods. Regression lines are represented by Equations (6) and (8). The mean values in the subperiods (stars) are taken from Table 2.

When data for the periods 1980–2006 and 2012–2018 were separated, the two relations were

$$\delta^{18}O_{1980\text{-}2006} = (0.336 \pm 0.013)\, T - (12.8 \pm 0.2),\ n = 320,\ r = 0.814, \tag{7}$$

$$\delta^{18}O_{2012\text{-}2018} = (0.312 \pm 0.036)\, T - (12.7 \pm 0.5),\ n = 74,\ r = 0.718. \tag{8}$$

Although Equations (7) and (8) indicate a slight change in the temperature coefficient, the difference is comparable to the statistical uncertainties in the a-values, and therefore it can be concluded that there was no change in the relation $\delta^{18}O$ versus T in the recent period compared to the older one. The two lines (Equations (7) and (8)) are practically indistinguishable (Figure 14). The new relation for the complete set of long-term data for Zagreb (Equation (6)) was not different (i.e., it is within uncertainties) from the same relation for the data for the 1980–1996 period ($\delta^{18}O = (0.325 \pm 0.016)\, T - (12.6 \pm 0.2)$, $r = 0.83$, $n = 183$) [62]. The slopes in Equations (7) and (8) were also comparable to the slope of 0.31‰ per °C, which was determined from long-term data from midlatitude stations in the Northern Hemisphere [4].

5. Conclusions

A 43-year-long record of data on the isotope composition of precipitation ($\delta^{18}O$, $\delta^{2}H$, d-excess, and the tritium activity concentration A), together with meteorological data (air temperature, precipitation amount) at a continental station in Zagreb, Croatia, was studied, divided into four almost equally long subperiods. However, due to some missing data on stable isotope composition, the first (1980–1985) and the last (2012–2018) subperiods were shorter than the two subperiods in between (1986–1996, 1996–2006).

A constant increase in mean annual temperature was observed at a rate of 0.07 °C per year. An increase in monthly mean temperature was observed in all months in 2012–2018 when compared to earlier subperiods, with larger differences in the spring–summer than in autumn. The most striking feature of the annual precipitation amount was larger variations in the last two subperiods compared to earlier. A shift of the month with the highest precipitation amount was observed, from June and August to September.

Annual mean δ^{18}O and δ^2H values in the whole long-term period showed an increase of 0.017‰ per year and 0.14‰ per year, respectively. When monthly mean values in different subperiods were compared, larger differences were observed in the first half of the year than in the second one. Both of these changes in the stable isotope composition resembled observed changes in air temperature.

Although mean annual *d*-excess remained constant over the whole long-term period, there was a tendency for a decrease in the *d*-excess value in the first half of the year and an increase in the second half due to the influence of air masses originating from the eastern Mediterranean. Together with a shift in the maximal monthly mean value from October to November and a significant increase in *d*-excess in November, these observations point to changes in the precipitation regime and circulation pattern of air masses during the most recent period.

Different regression methods for the calculation of the local meteoric water line for Zagreb gave very similar values for the slope and intercept, with a slight preference for the RMA method and with no difference between nonweighted and precipitation-weighted values. In addition, no significant difference in both LMWLs and the temperature dependence of δ^{18}O values was observed between the most recent period (2012–2018) and the earlier period (1980–2006). The observed temperature gradient of 0.33‰ per °C was comparable to that of other similar stations.

The tritium activity concentration in precipitation in Zagreb between 1976 and 1994 exhibited pronounced seasonal variations superposed on a generally decreasing trend with a half-life of about 6 years, which is typical for continental stations of the Northern Hemisphere. Since 1996, the mean annual *A* values have been almost constant, and the mean *A* value during the 2012–2018 period was 7.6 ± 0.8 TU. The tritium activity concentration in precipitation with no bomb-peak influence is worth further monitoring because of possible local contamination with technogenic tritium and also to enable studies of the solar cycle influence on the production of cosmogenic tritium.

The present analysis of long-term data on the isotope composition of precipitation may be useful for future comparisons to some other long-term records in nearby countries to obtain better knowledge on spatial and temporal variations across the wider region. It can also be a good basis for a comparison to some short-term records at other stations in Croatia. Last but not least, this analysis shows that climate changes are reflected in isotope compositions of precipitation, which means that further monitoring at stations with long-term records could be useful in studying the impact of climate changes on the environment, especially on water resources.

Supplementary Materials: The following are available online at http://www.mdpi.com/2073-4441/12/1/226/s1, Figure S1: Monthly δ^2H in precipitation at Zagreb, 1980–2018 period; Table S1: Stations from Croatia with available data on isotopes in precipitation; Table S2: Long-term monthly data (precipitation amount *P*, temperature *T*, δ^{18}O, δ^2H, *d*-excess, and tritium activity concentration *A*) for the station at Zagreb, Croatia; Table S3: Mean annual values δ^{18}O, δ^2H, *d*-excess, and *A*. Comparison of arithmetic and precipitation-weighted mean values.

Author Contributions: Conceptualization, I.K.B. and P.V.; data validation and interpretation, writing the manuscript—original draft preparation, final manuscript, visualization—most graphs, and communicating with the journal, I.K.B.; sample collection, tritium activity sample preparation, and discussion, J.B. and A.S.; data evaluation, database management, and LMWL Freeware software application, D.B.; stable isotope measurements, data reduction, and interpretation, P.V.; meteorological data analysis and visualization, I.L.M. All coauthors participated in discussions and reviewed the manuscript. All authors have read and agreed to the published version of the manuscript.

Funding: This research received no external funding. In the past, the isotope composition of precipitation was part of the following funders and projects: the Ministry of Science and Education (MSE) of the Republic of Croatia, project nos. 00980207 (natural radioisotopes and processes in gases, 1996–2002), 0098014 (natural isotopes of low-level activities and the development of instrumentation, 2002–2006), 098-0982709-2741 (natural radioisotopes in the investigation of Karst ecosystems and dating, 2007–2011); the IAEA, project nos. 11265 (Isotopic Composition of Precipitation in the Mediterranean Basin in Relation to Air Circulation Patterns and Climate Tritium and Stable Isotope Distribution in the Atmosphere in the Coastal Region of Croatia, 2000–2003), CRO/8/006 (the application of isotope techniques in the investigation of water resources and water protection in the Karst area of Croatia, 2005–2006), RER/8/012 (isotope methods for the management of drinking water resources in water-scarce areas, 2007–2008), CRO/8/007 (Using Isotope Tracers as a Tool for a Groundwater Vulnerability Assessment in the County of Split, Dalmatia, 2007–2008), RER/8/016 (Using Environmental Isotopes for an Evaluation of Streamwater/Groundwater Interactions in Selected Aquifers in the Danube Basin, 2009–2011), CRO/7/001 (Isotope Investigation of the Groundwater–Surface Water Interactions at the Well-Field Kosnica in the Area of the City of Zagreb, 2016–2017); EU project FP5 ICA2-CT-2002-10009 ANTHROPOL.PROT (a study of anthropogenic influence after the war and establishing protection measures in the National Park Plitvice and the Bihać Region at the border area between Croatia and Bosnia and Herzegovina, 2003–2005); and a project financed by the National Park Plitvice Lakes ("an investigation of the influence of forest ecosystems of the National Park Plitvice Lakes on the quality of water and lakes", 2003–2006). Cooperation between Croatian and Slovenian partners was funded in recent decades by MSE and the Slovenian Research Agency (SRA) as part of bilateral cooperation (BI-HR/01-03-011, BI-HR/04-05-13, and BI-HR/09-10-032) and the SRA research program P1-0143.

Acknowledgments: The authors are thankful to former Ruđer Bošković Institute laboratory staff members Bogomil Obelić, Nada Horvatinčić, Dušan Srdoč, Elvira Hernaus, and Božica Mustač and the present technician Anita Rajtarić, who all took part in sample and data collection. We are also thankful to colleagues from the Jožef Stefan Institute in Ljubljana (Jože Pezdič, Sonja Lojen, Nives Ogrinc, Silva Perko, Zdenka Trkov, and Stojan Žigon) for stable isotope measurements from the period 1980–2003 and to Jelena Parlov and Zoran Kovač from the Laboratory for Spectroscopy of the Faculty of Mining, Geology, and Petroleum Engineering, University of Zagreb, for stable isotope measurements from the period 2012–2018. Meteorological data were obtained on request (free of charge) from the Croatian Meteorological and Hydrological Service. We thank Catherine E. Hughes and Jagoda Crawford for help in getting LMWL Freeware software and for discussions of the formulae used.

Conflicts of Interest: The authors declare no conflicts of interest. The funders had no role in the design of the study; in the collection; analyses; or interpretation of data; in the writing of the manuscript; or in the decision to publish the results

List of Abbreviations

CMHS	Croatian Meteorological and Hydrological Service
GMWL	Global meteoric water line
GNIP	Global Network of Isotopes in Precipitation
GPC	Gas proportional counting
IAEA	International Atomic Energy Agency
IRMS	Isotope ratio mass spectrometry
LMWL	Local meteoric water line
LSC-EE	Liquid scintillation counting with electrolytic enrichment
MA	Major axis least squares regression
OLSR	Ordinary least squares regression
PWLSR	Precipitation-weighted OLSR
PWMA	Precipitation-weighted MA
PWRMA	Precipitation-weighted RMA
RBI	Ruđer Bošković Institute
RMA	Reduced major axis regression
rmSSEav	Average of the root mean square sum of squared errors
SMOW	Standard mean ocean water
TU	Tritium unit
WMO	World Meteorological Organization

References

1. Mook, W.G. *Environmental Isotopes in the Hydrological Cycle, Principles and Applications, Volumes I, IV and V*; Technical Documents in Hydrology No. 39; IAEA-UNESCO: Paris, France, 2001.
2. Dansgaard, W. Stable isotopes in precipitation. *Tellus* **1964**, *16*, 436–468. [CrossRef]
3. Maloszewski, P.; Zuber, A. Principles and practice of calibration and validation mathematical models for the interpretation of environmental tracer data in aquifers. *Adv. Water Resour.* **1993**, *16*, 173–190. [CrossRef]
4. Rozanski, K.; Araguás-Araguás, L.; Gonfiantini, R. Isotopic patterns in modern global precipitation. *Geophys. Monogr.* **1993**, *78*, 1–36. [CrossRef]
5. Rozanski, K.; Araguás-Araguás, L. Spatial and temporal variability of stable isotope composition of precipitation over the South American continent. *Bull. Inst. Fr. Études Andin.* **1995**, *24*, 379–390.
6. Adomako, D.; Gibrilla, A.; Maloszewski, P.; Ganyaglo, S.Y.; Rai, S.P. Tracing stable isotopes (δ^2H and δ^{18}O) from meteoric water to groundwater in the Densu River basin of Ghana. *Environ. Monit. Assess.* **2015**, *187*, 1–15. [CrossRef]
7. Vrzel, J.; Kip Solomon, D.; Blažeka, Ž.; Ogrinc, N. The study of the interactions between groundwater and Sava River water in the Ljubljansko polje aquifer system (Slovenia). *J. Hydrol.* **2018**, *556*, 384–396. [CrossRef]
8. Parlov, J.; Kovač, Z.; Nakić, Z.; Barešić, J. Using Water Stable Isotopes for Identifying Groundwater Recharge Sources of the Unconfined Alluvial Zagreb Aquifer (Croatia). *Water* **2019**, *11*, 2177. [CrossRef]
9. Kong, Y.; Wang, K.; Li, J.; Pang, Z. Stable Isotopes of Precipitation in China: A Consideration of Moisture Sources. *Water* **2019**, *11*, 1239. [CrossRef]
10. Rao, S.M. Injected radiotracer techniques in hydrology. *Proc. Indian Acad. Sci.* **1984**, *99*, 319–335.
11. Clark, I.; Fritz, P. *Environmental Isotopes in Hydrogeology*; Lewis Publ.: Boca Raton, FL, USA, 1997; p. 328.
12. Kendall, C.; McDonnell, J.J. *Isotope Tracers in Catchment Hydrology*; Elsevier Science: Amsterdam, The Netherlands, 1998; p. 840.
13. Mayr, C.; Langhamer, L.; Wissel, H.; Meier, W.; Sauter, T.; Laprida, C.; Massaferro, J.; Försterra, G.; Lücke, A. Atmospheric controls on hydrogen and oxygen isotope composition of meteoric and surface waters in Patagonia. *Hydrol. Earth Syst. Sci.* **2018**. [CrossRef]
14. Burnik Šturm, M.; Ganbaatar, O.; Voigt, C.C.; Kaczenskya, P. First field-based observations of δ^2H and δ^{18}O values of event-based precipitation, rivers and other water bodies in the Dzungarian Gobi, SW Mongolia. *Isot. Environ. Health Stud.* **2017**, *53*, 157–171. [CrossRef] [PubMed]
15. Lawrence, J.R.; White, J.R.C. The Elusive Climate Signal in the Isotopic Composition of Precipitation Stable Isotope Geochemistry: A Tribute to Samuel Epstein. In *Stable Isotope Geochemistry: A Tribute to Samuel Epstein*; Taylor, H.P., O'Neil, J.R., Jr., Kaplan, I.R., Eds.; Special Publication No.3; The Geochemical Society: Washington, DC, USA, 1991.
16. Gat, J.R. Some classical concepts of isotope hydrology. In *Isotopes in the Water Cycle: Past, Present and Future of a Developing Science*; Aggarwal, P.K., Gat, J.R., Fröhlich, K., Eds.; Springer: Dordrecht, The Netherlands, 2005; pp. 127–137.
17. Krklec, K.; Domínguez-Villar, D.; Lojen, S. The impact of moisture sources on the oxygen isotope composition of precipitation at a continental site in central Europe. *J. Hydrol.* **2018**, *561*, 810–821. [CrossRef]
18. Horvatinčić, N.; Krajcar Bronić, I.; Barešić, J.; Obelić, B.; Vidič, S. Tritium and stable isotope distribution in the atmosphere at the coastal region of Croatia. In *Isotopic Composition of Precipitation in the Mediterranean Basin in Relation to Air Circulation Patterns and Climate, IAEA-TECDOC-1453*; Gourcy, L., Ed.; IAEA: Vienna, Austria, 2005; pp. 37–50.
19. Giustini, F.; Brilli, M.; Patera, A. Mapping oxygen stable isotopes of precipitation in Italy. *J. Hydrol. Reg. Stud.* **2016**, *8*, 162–181. [CrossRef]
20. Longinelli, A.; Anglesio, E.; Flora, O.; Iacumin, P.; Selmo, E. Isotopic composition of precipitation in Northern Italy: Reverse effect of anomalous climatic events. *J. Hydrol.* **2006**, *329*, 471–476. [CrossRef]
21. Rozanski, K.L.; Gonfiantini, R. Isotopes in climatological studies: Environmental isotopes are helping us understand the world's climate. *IAEA Bull.* **1990**, *4*, 9–15.
22. Marchetti, D.W.; Marchetti, S.B. Stable isotope compositions of precipitation from Gunnison, Colorado 2007–2016: Implications for the climatology of a high-elevation valley. *Heliyon* **2019**, *5*, e02120. [CrossRef]

23. Ehleringer, J.R.; Cerling, T.E.; West, J.B.; Podlesak, D.W.; Chesson, L.A.; Bowen, G.J. Spatial considerations of stable isotope analyses in environmental forensics. In *Issues in Environmental Science and Technology*; Hester, R.E., Harrison, R.M., Eds.; Royal Society of Chemistry Publishing: Cambridge, UK, 2008; Volume 26, pp. 36–53. [CrossRef]

24. Bowen, G.J.; Wassenaar, L.I.; Hobson, K.A. Global application of stable hydrogen and oxygen isotopes to wildlife forensics. *Oecologia* **2005**, *143*, 337–348. [CrossRef]

25. West, A.G.; February, E.C.; Bowen, G.J. Spatial analysis of hydrogen and oxygen stable isotopes ("isoscapes") in ground water and tap water across South Africa. *J. Geochem. Explor.* **2014**, *145*, 213–222. [CrossRef]

26. Terzer, S.; Wassenaar, L.I.; Araguás-Araguás, L.J.; Aggarwal, P.K. Global isoscapes for $\delta^{18}O$ and δ^2H in precipitation: Improved prediction using regionalized climatic regression models. *Hydrol. Earth Syst. Sci.* **2013**, *17*, 4713–4728. [CrossRef]

27. IAEA/WMO. Global Network of Isotopes in Precipitation. The GNIP Database. Available online: https://nucleus.iaea.org/wiser (accessed on 22 November 2019).

28. Lewis, S.L.; Maslin, M.A. Defining the Anthropocene. *Nature* **2015**, *519*, 171–180. [CrossRef]

29. IPCC. The Intergovernmental Panel on Climate Change. Available online: https://www.ipcc.ch/sr15/ (accessed on 26 November 2019).

30. Climate Central. Available online: https://www.climatecentral.org/gallery/graphics/the-10-hottest-global-years-on-record (accessed on 22 November 2019).

31. NASA. Available online: https://www.nasa.gov/press-release/nasa-noaa-data-show-2016-warmest-year-on-record-globally (accessed on 22 November 2019).

32. Craig, H. Isotope variations in meteoric waters. *Science* **1961**, *133*, 1702–1703. [CrossRef] [PubMed]

33. Gat, J.R. Oxygen and hydrogen isotopes in the hydrologic cycle. *Annu. Rev. Earth Planet Sci.* **1996**, *24*, 225–262. [CrossRef]

34. Gat, R.J.; Mook, W.G.; Meijer, A.J. Atmospheric water. In *Environmental Isotopes in the Hydrological Cycle—rinciples and Applications*; Mook, W.G., Ed.; Technical Documents in Hydrology; IAEA: Vienna, Austria, 2001; No. 39; Volume 2, p. 113.

35. IAEA. *Stable Isotope Hydrology: Deuterium and Oxygen-18 in Water Cycle*; Gat, J.R., Gonfiantini, R., Eds.; Technical Reports Series 210; IAEA: Vienna, Austria, 1981; p. 339.

36. Coplen, T.B. New guidelines for the reporting of stable hydrogen, carbon, and oxygen isotope ratio data. *Geochim. Cosmochim. Acta* **1996**, *60*, 3359. [CrossRef]

37. Coplen, T.B.; Bohlke, J.K.; De Bievre, P.; Ding, T.; Holden, N.E.; Hopple, J.A.; Krouse, H.R.; Lamberty, A.; Peiser, P.S.; Revesz, K.; et al. Isotope-abundance variations of selected elements (IUPAC Technical Report). *Pure Appl. Chem.* **2002**, *74*, 1987–2017. [CrossRef]

38. Dunn, P.J.H.; Carter, J.F. (Eds.) *Good Practice Guide for Isotope Ratio Mass Spectrometry*, 2nd ed.; FIRMS: Bristol, UK, 2018; p. 84. ISBN 978-0-948926-33-4. Available online: http://www.forensic-isotopes.org/gpg.html (accessed on 26 November 2019).

39. Craig, H. Standard for Reporting Concentrations of Deuterium and Oxygen-18 in Natural Waters. *Science* **1961**, *133*, 1833–1834. [CrossRef]

40. Epstein, S.; Mayeda, T.K. Variations of ^{18}O content of waters from natural sources. *Geochim. Cosmochim. Acta* **1953**, *4*, 213–224. [CrossRef]

41. Brand, W.A.; Coplen, T.B.; Vogl, J.; Rosner, M.; Prohaska, T. Assessment of International Reference Materials for Isotope-Ratio Analysis (IUPAC Technical Report). *Pure Appl. Chem.* **2014**, *86*, 425–467. [CrossRef]

42. Dansgaard, W. The isotope composition of natural waters. *Medd. Grønland* **1961**, *165*, 120.

43. Gat, J.R.; Carmi, I. Evolution of the Isotopic Composition of Atmospheric Waters in the Mediterranean Sea Area. *J. Geophys. Res.* **1970**, *75*, 3039–3048. [CrossRef]

44. Cruz-San, J.; Araguas-Araguas, L.; Rozanski, K.; Benavente, J.; Cardenal, J.; Hidalgo, M.C.; Garcia-Lopez, S.; Martinez-Garrido, J.C.; Moral, F.; Olias, M. Sources of precipitation over South-Eastern Spain and groundwater recharge—An isotopic study. *Tellus B* **1992**, *44*, 226–236. [CrossRef]

45. IAEA. *Isotopic Composition of Precipitation in the Mediterranean Basin in Relation to Air Circulation Patterns and Climate: Final Report of a Coordinated Research Project 2000–2004*; IAEA-TECDOC-1453; IAEA: Vienna, Austria, 2005.

46. Gat, J.R.; Shemesh, A.; Tziperman, E.; Hecht, A.; Georgopoulos, D.; Basturk, O. The stable isotope composition of waters of the eastern Mediterranean Sea. *J. Geophys. Res.* **1996**, *101*, 6441–6451. [CrossRef]

47. Vreča, P.; Krajcar Bronić, I.; Horvatinčić, N.; Barešić, J. Isotopic characteristics of precipitation in Slovenia and Croatia: Comparisom of continental and maritime stations. *J. Hydrol.* **2006**, *330*, 457–469. [CrossRef]

48. Crawford, J.; Hughes, C.E.; Lykoudis, S. Alternative least squares methods for determining the meteoric water line, demonstrated using GNIP data. *J. Hydrol.* **2014**, *519*, 2331–2340. [CrossRef]

49. Hughes, C.E.; Crawford, J. A new precipitation weighted method for determining the meteoric water line for hydrological applications demonstrated using Australian and global GNIP data. *J. Hydrol.* **2012**, *464*, 344–351. [CrossRef]

50. Araguás-Araguás, L.; Fröhlich, K.; Rozanski, K. Deuterium and oxygen-18 isotope composition of precipitation and atmospheric moisture. *Hydrol. Process.* **2000**, *14*, 1341–1355. [CrossRef]

51. Peng, H.; Mayer, B.; Harris, S.; Krouse, H.R. A 10-yr record of stable isotope ratios of hydrogen and oxygen in precipitation at Calgary, Alberta, Canada. *Tellus B* **2004**, *56*, 147–159. [CrossRef]

52. Lucas, L.; Unterweger, M.P. Comprehensive review and critical evaluation of the half-life of tritium. *J. Res. Natl. Inst. Stand. Technol.* **2000**, *105*, 541–549. [CrossRef]

53. Nikolov, J.; Krajcar Bronić, I.; Todorović, N.; Stojković, I.; Barešić, J.; Petrović-Pantić, T. Tritium in water—Hydrology and health implications. In *Tritium—Advances in Research and Applications*; Janković, M.M., Ed.; NOVA Science Publishers: New York, NY, USA, 2018; pp. 157–213.

54. Tadros, C.V.; Hughes, C.E.; Crawford, J.; Hollins, S.E.; Chisari, R. Tritium in Australian precipitation: A 50 year record. *J. Hydrol.* **2014**, *513*, 262–273. [CrossRef]

55. Rozanski, K.; Gonfiantini, R.; Araguas-Araguas, L. Tritium in the global atmosphere: Distribution patterns and recent trends. *J. Phys. G Nucl. Part. Phys.* **1991**, *17*, 523–536. [CrossRef]

56. Galeriu, D.; Melintescu, A.; Beresford, N.A.; Crout, N.M.J.; Takeda, H. 14C and tritium dynamics in wild mammals: A metabolic model. *Radioprotection* **2005**, *40*, S351–S357. [CrossRef]

57. Hebert, D. Technogenic Tritium in Central European Precipitations. *Isot. Environ. Health Stud.* **1990**, *26*, 592–595. [CrossRef]

58. UN; ILO; WHO. *Selected Radionuclides—Environmental Health Criteria 25*; WHO: Geneva, Switzerland, 1983; ISBN 92-4-154085-0.

59. Nimac, I.; Perčec Tadić, M. New 1981–2010 climatological normals for Croatia and comparison to previous 1961–1990 and 1971–2000 normals. In Proceedings of the GeoMLA Conference, Belgrade, Serbia, 21–24 June 2016; University of Belgrade—Faculty of Civil Engineering: Belgrade, Serbia, 2016; pp. 79–85.

60. Köppen, W. Das geographische System der Klimate. In *Handbuch der Klimatologie*; Köppen, W., Geiger, G., Eds.; Gebrüder Borntraeger: Berlin, Germany, 1936; pp. 1–44.

61. Peel, M.C.; Finlyanson, B.L.; McMahon, T.A. Updated world map of the Köppen-Geiger climate classification. *Hydrol. Earth Syst. Sci.* **2007**, *11*, 1633–1644. [CrossRef]

62. Krajcar Bronić, I.; Horvatinčić, N.; Obelić, B. Two decades of environmental isotope records in Croatia: Reconstruction of the past and prediction of future levels. *Radiocarbon* **1998**, *40*, 399–416. [CrossRef]

63. Krajcar Bronić, I.; Vreča, P.; Horvatinčić, N.; Barešić, J.; Obelić, B. Distribution of hydrogen, oxygen and carbon isotopes in the atmosphere of Croatia and Slovenia. *Arch. Ind. Hyg. Toxicol.* **2006**, *57*, 23–29.

64. Krajcar Bronić, I.; Horvatinčić, N.; Barešić, J.; Obelić, B.; Vreča, P. Widening the Radiation Protection World, Tritium distribution in precipitation over Croatia and Slovenia. In Proceedings of the 11th International Congress of the International Radiation Protection Association, Madrid, Spain, 23–28 May 2004; IRPA: Madrid, Spain, 2014. ISBN 84-87078-05-2.

65. Horvatinčić, N.; Kapelj, S.; Sironić, A.; Krajcar Bronić, I.; Kapelj, J.; Marković, T. Investigation of water resources and water protection in the karst area of Croatia using isotopic and geochemical analyses. In Proceedings of the Advances in Isotope Hydrology and its Role in Sustainable Water Resources Management (IHS-2007), Vienna, Austria, 21–25 May 2007; IAEA: Vienna, Austria, 2007; Volume 2, pp. 295–304, ISBN 978-92-0-110207-2.

66. Horvatinčić, N.; Barešić, J.; Krajcar Bronić, I.; Karman, K.; Forisz, I.; Obelić, B. Study of the bank filtered groundwater system of the Sava River at Zagreb (Croatia) using isotope analyses. *Cent. Eur. Geol.* **2011**, *54*, 121–127. [CrossRef]

67. Babinka, S.; Obelić, B.; Krajcar Bronić, I.; Horvatinčić, N.; Barešić, J.; Kapelj, S.; Suckow, A. MRT of water from springs of the Plitvice Lakes and Una River area. In *Book of Abstracts IAEA-CN-151, Proceedings of the IAEA IHS International Symposium on Advances in Isotope Hydrology and its Role in Sustainable Water Resources Management, Vienna, Austria, 21–25 May 2007*; IAEA: Vienna, Austria, 2007; p. 45.

68. Mandić, M.; Bojić, D.; Roller-Lutz, Z.; Lutz, H.O.; Krajcar Bronić, I. Note on the spring region of Gacka River (Croatia). *Isot. Environ. Health Stud.* **2008**, *44*, 201–208. [CrossRef]

69. Pezdič, J. *Izotopi in Geokemijski Procesi*; Naravoslovnotehniška Fakulteta, Oddelek za Geologijo: Ljubljana, Slovenia, 1999; p. 269.

70. Vreča, P.; Krajcar Bronić, I.; Leis, A.; Brenčič, M. Isotopic composition of precipitation in Ljubljana (Slovenia). *Geologija* **2008**, *51*, 169–180. [CrossRef]

71. Vreča, P.; Krajcar Bronić, I.; Leis, A.; Demšar, M. Isotopic composition of precipitation at the station Ljubljana (Reaktor), Slovenia—Period 2007–2010. *Geologija* **2014**, *57*, 217–230. [CrossRef]

72. Vreča, P.; Krajcar Bronić, I.; Leis, A. Isotopic composition of precipitation in Portorož (Slovenia). *Geologija* **2011**, *54*, 129–138. [CrossRef]

73. Vreča, P.; Malenšek, N. Slovenian Network of Isotopes in precipitation (SLONIP)—A review of activities in the period 1981–2015. *Geologija* **2016**, *59*, 67–84. [CrossRef]

74. Horvatinčić, N.; Krajcar Bronić, I.; Obelić, B.; Bistrović, R. Long-time atmospheric tritium record in Croatia. *Acta Geol. Hung.* **1996**, *39*, 81–84.

75. Horvatinčić, N.; Krajcar Bronić, I. ^{14}C and ^{3}H as indicators of the environmental contamination. *RMZ Mater. Geoenviron.* **1998**, *45*, 56–60.

76. Barešić, J.; Horvatinčić, N.; Krajcar Bronić, I.; Obelić, B.; Vreča, P. Stable isotope composition of daily and monthly precipitation in Zagreb. *Isot. Environ. Health Stud.* **2006**, *42*, 239–249. [CrossRef] [PubMed]

77. Zaninović, K.; Gajić-Čapka, M.; Perčec Tadić, M.; Vučetić, M.; Milković, J.; Bajić, A.; Cindrić, K.; Cvitan, L.; Katušin, Z.; Kaučić, D.; et al. *Klimatski Atlas Hrvatske/Climate Atlas of Croatia 1961–1990, 1971–2000*; Državni Hidrometeorološki Zavod: Zagreb, Croatia, 2008.

78. Gehre, M.; Hoefling, R.; Kowski, P.; Strauch, G. Sample preparation device for quantitative hydrogen isotope analysis using chromium metal. *Anal. Chem.* **1996**, *68*, 4414–4417. [CrossRef]

79. Surić, M.; Roller-Lutz, Z.; Mandić, M.; Krajcar Bronić, I.; Juračić, M. Modern C, O, and H isotope composition of speleothem and dripwater from Modrič Cave, eastern Adriatic coast (Croatia). *Int. J. Speleol.* **2010**, *39*, 91–97. [CrossRef]

80. Coplen, T.B.; Wassenaar, L.I. LIMS for Lasers for achieving long-term accuracy and precision of δ^2H, δ^{17}O, and δ^{18}O of waters using laser absorption spectrometry. *Rapid Commun. Mass Spectrom.* **2015**, *29*, 2122–2130. [CrossRef]

81. Horvatinčić, N. Radiocarbon and tritium measurements in water samples and application of isotopic analyses in hydrology. *Fizika* **1980**, *12*, 201–218.

82. Krajcar Bronić, I.; Obelić, B.; Srdoč, D. The simultaneous measurement of tritium activity and background count rate in a proportional counter by the Povinec method: Three years experience at the Rudjer Bošković Institute. *Nucl. Instrum. Methods Phys. Res. B* **1986**, *17*, 498–500. [CrossRef]

83. Barešić, J.; Horvatinčić, N.; Krajcar Bronić, I.; Obelić, B. Comparison of two techniques for low-level tritium measurement—Gas proportional and liquid scintillation counting. In Proceedings of the Third European IRPA Congress, Helsinki, Finland, 14–18 June 2010; IRPA: Helsinki, Finland, 2010; pp. 1988–1992.

84. Barešić, J.; Krajcar Bronić, I.; Horvatinčić, N.; Obelić, B.; Sironić, A.; Kožar-Logar, J. Tritium activity measurement of water samples using liquid scintillation counter and electrolytical enrichment. In Proceedings of the 8th Symposium of the Croatian Radiation Protection Association, Krk, Croatia, 13–15 April 2011; Krajcar Bronić, I., Kopjar, N., Milić, M., Branica, G., Eds.; HDZZ: Zagreb, Croatia, 2011; pp. 461–467.

85. Krajcar Bronić, I.; Barešić, J.; Sironić, A.; Horvatinčić, N. Stability analysis of systems for preparation and measurement of ^{3}H and ^{14}C (Analiza stabilnosti sustava za pripremu i mjerenje ^{3}H i ^{14}C, in Croatian with English Abstract). In Proceedings of the 9th Symposium of the Croatian Radiation Protection Association, Krk, Croatia, 10–12 April 2013; Knežević, Ž., Majer, M., Krajcar Bronić, I., Eds.; CRPA: Zagreb, Croatia, 2013; pp. 495–501.

86. Stojković, I.; Todorović, N.; Nikolov, J.; Krajcar Bronić, I.; Barešić, J.; Kozmidic Luburić, U. Methodology of tritium determination in aqueous samples by Liquid Scintillation Counting techniques. In *Tritium—Advances in Research and Applications*; Janković, M.M., Ed.; NOVA Science Publishers: New York, NY, USA, 2018; pp. 99–156.

87. Rozanski, K.; Gröning, M. Tritium assay in water samples using electrolytic enrichment and liquid scintillation spectrometry. In *Quantifying Uncertainty in Nuclear Analytical Measurements IAEA-TECDOC-1401*; IAEA: Vienna, Austria, 2004; pp. 195–217.

88. Gröning, M.; Auer, R.; Brummer, D.; Jaklitsch, M.; Sambandam, C.; Tanwee, A.; Tatzber, H. Increasing the performance of tritium analysis by electrolytic enrichment. *Isot. Environ. Health Stud.* **2009**, *45*, 118–125. [CrossRef]

89. IAEA. *Statistical Treatment of Environmental Isotopes in Precipitation*; Technical Report Series 331; IAEA: Vienna, Austria, 1992; p. 781.

90. Australian Nuclear Science and Technology Organisation (ANSTO). Local Meteoric Water Line Freeware. 2012. Available online: https://openscience.ansto.gov.au/collection/879 (accessed on 28 November 2019).

91. Barešić, J.; Štrok, M.; Svetek, B.; Vreča, P.; Krajcar Bronić, I. Activity concentration of tritium (^{3}H) in precipitation—Long-term investigations performed in Croatia and Slovenia. In Proceedings of the Sixth International Conference on Radiation and Applications in Various Fields of Research, Ohrid, North Macedonia, 18–22 June 2018; Ristić, G., Ed.; RAD Association: Niš, Serbia, 2018; p. 186.

92. Krajcar Bronić, I. Environmental ^{14}C and ^{3}H levels in Croatia. In *Book of Abstracts, ENVIRA2017, Proceedings of the 4th International Conference on Environmental Radioactivity: Radionuclides as Tracers of Environmental Processes, Vilnius, Litva, 29 May–2 June 2017*; Lujaniene, L., Povinec, P., Eds.; ENVIRA: Vilnius, Litva, 2017; p. 152.

93. Palcsu, L.; Morgenstern, U.; Sültenfuss, J.; Koltai, G.; László, E.; Temovski, M.; Major, Z.; Nagy, J.T.; Papp, L.; Varlam, C.; et al. Modulation of Cosmogenic Tritium in Meteoric Precipitation by the 11-year Cycle of Solar Magnetic Field Activity. *Sci. Rep.* **2018**, *8*, 12813. [CrossRef]

Article

Application of a Self-Organizing Map of Isotopic and Chemical Data for the Identification of Groundwater Recharge Sources in Nasunogahara Alluvial Fan, Japan

Takeo Tsuchihara *, Katsushi Shirahata, Satoshi Ishida and Shuhei Yoshimoto

Institute for Rural Engineering, National Agriculture and Food Research Organization, Ibaraki 305-8609, Japan;
shirahatak@affrc.go.jp (K.S.); ishidast@affrc.go.jp (S.I.); shuy@affrc.go.jp (S.Y.)
* Correspondence: takeo428@affrc.go.jp

Received: 26 November 2019; Accepted: 16 January 2020; Published: 18 January 2020

Abstract: Paddy rice fields on an alluvial fan not only use groundwater for irrigation but also play an important role as groundwater recharge sources. In this study, we investigated the spatial distribution of isotopic and hydrochemical compositions of groundwater in the Nasunogahara alluvial fan in Japan and applied a self-organizing map (SOM) to characterize the groundwater. The SOM assisted with the hydrochemical and isotopic interpretation of the groundwater in the fan, and clearly classified the groundwater into four groups reflecting the different origins. Two groundwater groups with lower isotopic ratios of water than the mean precipitation values in the fan were influenced by the infiltration of river water flowing from higher areas in the catchments and were differentiated from each other by their Na^+ and Cl^- concentrations. A groundwater group with higher isotopic ratios was influenced by the infiltration of paddy irrigation water that had experienced evaporative isotopic enrichment. Groundwater in the fourth group, which was distributed in the upstream area of the fan where dairy farms dominated, showed little influence of recharge waters from paddy rice fields. The findings of this study will contribute to proper management of the groundwater resources in the fan.

Keywords: groundwater; self-organizing map; stable isotope ratios; radon; major ions; alluvial fan; paddy rice field

1. Introduction

Alluvial fans are important lowland areas in Japan because approximately 70% of the total land area of Japan is mountainous or hilly. Vast paddy rice fields are situated on alluvial fans, which occupy more than half (54%) of Japan's lowland area [1]. River water is usually diverted at the apexes of such alluvial fans and then delivered through a network of irrigation canals to each paddy rice field by gravity [2]. Alluvial fans are also vital aquifer systems that support local agricultural, industrial, and socio-economic development worldwide, wherever groundwater is utilized as an important water resource [3]. Aquifers of permeable sand and gravel formed in alluvial fans are vulnerable to contamination from the surface, and therefore interactions between surface water and groundwater can even affect hydrochemical processes (e.g., [4–7]). In addition, the infiltration of irrigation water contributes considerably to groundwater recharge [8] and also produces complex groundwater flows in paddy-dominant alluvial fans. For the sustainable use of groundwater resources in fans, it is important to understand these groundwater flows and recharge sources.

Environmental isotopes and hydrochemical data are powerful tools for investigating groundwater flow conditions (e.g., [9,10]), including hydrogeological investigations of alluvial fans (e.g., [8,11,12]). A multifaceted analysis using a suite of tracers (multi-tracer approach) can be effectively used to interpret complex groundwater flow conditions that cannot be observed directly. Multivariate analysis methods, such as principal component analysis and factor analysis, which allow for the dimensionality of multiple hydrochemical indicators to be reduced and features of groundwater flow conditions to be extracted, have been widely applied in hydrological system analysis (e.g., [13–15]). However, these multivariate analyses are generally based on linear principles and they cannot overcome difficulties arising from biases due to the complexity and nonlinearity of the datasets and from inherent correlations between variables [16]. Therefore, a self-organizing map (SOM) approach, a neural network–based pattern analysis with unsupervised learning, has recently been applied to map changes in groundwater levels and chemistry in space and time (e.g., [17–20]). An SOM, which can cluster a set of hydrochemical data into two or more independent groups, is superior to other statistical tools because it can: (1) deal with system nonlinearities, (2) be developed from data without requiring mechanistic knowledge of the system, (3) handle noisy or irregular data and be easily and quickly updated, and (4) be used to interpret and visualize information of multiple variables or parameters [16,21].

In this study, we performed an environmental isotopic and hydrochemical investigation of the Nasunogahara alluvial fan in central Japan, which was formed by several rivers and has a vast area (400 km^2), about 40% of which is used for rice production. The paddy rice fields are irrigated by both river water and groundwater. The river water is transmitted through the Nasu canal system, one of the three major canals for paddy field irrigation in Japan. The yield of groundwater pumped out of the fan for irrigation is the second largest for agricultural use in Japan [22]. The groundwater in this fan also emerges as numerous springs in the central to lower part of the fan. These springs support a unique ecosystem inhabited by freshwater fish (three-spined stickleback, *Gasterosteus aculeatus*) that has been designated as a natural monument [23]. The groundwater in the alluvial fan is recharged not only by direct percolation of precipitation but also by the infiltration of water from the rivers that flow across the fan and water from irrigated paddy rice fields [24]. Thus, a part of the irrigation water used in the alluvial fan, whether derived from surface water or groundwater, returns from the paddy rice fields to the aquifer.

Previous studies of groundwater in this area have focused on spatio-temporal variations in the groundwater quality [25,26], groundwater nitrate dynamics [27], and the groundwater budget, as simulated by a 2D horizontal model [28]. Stable isotope ratios of water have been effectively used to identify recharge sources of the groundwater in the central part of the fan [24], but the applicability of the method to the whole fan remains to be demonstrated, and how different recharge sources influence the flow of groundwater in the aquifer is still unclear. Therefore, further research is needed to provide information for farmers and other local residents to enable sustainable use of groundwater in the fan and conservation of the springs. Therefore, we measured multiple environmental isotopes and the hydrochemistry of groundwater in the Nasunogahara alluvial fan during two paddy cultivation periods, one with and one without irrigation. We then applied an SOM based on our results to identify the groundwater recharge sources and interpret groundwater flow through the aquifer system.

2. Study Area

2.1. Location, Land Use, and Water Use

The Nasunogahara alluvial fan in central Japan was formed by the Naka, Sabi, Kuma, and Houki rivers, which flow out of the mountainous region to the northwest of the fan (Figure 1). The alluvial fan extends over an area of approximately 400 km^2 and is bounded by the Naka River along its eastern edge and the Houki River along its western edge. The flow in the upper reaches of the Sabi and Kuma rivers is usually underground, but these rivers emerge above ground in the central part of the fan and join the Houki River in the toe of the fan. Then, the Houki and Naka rivers converge in the southeasternmost part of the fan. As a result, the Nasunogahara fan is spindle-shaped, different from most alluvial fans. The topography of the fan is mainly flat; its elevation ranges from 60 to 560 m a.s.l., although scattered hills dissected by numerous streams occupy the lower part of the fan. The slope near the fan apex is 3/100, and the slope in the central and lower parts of the fan is 1/100; the average slope is 1.6/100. Paddy rice fields cover about 40% of the fan surface [29] and are densely distributed in the central and lower parts (Figure 1). About one-quarter of the paddy rice fields are irrigated by surface water diverted from the rivers, and the rest are irrigated by groundwater. In 2008, 240 million m^3 of groundwater from the fan was used for agriculture [22]. Livestock farms (mostly dairy farms) are located mainly in the upper part of the fan, where there is less groundwater available for irrigation. The mean air temperature and annual precipitation from 1981 to 2010, measured at Kuroiso weather station (343 m a.s.l) (Figure 1), were 11.7 °C and 1526 mm, respectively.

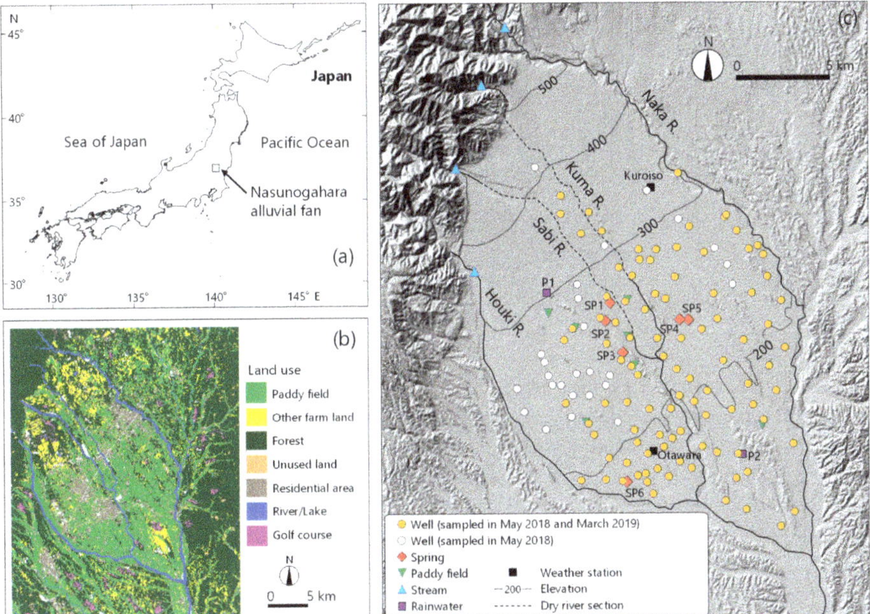

Figure 1. Maps showing the (**a**) location of the study site, (**b**) land use, and (**c**) water sampling sites. Groundwater was not sampled at the sites denoted by white circles in March 2019 because the wells were dry owing to a drop in the groundwater level. The land use map was created by using National Land Numerical Information [29]. The gray topographic base map in (**c**) was from the Geospatial Information Authority of Japan [30].

2.2. Hydrogeological Settings

The Nasunogahara alluvial fan consists mostly of Quaternary deposits overlying a basement composed of pre-Quaternary green tuff and mudstone (Shioya and Arakawa groups). In order of decreasing depth, the Sakaibayashi gravel formation, Tatenokawa tuff formation, Kuroiso volcanic breccia, Nabekake conglomerate, Kanemarubara gravel formation, Nasuno gravel formation, the Samui and Okuzawa gravel formations, and a flood plain gravel bed overlie the basement rocks (Figure 2).

An unconfined aquifer is in the flood plain gravel bed and the Samui and Okuzawa formations; in this aquifer, the groundwater table is higher during summer and disappears during winter. A second unconfined aquifer in the Nasuno gravel formation supplies most of the groundwater used for irrigation; it typically has a high hydraulic conductivity of 10^{-4} to 10^{-3} m/s [31,32]. The Tatenokawa tuff formation is impermeable and the gravels of overlying formations cover a paleotopography of hills and valleys on the upper surface of the Tatenokawa tuff. The Sakaibayashi gravel formation underlies the Tatenokawa tuff formation and is a confined aquifer. In this study, groundwater samples were collected from depths down to 30.3 m in wells installed in the first and second unconfined aquifers, as described in Section 3.1. The groundwater level of the fan begins to rise in April, reaches a maximum in October, and then declines through March, echoing changes in precipitation [24]. Groundwater levels are thus high from April to September, the paddy rice irrigation period, suggesting that the groundwater may be recharged by paddy irrigation water.

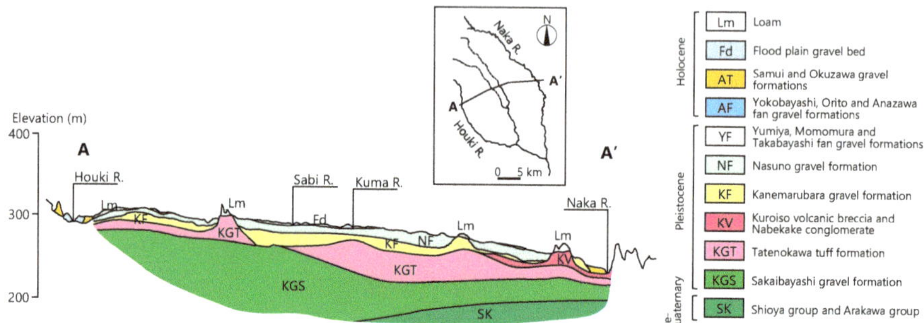

Figure 2. Geological cross section of the Nasunogahara alluvial fan (modified from the Kanto Regional Agricultural Administration Office [31]).

3. Methods

3.1. Sample Acquisition

Many hand-dug wells for paddy irrigation (as well as for upland crop irrigation and domestic use) are distributed across the Nasunogahara alluvial fan, although there are a few wells in the upper part of the fan (Figure 1). We collected samples for the analyses of stable isotopic composition, radon radioisotope (^{222}Rn), and major ion concentrations of water from wells, river waters, and paddy waters, both during the irrigation period (IP, 15–18 May 2018) and during the non-irrigation period (NP, 4–7 March 2019). In the IP, groundwater was collected from 124 wells, but in the NP, some of the wells had dried up because of a drop in the groundwater level (white circles in Figure 1), so groundwater was collected from only 98 wells. The ground elevation of the wells ranged from 134 to 417 m a.s.l., and the well depth ranged from 3.3 to 30.3 m, with an average depth of 9.6 m below ground level. Toward the apex of the fan, the wells were deeper, and toward the toe of the fan, they were shallower. The mean depths of groundwater in the wells in the IP and NP were 3.8 and 2.5 m, respectively. Therefore, isotopic and chemical compositions of groundwaters were considered to be the same regardless of water depth. An acrylic well bailer was used to collect groundwater samples from the wells. Paddy waters were collected from the irrigation water inlet and outlet of seven paddy rice fields only in the IP. Spring waters were collected at six sites (SP1–6) in the IP and three sites (SP4–6) in the NP. Some springs dried up in the NP because of the drop in the groundwater level. A polyethylene cup or EVA (ethylene-vinyl acetate copolymer) bucket were used to sample the paddy and spring waters. River water samples were collected from each of the four rivers (Naka, Kuma, Sabi, and Houki rivers), either from the bank or from a bridge, using an EVA bucket. A relatively low rainfall was observed and the maximum daily rainfalls in and before the IP and NP were 8 and 14 mm, respectively. Therefore, the collected river waters were close to baseflow conditions. Rainwater was collected at site P1 (321 m a.s.l.; Figure 1) from 14 March 2018 to 21 August 2019 (527 days) and site P2 (179 m a.s.l.) from 9 August 2018 to 21 August 2019 (377 days) at a mean interval of 33 days in rain collectors designed, following recommendations described in International Atomic Energy Agency (IAEA)/Global Network of Isotopes in Precipitation (GNIP) [33], to prevent water re-evaporation. The electrical conductivity (EC) of the sampled surface water and groundwater was measured by using a portable EC sensor (TOA-DKK, WM-32EP, Tokyo, Japan) immediately after the sampling. All water samples for chemical and isotopic analyses were transferred to clean polyethylene vials and brought to the laboratory of the Institute for Rural Engineering, National Agriculture, and Food Research Organization (Ibaraki, Japan), where they were filtered through syringe filters with a polytetrafluoroethylenemembrane (0.45 µm) and stored in a refrigerator until analysis.

3.2. Analysis of Environmental Isotopic and Hydrochemical Compositions

Major ions (Na^+, K^+, Mg^{2+}, Ca^{2+}, Cl^-, NO_3^-, SO_4^{2-}) were quantified using ion chromatography (TOA-DKK, ICA-2000, Tokyo, Japan) using multicomponent standard solutions. Nitrate (NO_3^-) is reported as the nitrate-nitrogen (NO_3-N) concentration. Alkalinity was determined using titration to pH 4.8 and is reported as HCO_3^-. For all water samples, errors in the cation–anion charge balance were less than 5%. ^{222}Rn, which is a radioactive gas generated by the decay of ^{226}Ra in geological strata, is soluble in water and has a half-life of 3.8 days. The ^{222}Rn concentration in groundwater finally reaches radioactive equilibrium. It reaches 92% of the value at secular equilibrium after two weeks underground. Using these characteristics, ^{222}Rn is often used as a target tracer to investigate the hydrological aspects of groundwater (e.g., estimation of infiltration from surface water to aquifers [34] and groundwater flow rate [35]), because far more ^{222}Rn is contained in groundwater than in surface water. The radioactivity of ^{222}Rn in samples was analyzed using a liquid scintillation counter (Packard, 2250 CA, CT, USA) after extraction with toluene within one week after collecting the water, and the concentrations of ^{222}Rn were calculated based on the count rate [36]. The stable isotopic compositions of the water were measured using a cavity ringdown spectrometer-based water isotope analyzer (Picarro, L2140-i, Santa Clara, CA, USA). They were reported with respect to Vienna Standard Mean Ocean Water (VSMOW) and are described here in "delta" notation as $\delta^{18}O$ and δ^2H (in parts per thousand, ‰). Three working standards calibrated by international measurement standards were used for the analysis of $\delta^{18}O$ and δ^2H. The analytical precision in this study was ±0.05‰ for $\delta^{18}O$ and ±0.50‰ for δ^2H. The deuterium excess (d-excess, used to indicate the kinetic fractionation effect of evaporation [37]) of each water sample was calculated as d-excess = $\delta^2H - 8\delta^{18}O$ [38]. The Mann–Whitney U test, which is applicable to non-normally distributed variables, was used to assess the independence of the measured isotopic and hydrochemical compositions between the IP and NP.

3.3. SOM

An SOM is an unsupervised training algorithm of an artificial neural network model developed by Kohonen [39]. An SOM can project high-dimensional target data onto a low-dimensional format (typically a two-dimensional grid map called a "feature map") [40]. Therefore, it is an effective and powerful analysis tool for the detection of data characteristics using pattern classification and visualization [16]. An SOM consists of an input layer and an output layer. The output layer often consists of a rectangular grid map with a hexagonal lattice ("nodes"). Each node in the output layer has a vector of coefficients associated with the input data, the so-called reference vector (also referred to as the weight vector and the codebook vector). The vectors can be obtained using iterative updates during a training phase consisting of three main procedures: competition between nodes, selection of a winning node, and updating of the reference vectors [40,41].

The determination of the SOM structure (calculation of the total number of nodes and the ratio of the number of map rows to the number of map columns) is an important step in the application of an SOM. If the number of nodes (i.e., map size) is too small, it might not explain some important differences that should be detected. Conversely, if the map size is too big, the differences between adjoining nodes are too small [42]. The optimal number of nodes in an SOM can be selected by using the heuristic rule $m = 5\sqrt{n}$, where m denotes the total number of nodes and n is the number of samples in the data set [40,43]. The ratio of the number of rows to the number of columns in the feature map is determined by calculating the square root of the ratio of the largest eigenvalue of the correlation matrix of the input data to the second-largest eigenvalue [19]. In this study, the eigenvalues were obtained using principal component analysis of the input dataset. Normalization of the data was necessary prior to the application of the SOM to ensure that all values of the chemical parameters were given similar importance. The results of the SOM application are sensitive to the method used for data pre-processing because the SOM is trained to organize itself according to the Euclidean distances between input data [40]. In this study, all variables of the input data were transformed to have a range of 0 to 1 such that they would have equal importance in the formation of the SOM.

For clustering the trained SOM, several methods, each with advantages and disadvantages, are available. Agglomerative clustering and partition clustering result in clear groupings on the trained map, whereas application of a U-matrix (unified distance matrix), which is commonly used to create a rough visualization of the classification, may not define clear boundaries (e.g., [44]). Ward's agglomerative hierarchical clustering (Ward's method) is an easy way to cluster nodes on the trained map because it is not necessary in this method to determine the optimum number of groups before performing the clustering calculations [42]. Ward's method has been commonly used to subdivide the trained map into several groups according to the similarity of the reference vectors of the nodes (e.g., [43,45]). Thus, Ward's method was applied in this study to identify groups on the trained map.

In this study, the input layer of the SOM consisted of 11 chemical and isotopic parameters (δ^2H, δ^{18}O, ^{222}Rn, and the major ion concentrations) of groundwater samples collected at 124 and 98 wells in the IP and NP, respectively. In accordance with the heuristic rules described above, a feature map with 72 (9 × 8) nodes was selected. The "kohonen" package in the R statistical software environment (version 3.5.1, Vienna, Austria) [46] was used to implement the SOM. Default values of the package parameters (e.g., learning rate and size of the neighborhood) were used. The observed hydrochemical and isotopic data of the groundwater in the fan were preprocessed by the proposed SOM, and then the groundwater data were spatially clustered into different homogeneous groups on the trained map. To identify the groundwater recharge sources, the hydrochemical and isotopic compositions of the groups were plotted on various types of diagrams (i.e., hexa-diagrams, a trilinear diagram, a δ^{18}O–δ^2H scatter diagram, and spatial distribution maps) and compared.

4. Results

4.1. Groundwater Level

We mapped groundwater levels, depth to the water table, and cross-sections of the water table in Figure 3. The groundwater table was almost parallel to the ground surface of the fan, which sloped toward the southeast, and the groundwater flowed in the same direction.

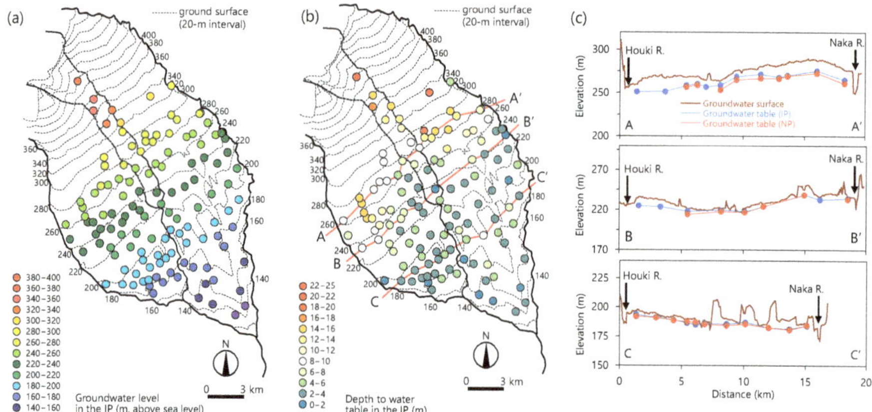

Figure 3. (**a**) Spatial distribution of the groundwater level in the irrigation period (IP), (**b**) depth to the water table in the IP, and (**c**) cross-sections of the groundwater table. The contour line of the ground surface was modified from the Geospatial Information Authority of Japan [30].

The depth to the groundwater table in the upper to middle part of the fan (from the apex to about 260 m above sea level) was relatively large. In this area, the Sabi and Kuma rivers were losing streams and hence recharged the aquifer. The terrace cliffs on the right bank of the Naka River along the eastern edge of the fan varied from several meters to several tens of meters in height (Figure 3c), and therefore, the river bed of the Naka River was lower than the groundwater table. This means that there was no recharge from the Naka river to the aquifer. The riverbed of the Houki River was about 10 m higher than the groundwater surface near the middle part of the fan. The Houki River recharged the aquifer in this part and the infiltrated river water flowed in the downstream direction in accordance with hydraulic gradient, whereas the groundwater table became higher toward the toe of the fan. The lower part of the fan, where the water level became closer to the ground surface, corresponded to the groundwater discharge area.

4.2. Chemical Compositions

Analytical results of the chemical and isotopic parameters of the groundwater, along with those of spring water, paddy water, river water, and rainwater, are shown in Table 1. Four groups of the groundwater were classified using the SOM, as described later in Section 4.4. The chemical compositions of the groundwater, spring water, and river water are shown as a Piper (trilinear) diagram in Figure 4 and as hexa-diagrams in Figure 5. The Ca-HCO$_3$ type typically indicates shallow fresh groundwater, while Na-HCO$_3$ type typically indicates weathering and ion exchange processes. As shown by the key diagram and hexa-diagrams, some groundwater samples had a Ca-HCO$_3$-to-Ca-SO$_4$-Cl-type composition, whereas the Ca-HCO$_3$-type composition was dominant. Groundwater samples were plotted near the center of the cation ternary diagram (Figure 4), suggesting little variability in the relative proportions of these cations. The anion ternary diagram indicated larger variability among samples: Groundwater samples were mostly HCO$_3{}^-$-type water, whereas some groundwater samples were scattered between HCO$_3{}^-$ and SO$_4{}^{2-}$ types.

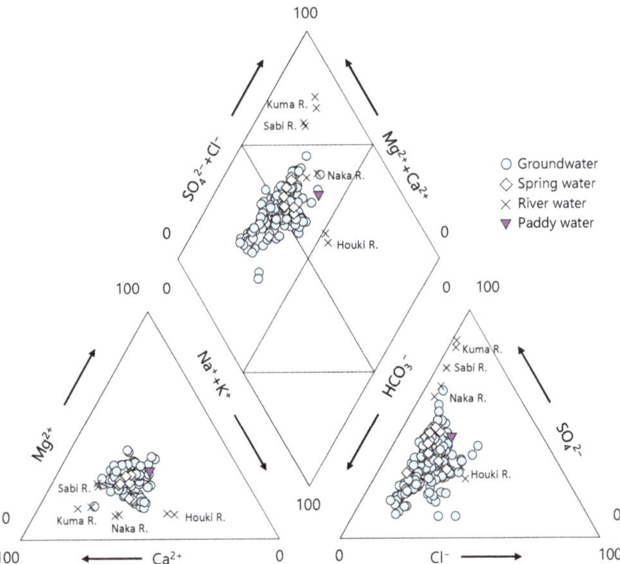

Figure 4. Trilinear diagram of groundwater, spring water, river water, and paddy water.

Table 1. Analytical results for groundwater, spring water, paddy water, river water, and rainwater.

Type	Group/Site	n	Period		EC	Na$^+$	K$^+$	Mg^{2+}	Ca^{2+}	HCO$_3^-$	Cl$^-$	SO$_4^{2-}$	NO$_3$-N	δ^{18}O	δ^2H	d-Excess	^{222}Rn
					mS/m	mg/L	mg/L	mg/L	mg/L	mg/L	mg/L	mg/L	mg/L	‰	‰	‰	Bq/L
Groundwater	All	222	IP, NP	Mean	17.3	9.0	0.3	1.3	4.9	42.0	8.7	20.1	2.6	−8.5	−56	12.5	12.4
				Median	17.4	8.6	0.3	1.2	4.6	40.3	8.1	20.7	2.5	−8.6	−56	12.6	12.9
				25th	15.1	7.1	0.2	0.9	3.9	30.4	6.1	18.2	1.9	−8.9	−57	11.4	10.7
				75th	19.7	10.1	0.3	1.6	5.8	50.8	10.2	22.8	3.1	−8.2	−54	13.4	14.5
	Group 1	34	IP, NP	Mean	20.3	13.9	2.0	4.1	16.8	46.4	15.4	24.0	1.9	−8.8	−57	12.9	11.6
				Median	20.0	13.8	2.1	3.8	16.4	46.9	15.4	23.5	1.6	−8.8	−57	13.0	12.4
				25th	19.5	13.3	1.9	3.6	15.8	44.4	13.8	22.8	1.2	−9.1	−59	12.4	10.1
				75th	20.6	15.1	2.3	4.1	17.5	50.3	16.5	24.4	2.0	−8.6	−56	13.3	13.5
	Group 2	73	IP, NP	Mean	17.2	8.6	1.4	5.1	13.5	39.0	8.6	20.2	3.0	−8.0	−53	11.0	12.5
				Median	17.0	8.5	1.4	5.2	13.0	38.4	8.1	20.6	3.0	−8.1	−54	11.1	12.9
				25th	16.2	7.6	1.1	4.6	12.1	33.6	7.1	18.6	2.6	−8.3	−55	10.2	10.6
				75th	18.0	9.1	1.5	5.7	14.6	45.2	9.8	23.5	3.4	−7.7	−52	11.8	14.6
	Group 3	72	IP, NP	Mean	14.4	6.8	1.1	3.9	10.7	29.6	5.5	19.2	2.1	−8.8	−57	13.5	12.2
				Median	14.0	6.7	1.0	3.9	10.8	28.9	5.2	20.4	2.0	−8.8	−57	13.5	12.9
				25th	13.2	6.1	0.7	3.5	9.7	24.9	4.3	18.3	1.7	−9.1	−58	12.8	10.4
				75th	15.2	7.3	1.4	4.2	11.7	33.7	6.4	21.3	2.3	−8.6	−56	14.5	15.0
	Group 4	43	IP, NP	Mean	20.0	9.7	1.0	6.7	18.2	64.4	9.1	18.3	3.4	−8.6	−56	12.9	13.2
				Median	20.0	9.8	1.0	6.6	17.8	62.8	8.9	18.5	3.2	−8.7	−56	12.9	13.0
				25th	19.3	9.2	0.8	6.0	16.4	56.3	8.3	15.8	2.8	−8.8	−57	12.3	11.2
				75th	21.2	10.3	1.2	7.1	19.8	69.7	9.8	20.7	4.0	−8.4	−55	13.4	14.8
Spring water	SP1	1	IP		11.8	5.3	0.6	2.8	9.0	19.4	3.6	19.1	1.4	−9.4	−60	15.1	13.3
	SP2	1	IP		11.3	5.3	0.7	2.8	9.5	24.4	3.8	20.9	1.5	−9.3	−60	14.3	8.4
	SP3	1	IP		13.6	5.9	0.8	3.4	10.5	22.0	5.1	21.0	1.9	−9.0	−58	13.8	11.3
	SP4	2	IP, NP	Mean	14.1	7.7	1.2	4.0	11.5	36.9	4.9	18.9	1.9	−8.9	−57	13.7	11.7
	SP5	2	IP, NP	Mean	14.4	7.9	1.1	4.3	11.0	31.7	6.0	19.1	2.2	−8.8	−57	13.4	12.6
	SP6	2	IP, NP	Mean	18.4	12.2	2.0	3.7	16.4	42.6	13.1	23.8	1.9	−8.9	−58	13.1	13.3
Paddy water		14	IP	Mean	11.7	6.2	3.3	2.9	7.1	19.1	6.9	20.5	0.5	−5.4	−45	−2.2	−
River water	Naka R.	2	IP, NP	Mean	15.7	10.2	0.2	1.6	15.9	25.3	2.7	42.7	0.1	−10.0	−63	17.3	0.3
	Kuma R.	2	IP, NP	Mean	14.8	5.0	0.2	2.0	17.6	9.6	1.0	52.5	0.2	−9.9	−62	16.4	0.2
	Sabi R.	2	IP, NP	Mean	11.1	3.6	0.4	2.5	10.6	11.9	1.2	32.5	0.2	−9.7	−62	15.4	0.5
	Houki R.	2	IP, NP	Mean	23.2	23.3	2.8	2.8	14.9	47.5	23.9	27.5	0.3	−10.0	−64	15.8	0.3
Rainwater	P1	16	All year	Mean	−	−	−	−	−	−	−	−	−	−8.5	−56	12.1	−
	P2	12	All year	Mean	−	−	−	−	−	−	−	−	−	−8.0	−52	12.1	−

EC: electrical conductivity, IP: irrigation period, NP: non-irrigation period, −: not analyzed, 25th and 75th: 25th and 75th percentiles, mean values of rainwater were weighted using the precipitation amount.

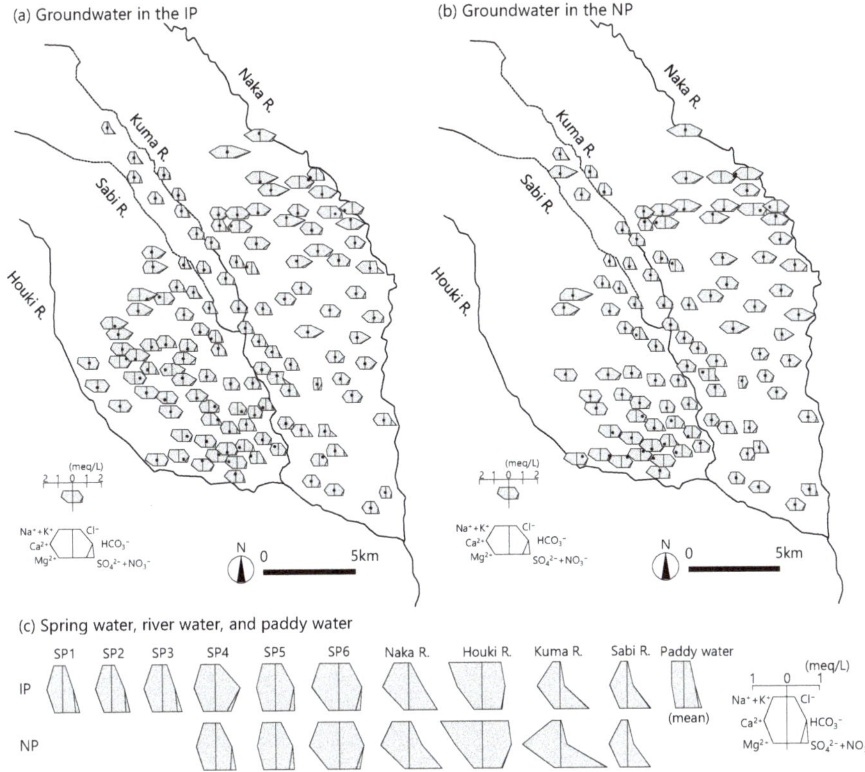

Figure 5. Hexa-diagrams of groundwater, spring water, river water, and paddy water.

Table 2 shows the matrix of the Pearson correlation coefficients for EC and major ions of all groundwater samples. Ca^{2+}, which is highly correlated with EC, is a major controlling factor of EC. A strong positive correlation between Ca^{2+} and HCO_3^- indicates that a dominant water type of the groundwaters in this aquifer was Ca-HCO_3 type. Concentrations of Ca^{2+} and HCO_3^- in aquifers of alluvial fans increased due to mineralization induced by a water–rock interaction during the groundwater flow process (e.g., [47]). Figure 6 shows the relationships between EC, Ca^{2+}, and HCO_3^- of the groundwater with groundwater level. These values did not increase with the downstream direction in the fan. The spatial distribution of the hexa-diagrams in Figure 5 indicates that the chemical compositions varied greatly depending on the location, especially the distances from the rivers.

Table 2. Pearson correlation coefficients for EC and major ions of all groundwater samples.

	EC	Na⁺	K⁺	Mg²⁺	Ca²⁺	HCO₃⁻	Cl⁻	SO₄²⁻	NO₃-N
EC	1								
Na⁺	0.67	1							
K⁺	0.24	0.45	1						
Mg²⁺	0.52	0.19	−0.19	1					
Ca²⁺	0.82	0.60	0.19	0.56	1				
HCO₃⁻	0.67	0.50	0.02	0.70	0.84	1			
Cl⁻	0.66	0.91	0.54	0.15	0.61	0.42	1		
SO₄²⁻	0.31	0.29	0.19	−0.01	0.26	−0.13	0.23	1	
NO₃-N	0.32	0.01	−0.07	0.69	0.27	0.28	0.08	−0.18	1

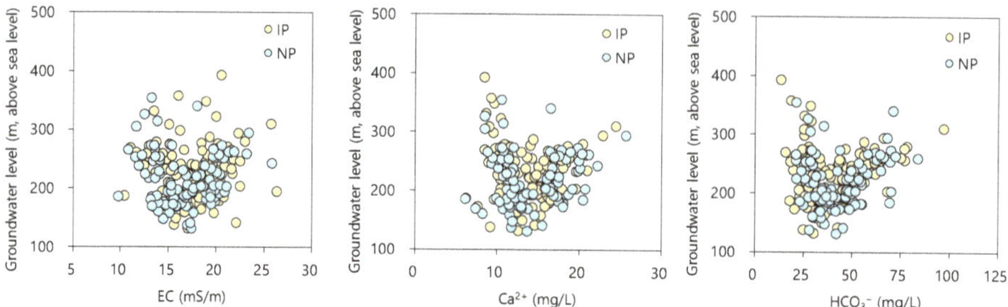

Figure 6. Relationships between EC, Ca^{2+}, and HCO_3^- with groundwater level.

Figure 7 illustrates the spatial distributions of EC and the concentrations of the three characteristic ions Ca^{2+}, SO_4^{2-}, and Cl^- of groundwater in the IP and NP. The spatial distribution of EC, which was low along the Sabi and Kuma rivers in the central part of the fan, was very consistent with previous survey results [25,48]. Along these two rivers, Ca^{2+}, like EC, was relatively low, whereas SO_4^{2-} was relatively high. In contrast, Cl^- concentrations were higher along the western edge of the fan near the Houki River than elsewhere in the fan. As is shown by the hexa-diagrams and ternary diagrams, the Naka, Sabi, and Kuma River waters were characterized by large SO_4^{2-} contents, whereas the Houki River water samples had higher Na^+ and Cl^- concentrations than the waters from the other rivers. Spring waters were plotted between the HCO_3^- and SO_4^{2-} types on the anion ternary diagram; SP1, SP2, and SP3 were characterized by larger SO_4^{2-} contents, as indicated by their hexa-diagram shapes. No significant seasonal difference between the IP and NP was observed in the chemical compositions of the river and spring waters.

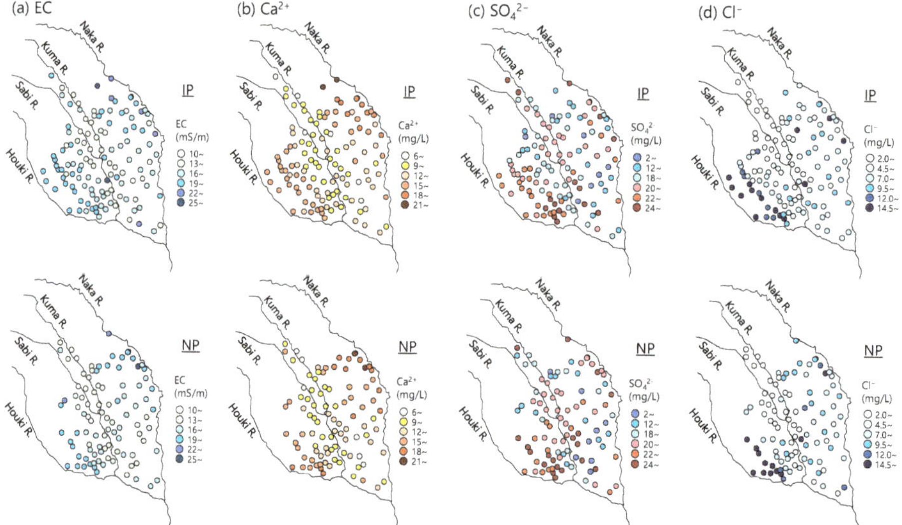

Figure 7. Distributions of (**a**) EC, (**b**) Ca^{2+}, (**c**) SO_4^{2-}, and (**d**) Cl^- concentrations in the groundwater. The upper and lower maps show the data in the IP and NP, respectively.

4.3. Isotopic Compositions

The relationship between $\delta^{18}O$ and δ^2H ($\delta^{18}O$–δ^2H scatter diagram) in all water samples is shown in Figure 8. In rainwater collected at sites P1 and P2, $\delta^{18}O$ and δ^2H varied over relatively wide ranges, from −12.9‰ to −3.1‰ and from −85‰ to −5‰, respectively. The local meteoric water line (LMWL) in this study area was determined to be $\delta^2H = 8.48\delta^{18}O + 20.09$. The slope of the LMWL was slightly larger than the global meteoric water line (GMWL) slope of 8 [49] and the slope of the meteoric water line on the Pacific Ocean side in Japan (7.8) [50]. The weighted means of $\delta^{18}O$ and δ^2H (weighted using the precipitation amount during the sampling time interval) were −8.3‰ and −54‰ at P1 and P2, respectively; these values, which were plotted below the LMWL, reflect the low d-excess value of rainwater in summer in Japan [51] because the region of the Nasunogahara fan receives more precipitation in summer than in winter.

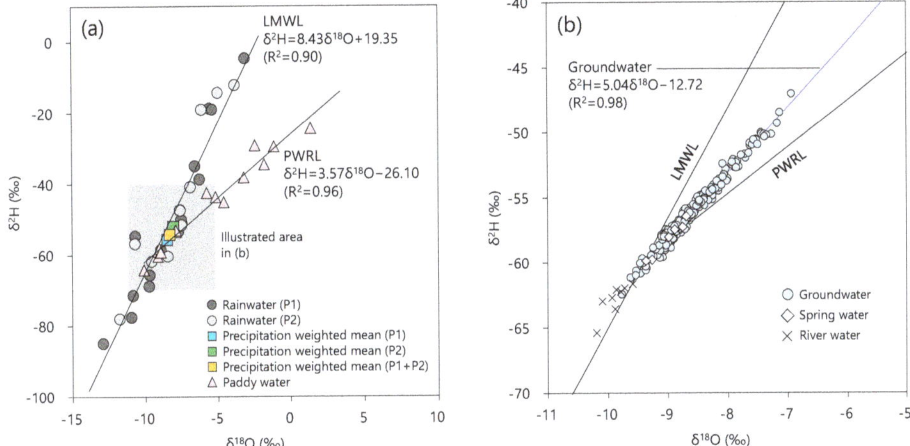

Figure 8. Relationship between $\delta^{18}O$ and δ^2H of (**a**) rainwater and paddy water and (**b**) groundwater. LMWL is the local meteoric water line identified by fitting a regression line to observations at sites P1 and P2. PWRL is the regression line fitted to the paddy water data.

The slope of the regression line between $\delta^{18}O$ and δ^2H of paddy waters, denoted as PWRL in Figure 8, was 3.6. The plots of the paddy water data deviated from the LMWL and the d-excess value of the paddy water (−2.2‰) was extremely low compared to those of the groundwater, rainwater, and river water (Table 1). Generally, kinetic fractionation of oxygen and hydrogen isotopes during evaporation causes the remaining water to become enriched in the heavier isotopes (^{18}O and ^2H); as a result, trends with slopes ranging from 4 to 7 are seen on the $\delta^{18}O$–δ^2H scatter diagram, whereas the equilibrium isotopic fractionation line lies near the LMWL and has a slope close to 8 [52]. If there is no rain, irrigation water is supplied to the paddy fields almost every day and is affected by evaporative enrichment, even in just one day. Paddy waters affected by evaporative isotopic enrichment would have higher isotopic ratios and lower d-excess values [53] and effects the isotopic compositions in groundwater [54]. In this study, paddy waters collected at paddy field outlets had higher isotopic ratios than those collected at the inlets because evaporative isotopic enrichment occurred during the residence time of the water in the paddy rice fields. The slope of the regression line fitted to all groundwater data was 5.04, between those of the LMWL and the PWRL (Figure 8b). The isotopic ratios of the river water samples were smaller than those of groundwater samples and plotted near the LMWL.

The spatial distributions of $\delta^{18}O$, d-excess, and the ^{222}Rn concentration in groundwater in the IP and NP are shown in Figure 9. The groundwater along the Sabi and Kuma rivers had comparatively low $\delta^{18}O$ (less than −8.5‰) and high d-excess values (more than 12.0‰), whereas groundwater in the southeastern part of the fan had high $\delta^{18}O$ (more than −8.0‰) and low d-excess values (less than 11.0‰). Compared with the IP, the domain with a high $\delta^{18}O$ and low d-excess values in the southeastern fan shrank in the NP. Spatial differences in the ^{222}Rn concentration in groundwater were indistinct, although concentrations in the NP were higher than those in the IP. The ^{222}Rn concentration in waters of the four rivers was close to zero (Table 1).

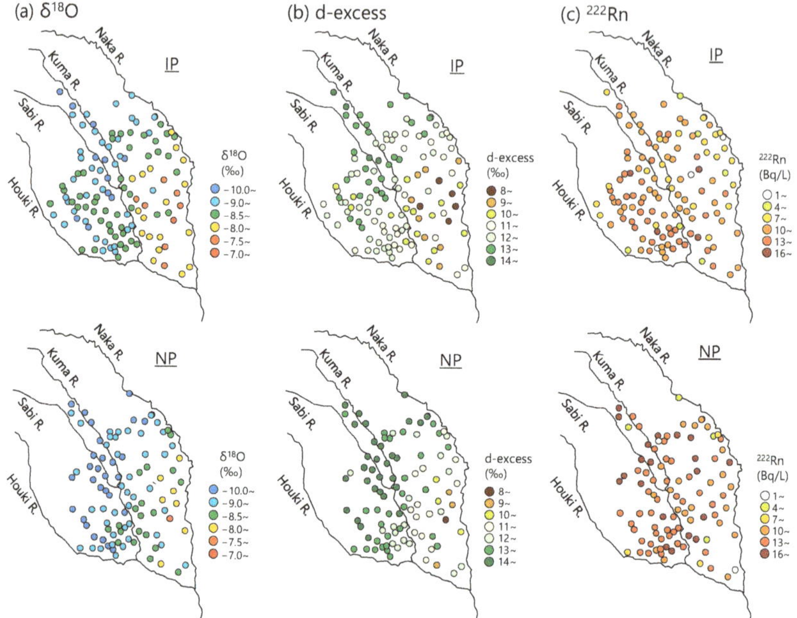

Figure 9. Distributions of (**a**) $\delta^{18}O$, (**b**) d-excess, and (**c**) the ^{222}Rn concentration in groundwater in the IP (upper maps) and NP (lower maps).

4.4. SOM and Clustering Results

Figure 10 shows the component planes of the hydrochemical and isotopic parameters finally obtained after iterative training of the SOM. Each component plane shows the value of one component of the reference vector in each of the 72 nodes. Visualization of the data using component planes is helpful for finding interrelationships among the different parameters [21]. For example, the component plane patterns of Na^+ and Cl^- were visibly similar; both had larger values in the lower-left and smaller values in the upper-left part of the plane (Figure 10). The component planes of several other parameter combinations (e.g., $\delta^{18}O$–δ^2H, Ca–HCO_3^-, and Mg^{2+}–NO_3-N) also showed similar patterns (Figure 10), whereas the component planes of SO_4^{2-}, K^+, and ^{222}Rn did not exhibit any similarity with those of other parameters.

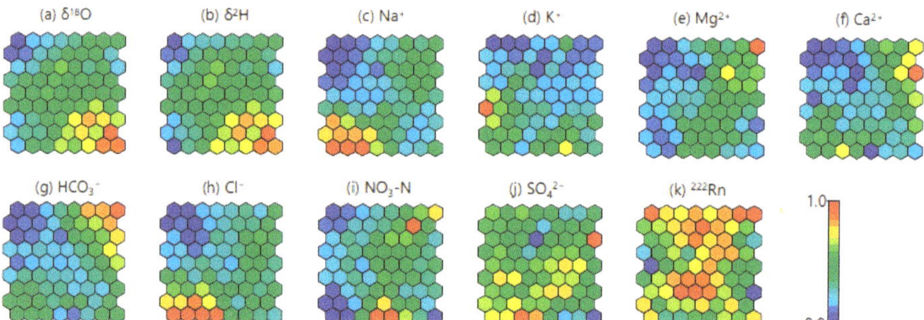

Figure 10. Component planes of the isotopic and hydrochemical parameters: (**a**) $\delta^{18}O$, (**b**) δ^2H, (**c**) Na^+, (**d**) K^+, (**e**) Mg^{2+}, (**f**) Ca^{2+}, (**g**) HCO_3^-, (**h**) Cl^-, (**i**) NO_3-N, (**j**) SO_4^{2-}, and (**k**) ^{222}Rn. Each plane shows the standardized value of each parameter (transformed to the range from 0 to 1) in each node.

The SOM nodes were subdivided into four groups (groups 1 to 4, comprising 34, 73, 72, and 43 groundwater samples, respectively) according to the dendrogram produced by the hierarchical cluster analysis (Figure 11). Figure 12 shows the spatial distributions of the color-coded hexa-diagrams of the four groups in the IP and NP. Group 1 was mostly distributed near the Houki River, which flows along the western edge of the fan. Group 2 was distributed in the central and lower parts of the fan. Group 3 was distributed along the Sabi and Kuma rivers. Group 4 was distributed across the middle part of the fan.

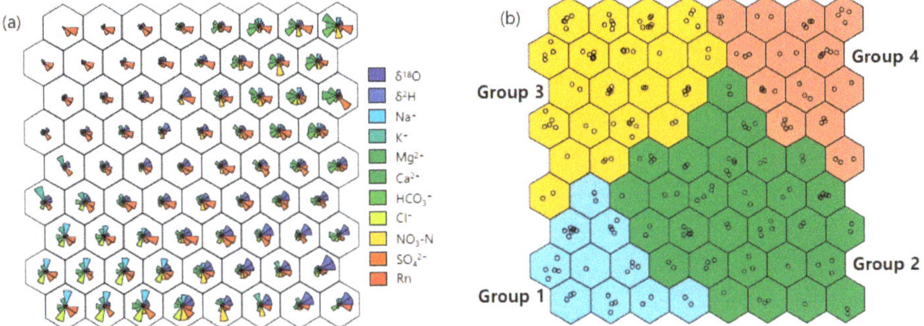

Figure 11. (**a**) All reference vectors in each of the 72 nodes and (**b**) classification of the self-organizing map (SOM) nodes into four groups. The circles on each map hexagon represent the groundwater samples assigned to that node.

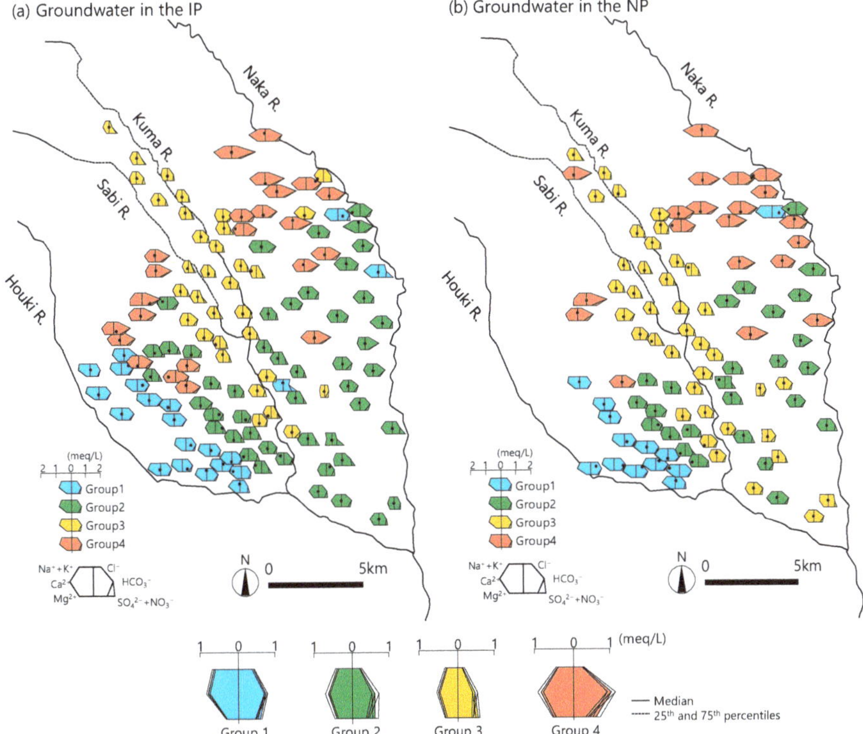

Figure 12. Spatial distribution of hexa-diagrams of groundwaters in four groups classified by the SOM in the (**a**) IP and (**b**) NP.

The EC of group 3 groundwater was lower than that of the other groups (Table 1), indicating that it contained lower amounts of dissolved components. Groundwater in groups 1 and 4 had higher EC than groundwater in groups 2 and 3. In particular, concentrations of Na^+ and Cl^- were notably higher in group 1, whereas those of Ca^{2+} and HCO_3^- were higher in group 4. The mean NO_3-N concentration was the highest in group 4, but all groundwater samples in this study had nitrate concentrations below the standard level of 10 mg/L NO_3-N in Japan. The width of each hexa-diagram is an approximate indication of the total ionic content. Thus, the hexa-diagram width of group 3, which had a low EC, was small, whereas the hexa-diagram widths of groups 1 and 4, which had higher ECs, were large (Figure 12). Chemical compositions of groundwaters in the four groups, spring water, and river water are shown as a trilinear diagram in Figure 13. Groundwater samples of group 4 were mostly HCO_3^--type water, whereas those of groups 1 to 3 tended to be scattered between HCO_3^- and SO_4^{2-} types. In addition, the relative abundance of Cl^- in group 1 samples was somewhat large.

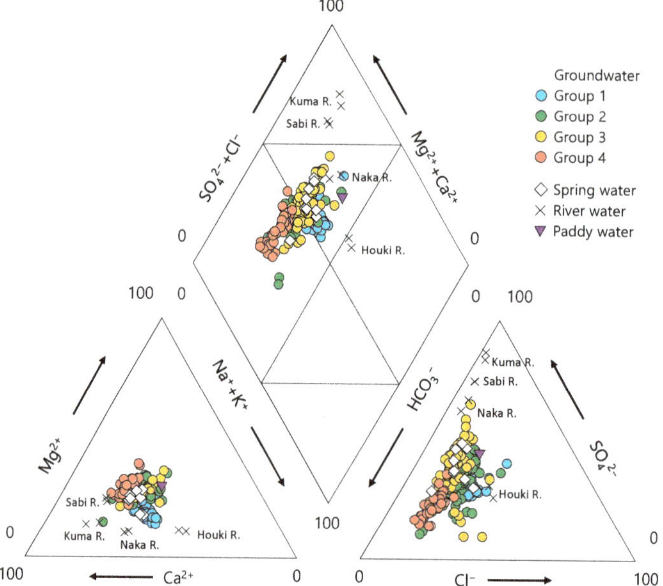

Figure 13. Trilinear diagram of the four groundwater groups.

The $\delta^{18}O$–δ^2H scatter diagram of groundwater in the four groups is shown in Figure 14. The groundwater in each of the four groups had characteristic $\delta^{18}O$ and δ^2H values: group 2 waters had the highest values, followed by waters of groups 4 and 1, and group 3 waters had the lowest values (Figure 14, Table 1). Group 2 waters had the lowest d-excess value and group 3 waters had the highest d-excess value.

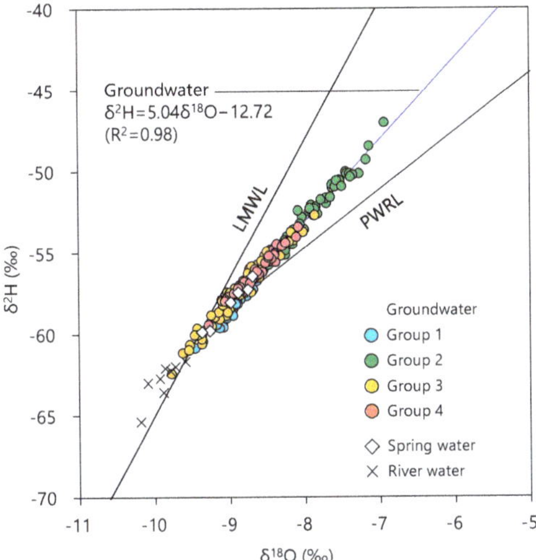

Figure 14. Relationship between $\delta^{18}O$ and δ^2H of the four groundwater groups. LMWL and PWRL are the local meteoric water line and the paddy water regression line, respectively, as shown in Figure 8.

5. Discussion

5.1. Influence of Recharge Sources on Groundwater Hydrochemical and Isotopic Compositions

It has been pointed out that shallow groundwater in the Nasunogahara alluvial fan is recharged by multiple sources, including the infiltration of rainwater, river water, and paddy irrigation water (e.g., [24]). The isotopic ratios of groundwater in this study, which were plotted as an elongated cluster between the LMWL and the PWRL in the $\delta^{18}O$–δ^2H diagram (Figure 8), also indicate a mixture of groundwaters recharged from three different sources. The fact that the isotopic ratios of the groundwater were plotted between the LMWL and PWRL is strong evidence that groundwater in the fan was influenced by the infiltration of not only rainwater but also paddy irrigation water (Figure 8). In addition, the comparison between groundwater levels and ground surface elevations indicated that there was infiltration from the Houki, Sabi, and Kuma rivers to the aquifer (Figure 3). The isotopic ratios of river waters (mean values of four rivers of −9.9‰ for $\delta^{18}O$ and −63‰ for δ^2H) were much less than the weighted means of rainwater (see Section 4.3) (Figure 8). The reason for the lower ratios of these river waters was that the rivers gathered precipitation that falls at high altitude in the mountains, which had lower isotopic ratios because of the isotopic altitude effect. The isotopic ratios of some groundwater samples were lower than the weighted mean rainwater $\delta^{18}O$ and δ^2H values, suggesting that these samples represented a mixture of waters with low isotopic ratios. We therefore inferred that the groundwaters in the western part and upper part of the fan were affected by the infiltration of waters from the Houki River, and the Kuma and Sabi rivers, respectively. The river waters contained baseflows supplied from the mountainous watersheds, but ^{222}Rn concentrations of river waters were low. This indicates that dissolved ^{222}Rn in river waters was lost through dispersion to the atmosphere and radioactive decay before reaching the fan. It had been expected that the ^{222}Rn concentration of the groundwater near the rivers decreased due to river water infiltration, but the concentrations were not lower than those far away from the rivers (Figure 9). This means that the residence time of infiltrated waters to reaching sampling wells was more than two weeks, which were required to reach radioactive equilibrium.

The hydrochemical data also implied that infiltration of river water recharged some groundwaters. The groundwater around the Sabi and Kuma rivers, the EC, and some dissolved ion concentrations, such as that of Ca^{2+}, were relatively low, but SO_4^{2-} was relatively high; thus, we can infer that the groundwaters were influenced by the infiltration of waters from these rivers, which had similar compositions, and this inference was supported by the low isotopic ratios and high d-excess values of the groundwaters. The groundwater in the western part of the fan had relatively high Na^+ and Cl^- concentrations, reflecting the infiltration of Houki River water, which was also characterized by relatively high Na^+ and Cl^- concentrations (Figures 5 and 7). Similarly, the relatively low isotopic ratios and high EC of the groundwaters suggested a large contribution of infiltration from the Houki River (Figures 7 and 9). Thus, groundwater in the fan reflected three different recharge sources and the fraction of the contribution from each source varied by location.

5.2. Characterization of Groundwater Using SOM

The SOM classified the groundwater in the fan into four groups with different hydrochemical and isotopic compositions. The groundwater of each group was inferred to be affected by different recharge sources. Group 1 groundwaters, distributed in the western part of the fan, had low isotopic ratios and high Na^+ and Cl^-, suggesting the infiltration of the Houki River. The relatively high isotopic ratios and low d-excess values of group 2 groundwaters indicated a large recharge contribution due to infiltration from paddy rice fields. Group 3 groundwaters around the Sabi and Kuma rivers were inferred to be influenced by infiltration of these rivers, which had similar chemical and isotopic compositions: low EC, low dissolved ion concentrations excluding SO_4^{2-}, and low stable isotopic ratios. Group 4 groundwaters, distributed at higher elevations than group 2 groundwaters, had lower isotopic ratios than the weighted means of rainwater, suggesting that these samples were affected by a mixture of Sabi and Kuma river waters. However, the relatively large EC and ionic contents indicated a smaller contribution of infiltration from these river waters compared to group 3 groundwaters. In the upper part of the fan, there are few paddy rice fields but many livestock farms (Figure 1). Reflecting this difference in land use, NO_3-N concentrations in group 4 groundwaters were higher than those in the other groups (Table 1). We inferred that these higher concentrations reflected the high nitrogen load associated with livestock farming (e.g., [27]). It can be readily seen that the contribution of infiltration from paddy rice field to groundwater recharge was smaller for group 4 than for group 2 groundwaters. In contrast, smaller ionic contents, including NO_3-N, in group 2 groundwaters indicate a dilution due to infiltration of paddy waters.

The application and results of the SOM assisted the interpretation of the difference in hydrochemical and isotopic compositions depending on the location and the influence of multiple recharge sources on the groundwater. Data that produce a scattered distribution in the trilinear diagram, hexa-diagrams, and $\delta^{18}O–\delta^2H$ diagram make it difficult to resolve ambiguous boundaries and to produce reasonable classification of the groundwaters influenced by multiple recharge sources, especially for the shallow groundwater with similar chemical and isotopic compositions. The SOM application could automatically establish the four different groups, independent of human-subjective criteria, and summarized the spatial distribution of each group with the readily understandable visualization (Figure 15).

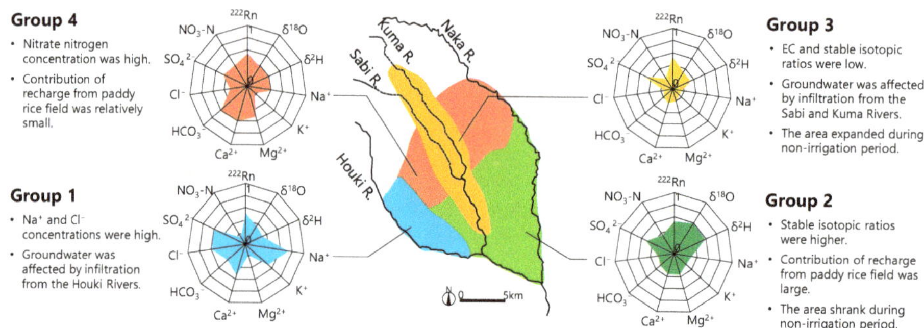

Figure 15. Characteristics of the Nasunogahara alluvial fan groundwaters in each of the four groups. Medians of all variables plotted on the radar charts were transformed to range from 0 to 1.

Figure 12 illustrates the differences in the spatial distribution of the four groundwater groups classified by the SOM between the IP and NP. We focus here on the central and eastern parts of the fan because we could not collect groundwater samples in the western part of the fan in the NP. One large seasonal change was that the distribution area of group 2 shrank in the NP, whereas that of group 3 expanded in the downstream direction. EC, δ^{18}O, δ^{2}H, and d-excess values and ^{222}Rn concentrations of groundwater at the observation sites classified into group 2 in the IP differed significantly between the IP and NP ($p < 0.05$) (Figure 16). Although the EC apparently decreased in the NP in the southeastern part of the fan (where only group 2 samples were distributed in the IP) (Figure 7), the EC in group 2 groundwaters did not differ significantly between the IP and NP ($p = 0.12$). The major ion concentrations in group 2 groundwaters also did not differ significantly between the two periods ($p > 0.10$). The median δ^{18}O and δ^{2}H values of group 2 groundwaters in the IP were −8.0‰ and −53‰, respectively, and the median values at the same sites in the NP decreased to −8.4‰ and −55‰, respectively, whereas the median d-excess value increased from 11.0‰ in the IP to 12.3‰ in the NP. These results indicated that, with respect to the relation between δ^{18}O and δ^{2}H, the group 2 groundwaters deviated more from the LMWL in the IP than in the NP. The high isotopic ratios of the group 2 groundwaters indicated that they were likely more greatly affected by recharge from paddy rice fields than groundwaters of the other groups. This inference is consistent with the higher isotopic ratios that were observed in the IP than in the NP. As a result of this seasonal difference, some sampling sites shifted from group 2 in the IP to group 3 in the NP, reflecting a reduced influence of recharge from paddy rice fields in the NP and a corresponding relative increase in the influence of infiltration of river waters at these sites. In addition, ^{222}Rn concentrations of groundwaters were lower in the IP than in the NP (Figure 16). Generally, ^{222}Rn concentrations in groundwater decrease during the irrigation period because of downward flow of soil water pushed out by irrigation water [55]. Accordingly, the difference in ^{222}Rn concentrations between the IP and NP also indicated a change in the relative contributions of infiltration from paddy rice fields and rivers to groundwater recharge.

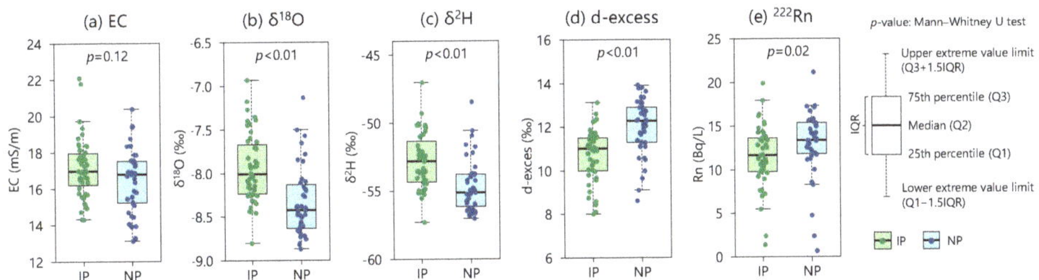

Figure 16. Comparison of groundwater EC values and isotopic compositions between the IP and NP at observation sites classified into group 2 in the IP.

6. Conclusions

The groundwater in the Nasunogahara alluvial fan, which is mostly covered by paddy rice fields, was investigated by using environmental isotopes and hydrochemical investigations and the application of an SOM. The obtained isotopic and hydrochemical data indicated that the groundwater in the fan was affected by three different recharge sources: precipitation, river waters, and paddy rice field irrigation water. Furthermore, the SOM clearly classified the groundwater in the fan into four groups (groups 1 to 4), reflecting different recharge sources. The characteristics of the groundwater of each group can be summarized as follows:

1. Group 1 groundwater, distributed around the Houki River, which flows along the western edge of the fan, had relatively low isotopic ratios, but high EC values and high Na^{+} and Cl^{-} concentrations. Group 1 groundwater was thus inferred to have been greatly affected by infiltration from the Houki River, which had water of a different chemical composition compared to the other rivers.

2. Group 2 groundwater, distributed in the central and lower part of the fan, had high isotopic ratios and was inferred to be recharged mainly by infiltration of paddy waters that had been affected by evaporative isotopic enrichment. However, the area of group 2 shrank during the non-irrigation period when the infiltration of paddy water did not occur.

3. Group 3 groundwater, distributed around the Sabi and Kuma Rivers, which flows down the center of the fan, had a low EC and low isotopic ratios, indicating a greater influence of infiltration of water from the two rivers compared to other recharge sources.

4. Group 4 groundwater, distributed on the upstream side of group 2 groundwater, had lower isotopic ratios than group 2 groundwater. In the upper part of the fan, few paddy rice fields but many livestock farms are distributed throughout this region; thus, recharge from paddy rice fields was relatively small. Furthermore, NO_3-N concentrations in group 4 groundwaters were higher than those in the other groups.

These results show that multiple tracers could be effectively used to evaluate the influence of different groundwater recharge sources, including paddy water in rice cultivation areas, and the application of an SOM can provide understandable and readily visualized results to assist in the interpretation of the characteristics of groundwater in the fan. The findings of this study can therefore contribute to proper planning for regional groundwater use and the preservation of spring waters in the fan.

Author Contributions: Design of the study framework, T.T. and S.I.; director of the field survey and analysis T.T.; water sample collection, all authors; writing, T.T.; reviewing and editing, K.S., S.I., and S.Y. All authors have read and agreed to the published version of the manuscript.

Funding: This study was partly supported by JSPS KAKENHI (grant number JP19K06301).

Acknowledgments: We express our gratitude to the Union Nasunogahara Land Improvement District (LID: farmers' water management organization), other LIDs, and the individual well owners for permission to sample groundwater for this study. Finally, we would like to thank the academic editors and four anonymous reviewers for their helpful comments and suggestions.

Conflicts of Interest: The authors declare no conflict of interest.

References

1. Seto, R. Land use by landform classification in Japan -Analysis with the data of grid square basis-. *Map* **1986**, *24*, 1–11, (In Japanese with English abstract).

2. Tabayashi, A. Irrigation systems in Japan. *Geogr. Rev. Jpn.* **1987**, *60*, 41–65. [CrossRef]

3. Li, F.; Pan, G.; Tang, C.; Zhang, Q.; Yu, J. Recharge source and hydrogeochemical evolution of shallow groundwater in a complex alluvial fan system, southwest of north China plain. *Environ. Geol.* **2008**, *55*, 1109–1122. [CrossRef]

4. Yu, H.L.; Chu, H.J. Understanding space-time patterns of groundwater system by empirical orthogonal functions: A case study in the Choshui River alluvial fan, Taiwan. *J. Hydrol.* **2010**, *381*, 239–247. [CrossRef]

5. Eastoe, C.J.; Hutchison, W.R.; Hibbs, B.J.; Hawley, J.; Hogan, J.F. Interaction of a river with an alluvial basin aquifer: Stable isotopes, salinity and water budgets. *J. Hydrol.* **2010**, *395*, 67–78. [CrossRef]

6. Bourke, S.A.; Cook, P.G.; Shanafield, M.; Dogramaci, S.; Clark, J.F. Characterisation of hyporheic exchange in a losing stream using radon-222. *J. Hydrol.* **2014**, *519*, 94–105. [CrossRef]

7. Karan, S.; Sebok, E.; Engesgaard, P. Air/water/sediment temperature contrasts in small streams to identify groundwater seepage locations. *Hydrol. Process.* **2017**, *31*, 1258–1270. [CrossRef]

8. Yoshioka, Y.; Nakamura, K.; Nakano, T.; Horino, H.; Shin, K.C.; Hashimoto, S.; Kawashima, S. Multiple-indicator study of groundwater flow and chemistry and the impacts of river and paddy water on groundwater in the alluvial fan of the Tedori River, Japan. *Hydrol. Process.* **2016**, *30*, 2804–2816. [CrossRef]

9. Doveri, M.; Mussi, M. Water isotopes as environmental tracers for conceptual understanding of groundwater flow: An application for fractured aquifer systems in the "Scansano-Magliano in Toscana" area (Southern Tuscany, Italy). *Water* **2014**, *6*, 2255–2277. [CrossRef]

10. Yeh, H.F.; Lin, H.I.; Lee, C.H.; Hsu, K.C.; Wu, C.S. Identifying seasonal groundwater recharge using environmental stable isotopes. *Water* **2014**, *6*, 2849–2861. [CrossRef]

11. Zhong, C.H.; Yang, Q.C.; Ma, H.Y.; Bian, J.M.; Zhang, S.H.; Lu, X.G. Application of environmental isotopes to identify recharge source, age, and renewability of phreatic water in Yinchuan Basin. *Hydrol. Process.* **2019**, *33*, 2166–2173. [CrossRef]

12. Liu, Y.; Yamanaka, T. Tracing groundwater recharge sources in a mountain-plain transitional area using stable isotopes and hydrochemistry. *J. Hydrol.* **2012**, *464*, 116–126. [CrossRef]

13. Valder, J.F.; Long, A.J.; Davis, A.D.; Kenner, S.J. Multivariate statistical approach to estimate mixing proportions for unknown end members. *J. Hydrol.* **2012**, *460–461*, 65–76. [CrossRef]

14. Cortes, J.E.; Muñoz, L.F.; Gonzalez, C.A.; Niño, J.E.; Polo, A.; Suspes, A.; Siachoque, S.C.; Hernãndez, A.; Trujillo, H. Hydrogeochemistry of the formation waters in the San Francisco field, UMV basin, Colombia—A multivariate statistical approach. *J. Hydrol.* **2016**, *539*, 113–124. [CrossRef]

15. Islam, M.M.; Lenz, O.K.; Azad, A.K.; Ara, M.H.; Rahman, M.; Hassan, N. Assessment of spatio-temporal variations in water quality of Shailmari River, Khulna (Bangladesh) using multivariate statistical techniques. *J. Geosci. Environ. Protect.* **2017**, *5*, 1–26. [CrossRef]

16. Dieng, N.M.; Orban, P.; Stumpp, C.; Faye, S.; Dassargues, A. Temporal changes in groundwater quality of the Saloum coastal aquifer. *J. Hydrol. Reg. Stud.* **2017**, *9*, 163–182. [CrossRef]

17. Chen, I.T.; Chang, L.C.; Chang, F.J. Exploring the spatio-temporal interrelation between groundwater and surface water by using the self-organizing maps. *J. Hydrol.* **2018**, *556*, 131–142. [CrossRef]

18. Choi, B.Y.; Yun, S.T.; Kim, K.H.; Kim, J.W.; Kim, H.M.; Koh, Y.K. Hydrogeochemical interpretation of South Korean groundwater monitoring data using Self-Organizing Maps. *J. Geochem. Explor.* **2014**, *137*, 73–84. [CrossRef]

19. Nakagawa, K.; Amano, H.; Kawamura, A.; Berndtsson, R. Classification of groundwater chemistry in Shimabara, using self-organizing maps. *Hydrol. Res.* **2017**, *48*, 840–850. [CrossRef]

20. Agoubi, B. Assessing hydrothermal groundwater flow path using Kohonen's SOM geochemical data and groundwater temperature cooling trend. *Environ. Sci. Pollut. Res.* **2018**, *25*, 13597–13610. [CrossRef]

21. An, Y.; Zou, Z.; Li, R. Descriptive Characteristics of Surface Water Quality in Hong Kong by a Self-Organising Map. *Int. J. Environ. Res. Public Health* **2016**, *13*, 115. [CrossRef] [PubMed]

22. Rural Development Bureau, Ministry of Agriculture, Forestry and Fisheries (MAFF). *Actual Use of Groundwater for Agriculture*; MAFF: Tokyo, Japan, 2011; pp. 1–13. (In Japanese)

23. Otawara City Habitat of Three-Spined Stickleback (Taya River). Available online: https://www.city.ohtawara.tochigi.jp/docs/2013082781284/ (accessed on 28 October 2019). (In Japanese).

24. Wakui, H.; Yamanaka, T. Source of groundwater recharge and their local differences in the central part of Nasu fan as revealed by stable isotopes. *J. Groundw. Hydrol.* **2006**, *48*, 263–277, (In Japanese with English abstract). [CrossRef]

25. Hiyama, T.; Suzuki, Y. Groundwater in the Nasuno basin -Spatial and seasonal changes in water quality-. *J. Jpn. Assoc. Hydrol. Sci.* **1991**, *21*, 143–154, (In Japanese with English abstract).

26. Babiker, I.S.; Mohamed, M.A.A.; Hiyama, T. Assessing groundwater quality using GIS. *Water Resour. Manag.* **2007**, *21*, 699–715. [CrossRef]

27. Somura, H.; Goto, A.; Matsui, H.; Elhassan, A.M. Impacts of nutrient management and decrease in paddy field area on groundwater nitrate concentration: A case study at the Nasunogahara alluvial fan, Tochigi Prefecture, Japan. *Hydrol. Process.* **2008**, *22*, 4752–4766. [CrossRef]

28. Elhassan, A.M.; Goto, A.; Mizutani, M. Combining a tank model with a groundwater model for simulating regional groundwater flow in an alluvial fan. *Trans. Jpn. Soc. Irrig.* **2001**, *215*, 21–29.

29. National Land Information Division, National Spatial Planning and Regional Policy Bureau, MILT of Japan. National Land Numerical Information Download Service. Available online: http://nlftp.mlit.go.jp/ksj/ (accessed on 4 September 2019).

30. Geospatial Information Authority of Japan. GSI Map. Available online: https://maps.gsi.go.jp/ (accessed on 4 September 2019).

31. Hydrogeological Map of Nasuno-ga-hara Area. *Kanto Regional Agricultural Administration Office*; MAFF of Japan: Tokyo, Japan, 1993. (In Japanese)

32. The Survey Report of Subsurface Dam in Kanto Area. *Kanto Regional Agricultural Administration Office*; MAFF of Japan: Tokyo, Japan, 1999. (In Japanese)

33. IAEA/GNIP Precipitation Sampling Guide (V2.02). September 2014. Available online: http://www-naweb.iaea.org/napc/ih/documents/other/gnip_manual_v2.02_en_hq.pdf (accessed on 31 October 2019).

34. Hoehn, E.; von Gunten, H.R. Radon in groundwater: A tool to assess infiltration from surface waters to aquifers. *Water Resour. Res.* **1989**, *25*, 1795–1803. [CrossRef]

35. Hamada, H. Estimation of groundwater flow rate using the decay of ^{222}Rn in a well. *J. Environ. Radioact.* **2000**, *47*, 1–13. [CrossRef]

36. Hamada, H.; Komae, T. Investigation on shallow groundwater in a small basin using natural radioisotopes. *Radioisotopes* **1996**, *45*, 71–81. [CrossRef]

37. Gat, J.R. Oxygen and hydrogen isotopes in the hydrologic cycle. *Annu. Rev. Earth Planet. Sci.* **1996**, *24*, 225–262. [CrossRef]

38. Dansgaard, W. Stable isotopes in precipitation. *Tellus* **1964**, *16*, 436–468. [CrossRef]

39. Kohonen, T. Self-organized formation of topologically correct feature maps. *Biol. Cybern.* **1982**, *43*, 59–69. [CrossRef]

40. Jin, Y.H.; Kawamura, A.; Park, S.C.; Nakagawa, N.; Amaguchi, H.; Olsson, J. Spatiotemporal classification of environmental monitoring data in the Yeongsan River basin, Korea, using self-organizing maps. *J. Environ. Monit.* **2011**, *13*, 2886–2894. [CrossRef] [PubMed]

41. Nguyen, T.T.; Kawamura, A.; Tong, T.N.; Nakagawa, N.; Amaguchi, H.; Gilbuena, R. Clustering spatio–seasonal hydrogeochemical data using self-organizing maps for groundwater quality assessment in the Red River Delta. *J. Hydrol.* **2015**, *522*, 661–673. [CrossRef]

42. Farsadnia, F.; Rostami Kamrood, M.; Moghaddam Nia, A.; Modarres, R.; Bray, M.T.; Han, D.; Sadatinejad, J. Identification of homogeneous regions for regionalization of watersheds by two-level self-organizing feature maps. *J. Hydrol.* **2014**, *509*, 387–397. [CrossRef]

43. Hentati, A.; Kawamura, A.; Amaguchi, H.; Iseri, Y. Evaluation of sedimentation vulnerability at small hillside reservoirs in the semi-arid region of Tunisia using the Self-Organizing Map. *Geomorphology* **2010**, *122*, 56–64. [CrossRef]

44. Vesanto, J.; Alhoniemi, R. Clustering of the self organizing map. *IEEE Trans. Neural Netw.* **2000**, *11*, 586–600. [CrossRef]

45. Faggiano, L.; Zwart, D.; García-Berthou, E.; Lek, S.; Gevrey, M. Patterning ecological risk of pesticide contamination at the river basin scale. *Sci. Total Environ.* **2010**, *408*, 2319–2326. [CrossRef]

46. Wehrens, R.; Buydens, L.M.C. Self-and super-organizing maps in r: The kohonen package. *J. Stat. Softw.* **2017**, *21*, 1–19.

47. Ogawa, R.; Yamanaka, M. Hydrogeochemical controlling on groundwater in the Hadano Basin, Kanagawa Prefecture: Processes of mineral weathering and dissolved inorganic carbon supply. In *Proceedings of the Institute of Natural Sciences. Sect*; Nihon University: Tokyo, Japan, 2018; Volume 53, pp. 125–134, (In Japanese with English abstract).

48. Yamanaka, T.; Tanaka, T.; Asanuma, J.; Hamada, Y. *Interaction Between Groundwater and River Water in the Nasu Fan, Tochigi*; Bull Terr Environment Research Ctr., University of Tsukuba: Tsukuba, Japan, 2003; Volume 4, pp. 51–59, (In Japanese with English abstract).

49. Craig, H. Isotopic variations in meteoric waters. *Science* **1961**, *133*, 1702–1703. [CrossRef]

50. Ichiyanagi, K.; Tanoue, M. Spatial analysis of annual mean stable isotopes in precipitation across Japan based on an intensive observation period throughout 2013. *Istopes Environ. Health Stud.* **2016**, *52*, 353–362. [CrossRef] [PubMed]

51. Yoshimura, K.; Ichiyanagi, K. A reconsideration of seasonal variation in precipitation deuterium excess over East Asia. *J. Jpn. Soc. Hydrol. Water Res.* **2009**, *22*, 262–276, (In Japanese with English abstract). [CrossRef]

52. Gibson, J.J.; Prepas, E.E.; McEachern, P. Quantitative comparison of lake throughflow, residency, and catchment runoff using stable isotopes: Modelling and results from a regional survey of Boreal lakes. *J. Hydrol.* **2002**, *262*, 128–144. [CrossRef]

53. Tsuchihara, T.; Shirahata, K.; Yoshimoto, S.; Ishida, S. National-scale variations in the stable isotopic compositions of irrigation-pond and spring waters across Japan. *Paddy Water Environ.* **2019**, *17*, 429–438. [CrossRef]

54. Mahindawansha, A.; Breuer, L.; Chamorro, A.; Kraft, P. High-frequency water isotopic analysis using an automatic water sampling system in rice-based cropping systems. *Water* **2018**, *10*, 1327. [CrossRef]

55. Hamada, H.; Komae, T. Analysis of recharge by paddy field irrigation using ^{222}Rn concentration in groundwater as an indicator. *J. Hydrol.* **1998**, *205*, 92–100. [CrossRef]

Article

Application of Stable Water Isotopes to Improve Conceptual Model of Alluvial Aquifer in the Varaždin Area

Tamara Marković [1,*], Igor Karlović [1], Melita Perčec Tadić [2] and Ozren Larva [1]

[1] Croatian Geological Survey, 10 000 Zagreb, Croatia; ikarlovic@hgi-cgs.hr (I.K.); olarva@hgi-cgs.hr (O.L.)
[2] Meteorological and Hydrological Service, 10 000 Zagreb, Croatia; melita.percec.tadic@cirus.dhz.hr
* Correspondence: tmarkovic@hgi-cgs.hr

Received: 9 December 2019; Accepted: 28 January 2020; Published: 30 January 2020

Abstract: To understand groundwater flow and geochemical processes within an aquifer, it is necessary to set up a conceptual model of the aquifer. To accomplish this, different methods are used, and one of them is an isotopic technique. The study area is located in the Varaždin area (NW Croatia). The aquifer represents the main source of potable water for the town of Varaždin and the surrounding settlements. The conceptual model of the alluvial aquifer has to be set up prior to creating a groundwater flow and transport model. Measurements of ratios $\delta^{18}O$ and $\delta^{2}H$ in ground- and surface waters and precipitation samples were carried out. The relationship between ratios $\delta^{18}O$, $\delta^{2}H$, and d-excess for local precipitation in the study area showed that precipitation originates from the Atlantic air masses, although during the colder periods of the year, influence of the Mediterranean air masses was not negligible. The monitored period was warmer and wetter than average. Evaporation was observed at all monitored surface waters, but the largest rate was at the location of a gravel pit in Šijanec. The isotopic composition of the precipitation and groundwater showed a good correlation due to the isotopic homogenization of groundwater along the flow path.

Keywords: stable isotopes; deuterium and oxygen-18; hydrogeological conceptual model; alluvial aquifer; Varaždin area

1. Introduction

To understand groundwater flow and geochemical processes within an aquifer, it is necessary to set up a conceptual model of it. Different methods are used, and one of them is an isotopic technique, which is often used successfully to help elucidate hydrological studies [1]. Knowledge about the isotopic ratios of oxygen ($\delta^{18}O$) and hydrogen ($\delta^{2}H$) in atmospheric precipitation and groundwater is important for hydrological, hydrogeological, climatological, and meteorological applications [1–8] because it can provide information on the mean recharge elevation of the aquifer, the mean residence time, water–rock interactions, etc.

The study area is located in the Varaždin area (NW Croatia). The aquifer represents the main source of potable water for the town of Varaždin and the surrounding settlements. The favorable climate, topography, and available groundwater have insured intensive agricultural practices involving the application of large amounts of synthetic fertilizers and manure that have subsequently led to high nitrate concentrations in the Varaždin aquifer. High concentrations of nitrate have caused the shutting down of the Varaždin pumping site. To determine the behavior of nitrates, a groundwater flow and transport model will be used. A conceptual model of the alluvial aquifer, therefore, has to be set up, and to improve this model, measurements of $\delta^{18}O$ and $\delta^{2}H$ in the ground- and surface waters and precipitation samples were carried out.

The research is still ongoing and, in this paper, only the results of two-year measurements are elaborated in detail. The main goal of this paper is to provide an overview of the spatial and temporal variability in $\delta^{18}O$ and $\delta^{2}H$ values in precipitation and in the ground and surface waters, and this information will be used to determine recharge areas.

2. Geological, Hydrogeological, and Climatic Setting of the Study Area

2.1. Geological and Hydrogeological Setting of the Study Area

The study area of the Varaždin aquifer system is located in NW Croatia, upstream of the town of Varaždin in the valley of the Drava River and it covers an area of approximately 200 km^2 (Figure 1). The Varaždin alluvial aquifer is composed of sediments of the Quaternary age, deposited during the Pleistocene and Holocene eras as a result of accumulation processes of the Drava River [9]. The alluvial deposits consist primarily of gravel and sand with occasional lenses and interbeds of silt and clay (Figure 2). The thickness of the aquifer varies from less than 5 m in the NW part to about 65 m in the SE part of the study area (Figure 2, cross-section A–A'). The Varaždin aquifer is an unconfined aquifer, and the groundwater is in direct contact with the surface water: the Drava River and the Plitvica stream, the derivation channel, the accumulation lake Varaždin (Varaždin Lake), and the gravel pit Šijanec (Figure 2). The direction of groundwater flow is generally NW–SE and is parallel to the direction of the Drava River flow. The covering layer of the aquifer is not continuously developed (Figure 2), which makes the aquifer vulnerable to contamination from the surface. Favorable hydrogeological conditions enabled the development of two pumping sites—Varaždin and Vinokovšćak (Figure 2). High concentrations of nitrate have caused the shutting down of the pumping site Varaždin, but Vinokovšćak is still in operation.

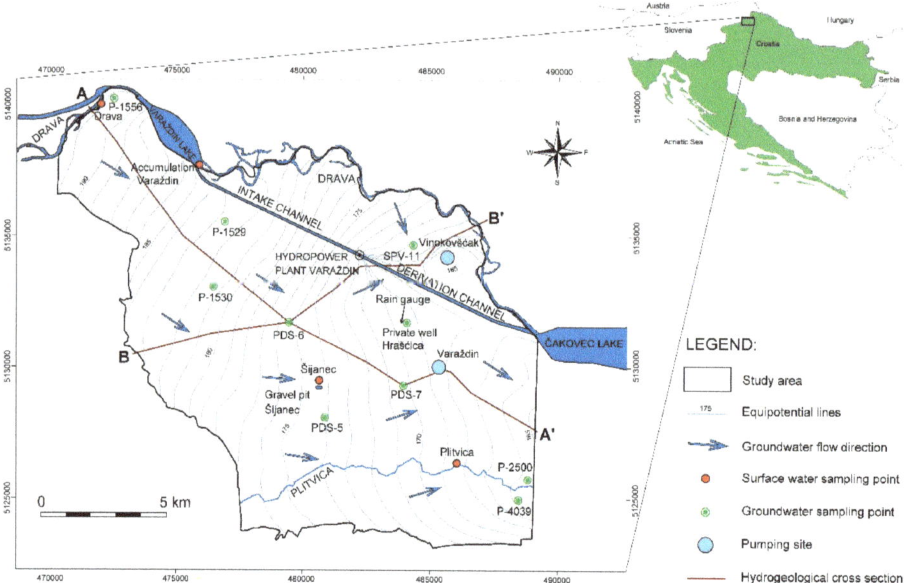

Figure 1. Geographical position of the study area with locations of the sampling points; the transects A–A' and B–B' correspond to the representative hydrogeological cross-sections shown in Figure 2. The general groundwater flow direction is defined by the water heads and stable isotopes in the study area.

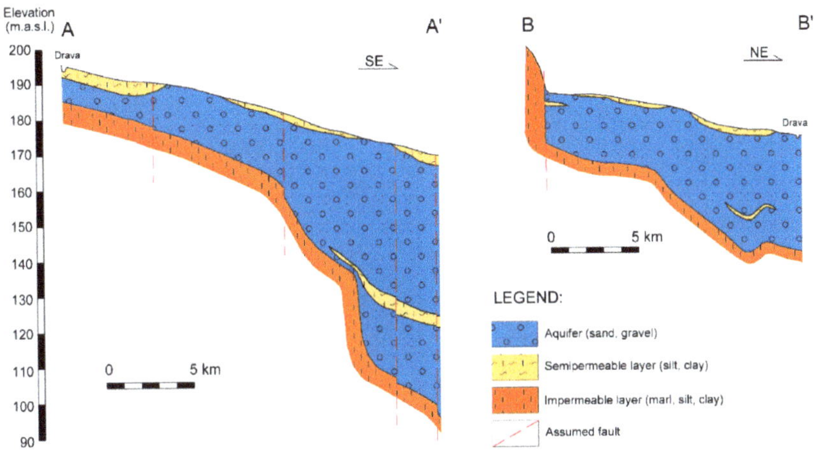

Figure 2. Schematic hydrogeological cross-sections across the study area (modified according to Reference [10]).

2.2. Climatic Setting of the Study Area

The Varaždin meteorological station (46.3° N, 16.137° E, 167 m a.s.l.) and the surroundings have a characteristic precipitation regime with more precipitation during the summer [11]. Because of these features, the local climate is categorized in the Cfb group of the Köppen–Geiger classification system, which is known as "warm-temperate climate" or "marine west coast climate." The study area is the former [12].

The coldest month is January, with average temperatures of 0.0 °C, while July and August are the warmest months with average temperatures of 20.9 and 20.1 °C, respectively (Table 1). Mean annual temperature is 10.6 °C. The inter-annual variability is the largest in February and January (standard deviation (sd) in Table 1), meaning that average differences from the mean are largest at the end of the winter.

Table 1. Monthly and annual average air temperature in Varaždin. Statistics in rows refer to maximum, mean, and minimum average monthly and annual temperature in the 1981–2010 period, and the standard deviation (sd).

Values	01	02	03	04	05	06	07	08	09	10	11	12	Annual
max (°C)	5.8	6.1	9.9	14.0	18.7	23.8	22.8	24.5	18.0	13.8	9.5	4.7	12.1
mean (°C)	0.0	1.6	6.1	10.9	16.0	19.1	20.9	20.1	15.6	10.6	5.3	1.1	10.6
sd (°C)	2.5	3.2	2.1	1.4	1.4	1.4	1.2	1.5	1.3	1.5	2.3	1.8	0.8
min (°C)	−6	−4.5	0.4	8.0	12.2	16.7	18.5	18.0	12.5	8.1	0.8	−2.7	9.0

According to the data from the last climate normal period (1981–2010), the lowest average precipitation amounts were during the cold part of the year, and in January the mean precipitation was 38.7 mm (Table 2). From June to September, the precipitation amounts were on average larger than 80 mm, with the highest amount in September (98.3 mm) and the second-highest in June (96.1 mm). The average annual precipitation was 832 mm. Inter-annual variability was largest in January (coefficient of variation (cv) = 0.81).

Table 2. The monthly and annual precipitation sum in Varaždin. Statistics in rows refer to maximum, mean, and minimum monthly and annual precipitation in the 1981–2010 period, the standard deviation (sd), and the coefficient of variation (cv).

Values	01	02	03	04	05	06	07	08	09	10	11	12	Annual
max (mm)	145.4	124.6	100.6	121.3	144.2	199.9	183.7	211.7	186.1	202	181.5	169.9	1200.3
mean (mm)	38.7	40.8	54.9	61.0	67.9	96.1	81.9	87.2	98.3	78.0	68.0	59.1	832.0
sd (mm)	31.4	26.7	26.6	34.1	31.1	40.3	41.0	53.9	46.3	48.0	42.0	35.8	131.4
cv	0.81	0.65	0.48	0.56	0.46	0.42	0.5	0.62	0.47	0.62	0.62	0.61	0.16
min (mm)	3.3	0.3	2.1	4.9	23.8	32.1	15.3	4.8	25.6	2.4	19.6	17.1	559.7

Based on the long-term data records for 1951–2018, the existence of seasonal trends was tested by a Mann–Kendall test at the 0.05 significance level, and Sen's slope was calculated to determine the trend value [13]. There is significant warming in all seasons, with Sen's slope ranging from 1.8 °C/100 years in autumn to 3.7 °C/100 years in summer (Table 3).

Table 3. Trend analysis of long-term temperature data for the period 1951–2019. A Mann–Kendall p-value of <0.05 indicates a significant trend. Sen's slope is a value of that trend. A Sen's slope p-value of <0.5 indicates a significant value of a trend.

Season	Mann–Kendall *p*-Value	Sen's Slope (°C/100y)	Sen's Slope *p*-Value
Autumn	1.26×10^{-2}	1.83×10^{-2}	1.26×10^{-2}
Spring	1.49×10^{-6}	3.30×10^{-2}	1.49×10^{-6}
Summer	2.89×10^{-9}	3.67×10^{-2}	2.89×10^{-9}
Winter	1.75×10^{-3}	3.10×10^{-2}	1.75×10^{-3}

There are no statistically significant changes in seasonal monthly precipitation in the 1951–2019 period (Table 4), even though there is a slight indication of drying in summer and in spring, while autumn and winter show a slight tendency of becoming wetter.

Table 4. Trend analysis of long-term precipitation data for the period 1951–2019. Parameters are the same as in Table 3.

Season	Mann–Kendall *p*-Value	Sen's Slope	Sen's Slope *p*-Value
Autumn	1.64×10^{-1}	8.00×10^{-1}	1.64×10^{-1}
Spring	4.88×10^{-1}	-2.66×10^{-1}	4.88×10^{-1}
Summer	8.46×10^{-2}	-9.72×10^{-1}	8.46×10^{-2}
Winter	9.46×10^{-1}	1.36×10^{-2}	9.46×10^{-1}

3. Materials and Methods

3.1. Water Sampling

Monthly composite precipitation was sampled in the Miko family courtyard in the village Hrašćica (46.3° N, 16.292° E; 177 m a.s.l.) in the period from June 2017 until June 2019. In the field, sample was poured into a 1 L plastic bottle with a tight-fitting cap.

In addition, ground- and surface water sampling was conducted on a monthly basis in the period from June 2017 to June 2019 for chemical and isotopic analyses. Ten observation wells (nine of them are located in the recharge area of the Varaždin pumping site and one in the recharge area of the Vinokovšćak pumping site) and four surface waters (Drava River, the Plitvica stream, Varaždin Lake, which is an accumulation, and a gravel pit in Šijanec) were sampled. The water depth was measured prior to pumping. The observation wells were sampled after stabilization of the parameters EC (electrical conductivity), T (water temperature), pH, and O_2 (dissolved oxygen content). Depths of the observation wells are given in Table 5. Samples were poured into a 50 mL plastic bottle with a tight-fitting cap. All samples were measured in the laboratory immediately upon returning from the field.

Table 5. Depths of the observation wells.

Observation Well	Elevation (m a.s.l.)	Depth (m)
Private well Hraščica	176.00	15.0
PDS-5	178.36	31.0
PDS-6	184.07	25.0
PDS-7	175.71	42.5
P-1529	187.32	8.0
P-1530	183.72	7.5
P-1556	193.03	15.6
P-2500	167.81	5.20
P-4039	167.76	8.0
SPV-11	177.69	40.0

Meteorological parameters (precipitation and air temperature) used here are from the main Varaždin meteorological station, located in the vicinity of the installed rain gauge.

3.2. Stable Isotope Analyses

The δ^{18}O and δ^{2}H were determined using Picarro L2130i (Santa, Clara, USA) in the Hydrochemical Laboratory of the Croatian Geological Survey. The instrument uses CRDS (Cavity Ring-Down Spectroscopy) technology [14]. All measurements were checked with Picarro's standards (Depleted $-29.6 \pm 0.2\ \delta^{18}$O; $-235 \pm 1.8\ \delta^{2}$H; Mid $-20.6 \pm 0.2\ \delta^{18}$O; $-159 \pm 1.3\ \delta^{2}$H; Zero $0.3 \pm 0.2\ \delta^{18}$O; $1.8 \pm 0.9\ \delta^{2}$H), which were checked periodically against the International Atomic Energy Agency (IAEA) standards: Vienna Standard Mean Ocean Water 2 (VSMOW2) and Standard Light Antarctic Precipitation 2 (SLAP2). Measurement precision was $\pm\ 0.3\ ^{o}/_{oo}$ for δ^{18}O and $\pm\ 1\ ^{o}/_{oo}$ for δ^{2}H.

It is generally known that all isotopic results are expressed as per the international measurement standard, VSMOW2 [15,16].

A global relationship between δ^{2}H and δ^{18}O has been observed [16] and called the Global Meteoric Water Line (δ^{2}H = 8·δ^{18}O + 10).

The deuterium excess (d-excess [17]) was calculated for each sample as follows:

$$\text{d-excess}\ (^{o}/_{oo}) = \delta^{2}\text{H} - 8\ \delta^{18}\text{O}.$$

In 2003, researchers verified this value by using global maps derived by interpolation from more than 340 stations [18] and, in 2005, this was reconfirmed using IAEA-GNIP datasets at 410 stations [19]. Higher values of d-excess are caused by intense evaporation of seawater in conditions of moisture deficit [20].

The local meteoric water line (LMWL) has been calculated using three types of linear regression analysis, two of which are recommended by the IAEA [21]: ordinary least squares regression (OLSR) and reduced major axis (RMA) regression. The new one takes into account the amount of precipitation, using precipitation weighted least squares regression (PWLSR) [22].

4. Results and Discussion

4.1. Precipitation and Temperature

The climatological conditions from June 2017 to June 2019, the period of the isotope sampling, were compared to a climate normal 1981–2010 (Figure 3).

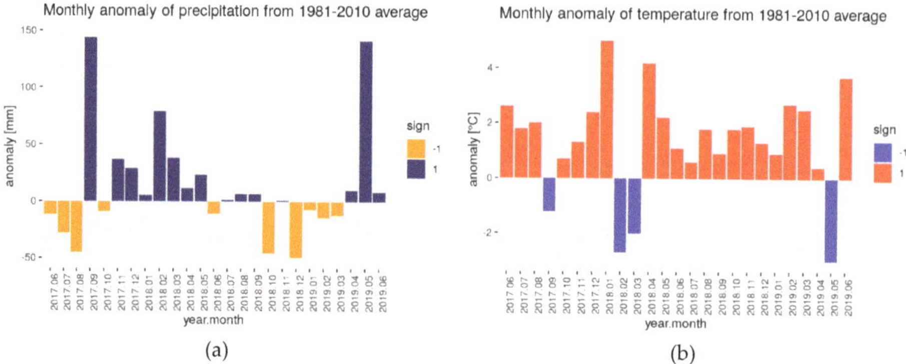

Figure 3. Monthly anomaly of precipitation (**a**) and air temperature (**b**) in the period from June 2017 to June 2019 based on 1981–2010 climate normal.

The summer of 2017 was drier and warmer from the average of 1981–2010. Autumn 2017 was mostly wetter than normal, with September 2017 being the wettest month of the period, and it was also colder than average in September. This period, with larger precipitation than average, continued until May 2018, and it was also warmer, except in February and March 2018. From June 2018 to March 2019, precipitation was mostly around or smaller than average, and it was warmer. May 2019 was the second wettest month during this period, and it was also colder than average. June 2019 was again warmer than average.

4.2. Stable Isotopes in Precipitation

The mean stable isotope $\delta^{18}O$ and δ^2H values and the associated d-excess are shown in Table 6.

Table 6. Yearly averaged isotopic composition of precipitation at the rain gauge in Hraščica (177 m a.s.l.).

Year	$\delta^{18}O$ ($^o/_{oo}$)	δ^2H ($^o/_{oo}$)	d-excess ($^o/_{oo}$)
2017 (VI-XII)	−7.78	−52.9	8.7
2018 (I-VI)	−9.96	−70.4	9.3
2018 (VII-XII)	−8.09	−57.1	7.6
2019 (I-VI)	−9.68	−67.3	10.1

The stable isotope $\delta^{18}O$ values varied from −14.91 to −4.5 $^o/_{oo}$, and δ^2H varied from −108.1 to −25.1 $^o/_{oo}$ (Figure 4, Table S1). The lowest $\delta^{18}O$ and δ^2H values were observed in winter and highest were observed in summer. The d-excess varied from 4.1 to 12.9 $^o/_{oo}$. The d-excess shows the influence of the Atlantic air masses. Nevertheless, the influence of the Mediterranean air masses in the study area was observed during the autumn and winter months (Figure 4). The Mediterranean air masses (precipitation) are characterized by a higher d-excess than the Atlantic air masses [20]. This was observed in References [23,24] in the continental part. Atypical climatological conditions during the observed period had influenced variations of monthly isotopic composition. A sudden change in the air temperature and/or precipitation amount during the season influenced the variation of the monthly isotopic composition of the rain. For example, May 2019 was colder and wetter than average (even than May 2018), and $\delta^{18}O$ and δ^2H values were automatically more negative. In addition, the lowest $\delta^{18}O$ and δ^2H values were measured in the coldest month, which was February 2018 (Figures 3 and 4). It was observed that the isotopic composition of the precipitation in the study area reflects climatological conditions well.

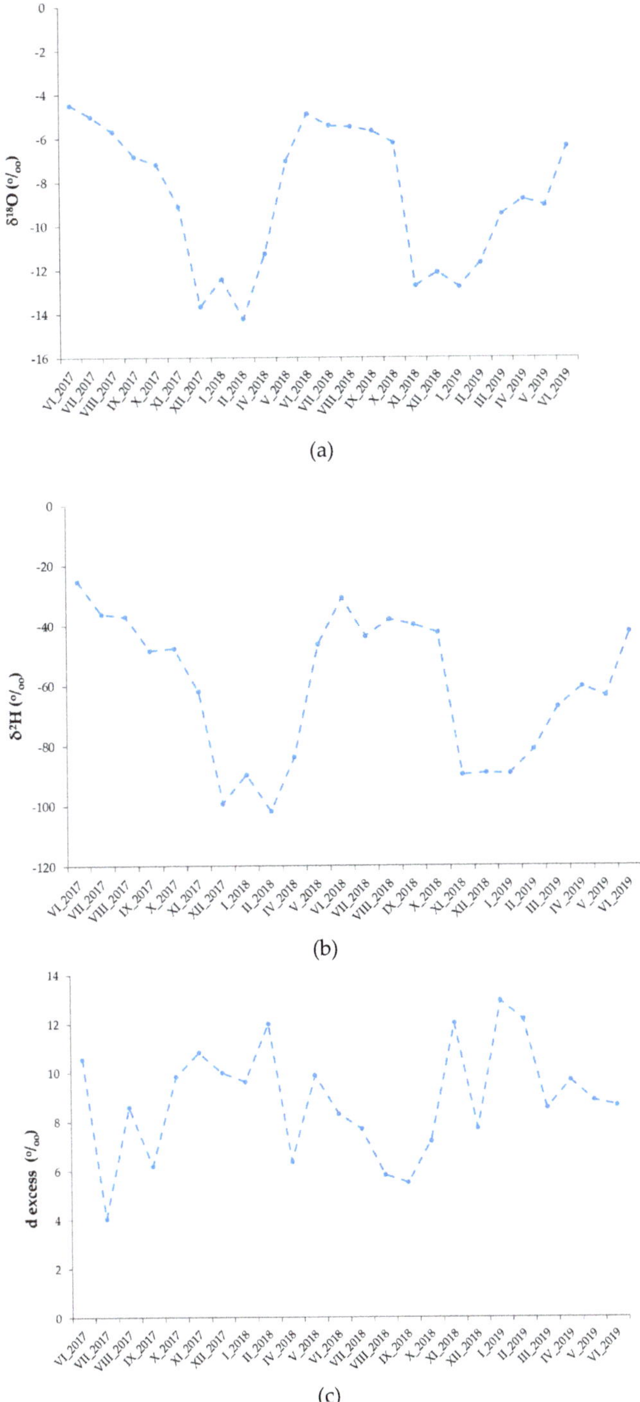

(a)

(b)

(c)

Figure 4. Monthly variations of (**a**) $\delta^{18}O$ ($^o/_{oo}$) values, (**b**) δ^2H ($^o/_{oo}$) values, and (**c**) d-excess ($^o/_{oo}$) in precipitation at the rain gauge in Hrašćica.

The measured stable isotope $\delta^{18}O$ and δ^2H values were weighted by the amount of precipitation at the Varaždin meteorological station for the observed period. However, there were no large differences between the measured stable isotope $\delta^{18}O$ and δ^2H values and weighted by the amount of precipitation. Because of this, they are not discussed here.

The calculated LMWL for the period from June 2017 to June 2019 is:

$$\text{OLSR } \delta^2H = (7.54 \pm 0.12)\, \delta^{18}O + (5.00 \pm 1.00),\ n = 23$$

$$\text{RMA } \delta^2H = (7.56 \pm 0.11)\, \delta^{18}O + (5.17 \pm 1.04),\ n = 23$$

$$\text{PWLSR } \delta^2H = (7.55 \pm 0.13)\, \delta^{18}O + (4.85 \pm 1.13),\ n = 23.$$

It was observed that all three methods yielded a very similar slope value and axis intercept (b value) of the LMWL, which was supported by very similar measured and weighted values.

In addition, calculated data were compared with data published in Reference [25]. There is a difference between these two slopes values for 0.09 $^o/_{oo}$, and it can be concluded that the OLSR values calculated from the measured data and the published meteoric water line of the study area are not different in terms of slope. However, there is a large difference between these two lines in the axis intercept values for 2.6 $^o/_{oo}$, and the published LMWL is slightly below the measured one (Figure 5). There are several reasons for that: shorter monitored period in our research; very untypical and extreme climatological conditions during our monitored period; and different measurement techniques (our samples were measured using CRDS technology and published were measured using Isotope Ratio Mass Spectrometry (IRMS) technology).

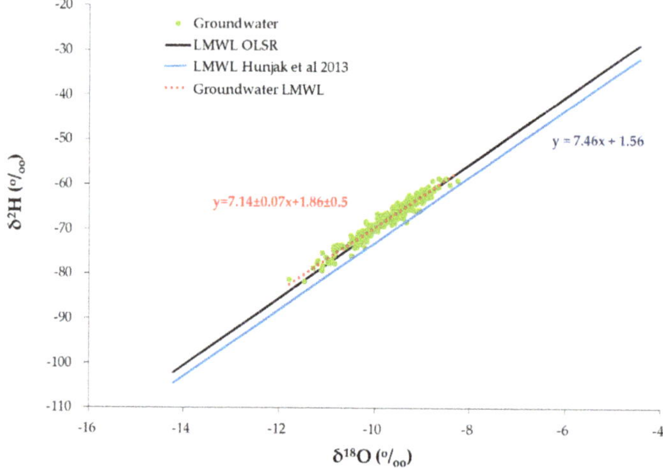

Figure 5. Distribution of groundwater $\delta^{18}O$ ($^o/_{oo}$) and δ^2H ($^o/_{oo}$) values around the local meteoric water line (LMWL).

4.3. Stable Isotopes in Ground and Surface Waters

The minimum, maximum, and average isotopic composition of the surface- and groundwaters are given in Tables 7 and 8 together with their average d-excess.

Table 7. Minimum, maximum, and averaged isotopic composition of the groundwater by sampled well.

Observation Well	$\delta^{18}O$ (°/_{oo})				δ^2H (°/_{oo})				d-excess (°/_{oo})
	min	max	average	sd	min	max	average	sd	average
Private well Hraščica	−11.47	−9.06	−10.00	0.65	−81.7	−65.4	−70.4	4.7	9.6
P-1529	−11.20	−8.87	−9.76	0.64	−77.6	−61.2	−68.2	4.4	9.9
P-1556	−10.85	−8.62	−9.62	0.64	−76.0	−58.8	−66.3	4.8	10.2
SPV-11	−11.79	−9.13	−10.32	0.63	−81.2	−62.3	−71.9	4.1	10.7
P-2500	−10.51	−8.42	−9.31	0.56	−71.3	−58.6	−64.1	3.5	10.4
P-4039	−10.33	−8.26	−9.46	0.54	−71.9	−58.5	−65.4	3.9	10.2
PDS-5	−10.97	−8.76	−9.74	0.61	−77.2	−61.0	−67.4	4.3	10.5
PDS-6	−10.93	−8.95	−9.79	0.55	−77.7	−61.1	−68.2	4.6	10.1
PDS-7	−10.97	−8.91	−9.79	0.58	−77.5	−60.9	−68.1	4.5	10.2

Table 8. Minimum, maximum, and averaged isotopic composition of surface water.

Observation Point	$\delta^{18}O$ (°/_{oo})				δ^2H (°/_{oo})				d-excess (°/_{oo})
	min	max	average	sd	min	max	average	sd	average
Plitvica stream	−10.15	−7.86	−9.11	0.52	−69.2	−56.9	−63.3	2.8	9.6
Drava River	−11.64	−8.99	−10.21	0.73	−80.1	−61.1	−70.7	4.8	11.0
Varaždin Lake	−12.12	−8.10	−10.33	0.87	−81.7	−59.1	−72.4	5.2	10.2
Gravel pit in Šijanec	−9.47	−3.36	−6.67	1.31	−65.7	−38.8	−51.1	6.5	2.3

The measured $\delta^{18}O$ values in the groundwater varied from −11.47 to −8.26 °/_{oo}, and the δ^2H values varied from −81.7 to −58.5 °/_{oo} (Table 7). The measured $\delta^{18}O$ values in the surface water varied from −12.12 to −3.36 °/_{oo}, and the δ^2H values varied from −81.7 to −38.8 °/_{oo} (Table 8).

The correlation between the $\delta^{18}O$ and δ^2H measured values of the groundwater is shown in Figure 5 and indicates that this relationship has a slope of 7.14. Using a Student's *t*-test according to Reference [26], a good relationship between groundwater and precipitation was observed. Generally, an isotope relationship between $\delta^{18}O$ and δ^2H with a slope of about 8 is normally observed for precipitation [16]. Since the relationship between the isotopic composition of precipitation and groundwater is good, it can be concluded that groundwater is recharged by precipitation. Values that are slightly more negative were measured in the SPV-11 well and the private well, while at observation wells PDS-5, PDS-6, PDS-7, and P-1529, values are almost identical (Table 7). The highest $\delta^{18}O$ and δ^2H values were measured at observation wells P-4039 and P-2500 (Table 7). The calculated average d-excess values varied from 9.6 to 10.7 °/_{oo}, indicating the influence of recharge by precipitation with signatures of the Atlantic air masses and good homogenization of groundwater along the flow path. This was observed at SPV-11, PDS-6, PDS-7, the private well, P-1530, and P-1529. However, depending on hydrodynamic conditions (low/high water levels), the vicinity of the river or lake, and the depth of the observation well, it was observed that wells, especially the shallower ones and/or those closer to the river and lake, showed high variation in d-excess values. These values were higher than 11 °/_{oo}, indicating recharge by surface waters and faster recharge by precipitation. This was observed at P-1556, PDS-5, P-2500, and P-4039.

The measured $\delta^{18}O$ and δ^2H values of the surface waters distributed around the LMWL shown in Figure 6 indicate a relationship between $\delta^{18}O$ and δ^2H for surface waters with slopes of 5.73 at Varaždin Lake, 6.32 at Drava River, and 4.77 at the gravel pit in Šijanec, indicating an influence of evaporation. A slope from 4 to 6 is attributed to waters with a significant rate of evaporation relative to the input [16]. It was observed that the evaporation process was strongest at the location gravel pit in Šijanec (Figure 7). Nevertheless, the gravel pit was used for fish farming. Because of this activity (resulting in an extra nutrient load due to fish feeding), a low water level, a high load of nutrients, high temperatures, and algae bloom occurred every summer, which had a significant influence on the isotopic and chemical features of this water. The winter-measured values of the δ^2H and $\delta^{18}O$ of the Drava River and the summer-measured values of Varaždin Lake are above the LMWL (Figure 6). For the Drava River, this can be explained by the fact that the larger part of the recharge area of the Drava River is situated far upstream of the study area and is under the influence of a different climate, and the influence of the recharge area in the study area is small. Varaždin Lake is recharged by the Drava River, especially in the late winter and springtime when isotopic values are more negative in the river. Since the lake has a high volume, turnover in the lake takes some time. In addition, the Plitvica stream, like the Drava River, has its recharge area in a mountain area where the climate is different, and because of that, more negative values are measured in the wintertime. During the late spring/summer period, the discharge of the stream is low. In the watercourse of the stream, small connected ponds are formed where evapotranspiration is present and, because of that, some values are below the LMWL.

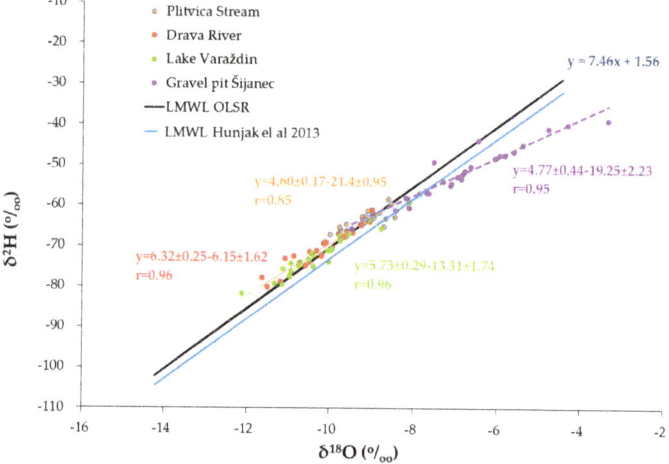

Figure 6. Distribution of surface waters $\delta^{18}O$ ($^o/_{oo}$) and δ^2H ($^o/_{oo}$) values around the LMWL.

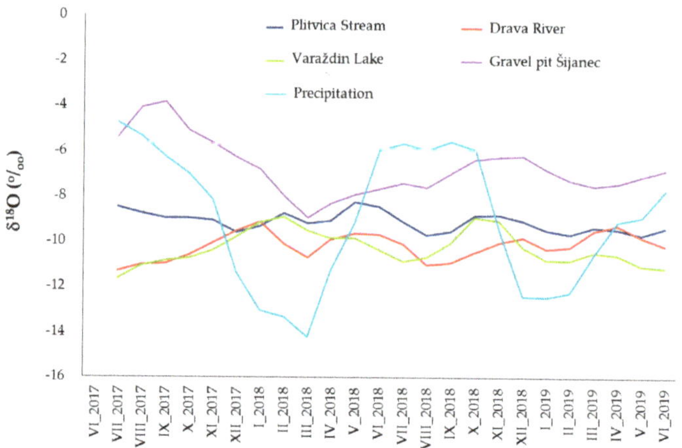

Figure 7. Distribution of $\delta^{18}O$ ($^o/_{oo}$) values in surface waters over the monitored time.

To connect the measured values, a simplified statistical correlation method was used, and results are shown in the correlation matrix in Table 9.

Table 9. Correlation matrix between ground and surface waters for $\delta^{18}O$ (‰) values.

Matrix	Private Well	P-1529	P-1556	SPV-11	P-2500	P-4039	PDS-5	PDS-6	PDS-7	Plitvica Stream	Drava River	Varaždin Lake	Garvel Pit Šijanec
Private well	1.00												
P-1529	0.93	1.00											
P-1556	0.96	0.91	1.00										
SPV-11	0.76	0.76	0.80	1.00									
P-2500	0.91	0.94	0.89	0.73	1.00								
P-4039	0.75	0.78	0.78	0.66	0.88	1.00							
PDS-5	0.93	0.94	0.93	0.78	0.92	0.78	1.00						
PDS-6	0.95	0.98	0.92	0.73	0.93	0.77	0.93	1.00					
PDS-7	0.93	0.97	0.95	0.70	0.93	0.82	0.96	0.96	1.00				
Plitvica Stream	0.04	0.17	0.17	0.00	0.07	0.03	0.18	0.17	0.22	1.00			
Drava River	0.67	0.59	0.58	0.36	0.48	0.45	0.48	0.62	0.53	−0.04	1.00		
Varaždin Lake	0.56	0.48	0.57	0.85	0.55	0.53	0.51	0.47	0.54	−0.31	0.51	1.00	
Gravel pit Šijanec	-0.63	−0.62	−0.67	−0.67	−0.63	−0.48	−0.66	−0.56	−0.62	−0.29	−0.20	−0.29	1.00

No statistical connection was observed between either the groundwater, the waters of Drava River, Varaždin Lake, or the Plitvica stream with water from the gravel pit in Šijanec. This is partly because there was no observation immediately downstream from the gravel pit. Moreover, it has a very small volume, and its influence is thus limited to its immediate surroundings. Furthermore, the gravel pit is not connected to the river, stream, or lake; consequently, the isotopic composition differs (Figure 1). In addition, a very weak correlation was observed between the waters from the P-2500 and P-4039 wells, which are in the vicinity of the Plitvica stream. This weak correlation is attributed to the drainage roll of the Plitvica stream in this part of the aquifer. A higher correlation was observed between both Varaždin Lake and the Drava River waters and the groundwater of the observation wells. A high correlation between observation wells on the right side of the intake/drain channels indicates a homogenization of the groundwater source (a mixture of the precipitation, river, and lake waters). The left side is different because of the influence of Varaždin Lake. Namely, the Drava River flows from the lake and has the same isotopic composition as the lake water (Figure 1). The river recharges the aquifer as a consequence of groundwater abstraction at the Vinokovšćak pumping site (Figure 1), and the influence of the local precipitation is minor.

5. Conclusions

Although $\delta^{18}O$ and δ^2H values of ground- and surface waters and precipitation were measured for only 24 months, the following conclusions are proposed:

(a) Meteorological conditions during the observed period were quite different from the average. The summer of 2017 was drier and warmer, while the autumn was wet and cold. This period had more precipitation than average, and this continued until May 2018. It was also warmer, except in February and March 2018. From June 2018 to March 2019, precipitation was mostly around or smaller than average, and it was warmer. May 2019 was the second wettest month during this period and was also colder than average. June 2019 was again warmer than average.

(b) The isotopic composition of the local precipitation in the study area varied according to the variable meteorological conditions and relationship between $\delta^{18}O$ and δ^2H, and the d-excess showed that precipitation originated from the Atlantic air masses, although during the colder parts of the year, the influence of the Mediterranean air masses was not negligible.

(c) Evaporation was observed at all surface waters, but the largest rate was at the location of the gravel pit in Šijanec. Moreover, it has a very small volume; therefore, its influence is limited to the immediate surroundings of the aquifer.

(d) The isotopic composition of precipitation and groundwater showed a good correlation due to the isotopic homogenization of groundwater along the flow path.

(e) The isotopic composition of the groundwater source on the right side of the intake/drain channels indicates a homogenization of the groundwater source (a mixture of precipitation, river, and lake waters), whereas the left side is different because of the higher influence of Varaždin Lake.

Supplementary Materials: The following are available online at http://www.mdpi.com/2073-4441/12/2/379/s1, Table S1: Determined isotopic composition of the precipitation by month for monitored period.

Author Contributions: T.M. contributed to conceptualization, isotopic data interpretation, and writing. I.K. contributed to the hydrogeology and geology description of the study area and graphical editing. M.P.T. contributed to statistical analysis and visualization of climatological data. O.L. contributed by polishing the article. All authors have read and agreed to the published version of the manuscript.

Funding: This research was funded by the Croatian Scientific Foundation (HRZZ), grant number HRZZ-IP-2016-06-5356.

Acknowledgments: The authors would like to thank the Miko family for taking care of the rain gauge, as well as the Croatian Meteorological and Hydrological Service for providing climatological data for the Varaždin meteorological station.

Conflicts of Interest: The authors declare that there is no conflict of interest.

References

1. Dotsika, E.; Lykoudis, S.; Poutoukis, D. Spatial distribution of the isotopic composition of precipitation and spring water in Greece. *Global Planet. Change* **2010**, *71*, 141–149. [CrossRef]
2. Rozanski, K.; Araguas-Araguas, L.; Gonfiantini, R. Isotopic patterns in modern global precipitation. Continental Isotopic Indicators of Climate. *Am. Geophys. Union Monogr.* **1993**, *78*, 1–36.
3. Vreča, P.; Krajcar Bronić, I.; Horvatinčić, N.; Barešić, J. Isotopic characteristics of precipitation in Slovenia and Croatia: Comparison of continental and maritime stations. *J. Hydrol.* **2006**, *330*, 457–469. [CrossRef]
4. Kanduč, T.; Mori, N.; Kocman, D.; Stibilj, V.; Grassa, F. Hydrogeochemistry of Alpine springs from North Slovenia: Insights from stable isotopes. *Chem. Geol.* **2012**, *300/301*, 40–54. [CrossRef]
5. Marković, T.; Brkić, Ž.; Larva, O. Using hydrochemical data and modelling to enhance the knowledge of groundwater flow and quality in an alluvial aquifer of Zagreb. Croatia. *Sci. of the Tot. Environ.* **2013**, *458–460*, 508–516. [CrossRef] [PubMed]
6. Lukač Reberski, J.; Marković, T.; Nakić, Z. Definition of the river Gacka springs subcatchment areas on the basis of hydrogeological parameters. *Geologia Croatica.* **2013**, *66*, 39–53. [CrossRef]
7. Gross, E.; Andrews, S.; Bergamaschi, B.; Downing, B.; Holleman, R.; Burdick, S.; Durand, D. The Use of Stable Isotope-Based Water Age to Evaluate a Hydrodynamic Model. *Water* **2019**, *11*, 2207. [CrossRef]
8. Cervi, F.; Dadomo, A.; Martinelli, G. The Analysis of Short-Term Dataset of Water Stable Isotopes Provides Information on Hydrological Processes Occurring in Large Catchments from the Northern Italian Apennines. *Water.* **2019**, *11*, 1360. [CrossRef]
9. Prelogović, E.; Velić, I. Kvartarna tektonska aktivnost u zapadnom dijelu Dravske potoline. *Geol. vjesnik* **1988**, *41*, 237–253.
10. Larva, O. Ranjivost vodonosnika na priljevnom području varaždinskih crpilišta (Aquifer vulnerability at catchment area of Varaždin pumping sites—in Croatian). Ph.D. Thesis, University of Zagreb, Zagreb, Croatia, December 2008.
11. Zaninović, K.; Gajić-Čapka, M.; Perčec Tadić, M. *Klimatski atlas Hrvatske. Climate atlas of Croatia: 1961–1990, 1971–2000*; Državni hidrometeorološki zavod: Zagreb, Croatia, 2008; p. 200.
12. Nimac, I.; Perčec Tadić, M. *New 1981–2010 Climatological Normals for Croatia and Comparison to Previous 1961–1990 and 1971–2000 Normals. Proceedings from GeoMLA Conference*; University of Belgrade—Faculty of Civil Engineering: Belgrade, Serbia, 2016; pp. 79–85.
13. Wilks, D.S. *Statistical Methods in the Atmospheric Sciences*, 2nd ed.; Academic Press: London, UK, 2006; p. 704.
14. Available online: http://www.picarro.com/technology/cavity_ring_down_spectroscopy (accessed on 3 June 2019).
15. Clark, I.D.; Fritz, P. *Environmental Isotopes in Hydrogeology*; CRC Press: Boca Raton, FL, USA, 2013; ISBN 978-1-4822-4291-1.
16. Craig, H. Isotopic Variations in Meteoric Waters. *Science* **1961**, *133*, 1702–1703. [CrossRef] [PubMed]
17. Dansgaard, W. Stable isotopes in precipitation. *Tellus* **1964**, *16*, 436–468. [CrossRef]
18. Bowen, G.J.; Revenaugh, J. Interpolating the isotopic composition of modern meteoric precipitation. *Water Resour. Res.* **2003**, *39*, 10. [CrossRef]
19. Gourcy, L.L.; Groening, M.; Aggarwal, P.K. Stable oxygen and hydrogen isotopes in precipitation. *Isotopes in the Water Cycle: Past. Present and Future of Developing Science.* Dordrecht, The Netherlands, 2005 Aggarwal, P.K.; Gat, J.R. and Froehlich, K.F.O., Eds.; Springer, 39–51. Available online: https://www.springer.com/gp/book/9781402030109 (accessed on 30 January 2020). [CrossRef]
20. Gat, J.R.; Carmi, I. Evolution of the isotopic composition of the atmospheric water in the Mediterranean Sea area. *J. Geophys. Res.* **1970**, *75*, 3039–3048.
21. IAEA. Statistical treatment of data on environmental isotopes in precipitation. *Technical Report Series 331. International Atomic Energy Agency, Vienna.* 1992. Available online: https://www.iaea.org/publications/1435/statistical-treatment-of-data-on-environmental-isotopes-in-precipitation (accessed on 30 January 2020).
22. Hughes, C.E.; Crawford, J. 2012: A new precipitation weighted method for determining the meteoric water line for hydrological applications demonstrated using Australian and global GNIP data. *J. Hydrol.* **2012**, *464–465*, 344–351. [CrossRef]
23. Vreča, P.; Krajcar Bronić, I.; Leis, A. Demšar. M. Isotopic composition of precipitation at the station Ljubljana (Reaktor). Slovenia—period 2007–2010. *Geologija* **2014**, *57*, 217–230.

24. Bottyán, E.; Czuppon, G.; Weidinger, T.; Haszpra, L.; Kármán, K. Moisture source diagnostics and isotope characteristics for precipitation in east Hungary: implications for their relationship. *Hydrol. Sci. J.* **2017**, *62*, 2049–2060.

25. Hunjak, T.; Lutz, H. O.; Roller-Lutz, Z. Stable isotope composition of the meteoric precipitation in Croatia. *Isot. Environ. Health Stud.* **2013**, *49*, 336–345. [CrossRef] [PubMed]

26. Boschetti, T.; Cifuentes, J.; Iacumin, P.; Selmo, E. Local Meteoric Water Line of Northern Chile (18 S–30 S): An Application of Error-in-Variables Regression to the Oxygen and Hydrogen Stable Isotope Ratio of Precipitation. *Water* **2019**, *11*, 791. [CrossRef]

 *water*

Article

Can Soil Hydraulic Parameters be Estimated from the Stable Isotope Composition of Pore Water from a Single Soil Profile?

Alexandra Mattei [1,2,*], Patrick Goblet [2], Florent Barbecot [1], Sophie Guillon [2], Yves Coquet [3] and Shuaitao Wang [2]

[1] GEOTOP, Département des sciences de la Terre et de l'atmosphère—Université du Québec à Montréal, CP8888 succ Centre-Ville, Montréal, QC H3C 3P8, Canada; barbecot.florent@uqam.ca

[2] MINES ParisTech, PSL University, Centre for geosciences and geoengineering, 35 rue Saint-Honoré, 77300 Fontainebleau, France; patrick.goblet@mines-paristech.fr (P.G.); sophie.guillon@mines-paristech.fr (S.G.); shuaitao.wang@mines-paristech.fr (S.W.)

[3] Université Paris-Saclay, INRAE, AgroParisTech, UMR ECOSYS, 78850 Thiverval-Grignon, France; yves.coquet@agroparistech.fr

* Correspondence: alexandra.mattei@mines-paristech.fr

Received: 4 October 2019; Accepted: 30 January 2020; Published: 1 February 2020

Abstract: Modeling water and solute transport in the vadose zone for groundwater resource management requires an accurate determination of soil hydraulic parameters. Estimating these parameters by inverse modeling using in situ observations is very common. However, little attention has been given to the potential of pore water isotope information to parameterize soil water transport models. By conducting a Morris and Sobol sensitivity analysis, we highlight the interest of combining water content and pore water isotope composition data in a multi-objective calibration approach to constrain soil hydraulic property parameterization. We then investigate the effect of the sampling frequency of the observed data used for model calibration on a synthetic case. When modeling is employed in order to estimate the annual groundwater recharge of a sandy aquifer, it is possible to calibrate the model without continuous monitoring data, using only water content and pore water isotopic composition profiles from a single sampling time. However, even if not continuous, multi-temporal data improve model calibration, especially pore water isotope data. The proposed calibration method was validated with field data. For groundwater recharge estimate studies, these results imply a significant reduction in the time and effort required, by avoiding long-term monitoring, since only one sampling campaign is needed to extract soil samples.

Keywords: $\delta^2 H$; inverse modeling; vadose zone; sensitivity analysis; soil hydraulic parameters estimation; groundwater recharge

---

1. Introduction

Understanding water and solute movement in the vadose zone is fundamental to many environmental and natural resource issues. This includes the protection of groundwater, which is the main source of drinking water in many regions of the world [1–3], in terms of both quantity and quality. Numerical physically-based models are widely used to explore and predict water flow and solute transport in the vadose zone. The governing flow and transport equations are well-established, and a large number of analytical or numerical solutions have been published over the last four decades [4]. However, a number of numerical and conceptual difficulties remain to be solved, especially for field-scale processes subject to natural boundary conditions. This is typically the case for the estimation of soil hydraulic parameters, because of the pronounced spatial heterogeneity of the vadose zone [5].

Some of these parameters can be measured directly for soil samples in the laboratory. However, measured laboratory-scale properties have little relevance to describing field-scale water flow and transport processes under natural conditions [6,7]. This is due to spatial variability at the field scale which is not captured by limited sampling volumes, and to specific boundary conditions in the laboratory that differ from real-world conditions [8–10]. The inverse modeling approach is an established method that fits model simulations to observed data and results in effective parametrization, which lumps the subscale heterogeneity of the system and describes its behavior at the targeted scale [11]. A key question regarding the inverse modeling approach is whether the measurements are accurate enough and contain enough information to estimate the effective soil hydraulic parameters with the required accuracy.

The majority of field-scale inverse modeling studies have only used information on the water content (e.g., [12,13]). However, in situ observations of water content do not necessarily provide sufficient information to accurately parameterize field-scale soil hydraulic properties [14]. Combining different types of observed variables to calibrate water flow and transport models (e.g., water content and matric potential [15]; bromide concentration and water content [16]; water content and oxygen isotope composition [17]; water content and hydrogen isotope composition [18]; and water content, oxygen isotope composition, and matric potential [10]) has been found to improve parameterization.

In particular, water stable isotope (deuterium (^{2}H) and oxygen-18 (^{18}O)) profiles have been shown to provide valuable insights into hydrological processes in the vadose zone since several hydrological processes, such as infiltration [19], evaporation [20], and root water uptake [21], control the shape of the pore water-stable isotope profiles. The fact that stable isotopes are part of the water molecule, and are therefore extracted (without fractionation [22]) via root water uptake, is particularly helpful for constraining transpiration, which would not be possible with an artificial tracer. [18] demonstrated that water-stable isotope profiles, in combination with soil moisture time series, allow soil models to be calibrated and time-varying site-specific transit time distributions in the vadose zone to be derived.

To date, water content and pore water isotope composition profiles from a single sampling time have not been rigorously tested for their potential to calibrate soil hydraulic properties in the vadose zone in a humid climate. If possible, such an approach would reduce the time and effort required for long-term soil water content measurements, since only one sampling campaign would be necessary to obtain the soil samples required. The main objective of this study is to determine whether the inclusion of pore water isotope data allows a realistic parameterization of soil water transport models without the need for continuous monitoring data.

The METIS code [23], a vadose zone unsaturated/saturated transport model including isotope transport and isotopic fractionation due to evaporation, was used in this study to simulate soil water content and δ^2H isotope data. A sensitivity analysis based on the methods of Morris [24] and Sobol [25] was performed to understand the interactions between soil hydraulic parameters and their impacts on modeled water content and isotope profiles and groundwater recharge calculation, and to highlight the interest of combining different observation types in model calibration. A synthetic case permitted insight into and a discussion of the performances of the calibration methods proposed here, compared with a calibration carried out using continuous monitoring data. Finally, the proposed calibration method was tested on field data. Using soil moisture and isotope profiles from a unique field campaign in a multi-objective approach to optimize model parameterization, a best set of soil hydraulic parameters was determined for two sites in southern Quebec, Canada. The accuracy of the parameterization for each site was assessed based on its ability to reproduce soil moisture and isotope profiles measured during a second campaign, one year later.

2. Materials and Methods

2.1. Site Description and Data Availability

The two sampling sites (SLA and SLB) were located in the Saint-Lawrence Lowlands, 60 km southwest of Montreal, in the Vaudreuil-Soulanges area (eastern Quebec, Canada). The study area has a humid climate with short, hot, and humid summers; cold and snowy winters; and rainy springs and autumns. The average annual precipitation in this area is 960 mm [26], with the average monthly temperature ranging from −11 to 23 °C [26]. The sites were located in a flat area in the southern part of Saint-Lazare (45°23′5.388″ N/74°11′50.316″ W), in the medium sands of the Saint-Lazare glacio-fluvial complex, which is a locally unconfined aquifer. The site SLA is located in grasslands 55 m from the site SLB, which is located in a pine forest. Further site characteristics are presented in Table 1.

At both sites, the unsaturated zone was sampled on 23 November 2017 and 9 November 2018. Approximately 800 g of soil samples was taken with a spatula, with a spacing of 5 cm between the soil surface to a 30 cm depth, and then 10 cm down to a 200 cm depth. Soil samples were stored in air-tight glass jars until water-stable isotope measurements were performed by laser spectrometry (TIWA-DLP-EP) at GEOTOP-UQAM using the direct vapor equilibration method [27]. The gravimetric water content and solid phase density were measured in the laboratory for the bulk samples.

Daily weather data (minimum and maximum temperature, precipitation, and relative humidity) with a 32 km grid spacing were obtained from the North American Regional Reanalysis (NARR) [28]. Monthly precipitation isotope composition values were measured at Mont-Saint Bruno (75 km from the study site) from January 2015 to November 2018.

A piezometer was installed in 2015 between the SLA and SLB sites. During the studied period (1 January 2016 to 31 December 2018), the water table level fluctuated between a 6.5 m and 4.5 m depth.

Table 1. Characteristics of the two study sites.

		SLA	SLB
Location		45°23′5.388″ N/74°11′50.316″ W	45°23′5.390″ N/74°11′50.320″ W
Elevation (m)		104	104
Geology		Cambrian formation	Cambrian formation
Soil depth (cm)	Horizon 1	0–20	0–30
	Horizon 2	20–200	30–200
Soil texture *	Horizon 1	Medium sand	Medium sand
	Horizon 2	Medium sand	Medium sand
Organic matter (%) **	Horizon 1	3	6
	Horizon 2	0	<1
Soil particle density (g cm^{-3})	Horizon 1	2.1	1.2
	Horizon 2	2.4	2.4
Land use		Grassland	Pine forest
Maximum rooting depth (cm)		10	20

* Texture classification system of Folk and Ward (1957) [29]. ** Large uncertainty exists.

2.2. Model Description

2.2.1. Water Flow

Transient water flow within the unsaturated soil profile was simulated by numerically solving the Richards equation using the finite-element code, METIS (1D), developed by the Geosciences Department of MINES ParisTech [23]:

$$\mathrm{div}(K_s\, K_r \mathrm{grad}\, H) = (F\, C(h)) \frac{\partial H}{\partial t}, \tag{1}$$

where K_s is the saturated hydraulic conductivity (L T^{-1}); K_r is the relative hydraulic conductivity (-); H is the hydraulic head (L); F is the porosity (L^3 L^{-3}); and C(h) is the specific retention capacity specified by S, the saturation ratio (L^3 L^{-3}), and h, the matric head (L):

$$C(h) = \frac{\partial S}{\partial h} \tag{2}$$

The behavior of the unsaturated medium depends on two essential characteristics of the material: the relative permeability curve and the suction curve. These relationships are determined by the residual and maximal saturation ratio (S_{min} (L^3 L^{-3}) and S_{max} (L^3 L^{-3}), respectively); three empirical parameters defining the shape of the retention curves: α (L^{-1}), n (–), and m (–) (with m = 1 – 1/n); and K_s, the saturated hydraulic conductivity (L T^{-1}), following the Mualem–van Genuchten (MvG) model [30,31]:

$$K_r(h) = \sqrt{\frac{S(h) - S_{min}}{S_{max} - S_{min}}} \left[1 - \left(1 - \left(\sqrt{\frac{S(h) - S_{min}}{S_{max} - S_{min}}} \right)^{\frac{1}{n}} \right)^n \right]^2, \tag{3}$$

$$\frac{S(h) - S_{min}}{S_{max} - S_{min}} = \frac{1}{\left[1 + (\alpha|h|)^n \right]^m}. \tag{4}$$

In order to account for root water uptake, a sink term was defined according to [32]. The required matric potential that describes the minimal pressure head for the root water uptake should be calculated using the Kelvin equation [33]. However, for sandy soil or gravel, due to numerical constraints, one needs to use high values. Here, the matric potential was set to −1000 cm. Root water uptake was parameterized by the site-specific rooting depth (Table 1) and a uniform root distribution.

Potential evapotranspiration (PET) was estimated using the Hargreaves and Samani formula [34] as a function of extraterrestrial radiation and daily maximum and minimum air temperature. As evaporation changes the isotopic composition of the pore water remaining in the soil by fractionation, whereas transpiration does not [22], potential evaporation and potential transpiration must be considered separately in the model. PET is therefore split into potential evaporation and potential transpiration according to Beer's law [35], which is a function of the leaf area index (LAI) and the canopy radiation extinction factor (set to 0.463, as per [36]).

Snow is typical during the winter season in the study area. Snow cover was modeled using the two-parameter semi-distributed Snow Accounting Routine model, CEMANEIGE [37]. Precipitation during periods with temperatures of less than 0 °C was assumed to take the form of snow, and did not immediately infiltrate the soil. Snow melt was then simulated by assuming it to be proportional to the air temperature above 0 °C, using a degree-day snow melting constant of 0.5 mm day^{-1} °C^{-1} and a cold content factor of 0.1. Groundwater recharge was calculated as downward water flux (L T^{-1}) at the bottom of the profile.

2.2.2. δ^2H Transport

In the METIS code, ^{2}H transport in soil is calculated according to the widely used advection–dispersion model, with dispersivity (λ (L)) as the only parameter [38]. The absolute value of δ (in parts per thousand Vienna Standard Mean Ocean Water (VSMOW)) is used to represent the isotopic concentration. Isotopic enrichment due to fractionation processes during evaporation was included in the model based on the formulation introduced by Gonfiantini (1986) [39] for isotope mass balance, as follows:

$$\delta_s = \delta^* + \left(\delta_p - \delta^* \right)(1 - f)^m, \tag{5}$$

where δ_s is the isotopic signal of the soil water in the upper centimeter, f is the fraction of evaporated water, δ_p is the isotopic signal of the water at the previous time step, δ^* is the limiting isotopic enrichment factor, and m is the enrichment slope, as described below. For more details, see [40].

δ^* is a function of air humidity, a; the isotopic composition of ambient air, δ_A; and a total enrichment factor, ε [41]:

$$\delta^* = \frac{a \times \delta_A + \varepsilon}{a - \frac{\varepsilon}{100}}. \tag{6}$$

δ_A is calculated based on the stable isotopic signal of precipitation, δ_{rain}, and the isotope fractionation factor, $\varepsilon+$, according to [42]

$$\delta_A = \frac{\delta_{rain} + \varepsilon^+}{\alpha^+}, \tag{7}$$

with α^+ being the equilibrium isotope fractionation factor defined by Horita and Wesolowski (1994) [43]. The temperature-dependent equilibrium isotope fractionation factor, ε^+, is defined as

$$\varepsilon^+ = \left(\alpha^+ - 1\right) \times 1000. \tag{8}$$

The total fractionation factor, ε, is equal to the sum of the equilibrium isotope fractionation factor, ε^+, determined above (Equation (8)), and the kinetic isotope fractionation factor, ε_K, defined by Gat (1995) [44]:

$$\varepsilon_k = (1 - a) \times C_k, \tag{9}$$

where C_k is the kinetic fractionation constant equal to 12.5‰ for $\delta^2 H$.

The enrichment slope m is given by

$$m = \frac{a - \frac{\varepsilon}{1000}}{1 - a + \frac{\varepsilon_k}{1000}}. \tag{10}$$

2.2.3. Initial and Boundary Conditions

For all study sites, the depth of the soil profiles was set to 600 cm and discretized into 601 nodes. The profiles were divided into two different horizons (horizon 1 and horizon 2) according to the field observations (Table 1).

The results of steady-state simulations of water content and a constant pore water isotope composition, representing the weighted average concentration in precipitation (−85‰ for $\delta^2 H$, Figure S1, in the supplementary material), were used as site-specific initial conditions. The modeled period extended from 1 January 2016 to 31 December 2018. The effect of the initial conditions on the calibration can be neglected, as a spin-up period of almost 2 years (1 January 2016 to 23 November 2017) was simulated prior to calibration.

The upper boundary condition was defined by variable atmospheric conditions that govern the loss of water and deuterium by evaporation, and the input of water due to infiltration and the accompanying flux concentrations of deuterium. The bottom boundary condition was set to zero-potential to represent the average water table level observed in the field.

2.3. Sensitivity Analyses

Environmental optimization studies are often affected by the equifinality problem [45], whereby multiple sets of parameters can produce similar results. This problem is exacerbated with large numbers of parameters and when only limited information about their interactions and their effects on the output is available. In the METIS code, six parameters have to be optimized for each horizon of the soil profiles to simulate water flow in the unsaturated zone: S_{min}, S_{max}, α, n, and K_s, describing the water retention and hydraulic conductivity characteristics in accordance with the MvG model and F soil porosity. The dispersivity, λ, also has to be optimized to simulate $\delta^2 H$ value dispersion in soil. The same dispersivity value was assumed for both horizons [38].

Water and δ^2H transport equations are known to be highly non-linear; however, little attention has been given to quantifying the influence of each parameter, individually and through interaction, on groundwater recharge estimates. To fill this gap, sensitivity analyses were performed in order to identify the most influential soil hydraulic parameters and their interactions, and to determine how these parameters affect groundwater recharge estimation. Additionally, in order to demonstrate if information contained in the isotope and water content profiles from a single sampling time are complementary in their abilities to optimize soil hydraulic parameters by inverse modeling, sensitivity analyses were performed to understand how soil hydraulic parameters affect the modeled water content and isotope profiles at a specific time.

Sensitivity analyses presented below were performed for a synthetic case. Simulated recharge, water content, and isotope profiles were obtained using the boundary conditions described above (Section 2.2.3) and the soil hydraulic parameters presented in Table 2, for a period of 3 years (1 January 2016 to 31 December 2018).

Table 2. Comparison of soil hydraulic parameters obtained using different types of data in the optimization procedure. The set of reference parameters is the "synthetic case". To calibrate the model, the other cases rely on one water content profile at a single time (case 1); one water content profile and one pore water isotope composition profile at a same single time (case 2); monthly monitoring of the water content and pore water isotope composition at a 15 cm depth, plus one water content and one pore water isotope composition profile at a single time (case 3); and daily water content monitoring at 10, 20, 50, and 100 cm depths, plus one water content and pore water isotope composition profile at a single time (case 4).

		n	α	K_s	F	S_{min}	S_{max}	λ
		(-)	(m^{-1})	$(m\ s^{-1})$	(-)	(-)	(-)	(m)
Synthetic case	Horizon 1	2.00	5.00	1.00×10^{-3}	0.40	0.02	0.60	0.01
	Horizon 2	3.00	10.00	1.00×10^{-4}	0.35	0.05	0.55	0.01
Case 1	Horizon 1	1.78	4.97	2.33×10^{-3}	0.32	0.12	0.54	0.01
	Horizon 2	2.37	21.36	8.63×10^{-3}	0.23	0.09	0.29	0.01
Case 2	Horizon 1	1.31	8.91	2.94×10^{-3}	0.37	0.05	0.51	0.01
	Horizon 2	2.12	27.48	7.61×10^{-3}	0.37	0.08	0.60	0.01
Case 3	Horizon 1	1.46	10.49	2.69×10^{-4}	0.39	0.01	0.50	0.01
	Horizon 2	2.37	23.80	1.25×10^{-4}	0.20	0.04	0.54	0.01
Case 4	Horizon 1	1.46	10.49	2.69×10^{-4}	0.39	0.01	0.50	0.01
	Horizon 2	2.37	23.80	1.25×10^{-4}	0.20	0.04	0.54	0.01

2.3.1. The Morris Method

In this study, the modified version of the Morris method proposed by [24] was used to investigate input soil hydraulic parameters to which groundwater recharge is most sensitive. The Morris method performs "One-At-a-Time" (OAT) analyses, whereby each parameter is varied one after another and the relative variation in model output (referred to as the Elementary Effect, E) is measured. Each parameter, X, is randomly selected in the input space, and the variation direction is also random. The repetition (r times) of OAT analyses allows the full input parameter space to be scanned. Along each trajectory, the so-called elementary effect of parameter X_j for the i-th repetition is defined as

$$E_j^i = \frac{Y(X + \Delta e_j) - Y(X)}{\Delta},$$
(11)

where Y is the model output with parameter set X, Δ is a value chosen in the range of $\{1/(p-1),..,1-1/p-1)\}$ by the user, p is the number of discretization levels, and e_j is a vector of the canonical basis. [24] recommends that Δ equals $p/2(p-1)$. For any X_j selected in the input space, $X + \Delta e_j$ is always in the input space.

To assess the sensitivity of the input factors, two indices, σ and μ^*, are computed as follows:

$$\mu_j^* = \frac{1}{r}\sum_{i=1}^{r}\left|E_j^i\right|,\tag{12}$$

$$\sigma_j = \sqrt{\frac{1}{r}\sum_{i=1}^{r}\left(E_j^i - \frac{1}{r}\sum_{i=1}^{r}E_j^i\right)^2}.\tag{13}$$

μ_j^* estimates the overall impact of the parameter X_j on the model output, and σ_j measures the non-linear and/or interaction effects with other parameters.

2.3.2. The Sobol Method

A modified version of the Sobol sensitivity analysis [46] was performed for the most influent input soil hydraulic parameters identified by the Morris method in order to quantify the amount of variance that each parameter contributes to the unconditional variance of the model output. These amounts are represented by Sobol's sensitivity indices (SIs). The SIs give quantitative information on the variance associated with a single parameter or related to interactions of multiple parameters. For a more complete explanation of the Sobol method, please refer to Sobol (2001) [25]. Sobol's sensitivity indices are expressed as follows:

$$\text{First-order indices}: S_{1,i} = \frac{V_i}{V},\tag{14}$$

$$\text{Second-order indices}: S_{i,j} = \frac{V_{ij}}{V},\tag{15}$$

$$\text{Total indices}: S_{Ti} = S_{1,i} + \sum_{j\neq i}S_{ij},\tag{16}$$

where V_i is the variance associated with the ith parameter and V is the total variance.

The first order index, $S_{1,j}$, represents the individual variance contribution of the parameter X_i to the total unconditional variance, also referred to as the "main effect". The second order index, $S_{i,j}$, explains the interaction effect between parameters X_i and X_j. The overall impact of parameter X_i, including the main effect and all its interactions with other parameters, is given by the total index S_{Ti}. If the sum of all first-order indices is less than 1, the model is non-additive.

Sobol sensitivity analysis was performed for groundwater recharge, which is ultimately the targeted output variable, as well as for soil water content and pore water isotope profiles simulated at a single sampling time, in order to investigate the benefit of combining soil water content and tracer data in the optimization procedure.

2.3.3. Implementation of the Sensitivity Analyses

The Python programming language and the Sensitivity Analysis Library (SALib) [47] in particular were used to conduct the sensitivity analyses. Sensitivity indices are calculated for scalar outputs obtained from objective functions. Here, the modified Kling–Gupta efficiency (KGE) index, as defined by [48], was used as the objective function (Equation (17)). This dimensionless index compares simulated and observed data with regards to their correlation (r), the ratio of their mean values (bias ratio, β), and the ratio of their coefficient of variation (variability ratio, γ) as follows:

$$KGE = 1 - [(1-r)^2 + (1-\beta)^2 + (1-\gamma)^2]^{0.5}.\tag{17}$$

Values of the objective function are stored in a one-dimensional array for the subsequent computation of the sensitivity indices. If the METIS code needs to reduce the computation step to less than one second, it is considered to be non-convergent; the script then terminates the simulation and attributes a large negative value to the objective function (-100). The same negative value is attributed when the duration of the modeled groundwater recharge is shorter than one year, which indicates that the run was unsuccessful.

2.4. Model Calibration

2.4.1. Data Types

This study focuses on the type of data used for model calibration and the implications when using the model to predict groundwater recharge. Four data types have been tested here using different approaches:

- The **one-profile approach** uses a single depth profile of the water content at a single sampling time (case 1) or one depth profile of both the water content and pore water isotope composition at a given time (case 2) to calibrate the soil hydraulic parameters. This approach does not require continuous monitoring data as it is only based on profiles at a given sampling time. Such a method facilitates model calibration by avoiding the time, cost, and effort associated with long-term soil water content measurements, since only one sampling campaign is needed to obtain the soil samples;
- The **monthly approach** (case 3) uses the monthly water content and pore water isotope composition at a 15 cm depth, plus one depth profile of both the water content and pore water isotope composition at a given time (as in case 2), to calibrate the soil hydraulic parameters;
- The **daily approach** (case 4) uses daily monitoring of the water content at four different depths (10, 20, 50, and 100 cm), and one depth profile of both the water content and pore water isotope composition at a single time to calibrate the soil hydraulic parameters. [18] used this approach.

As neither monthly water content or pore water isotope composition monitoring data from a 15 cm depth nor daily water content monitoring data from 10, 20, 50, and 100 cm depths were available for our study sites, we used the synthetic case to discuss the performances of the optimization procedures. Simulated water content and isotope profiles were obtained using the boundary conditions described above (Section 2.2.3) and the soil hydraulic parameters presented in Table 2, for a period of 3 years (1 January 2016 to 31 December 2108). These simulated data were subsampled with a spacing of 5 cm between the soil surface to a 30 cm depth, and then 10 cm down to a 200 cm depth. These subsampled data were used in place of observed data. The realism of the parameterization obtained with the different calibration approaches was assessed based on their abilities to reproduce observed groundwater recharge, both in terms of quantity and dynamics.

2.4.2. Multi-Objective Optimization Procedure

An optimization procedure was developed to calibrate soil hydraulic parameters of a two-layer model (i.e., α_1, n_1, S_{min1}, S_{max1}, K_{s1}, F_1, α_2, n_2, S_{min2}, S_{max2}, Ks_2, F_2, and λ) based on the difference between the measured and simulated soil moisture and pore water isotope composition. The Latin Hypercube Sampling (LHS) method [49] was used to sample soil hydraulic parameters in the 13-dimension parameter space. The range (min, max) for each parameter was chosen according to expert knowledge [21] and 5000 values were randomly chosen from within this range (Table 2). The LHS method ensures that parameter ranges are equally sampled and fully explored. Each computation is initiated with a different set of soil hydraulic parameters, and soil moisture and pore water isotope data are computed with these different parameter combinations. For each simulation, three KGE were calculated: soil moisture (KGE_θ), isotope data (KGE_c), and the average of both (KGE_{tot}). Sequential and simultaneous multi-objective approaches have already been tested elsewhere [10,18] and the simultaneous multi-objective approach was shown to be a time-efficient calibration procedure, resulting in a good representation of water flow and isotope transport. The set of soil hydraulic parameters leading to the best KGE_{tot} was retained.

3. Results and Discussion

3.1. Sensitivity Analyses

Morris sensitivity analysis method results produced for groundwater recharge (obtained from the synthetic case) are reported in Figure 1. To interpret the results by simultaneously taking into account both sensitivity measures, we used representation in the (μ^*-σ) plane, which allows soil hydraulic parameters to be classified into three types: parameters having negligible effects, with small μ^* and σ; parameters having non-linear effects and/or interactions with other parameters, with large μ^* and σ; and parameters having linear and additive effects, with large μ^* and small σ. Parameters α_1 and n_1 present particularly strong non-linear effects and/or interactions with other parameters. All factors with a high μ^* value also have a high σ value, indicating that none of the parameters have a purely linear effect. λ is the only parameter exhibiting small μ^* and σ values, meaning that it has a very limited effect on the output, and therefore can be fixed. Since all other soil hydraulic parameters exhibit significant μ^* and σ values, it was not possible to fix these and thereby reduce the dimensionality of the problem without affecting the quality of the simulation. Five parameters therefore had to be optimized for each horizon of the soil profiles. λ was set to 1 cm [38].

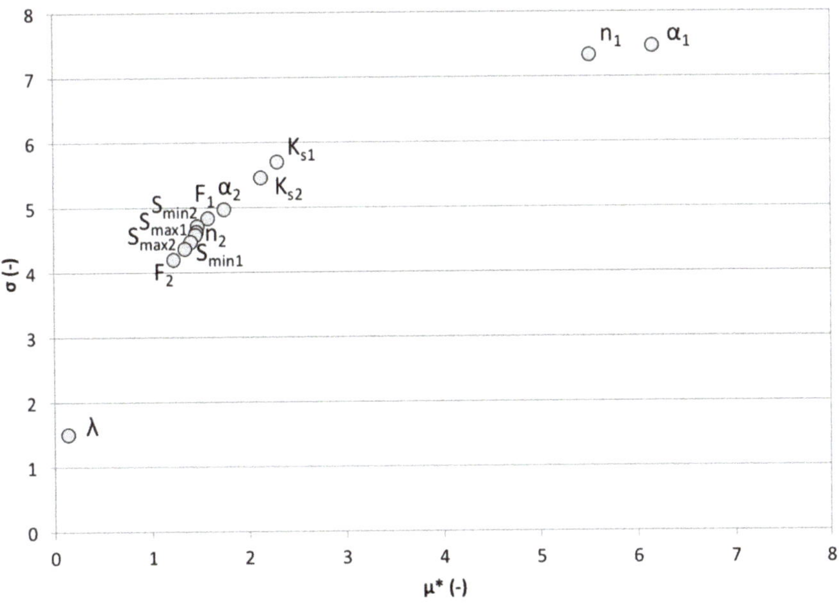

Figure 1. Results of the Morris method sensitivity analysis for output groundwater recharge.

Sobol's first-order index, S_1, which represents the individual variance contribution of a given parameter to the total unconditional variance, and Sobol's total index, S_T, which represents the total impact of a given parameter, including its main effect and all its interaction effects with other parameters, were calculated for the groundwater recharge calculated in the synthetic case (Figure 2a). Only the retention curve shape parameters n_1 and α_1 exhibit a high S_1 value, and thus have a significant direct influence on groundwater recharge variance. All the other soil hydraulic parameters have a first-order index of less than 1%, which indicates that their main effect on the output variance is negligible. The sum of all first-order indices is less than one, which means that the model is non-additive. Only 62% of the variance is attributable to the first-order effects, with interactions between soil hydraulic parameters playing a fundamental role. Almost 86% of the variance in simulated groundwater recharge is caused by n_1, either by variation in the parameter itself (49%) or by its interactions with other parameters. The effect of the second most influential parameter, α_1, represents only half of the n_1 effect (47%). It can be noted that all the other soil hydraulic parameters have a very low main effect, but a relatively high total effect. This indicates that they have a limited direct effect on the variance of the objective function, but an interaction effect with other parameters. All soil hydraulic parameters therefore influence the output variance either directly or through their interactions. No parameter other than dispersivity can be fixed without affecting the uncertainty of the output.

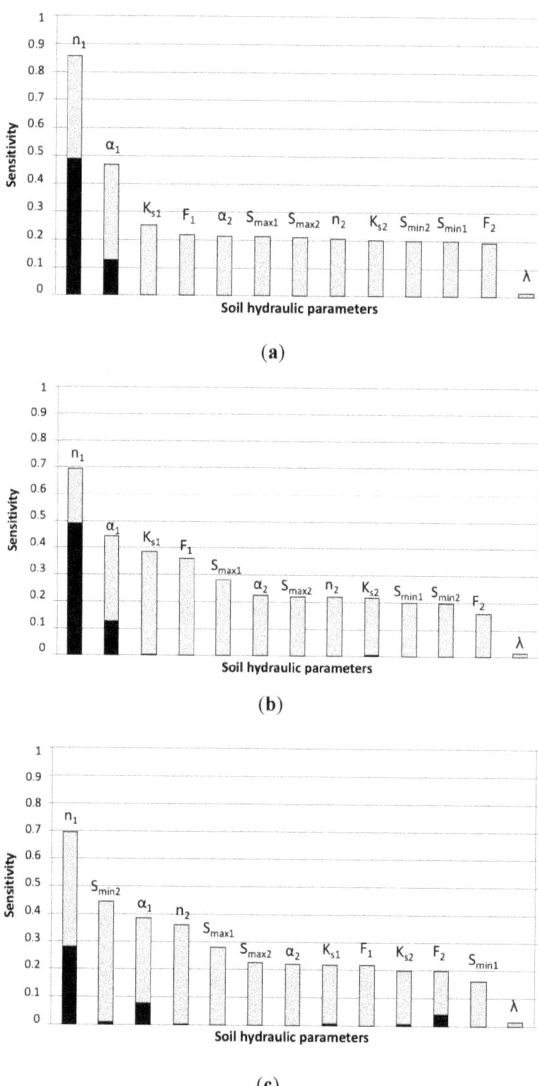

Figure 2. Results of the Sobol method sensitivity analysis for different model outputs: (**a**) groundwater recharge (1096 days of simulation); (**b**) water content profile after 1044 days of simulation*; and (**c**) pore water isotope profile after 1044 days of simulation*. S_1 is shown in black and S_T is shown in gray. Soil hydraulic parameters are ranked by decreasing influence. * 1044 days of simulation correspond to 9 November 2018. Results were checked to be independent of the chosen day.

The sensitivity analysis of the model reveals that the effect of horizon 1, the surface layer, strongly impacts the output. Within this layer, precipitation is partitioned into a vapor flux back into the atmosphere through evaporation. The accuracy of the upper boundary (e.g., precipitation and actual evaporation) and initial (e.g., water content and isotope content) conditions is therefore of great importance. However, these data are often not available or are associated with high uncertainties in outdoor experiments [50,51]. For example, various studies have shown that standard devices for precipitation measurement frequently underestimate the amount of rainfall [52,53], affecting the soil hydraulic property estimates [54]. [10] studied the impact of a less accurately defined upper boundary condition on simulated state variables (water content, matric potential, and $\delta^{18}O$) by comparing simulations that used daily precipitation measured with a rain gauge (tipping bucket method) with simulations in which precipitation was derived from lysimeter weights. The authors demonstrated that using less accurately defined boundary conditions clearly decreased the ability of the calibrated vadose zone model to simulate water content, matric potential, and drainage. Efforts are therefore needed to understand to what extent the collected records are consistent and representative of real field conditions.

Sobol's indices were calculated for one water content (Figure 2b) and pore water isotope composition (Figure 2c) depth profile collected at a single time (obtained from the synthetic case). The Sobol sensitivity analysis performed for the water content profile (Figure 2b) and groundwater recharge (Figure 2a) showed similar results in terms of the order of the most influential soil hydraulic parameters, and the main and total effects. Sobol's sensitivity analysis results for the isotope profile (Figure 2c) are significantly different. Even if n_1 remains the most influent parameter, S_{min2}, which plays a very minor role for the water content profile, is the second most influent parameter here. Inversely, Ks_1, which was the third most influent parameter for the water content profile, has only a minor influence on the isotope profile. Information contained in the isotope and water content profiles is therefore complementary in its abilities to optimize soil hydraulic parameters.

3.2. Model Parametrization

The synthetic case was used in order to test four data types for model calibration: one water content profile at a single sampling time (case 1); one water content profile and one pore water isotope composition profile at a single sampling time (case 2); monthly water content and pore water isotope composition monitoring at a 15 cm depth, plus one depth profile of both the water content and pore water isotope composition at a single time (case 3); and daily water content monitoring at 10, 20, 50, and 100 cm depths, plus one depth profile of both the water content and pore water isotope composition at a given time (case 4). Table 2 presents the best set of soil hydraulic parameters obtained for each calibration procedure, depending on the type of data used. n_1 appears to be well-estimated, which is consistent with the Sobol sensitivity analysis results. However, if we compare each set of soil hydraulic parameters with those of the "synthetic case", we observe that no approach allows it to be accurately simulated, even if KGE_{tot} is superior to 0.9 in each case. This result is not surprising considering, on the one hand, the poor main effect of the parameters and their high interactions and, on the other hand, the relatively low number of sets of parameters tested (5000). Nonetheless, this highlights the existence of local minima and the difficulty of accurately calibrating soil models through a local inversion approach.

Cases 3 and 4 lead to the same best set of soil hydraulic parameters. Daily values of water content at different depths do not contain more information for the inversion procedure than one monthly measurement of the water content and pore water isotope composition in the first horizon.

It is difficult to determine which data type is most appropriate for groundwater recharge estimates when only looking at soil hydraulic parameter values. In Table 3, we compare the annual groundwater recharge values obtained through the different approaches for two simulation years (2017 and 2018) to the value obtained in the synthetic case. The use of pore water isotopic composition profile measurements for the calibration significantly improves the estimate accuracy. Comparing cases 1 and 2, the error for the annual value is reduced from 15% to 7% in 2017 and from 32% to 16% in 2018. The use of temporal data for calibration also improves the accuracy of the groundwater recharge estimates, but less so; error is reduced from 16% to 9% in 2018 and is in the same range of values for 2017 (7%) based on cases 2 and 3, respectively. However, we can observe an overestimation of groundwater recharge for each case. If modeling is undertaken with the objective of estimating annual recharge values, using only water content profile and pore water isotopic composition profile data to calibrate the model leads to consistent results, even if the use of temporal data still remains the best approach.

Table 3. Comparison of groundwater recharge values obtained for two simulation years (2017 and 2018) using different types of data in the optimization procedure. The set of reference soil hydraulic parameters is the "synthetic case". To calibrate the model, the other cases rely on one water content profile at a single time (case 1); one water content profile and one pore water isotope composition profile at a same single time (case 2); monthly water content and pore water isotope composition monitoring at a 15 cm depth, plus one water content and pore water isotope composition profile at a single time (case 3); and daily water content monitoring at 10, 20, 50, and 100 cm depths, plus one water content and pore water isotope composition profile at a single time (case 4). Results are given in millimeters.

	Synthetic Case	Case 1	Case 2	Case 3	Case 4
2017	278	319	298	302	302
2018	152	201	177	166	166

We were also interested in the dynamics of groundwater recharge. As the annual time step is most relevant for groundwater recharge studies, we present annual cumulative curves in Figure 3. For each case, the annual groundwater recharge cumulative curve obtained is compared with the target curve (i.e., obtained with the set of soil hydraulic parameters from the synthetic case). The groundwater recharge dynamics are well-reproduced in all the tested cases (KGE > 0.7). The two groundwater recharge periods (spring and autumn) can be clearly identified. However, groundwater recharge always appears to be overestimated in autumn, especially in 2017. This might be caused by two important precipitation events (40 and 32 mm) occurring in October and only 21 days apart. The use of pore water isotopic composition profile measurements for the calibration significantly improves the accuracy of the simulated dynamics (from 0.90 to 0.97 in 2017 and from 0.71 to 0.83 in 2018 based on cases 1 and 2). The use of temporal data for calibration does not systematically improve the accuracy of groundwater recharge dynamics. If modeling is undertaken with the objective of estimating annual recharge values and identifying the seasonality of groundwater recharge, using only water content and pore water isotopic composition profile to calibrate the model is a reliable approach. Nonetheless, if daily groundwater recharge dynamics are needed for the study, the situation is different: only calibration methods including temporal data lead to reliable results. This highlights the need to select the data used for calibration based on the specific modeling objective.

In order to quantify the spread of groundwater recharge values around the optimal value (i.e., that obtained using the set of soil hydraulic parameters leading to the greatest KGE_{tot}), we also present groundwater recharge curves obtained with the five best sets of parameters (all KGE_{tot} are above 0.9) in Figure 3. The narrowest spread of groundwater recharge estimates is obtained for case 2, which means that the model response is constrained enough and strengthens the argument that this approach is suitable for calibrating soil models.

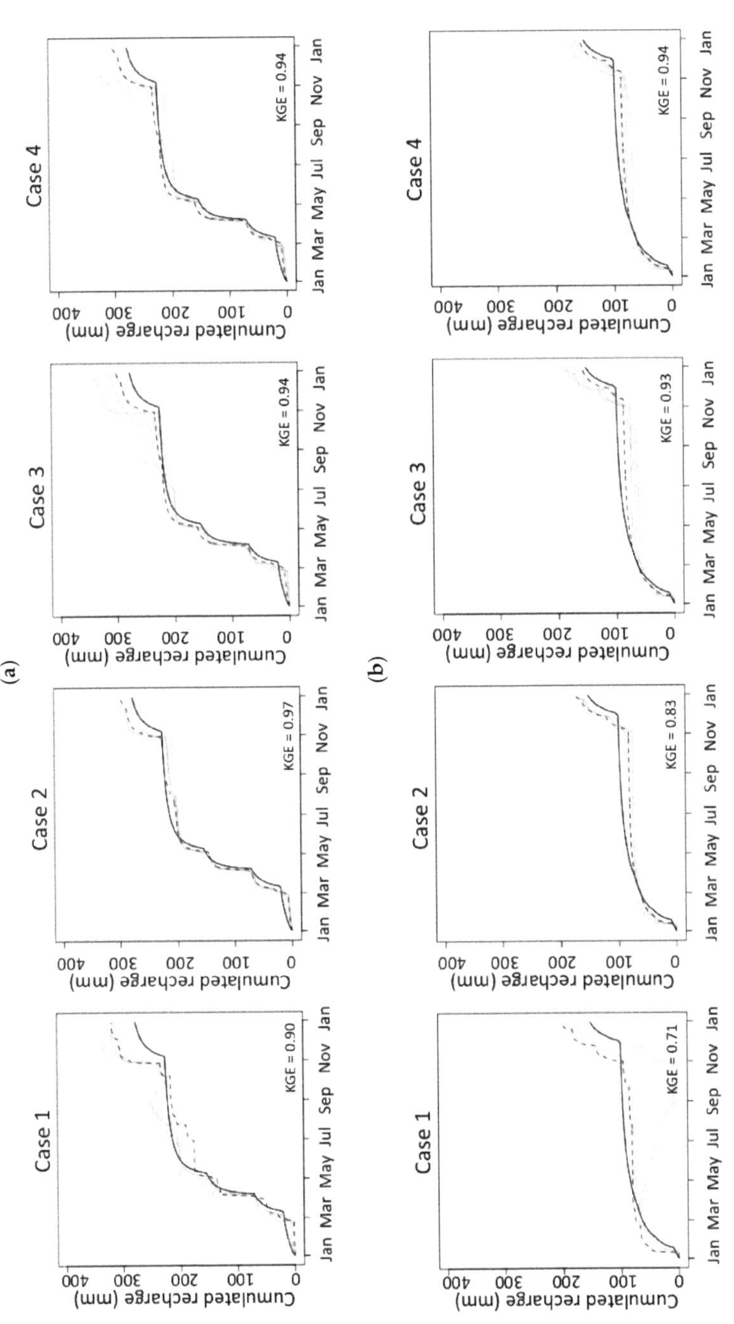

Figure 3. Annual cumulated groundwater recharge values for 2017 (**a**) and 2018 (**b**). Results of the synthetic case are presented as continuous black lines. The five best sets of parameter results are presented as gray lines. Kling–Gupta efficiency (KGE) is calculated for the synthetic case results and the best set of soil hydraulic parameter results (black dashed line).

3.3. Field Case Study

Under real conditions, using one water content and pore water isotope composition depth profile from a single time to calibrate the model was also found to be a reliable approach in this work. Water content profiles (KGE$_\theta$ from 0.56 to 0.77) and δ^2H profiles (KGE$_c$ from 0.57 to 0.71) were satisfactorily reproduced (Figures 4 and 5). The KGE$_{tot}$ values that were obtained for the validation model period (i.e., 2018) are of the same order of magnitude as those for the calibration period (i.e., 2017), for sites SLA and SLB. For SLA, if the first three measurements (which represent the first 15 cm below the surface level) are not considered, the KGE$_{tot}$ increases from 0.57 to 0.75 in 2017 and from 0.69 to 0.83 in 2018, mainly due to KGE$_c$ increasing. We suspect that the depleted values measured at 5 and 15 cm depths correspond to two small snowfall events (the first totally melted and the second partially melted on SLA), which occurred on 19 November and 16 November 2017, respectively, a few days before soil profile sampling. The monthly collection of precipitation isotope data does not allow the conditions in the top centimeters of the soil profile to be accurately reproduced. With an increasing infiltration depth, water gets mixed with the pore water previously present in the soil, leading to the following loss of its initial isotopic signal and the convergence toward a mean value, which explains the better results obtained at a greater depth. For SLB, we do not observe this increase in KGE$_c$. If we look more closely at the pore water isotope profile (Figure 5), the two snowfall events mentioned above are not evident; rather, measurements in the upper 15 cm are very close to the average precipitation composition in November (δ^2H = −75‰). SLB is located in a pine forest. Interception by the trees, and the needles and biomass cover on the soil could explain these results, by their not allowing small precipitation events to directly infiltrate into the soil.

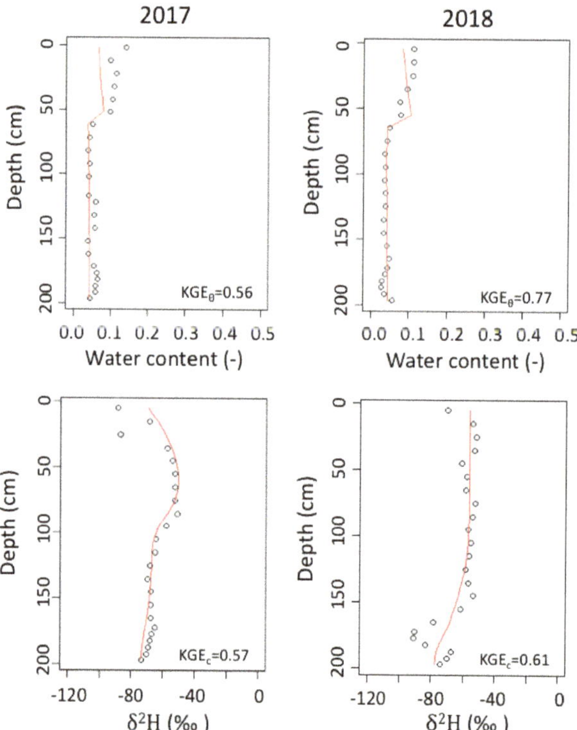

Figure 4. Observed (**circles**) and simulated (**lines**) water content (**top**) and δ^2H (**bottom**) soil profiles at site SLA, with the best parameter set.

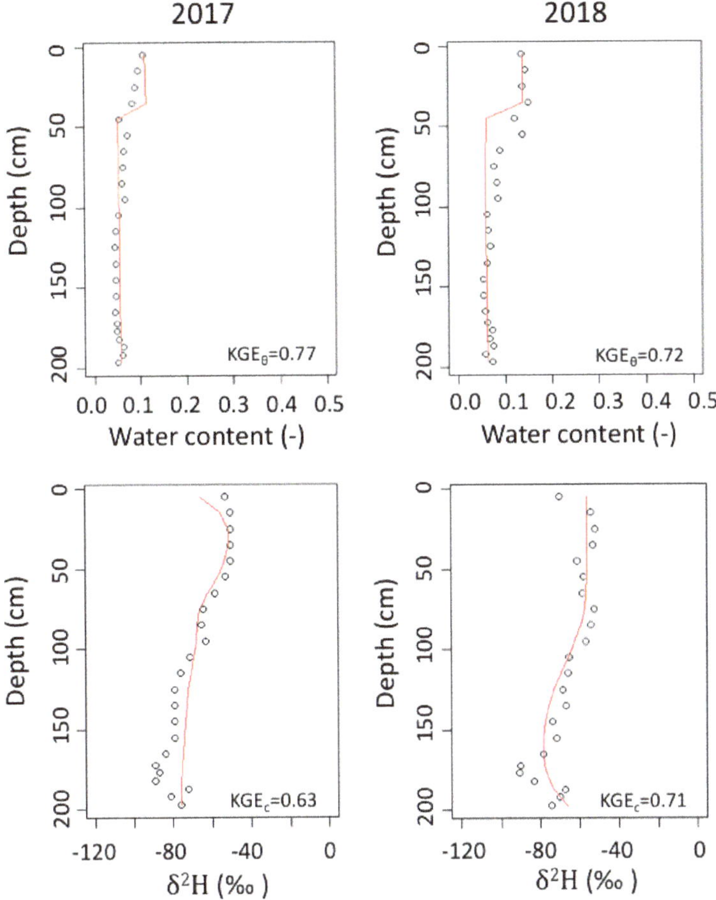

Figure 5. Observed (**circles**) and simulated (**lines**) water content (**top**) and δ^2H (**bottom**) soil profiles at site SLB with the best parameter set.

The soil hydraulic parameters obtained for sites SLA and SLB are different for the two horizons considered (Table 4), even if, for both sites, horizon 2 had no organic matter and the same granulometry. Vegetation affects the soil structure and its stability at different scales and through various direct and indirect mechanisms [55]. By penetrating the soil, roots form macropores which favor fluid transport. They also create failure zones, which contribute to fragmentation of the soil and the formation of aggregates, thus modifying soil hydraulic properties.

Table 4. Best set of soil hydraulic parameters obtained for sites SLA and SLB.

		n	α	K_s	F	S_{min}	S_{max}	λ
		(-)	(m^{-1})	(ms^{-1})	(-)	(-)	(-)	(m)
SLA	Horizon 1	1.87	11.32	5.46×10^{-3}	0.21	0.02	0.44	0.01
	Horizon 2	1.89	26.3	2.74×10^{-3}	0.36	0.02	0.24	0.01
SLB	Horizon 1	1.51	20.35	8.26×10^{-3}	0.28	0.04	0.54	0.01
	Horizon 2	2.16	10.61	9.49×10^{-3}	0.39	0.04	0.37	0.01

The mean annual groundwater recharge values obtained for sites SLA and SLB using the proposed method (Table 5), 345 mm and 357 mm, respectively, are consistent with previous studies [24]. As expected, recharge values and the repartition between evaporation and transpiration are different between the sites due to differences in vegetation. The role of evaporation is greater at site SLA, which is located in grassland, and the role of transpiration is greater at site SLB, which is located in a pine forest.

Table 5. Annual simulated groundwater recharge, evaporation, and transpiration values for sites SLA and SLB.

		Recharge (mm)	Evaporation (mm)	Transpiration (mm)
SLA	2017	429	307	117
	2018	261	220	75
SLB	2017	456	215	217
	2018	257	181	118

4. Conclusions

In this study, we investigated the possibility of calibrating a soil water transport model to estimate groundwater recharge using only one water content and pore water isotope composition depth profile from a single sampling time. By conducting sensitivity analyses, we highlighted the difficulties in accurately calibrating the soil model. Indeed, the interactions between soil hydraulic parameters play a fundamental role, and various combinations can lead to similar groundwater recharge simulations. Combining water content and tracer data in a multi-objective calibration approach helped to constrain soil hydraulic property determination, as the influence of soil hydraulic parameters for the two types of data was found to differ. The value of pore water isotope information for appropriate soil water transport model parameterization was demonstrated. For sandy soils, accurate calibration is possible without temporally-continuous monitoring data, using only water content and pore water isotopic composition profiles measured on a single date. However, even if not continuous, multi-temporal data improve model calibration, especially pore water isotope data. Indeed, monthly water content and pore water isotope composition monitoring in the first horizon provides as much information as daily water content monitoring in both horizons. We therefore encourage field monitoring methods to develop in this direction. More generally, it is important that the choice of data used for soil hydraulic parameter calibration be made in accordance with the objectives of the model.

Supplementary Materials: The following are available online at http://www.mdpi.com/2073-4441/12/2/393/s1, Figure S1: Measured δ^2H in precipitation at Mont Saint-Bruno, Quebec, Canada.

Author Contributions: Conceptualization, A.M., P.G., F.B., Y.C. and S.G.; data curation, F.B. and P.G.; methodology, A.M., P.G., F.B., S.W. and S.G.; investigation, A.M., F.B., P.G. and S.G.; resources, A.M., P.G. and S.W.; software, A.M., P.G. and S.W.; writing—original draft preparation, A.M.; supervision, P.G., Y.C., F.B. and S.G. All authors have read and agreed to the published version of the manuscript.

Funding: This research was funded by an NSERC discovery grant to Florent Barbecot.

Acknowledgments: We thank V. Horoi for help with the field work and T. Romary for the interesting discussions. We also thank the four anonymous referees and the editors, Polona Vreča and Zoltan Kern, for their careful reading of the manuscript and their many insightful comments and suggestions.

Conflicts of Interest: The authors declare no conflicts of interest.

References

1. Aeschbach, W.; Gleeson, T. Regional strategies for the accelerating global problem of groundwater depletion. *Nat. Geosci.* **2012**, *5*, 853–861. [CrossRef]

2. Taylor, R.; Scanlon, B.; Doell, P.; Rodell, M.; van Beek, R.; Wada, Y.; Longuevergne, L.; Leblanc, M.; Famiglietti, J.S.; Edmunds, M.; et al. Ground water and climate change. *Nat. Clim. Chang.* **2013**, *3*, 322–329. [CrossRef]

3. Famiglietti, J. The global groundwater crisis. *Nat. Clim. Chang.* **2014**, *4*, 945–948. [CrossRef]

4. Farthing, M.W.; Ogden, F. Numerical Solution of Richards' Equation: A Review of Advances and Challenges. *Soil Sci. Soc. Am. J.* **2017**, *81*, 1257–1269. [CrossRef]

5. Corwin, D.L.; Hopmans, J.; de Rooij, G.H. From Field- to Landscape-Scale Vadose Zone Processes: Scale Issues, Modeling, and Monitoring. *Vadose Zo. J.* **2006**, *5*, 129–139. [CrossRef]

6. Mermoud, A.; Xu, D. Comparative Analysis of Three Methods to Generate Soil Hydraulic Functions. *Soil Tillage Res.* **2006**, *87*, 89–100. [CrossRef]

7. Isch, A.; Montenach, D.; Hammel, F.; Ackerer, P.; Coquet, Y. A Comparative Study of Water and Bromide Transport in a Bare Loam Soil Using Lysimeters and Field Plots. *Water* **2019**, *11*, 1199. [CrossRef]

8. Ritter, A.; Hupet, F.; Muñoz-Carpena, R.; Lambot, S.; Vanclooster, M. Using inverse methods for estimating soil hydraulic properties from field data as an alternative to direct methods. *Agric. Water Manag.* **2003**, *59*, 77–96. [CrossRef]

9. Mertens, J.; Stenger, R.; Barkle, G.F. Multiobjective Inverse Modeling for Soil Parameter Estimation and Model Verification. *Vadose Zo. J.* **2006**, *5*, 917–933. [CrossRef]

10. Groh, J.; Stumpp, C.; Lücke, A.; Pütz, T.; Vanderborght, J.; Vereecken, H. Inverse Estimation of Soil Hydraulic and Transport Parameters of Layered Soils from Water Stable Isotope and Lysimeter Data. *Vadose Zo. J.* **2018**, *17*. [CrossRef]

11. Pachepsky, Y.; Smettem, K.; Vanderborght, J.; Herbst, M.; Vereecken, H.; Wösten, H. Reality and Fiction of Models and Data in Soil Hydrology. In *Unsaturated-zone Modeling: Progress, Challenges and Applications*; Feddes, R.A., de Rooij, G.H., van Dam, J.C., Eds.; Springer: New York, NY, USA, 2004; Volume 6, pp. 233–260.

12. Lu, X.; Jin, M.; van Genuchten, M.T.; Wang, B. Groundwater Recharge at Five Representative Sites in the Hebei Plain, China. *Groundwater* **2011**, *49*, 286–294. [CrossRef] [PubMed]

13. Wang, T.; Franz, T.E.; Yue, W.; Szilagyi, J.; Zlotnik, V.A.; You, J.; Chen, X.; Shulski, M.D.; Young, A. Feasibility analysis of using inverse modeling for estimating natural groundwater recharge from a large-scale soil moisture monitoring network. *J. Hydrol.* **2016**, *533*, 250–265. [CrossRef]

14. Wohling, T.; Vrugt, J.A.; Barkle, G.F. Comparison of Three Multiobjective Optimization Algorithms for Inverse Modeling of Vadose Zone Hydraulic Properties. *Soil Sci. Soc. Am. J.* **2008**, *72*, 305–319. [CrossRef]

15. Groh, J.; Puhlmann, H.; Wilpert, K. Calibration of a soil-water balance model with a combined objective function for the optimization of the water retention curve. *Hydrol. Wasserbewirtsch.* **2013**, *57*, 152–162.

16. Abbasi, F.; Simunek Jirka, J.; Feyen, J.; Shouse, P.J. Simultaneous Inverse Estimation of Soil Hydraulic and Solute Transport Parameters from Transient Field Experiments: Homogeneous Soil. *Am. Soc. Agric. Eng.* **2003**, *46*, 1085–1095. [CrossRef]

17. Stumpp, C.; Stichler, W.; Kandolf, M.; Šimůnek, J. Effects of Land Cover and Fertilization Method on Water Flow and Solute Transport in Five Lysimeters: A Long-Term Study Using Stable Water Isotopes. *Vadose Zone J.* **2012**, *11*. [CrossRef]

18. Sprenger, M.; Volkmann, T.H.M.; Blume, T.; Weiler, M. Estimating flow and transport parameters in the unsaturated zone with pore water stable isotopes. *Hydrol. Earth Syst. Sci.* **2015**, *19*, 2617–2635. [CrossRef]

19. Koeniger, P.; Leibundgut, C.; Link, T.; Marshall, J.D. Stable isotopes applied as water tracers in column and field studies. *Org. Geochem.* **2010**, *41*, 31–40. [CrossRef]

20. Zimmermann, U.; MüNnich, K.O.; Roether, W. Downward Movement of Soil Moisture Traced by Means of Hydrogen Isotopes. In *Isotopes in Hydrology*; IAEA, Isotopes in Hydrology: Vienna, Austria, 14–18 November 1966; International Atomic Energy Agency: Vienna, Austria, 1967; pp. 567–585.

21. Gehrels, J.C.; Peeters, J.E.M.; Vries, J.J.D.E.; Dekkers, M. The mechanism of soil water movement as inferred from 18O stable isotope studies. *Hydrol. Sci. J.* **1998**, *43*, 579–594. [CrossRef]

22. Souchez, R.; Lorrain, R.; Tison, J.L. Stable water isotopes and the physical environment. *Belgeo. Revue Belge Géographie* **2002**, *2*, 133–144. [CrossRef]

23. Goblet, P. *Simulation D'écoulement et de Transport Miscible en Milieu Poreux et Fracturé*; Manual: Fontainebleau, France, 2010.

24. Campolongo, F.; Cariboni, J.; Saltelli, A. An effective screening design for sensitivity analysis of large models. *Environ. Model. Softw.* **2007**, *22*, 1509–1518. [CrossRef]

25. Sobol, I.M. Global sensitivity indices for nonlinear mathematical models and their Monte Carlo estimates. *Math. Comput. Simul.* **2001**, *55*, 271–280. [CrossRef]

26. Larocque, M.; Meyzonnat, G.; Ouellet, M.A.; Graveline, M.H.; Gagné, S.; Barnetche, D.; Dorner, S. *Projet de connaissance des eaux souterraines de la zone de Vaudreuil-Soulanges*; Rapport Final déposé au Ministère du Développement durable, de l'Environnement et de la Lutte Contre les Changements Climatiques: Quebec, QC, Canada, 2015.

27. Mattei, A.; Barbecot, F.; Guillon, S.; Goblet, P.; Hélie, J.-F.; Meyzonnat, G. Improved accuracy and precision of water stable isotope measurements using the direct vapour equilibration method. *Rapid Commun. Mass Spectrom.* **2019**, *33*, 1613–1622. [CrossRef] [PubMed]

28. Mesinger, F.; DiMego, G.; Kalnay, E.; Mitchell, K.; Shafran, P.C.; Ebisuzaki, W.; Jović, D.; Woollen, J.; Rogers, E.; Berbery, E.H.; et al. North American Regional Reanalysis. *Bull. Am. Meteorol. Soc.* **2006**, *87*, 343–360. [CrossRef]

29. Folk, R.L.; Ward, W.C. Brazos River bar: a study in the significance of grain size parameters. *J. Sediment. Petrol.* **1957**, *27*, 3–26. [CrossRef]

30. Mualem, Y. A new model for predicting the hydraulic conductivity of unsaturated porous media. *Water Resour. Res.* **1976**, *12*, 513–522. [CrossRef]

31. Van Genuchten, M.T. A Closed Form Equation for Predicting the Hydraulic Conductivity of Unsaturated Soils 1. *Soil Sci. Soc. Am. J.* **1980**, *44*, 892–898. [CrossRef]

32. Feddes, R.A.; Kowalik, P.; Kolinska-Malinka, K.; Zaradny, H. Simulation of field water uptake by plants using a soil water dependent root extraction function. *J. Hydrol.* **1976**, *31*, 13–26. [CrossRef]

33. Šimůnek, J.; Šejna, M.; Saito, H.; Sakai, M.; van Genuchten, M.T. *The HYDRUS-1D Software Package for Simulating the One-Dimensional Movement of Water, Heat, and Multiple Solutes in Variably-Saturated Media*; University of California Riverside: Riverside, CA, USA, 2013.

34. Hargreaves, G.H.; Samani, Z.A. Estimating Potential Evapotranspiration. *Appl. Eng. Agric.* **1982**, *1*, 96–99.

35. Ritchie, J.A. Model for Predicting Evaporation From a Low Crop With Incomplete Cover. *Water Resour. Res.* **1972**, *8*, 1204–1213. [CrossRef]

36. Sutanto, S.; Wenninger, J.; Coenders-Gerrits, M.; Uhlenbrook, S. Partitioning of evaporation into transpiration, soil evaporation and interception: A comparison between isotope measurements and a HYDRUS-1D model. *Hydrol. Earth Syst. Sci.* **2012**, *16*, 2605–2616. [CrossRef]

37. Valéry, A. Modélisation Précipitations–Débit sous Influence Nivale: Elaboration d'un Module Neige et Evaluation sur 380 Bassins Versants. Ph.D. Thesis, AgroParisTech, Antony, France, 2010.

38. Vanderborght, J.; Vereecken, H. One-Dimensional Modeling of Transport in Soils with Depth-Dependent Dispersion, Sorption and Decay. *Vadose Zone J.* **2007**, *6*, 140–148. [CrossRef]

39. Gonfiantini, R. Chapter 3—Environmental isotopes in lake studies. In *The Terrestrial Environment, B*; Fritz, P., Fontes, J.C., Eds.; Handbook of Environmental Isotope Geochemistry; Elsevier: Amsterdam, The Netherlands, 1986; pp. 113–168. ISBN 978-0-444-42225-5.

40. Skrzypek, G.; Mydłowski, A.; Dogramaci, S.; Hedley, P.; Gibson, J.J.; Grierson, P.F. Estimation of evaporative loss based on the stable isotope composition of water using Hydrocalculator. *J. Hydrol.* **2015**, *523*, 781–789. [CrossRef]

41. Gat, J.R.; Levy, Y. Isotope hydrology of inland Sabkhas in the Bardawil area. *Sinai. Limnol. Oceanogr.* **1978**, *23*, 841–850. [CrossRef]

42. Gibson, J.J.; Birks, S.J.; Edwards, T.W.D. Global prediction of δA and $\delta 2H$-$\delta 18O$ evaporation slopes for lakes and soil water accounting for seasonality. *Glob. Biogeochem. Cycles* **2008**, *22*, GB2031. [CrossRef]

43. Horita, J.; Wesolowski, D.J. Liquid-vapor fractionation of oxygen and hydrogen isotopes of water from the freezing to the critical temperature. *Geochim. Cosmochim. Acta* **1994**, *58*, 3425–3437. [CrossRef]

44. Gat, J.R. *Stable Isotopes of Fresh and Saline Lakes BT—Physics and Chemistry of Lakes*; Lerman, A., Imboden, D.M., Gat, J.R., Eds.; Springer: Berlin/Heidelberg, Germany, 1995; pp. 139–165. ISBN 978-3-642-85132-2.

45. Beven, K. A manifesto for the equifinality thesis. *J. Hydrol.* **2006**, *320*, 18–36. [CrossRef]

46. Saltelli, A. Sensitivity Analysis for Importance Assessment. *Risk Anal.* **2002**, *22*, 579–590. [CrossRef]

47. Usher, W.; Herman, J.; Mutel, C. SALib: Sensitivity Analysis Library in Python (Numpy). Contains Sobol, Morris, Fractional Factorial and FAST methods. 2015. Available online: http://SALib.github.io/SALib/ (accessed on 1 March 2019).

48. Kling, H.; Fuchs, M.; Paulin, M. Runoff conditions in the upper Danube basin under an ensemble of climate change scenarios. *J. Hydrol.* **2012**, *424–425*, 264–277. [CrossRef]

49. McKay, M.D.; Beckman, R.J.; Conover, W.J. A Comparison of Three Methods for Selecting Values of Input Variables in the Analysis of Output from a Computer Code. *Technometrics* **1979**, *21*, 239–245.

50. Vrugt, J.A.; ter Braak, C.J.F.; Clark, M.P.; Hyman, J.M.; Robinson, B.A. Treatment of input uncertainty in hydrologic modeling: Doing hydrology backward with Markov chain Monte Carlo simulation. *Water Resour. Res.* **2008**, *44*, W00B09. [CrossRef]

51. Mannschatz, T.; Dietrich, P. Model Input Data Uncertainty and Its Potential Impact on Soil Properties. In *Sensitivity Analysis in Earth Observation Modelling*; Petropoulos, G., Srivastava, P.K., Eds.; Elsevier: Oxford, UK, 2017; pp. 25–52.

52. Gebler, S.; Franssen, H.-J.; Pütz, T.; Post, H.; Schmidt, M.; Vereecken, H. Actual evapotranspiration and precipitation measured by lysimeters: A comparison with eddy covariance and tipping bucket. *Hydrol. Earth Syst. Sci.* **2015**, *19*, 2145–2161. [CrossRef]

53. Hoffmann, M.; Schwartengräber, R.; Wessolek, G.; Peters, A. Comparison of simple rain gauge measurements with precision lysimeter data. *Atmos. Res.* **2016**, *174–175*, 120–123. [CrossRef]

54. Peters-Lidard, C.; Mocko, D.M.; Garcia, M.; Santanello, J.; Tischler, M.A.; Susan Moran, M.; Wu, Y. Role of precipitation uncertainty in the estimation of hydrologic soil properties using remotely sensed soil moisture in a semiarid environment. *Water Resour. Res.* **2008**, *44*, W05S18. [CrossRef]

55. Angers, D.A.; Caron, J. Plant-induced Changes in Soil Structure: Processes and Feedbacks. *Biogeochemistry* **1998**, *42*, 55–72. [CrossRef]

Article

Isotopic 'Altitude' and 'Continental' Effects in Modern Precipitation across the Adriatic–Pannonian Region

Zoltán Kern [1,*], István Gábor Hatvani [1], György Czuppon [1,2], István Fórizs [1], Dániel Erdélyi [1,3], Tjaša Kanduč [4], László Palcsu [2] and Polona Vreča [4]

1 Institute for Geological and Geochemical Research, Research Centre for Astronomy and Earth Sciences, MTA Centre for Excellence, Budaörsi út 45, H-1112 Budapest, Hungary; hatvaniig@gmail.com (I.G.H.); czuppon@geochem.hu (G.C.); forizs@geochem.hu (I.F.); danderdelyi@gmail.com (D.E.)
2 Isotope Climatology and Environmental Research Centre (ICER), Institute for Nuclear Research, Bem tér 18/c, H-4026 Debrecen, Hungary; palcsu.laszlo@atomki.mta.hu
3 Centre for Environmental Sciences, Department of Geology, Eötvös Loránd University, Pázmány Péter stny 1, H-1117 Budapest, Hungary
4 Department of Environmental Sciences, Jožef Stefan Institute, Jamova cesta 39, 1000 Ljubljana, Slovenia; tjasa.kanduc@ijs.si (T.K.); polona.vreca@ijs.si (P.V.)
* Correspondence: zoltan.kern@gmail.com; Tel.: +36-70-253-8188

Received: 17 April 2020; Accepted: 19 June 2020; Published: 24 June 2020

Abstract: It is generally observed that precipitation is gradually depleted in ^{18}O and ^{2}H isotopes as elevation increases ('altitude' effect) or when moving inland from seacoasts ('continental' effect); the regionally accurate estimation of these large-scale effects is important in isotope hydrological or paleoclimatological applications. Nevertheless, seasonal and spatial differences should be considered. Stable isotope composition of monthly precipitation fallen between January 2016 and December 2018 was studied for selected stations situated along an elevation transect and a continental transect in order to assess the isotopic 'altitude' and 'continental' effects in modern precipitation across the Adriatic–Pannonian region. Isotopic characteristics argue that the main driver of the apparent vertical depletion of precipitation in heavy stable isotopes is different in summer (raindrop evaporation) and winter (condensation), although, there is no significant difference in the resulting 'altitude' effect. Specifically, an 'altitude' effect of −1.2‰/km for $\delta^{18}O$ and −7.9‰/km for $\delta^{2}H$ can be used in modern precipitation across the Adriatic–Pannonian region. Isotopic characteristics of monthly precipitation showed seasonally different patterns and suggest different isotope hydrometeorological regimes along the continental transect. While no significant decrease was found in $\delta^{18}O$ data moving inland from the Adriatic from May to August of the year, a clear decreasing trend was found in precipitation fallen during the colder season of the year (October to March) up to a break at ~400 km inland from the Adriatic coast. The estimated mean isotopic 'continental' effect for the colder season precipitation is −2.4‰/100 km in $\delta^{18}O$ and −20‰/100 km in $\delta^{2}H$. A prevailing influence of the Mediterranean moisture in the colder season is detected up to this breakpoint, while the break in the $\delta^{18}O$ data probably reflects the mixture of moisture sources with different isotopic characteristics. A sharp drop in the d-excess (>3‰) at the break in precipitation $\delta^{18}O$ trend likely indicates a sudden switch from the Mediterranean moisture domain to additional (mainly Atlantic) influence, while a gradual change in the d-excess values might suggest a gradual increase of the non-Mediterranean moisture contribution along the transect.

Keywords: stable isotopes; oxygen; hydrogen; d-excess; elevation effect; altitude effect; continental effect; Slovenia; Hungary

---

1. Introduction

Precipitation is a key element of the atmospheric branch of the water cycle, and the study of precipitation is essential for improving the understanding of the Earth's water balance on both regional and global scales [1]. An important method for tracing the water cycle is that of comparing the ratios of heavy to light stable isotopes of hydrogen and oxygen (^{2}H/^{1}H; ^{18}O/^{16}O) in the water molecule [2–4]. The ratio of heavy to light stable isotopes is traditionally expressed as per mil (‰) deviation relative to Vienna Standard Mean Ocean Water (VSMOW), and delta notation (δ^{18}O and δ^2H) is commonly used to report the measured isotope variations [5].

It is generally observed that precipitation is gradually depleted in ^{18}O and ^{2}H isotopes as elevation increases [6]. The phenomenon is commonly called the 'altitude' effect and primarily results from the cooling of the air masses as they ascend on a mountain, accompanied by the rainout of the excess moisture [7,8]. In case of precipitation δ^{18}O a gradient with elevation has a global average of −2.8‰/km, ranging from −1.7 to −5.0‰/km; the European average is −2.1‰/km [9]. A similar general observation is that precipitation is gradually depleted in ^{18}O and ^{2}H isotopes when moving inland from seacoasts. This phenomenon is called the 'continental' effect, and it reflects the gradual depletion of the residual marine moisture in heavy isotopes during the sequential rainout as air masses travel from the marine moisture source to the center of the continents [6]. In the case of Europe, for example, a gradient in the long-term mean precipitation δ^{18}O shows an ~8‰ decrease over a distance of approximately 4000 km from the Irish coasts to the foothills of the Ural Mountains [10]. Besides the above-mentioned ones, there are a few other empirical relationships observed between environmental parameters and the stable hydrogen and oxygen isotope composition of precipitation, such as 'temperature', 'latitude' and 'amount' effects [6,10].

These large-scale effects are also important from a practical point of view because they have enabled useful applications in isotope hydrology (e.g., identification of the mean elevation of groundwater recharge [11,12]), paleoclimatology [13], paleoaltimetry [14], food authenticity [15], etc. However, numerous studies have reported seasonal and spatial differences in both 'altitude' [16–19] and 'continental' effects [10,20,21]. Therefore, the regional estimation of these empirical relations in modern precipitation is critically important for underpinning the application of appropriate gradient values to subcontinental scale (e.g., river basin) practical problems (e.g., estimation of mean recharge area [22]) or paleoclimatological comparisons [13].

The aim of the study is to assess the isotopic 'altitude' and 'continental' effects in modern precipitation across the Adriatic–Pannonian region, specifically to (i) check the potential seasonal changes and (ii) determine the characteristic gradients specifically for this region which can be applied in isotope hydrological/hydrogeological studies and other research in the future.

2. Study Area and Main Moisture Sources

The Adriatic–Pannonian region encompasses the northern part of Southeast Europe and the eastern part of Central Europe. The Mediterranean Sea was identified as the main marine moisture source for Central and Southeastern Europe [23–25], and changes in its moisture supply have a decisive role on dryness conditions in this region [26]. More specifically, the Western and Central Mediterranean are the dominant marine moisture source regions throughout the year, although locally recycled continental moisture is the dominant source of the atmospheric moisture in the summer [24]. At the northern part of the study region, the westerlies transport Atlantic marine moisture mixed with recycled terrestrial moisture from Western and West-Central Europe, in addition to the Mediterranean moisture contribution [27,28]. Studies dealing with single precipitation events in Eastern Hungarian localities found occasional moisture contributions from eastern and northern directions mostly in the October to March period [27,28].

The surface water of the Mediterranean Sea is more enriched in heavy isotopes than Atlantic surface water [29]. This feature and the distinct evaporative conditions in these regions imprint the resultant atmospheric moisture with a source-specific isotopic signature (e.g., see [30,31]; more details in Section 4.2).

3. Data and Methods

Stable hydrogen and oxygen isotope compositions (δ^2H and δ^{18}O) of monthly aggregated precipitation were gathered from the national precipitation monitoring networks of Austria [32], Slovenia [33,34] and Hungary [35] and from additional individual stations with accessible data, namely Debrecen [36] and Hrašćica [37], from three consecutive years (2016, 2017, 2018). Subsets suitable for assessing the 'altitude' and 'continental' effects have been selected from the gathered data set (Figure 1). The sampling methods might be different among the stations, but they always employed a collector configuration recommended for precipitation isotope analysis [38]. The analytical techniques may also differ between data sources (e.g., mass spectrometry for Slovenian samples, laser spectroscopy for the Hungarian and Croatian samples); however, the raw values were converted by all laboratories to the same international reference scale (VSMOW/SLAP), ensuring comparability and joint assessment of the data. In addition, measurements were carried out together with laboratory reference materials that are calibrated periodically against primary IAEA calibration standards to VSMOW/SLAP scale. The highest estimated uncertainties of the data are ±1‰ and ±0.3‰ for δ^2H and δ^{18}O, respectively.

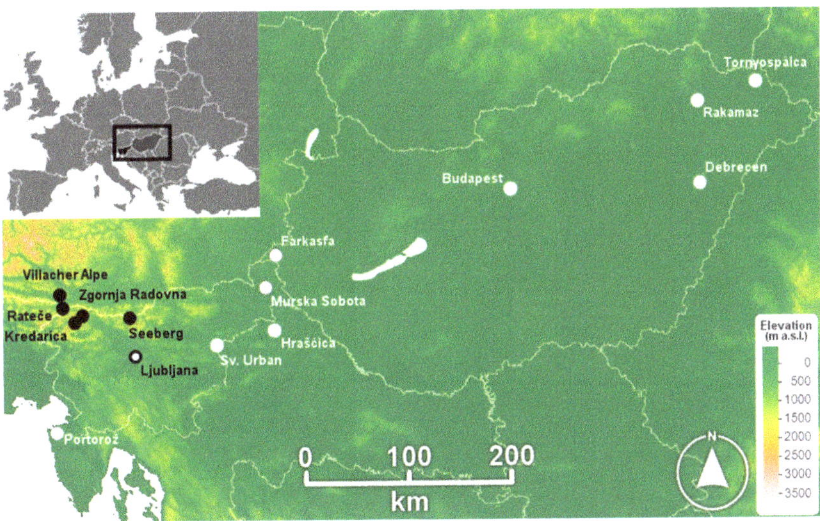

Figure 1. The study area and the location of the precipitation stable isotope monitoring stations. The sites used in the evaluation of the 'continental' effect are indicated by white dots, while the ones used in studying the 'altitude' effect are indicated by black dots; Ljubljana was used in both. Site details can be found in Table 1. The black rectangle in the inset map of Europe shows the position of the studied region. Country map of Europe by FreeVectorMaps.com. Digital elevation model taken from http://srtm.csi.cgiar.org/.

Stations situated within a relatively close distance but spanning a pronounced elevation range (from 282 to 2515 m a.s.l.) around the Slovenian–Austrian border (Figure 1 and Table 1) offer the opportunity to estimate the regional 'altitude' effect in modern precipitation in the Adriatic–Pannonian region. Due to the restricted latitudinal extension of this elevation transect, latitudinal or continental dependence are assumed to have only a negligible effect on the stable isotope composition of precipitation and a similarly uniform moisture source can be plausibly assumed for these closely located stations. The elevation range of this transect practically comprises the elevation extent of the entire Adriatic–Pannonian region, so the derived empirical relation should be applicable to the entire region.

To study the continental effect, stations situated along a ~750 km transect, stretching in a SW–NE direction from the Adriatic coast (Istria) to the NE corner of the Great Hungarian Plain, were selected (Figure 1 and Table 1). To eliminate potential interference with 'altitude' effect, only stations below 350 m a.s.l. were considered. This transect was chosen due to the predominance of the Mediterranean as the moisture source over the studied region [25,26].

Table 1. Basic geographical information of precipitation stable isotope monitoring stations operating in East-Central Europe between January 2016 and December 2018. Country codes follow the ISO-3166-1 ALPHA-2 standard. Latitude and longitude are in WGS84 projection; EPSG: 4326.

Name	Latitude (°)	Longitude (°)	Elevation (m a.s.l.)	No. of Monthly Data (δ^2H; δ^{18}O)	Country	Used in
Villacher Alpe	46.603	13.672	2164	35; 35	AT	Alt
Seeberg	46.417	14.533	940	11; 11	AT	Alt
Hrašćica	46.3	16.292	177	18; 18	HR	Cont
Farkasfa	46.910	16.309	312	35; 35	HU	Cont
Budapest	47.432	19.187	139	36; 36	HU	Cont
Rakamaz	48.128	21.470	103	24; 24	HU	Cont
Debrecen	47.475	21.494	110	36; 34	HU	Cont
Tornyospálca	48.273	22.177	108	24; 24	HU	Cont
Portorož	45.475	13.616	2	35; 35	SI	Cont
Rateče	46.497	13.713	864	32; 33	SI	Alt
Kredarica	46.379	13.849	2514	34; 34	SI	Alt
Zg. Radovna	46.428	13.943	750	35; 35	SI	Alt
Ljubljana	46.095	14.597	282	35; 35	SI	Alt/Cont
Sv. Urban	46.184	15.591	283	34; 34	SI	Cont
Murska Sobota	46.652	16.191	186	33; 32	SI	Cont

To statistically investigate the 'altitude' and 'continental' effects, the monthly precipitation stable isotope values were plotted against station elevation and distance from the Adriatic coast, respectively. Next, linear trendlines were fitted with least squares regression and the fits' statistics were investigated (slope (coefficients), p, adjusted r^2). Only the slope coefficients with their significance at the usually applied significance thresholds are presented. Besides the overall trends, in the seasonal (winter and summer half-years) difference in 'altitude' effect was explored with an independent sample Mann–Whitney U-test (MW U-test) [39].

Deuterium excess (d-excess), defined as d = δ^2H − 8 × δ^{18}O [6] was also calculated. The uncertainty of d-excess is ≤1.3‰, as propagated from the highest uncertainties of the primary isotopic parameters. This second-order isotopic parameter is a useful tracer of the origin of moisture because it is relatively invariant during transport and during the formation of condensate accompanied by equilibrium isotopic fractionation [40] but is an informative indicator of non-equilibrium isotopic effects [6] and the mixing of moistures of different origin [41].

4. Results and Discussion

In δ-δ space, the global meteoric water line [42] fits nicely to the used data (Figure S1), certifying that the isotopic compositions correspond to precipitation origin. Due to the strong linear relationship between δ^2H and δ^{18}O in precipitation (Figure S1), the results for 'altitude' and 'continental' effects on the monthly scale are only presented graphically for δ^{18}O for the studied 36-month period between 2016 January and 2018 December.

4.1. 'Altitude' Effect

The analysis of the precipitation δ^{18}O gradient indicated a decreasing trend in most of the months, ranging from 0.1 to −3.5‰/km (Figure 2a). The distribution of these monthly δ^{18}O lapse rates tends to be higher than the reported global range (from −1.7 to −5‰/km [9]), although overlapping with the upper portion of the global range, and corresponds well with the range of other large-scale compilations [8]. The fitted linear slope was significant in roughly half of the cases, and most of the insignificant ($p > 0.1$) trends were observed in the October–March half-year (59% in the case of δ^{18}O, 62.5% in the case of δ^2H).

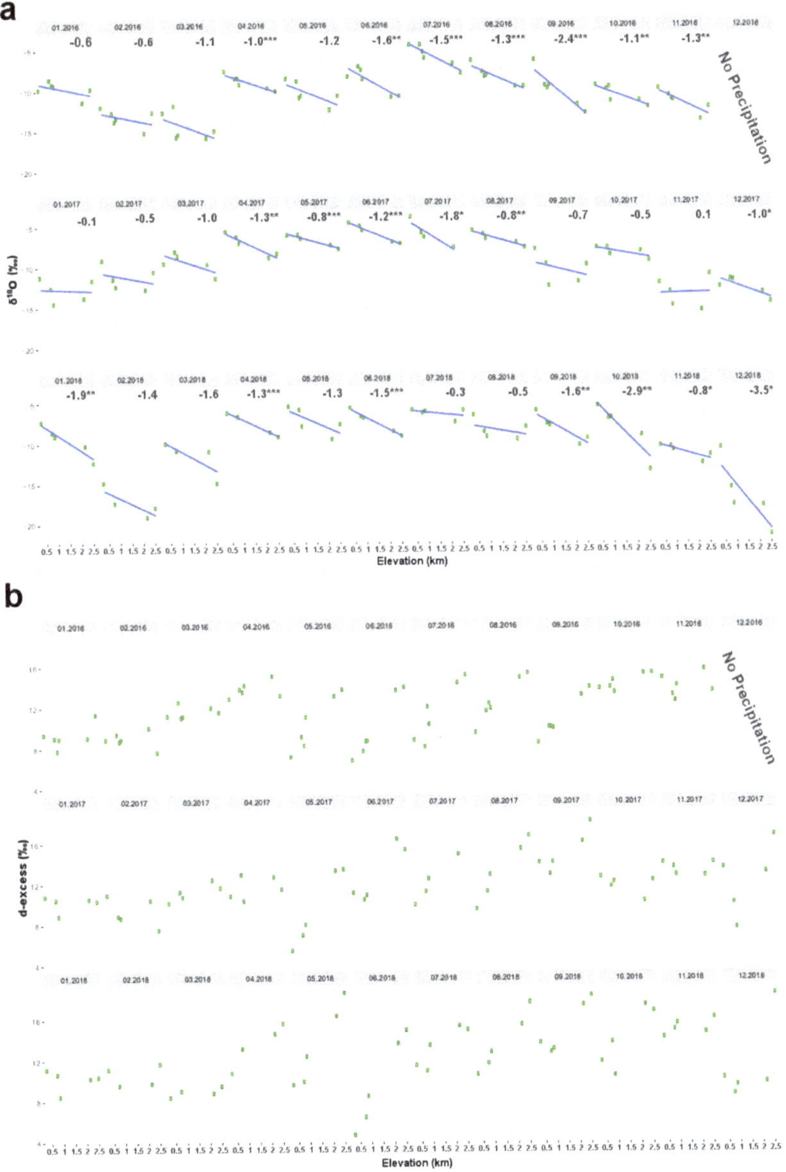

Figure 2. Stable isotope characteristics of monthly precipitation along a representative elevation transect in the Adriatic–Pannonian region: δ^{18}O (**a**) and deuterium excess (d-excess) (**b**) vs. elevation. Best fit lines (blue) for precipitation δ^{18}O data and elevation are plotted. The slope and significance of the linear fit are coded as α-values of 0.01, 0.05 or 0.1 and are marked with (***), (**) or (*), respectively.

Corresponding d-excess values along the elevation transect show a different vertical pattern when comparing colder (October to March) and warmer (April to September) months (Figure 2b). While d-excess values are scattered in a relatively narrow range without any systematic shift along the elevation gradient in the colder season, steep gradients with low d-excess values for low elevation stations can be seen in the warmer season (Figure 2b). This suggests distinct isotopic precipitation regimes for the warmer and colder parts of the year.

According to the classical concept of Rayleigh rainout, process whatever isotopic properties acquired by the cloud before, precipitation is removed under isotopic equilibrium conditions [6]. Under this ideal Rayleigh rainout scenario, the atmospheric moisture is depleted in heavy isotope content commensurate with the loss of moisture by precipitation; the d-excess during this process remains invariant [43], even though there is a tendency for an increase in this value at higher elevations [8].

The remarkably constant d-excess data along the elevation transect suggests a single dominant moisture source during the colder season [19]. Thus, in winter conditions, Rayleigh rainout effect can be taken as the predominant one, with a single dominating moisture source and with the moisture mass ascending to form orographic precipitation which is continuously depleted in heavy isotopes with increasing elevation [8,19]. However, the change of d-excess values along the elevation transect in the warmer season suggests that other factors which change the isotope composition besides the basic Rayleigh effect need to be considered. Evaporative enrichment of isotopes in raindrops during the fall beneath the cloud base reduces the d-excess [6,44]. The amplification of this effect is indeed expected in the warmer season; and it is expected to be larger at low elevations, conforming to the observed vertical d-excess trend (Figure 2b). Therefore, this appears to be a 'pseudo-altitude effect' [45] when vertical change of stable isotopes in precipitation is not due to the rainout of vapor of a single origin [8].

It should be noted that precipitation is dominantly rain in the warmer season over the studied elevation transect; while in the colder season, only snow is expected above ~1000 m, and a mixture of snow and rain is expected in the lower part of the transect. Since vapor → ice fractionation is slightly larger than vapor → water fractionation around the altitude where the regime switches from rain to snow (roughly corresponding to the 0 °C isotherm), the condensed hydrometeor is expected to be less depleted in heavy isotopes than it would be if the vapor → water fractionation was still in action. This phenomenon might be an additional effect that influences the monotony of the vertical upward decrease in precipitation $\delta^{18}O$ and causes the fitted linear regression to be frequently non-significant in the colder season.

Comparing the frequency distribution of monthly estimated 'altitude' effect for the colder and warmer seasons for the 2016–2018 period (Figure 3), no significant difference is found in the seasonal distributions ($p = 0.351$ for $\delta^{18}O$ and $p = 0.832$ for δ^2H according to the MW U-Test) despite the undeniably different responsible processes dictating the vertical trend of precipitation stable isotopes in the warmer and colder seasons. Hence, from a practical point of view, the mean of the monthly gradients calculated for the studied 3-year-long period ($\delta^{18}O$: −1.2‰/km; δ^2H: −7.9‰/km) can be considered as empirical isotopic 'altitude' effect in modern precipitation across the Adriatic–Pannonian region.

The estimated 'altitude' effect based on spring water $\delta^{18}O$ values (−1.1‰/km [22]) investigated between 2005 and 2007 in the vicinity of the studied elevation transect (in the valley of the Radovna River in NW Slovenia) is in good agreement with the current results. Comparing these results to former studies from the region, which typically employed fewer stations and/or a narrower elevation range, one can say that the magnitude of the determined isotopic 'altitude' effect tends to be smaller than the $\delta^{18}O$ vertical gradient calculated to the south of the study area, e.g., near Zagreb (−2.8‰/km) [46] or in the northern Adriatic (−3.0‰/km) [47]. However, at the northern border of the study area, in the Western Carpathians, the 'altitude' effect estimated both from spring water $\delta^{18}O$ values (−1.0 to −1.5‰/km) [48] and from precipitation $\delta^{18}O$ values (−2.1‰/km) [49] is relatively closer to the empirical $\delta^{18}O$ lapse rate obtained here for modern precipitation. At the eastern border of the study area, in the Eastern Carpathians, the estimated long-term 'altitude' effect based on spring water $\delta^{18}O$ values also returned comparable results (−1.5‰/km [50]).

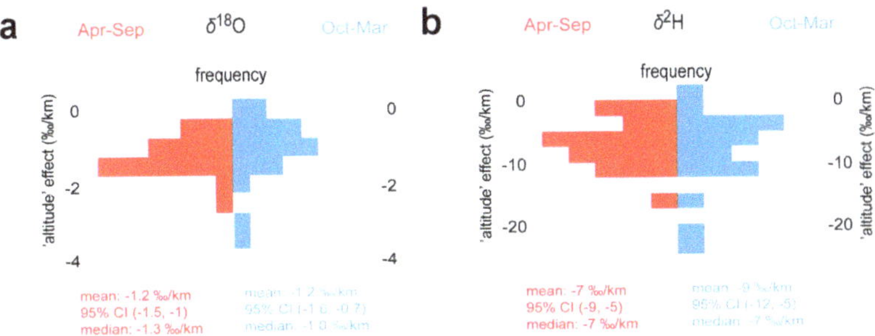

Figure 3. Frequency distribution of 'altitude' effects estimated from monthly precipitation $\delta^{18}O$ (**a**) and δ^2H (**b**) data for colder (October to March) and warmer (April to September) seasons.

4.2. 'Continental' Effect

As an initial observation, one can see that the seasonal change in precipitation stable isotopes is well reflected by the shifting of the whole data set, with more negative $\delta^{18}O$ values seen in winter and less negative $\delta^{18}O$ values in summer in each studied year. The monthly patterns in precipitation $\delta^{18}O$ values along the transect streching from the Adriatic coast to the northeastern corner of the Great Hungarian Plain did not indicate any trend in precipitation $\delta^{18}O$ data from May to August (Figure 4a). However, a generally negative gradient is outlined for the dataset from April to September, resembling an expected 'continental' effect.

A similar seasonal contrast has been reported for the long-term mean δ^2H gradient along the Atlantic stormtrack in Western Europe and was explained by enhanced raindrop evaporation in summertime at inland locations [20]. Occasionally, a remarkable shift (~10‰) can be seen (e.g., June and July 2017) toward lower d-excess values for single inland stations, which might indicate a kinetic isotope effect such as raindrop evaporation (Figure 4b). However, the situation seems to be more complicated here because usually there is no visible shift toward lower d-excess values during the summer months (see e.g., 2018) or, at most, a wide scattering (e.g., May 2016) can be observed (Figure 4b). The absence of a 'continental' effect was reported over the Amazon Basin [51], which was explained by the intense terrestrial moisture recycling [52]. In the studied region, summer is the high season for vegetation activity, accompanied by the highest transpiration flux within the year. Transpiration is essentially a non-fractionation process [53] and thus re-introduces isotopically unchanged water to the atmosphere through evapotranspiration. The large-scale dominance of transpired moisture might explain the coupled lack of trend of $\delta^{18}O$ and d-excess values, while the variable importance of evaporation and transpiration might spoil the empirical isotopic 'continental' effect and cause an accompanying wide scattering in d-excess values in precipitation. Supporting this explanation, Lagrangian moisture source modeling clearly indicated the dominant contribution of locally recycled continental moisture in the summer precipitation for Southeastern Europe [24].

A strikingly different pattern is discernible from September to April. Precipitation $\delta^{18}O$ values remarkably decrease with increasing distance from the Adriatic coast over the first part of the studied profile, with a break in this trend at ~400 km (Figure 4a). The $\delta^{18}O$ values beyond this point are not as depleted as would be expected from the linear trend fitted to the data of the first segment of the profile. The largest gradients were observed from November to March, while the estimated effect is smaller and usually not significant ($p > 0.1$) in April and September (Figure 4a). It is likely that the transition between the winter regime and the summer regime discussed above occurs during these months. An apparent nonlinearity of the 'continental' effect in the winter season was also reported along the distillation route of Atlantic moisture, with a gradient larger by a factor of two from the Irish coast to Central Europe compared to the rest of the transect to the foothills of the Ural Mountains. [10]. The decreased gradient was explained by additional moisture supply (e.g., Black Sea, Caspian Sea) substantially contributing to the water balance over Eastern Europe.

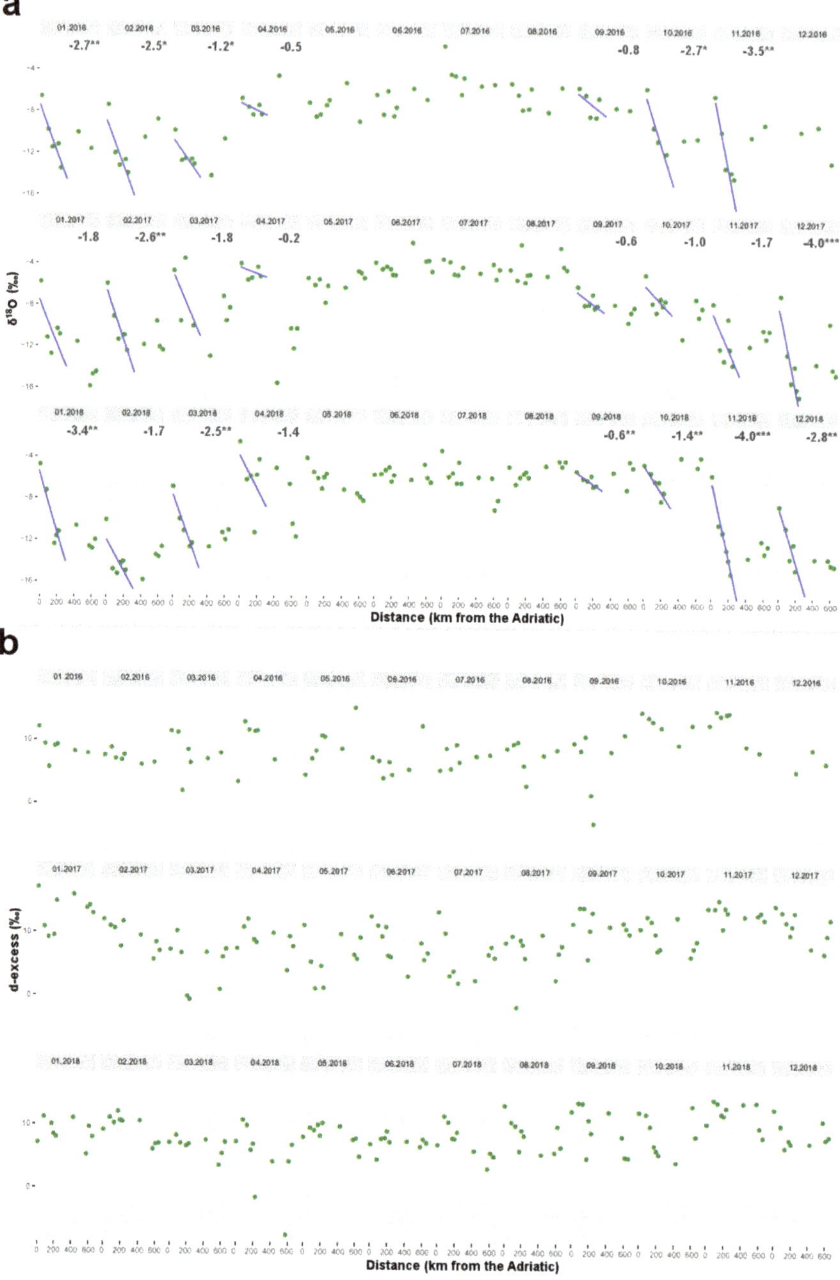

Figure 4. Stable isotope characteristics of monthly precipitation from the Adriatic coast to the northeastern corner of the Great Hungarian Plain: $\delta^{18}O$ (**a**) and deuterium excess (d-excess) (**b**) vs. distance from the Adriatic coast. Best fit lines (blue) for precipitation $\delta^{18}O$ data and distance are plotted over the first half of the continental transect. The significance of the linear fit is coded as α-values of 0.01, 0.05 or 0.1 that are marked with (***), (**) or (*), respectively.

Monthly d-excess patterns along the continetal transect across the Adriatic–Pannonian region also suggest additional moisture supply for the winter precipitation over the most inland section of the transect. Usually a shift can be observed in the d-excess values at the place where the break appears in the decrease of the $\delta^{18}O$ values (Figure 4b). Deuterium-excess values are grouped around a relatively higher mean for the first part of the transect accompanying the prevailing $\delta^{18}O$ gradient, while the d-excess values of the stations drop and usually group around a lower mean beyond this break. Characteristic examples are November 2016, December 2017 and February 2018, when precipitation d-excess values along the first part of the profile are typically at least 3‰ higher compared to the last part of the profile (Figure 4b). The average October-to-March d-excess value of the stations with a distance < 350 km to the Adriatic coast is 1.2‰, higher compared to the average for those stations situated further than 450 km from the sea.

This isotopic pattern agrees with the results of Lagrangian moisture source modeling for the winter (wet) season in pointing out the Western and Central Mediterranean as the dominant moisture source regions in Southeast Europe [24], while the northern sector of the study area is connected to the Northern Alps where the North Atlantic was found to be the primary marine moisture source [23]. Water vapor originating from the Western and Central Mediterranean has characteristically higher d-excess values [30] compared to the water vapor of the Atlantic atmospheric marine boundary layer [31]. Therefore, it is likely that the lower d-excess values in the northern part of the study area confirm the influence of the Atlantic moisture supply. Indeed, studies on the stable isotope composition of precipitation in Hungary (representing the inland sector of the studied continental transect; see Figure 1) found characteristically higher d-excess values in the case of precipitation arriving from the Mediterranean compared to precipitation arriving from the Atlantic directions [27,28]. The break in the inland-trend of precipitation $\delta^{18}O$ values and the accompanying sharp shift in the d-excess values probably indicate a sudden switch from the Mediterranean dominated regime to the Atlantic influence, while the gradual change in d-excess values (e.g., January 2016, February 2017) suggests gradual mixing of moisture delivered from both sources (Figure 4). Occasional contribution from additional moisture sources (e.g., from eastern and northern directions) between October and March can further complicate the isotope hydrometeorological situation in the NE sector of the Great Hungarian Plain [27,28].

Taking all of these results into consideration, a classical isotopic 'continental' effect can be considered only up to ~400 km inland from the Adriatic coast, and the mean gradients estimated from the monthly $\delta^{18}O$ and $\delta^{2}H$ data are −2.4 and −20 ‰/100 km, respectively. Comparing these results to former studies, we can conclude that this regional gradient is substantially larger than that estimated for Western Europe (−0.38‰/100 km) from the Irish coast to Central Europe for $\delta^{18}O$ [10] and the −3.3‰/100 km estimated for $\delta^{2}H$ [20]. This observation of a much steeper gradient along the Adriatic moisture source pathway can be explained by the stronger temperature gradient between the Adriatic coast and the center of the Pannonian Basin (i.e., mild Mediterranean winter vs. cold continental winter) compared to the temperature gradient prevailing over Western Europe in the winter season. The stronger temperature gradient likely forces a higher degree of condensation and a higher degree of fractionation over the same distance.

5. Conclusions

Assessing the stable isotope composition of monthly precipitation fallen between January 2016 and December 2018 for selected stations situated along an elevation transect and a continental transect provided empirical estimates for the isotopic 'altitude' and 'continental' effects in modern precipitation across the Adriatic–Pannonian region.

Isotopic characteristics suggest that the main driver of the apparent vertical depletion of heavy stable isotopes in precipitation differs in warmer (raindrop evaporation) and colder (condensation) seasons, although there is no significant difference in the resulting 'altitude' effect. Therefore, an 'altitude' effect of −1.2‰/km for $\delta^{18}O$ and −7.9‰/km for $\delta^{2}H$ can be recommended to be used for modern precipitation in the Adriatic–Pannonian region.

The monthly patterns in precipitation $\delta^{18}O$ values along a ~750 km long transect stretching from the Adriatic coast to the norteastern corner of the Great Hungarian Plain did not indicate any significant trend in precipitation $\delta^{18}O$ data from May to August. This can be explained by the overwhelming influence of terrestrial recycled moisture by evapotranspiration. The recycled moisture negates the effect of the classical Rayleigh distillation process, erasing the expected decreasing trend from the precipitation stable isotope data moving inland from the Adriatic coast. A significant 'continental' effect emerged in the stable isotope composition of winter precipitation along the same transect. However, isotopic characteristics argue for the prevailing influence of the Mediterranean moisture source in winter precipitation only up to ~400 km distance inland from the Adriatic coast. The classical concept of the 'continental' effect is applicable practically from the Adriatic coast to the Danube, and the estimated mean isotopic 'continental' effect for the winter precipitation is $-2.4‰/100$ km in $\delta^{18}O$ and $-20‰/100$ km in δ^2H. A break in the spatial trend of isotopic characteristics in winter precipitation at ~400 km probably reflects the mixture of moisture sources that travel along different distillation routes and delivering moisture with different isotopic composition. A sharp drop in the d-excess (>3‰) corresponding in space with the break in precipitation $\delta^{18}O$ trend likely indicates a sudden switch from the Mediterranean moisture domain to additional (mainly Atlantic) influence, while a gradual change in the d-excess values might suggest a gradual increase of the non-Mediterranean moisture contribution along the transect.

The regionally determined isotopic 'altitude' and 'continental' effects of modern precipitation can be recommended for use in future isotope hydrological or paleoclimatological applications in the Adriatic–Pannonian region.

Supplementary Materials: The following are available online at http://www.mdpi.com/2073-4441/12/6/1797/s1, Figure S1: Crossplot of the $\delta^{18}O$ vs. δ^2H values used in the present study, Table S1: Stable isotopic data used in the study.

Author Contributions: Conceptualization, Z.K.; methodology, Z.K. and I.G.H.; formal analysis, I.G.H. and Z.K.; resources, G.C., L.P., P.V. and T.K.; data curation, P.V. and D.E.; writing—original draft preparation, Z.K. and I.G.H.; writing—review and editing, P.V., I.F. and G.C.; visualization, I.G.H., D.E. and Z.K.; project administration, Z.K. and P.V.; funding acquisition, Z.K., P.V. and L.P. All authors have read and agreed to the published version of the manuscript.

Funding: This work was supported by the National Research, Development and Innovation Office under Grants SNN118205 and PD121387 and by the Slovenian Research Agency ARRS under Grants N1-0054, J4-8216 and P1-0143. The research was partly supported by the European Union and the State of Hungary, co-financed by the European Regional Development Fund in the project of GINOP-2.3.2-15-2016-00009 'ICER'.

Acknowledgments: The authors express their gratitude to the staff of the Slovenian Environment Agency at the meteorological stations and I. Kanduč for their valuable help in sampling and to S. Žigon for assistance with laboratory work. The results of this study have been discussed within the COST Action: "WATSON" CA19120. The authors applied the FLAE approach for the sequence of authors. See https://doi.org/10.1371/journal.pbio. 0050018 for further details. This is contribution No. 73 of the 2ka Paleoclimatology Research Group.

Conflicts of Interest: The authors declare no conflict of interest.

References

1. Peixóto, J.P.; Oort, A.H. The Atmospheric Branch of the Hydrological Cycle and Climate. In *Variations in the Global Water Budget*; Street-Perrott, A., Beran, M., Ratcliffe, R., Eds.; Springer: Dordrecht, The Netherlands, 1983; pp. 5–65.

2. Fórizs, I. Isotopes As Natural Tracers In The Water Cycle: Examples From The Carpathian Basin. *Stud. Phys.* **2003**, *1*, 69–77.

3. Yoshimura, K. Stable Water Isotopes in Climatology, Meteorology, and Hydrology: A Review. *J. Meteorol. Soc. Jan. Ser. II* **2015**, *93*, 513–533. [CrossRef]

4. Gat, J.R.; Mook, W.G.; Meijer, H.A. *Environmental Isotopes in the Hydrological Cycle*; International Atomic Energy Agency: Paris, France, 2001; Volume 2, p. 73.

5. Coplen, T.B. Reporting of stable hydrogen, carbon and oxygen isotopic abundances. *Pure App. Chem.* **1994**, *66*, 273–276. [CrossRef]

6. Dansgaard, W. Stable isotopes in precipitation. *Tellus* **1964**, *16*, 436–468. [CrossRef]

7. Ambach, W.; Dansgaard, W.; Eisner, H.; Møller, J. The altitude effect on the isotopic composition of precipitation and glacier ice in the Alps. *Tellus* **1968**, *20*, 595–600. [CrossRef]

8. Gonfiantini, R.; Roche, M.-A.; Olivry, J.-C.; Fontes, J.-C.; Zuppi, G.M. The altitude effect on the isotopic composition of tropical rains. *Chem. Geol.* **2001**, *181*, 147–167. [CrossRef]

9. Poage, M.A.; Chamberlain, C.P. Empirical Relationships Between Elevation and the Stable Isotope Composition of Precipitation and Surface Waters: Considerations for Studies of Paleoelevation Change. *Am. J. Sci.* **2001**, *301*, 1–15. [CrossRef]

10. Rozanski, K.; Araguás-Araguás, L.; Gonfiantini, R. Isotopic patterns in modern global precipitation. In *Climate Change in Continental Isotopic Records*; American Geophysical Union: Washington, DC, USA, 1993; pp. 1–36.

11. Clark, I.D.; Fritz, P. *Environmental Isotopes in Hydrogeology*; CRC Press Taylor & Francis Group: Boca Raton, FL, USA; London, UK; New York, NY, USA, 1997; p. 342.

12. Sappa, G.; Vitale, S.; Ferranti, F. Identifying Karst Aquifer Recharge Areas using Environmental Isotopes: A Case Study in Central Italy. *Geosciences* **2018**, *8*, 351. [CrossRef]

13. Jasechko, S.; Lechler, A.; Pausata, F.S.R.; Fawcett, P.J.; Gleeson, T.; Cendón, D.I.; Galewsky, J.; LeGrande, A.N.; Risi, C.; Sharp, Z.D.; et al. Late-glacial to late-Holocene shifts in global precipitation δ^{18}O. *Clim. Past* **2015**, *11*, 1375–1393. [CrossRef]

14. Blisniuk, P.M.; Stern, L.A. Stable isotope paleoaltimetry: A critical review. *Am. J. Sci.* **2005**, *305*, 1033–1074. [CrossRef]

15. Kelly, S.; Heaton, K.; Hoogewerff, J. Tracing the geographical origin of food: The application of multi-element and multi-isotope analysis. *Trends Food Sci. Technol.* **2005**, *16*, 555–567. [CrossRef]

16. Niewodnizański, J.; Grabczak, J.; Barański, L.; Rzepka, J. The Altitude Effect on the Isotopic Composition of Snow in High Mountains. *J. Glaciol.* **1981**, *27*, 99–111. [CrossRef]

17. Kern, Z.; Kohán, B.; Leuenberger, M. Precipitation isoscape of high reliefs: Interpolation scheme designed and tested for monthly resolved precipitation oxygen isotope records of an Alpine domain. *Atmos. Chem. Phys.* **2014**, *14*, 1897–1907. [CrossRef]

18. Kong, Y.; Pang, Z. A positive altitude gradient of isotopes in the precipitation over the Tianshan Mountains: Effects of moisture recycling and sub-cloud evaporation. *J. Hydrol.* **2016**, *542*, 222–230. [CrossRef]

19. Jiao, Y.; Liu, C.; Liu, Z.; Ding, Y.; Xu, Q. Impacts of moisture sources on the temporal and spatial heterogeneity of monsoon precipitation isotopic altitude effects. *J. Hydrol.* **2020**, *583*, 124576. [CrossRef]

20. Rozanski, K.; Sonntag, C.; Münnich, K.O. Factors controlling stable isotope composition of European precipitation. *Tellus* **1982**, *34*, 142–150. [CrossRef]

21. Winnick, M.J.; Chamberlain, C.P.; Caves, J.K.; Welker, J.M. Quantifying the isotopic 'continental effect'. *Earth Planet. Sci. Lett.* **2014**, *406*, 123–133. [CrossRef]

22. Torkar, A.; Brenčič, M.; Vreča, P. Chemical and isotopic characteristics of groundwater-dominated Radovna River (NW Slovenia). *Environ. Earth Sci.* **2016**, *75*, 1296. [CrossRef]

23. Sodemann, H.; Zubler, E. Seasonal and inter-annual variability of the moisture sources for Alpine precipitation during 1995–2002. *Int. J. Climatol.* **2010**, *30*, 947–961. [CrossRef]

24. Gómez-Hernández, M.; Drumond, A.; Gimeno, L.; Garcia-Herrera, R. Variability of moisture sources in the Mediterranean region during the period 1980–2000. *Water Resour. Res.* **2013**, *49*, 6781–6794. [CrossRef]

25. Ciric, D.; Nieto, R.; Losada, L.; Drumond, A.; Gimeno, L. The Mediterranean Moisture Contribution to Climatological and Extreme Monthly Continental Precipitation. *Water* **2018**, *10*, 519. [CrossRef]

26. Stojanovic, M.; Drumond, A.; Nieto, R.; Gimeno, L. Variations in Moisture Supply from the Mediterranean Sea during Meteorological Drought Episodes over Central Europe. *Atmosphere* **2018**, *9*, 278. [CrossRef]

27. Bottyán, E.; Czuppon, G.; Weidinger, T.; Haszpra, L.; Kármán, K. Moisture source diagnostics and isotope characteristics for precipitation in east Hungary: Implications for their relationship. *Hydrol. Sci. J.* **2017**, *62*, 2049–2060. [CrossRef]

28. Czuppon, G.; Bottyán, E.; Krisztina, K.; Weidinger, T.; Haszpra, L. Significance of the air moisture source on the stable isotope composition of the precipitation in Hungary. In Proceedings of the Conference The EGU General Assembly 2017, Vienna, Austria, 23–28 April 2017; p. 13458.

29. LeGrande, A.N.; Schmidt, G.A. Global gridded data set of the oxygen isotopic composition in seawater. *Geophys. Res. Lett.* **2006**, *33*, L12604. [CrossRef]

30. Gat, J.R.; Klein, B.; Kushnir, Y.; Roether, W.; Wernli, H.; Yam, R.; Shemesh, A. Isotope composition of air moisture over the Mediterranean Sea: An index of the air-sea interaction pattern. *Tellus B: Chem. Phys. Meteorol.* **2003**, *55*, 953–965. [CrossRef]

31. Benetti, M.; Steen-Larsen, H.C.; Reverdin, G.; Sveinbjörnsdóttir, Á.E.; Aloisi, G.; Berkelhammer, M.B.; Bourlès, B.; Bourras, D.; de Coetlogon, G.; Cosgrove, A.; et al. Stable isotopes in the atmospheric marine boundary layer water vapour over the Atlantic Ocean, 2012–2015. *Sci. Data* **2017**, *4*, 160128. [CrossRef]

32. Umweltbundesamt, H.O.F. Bundesministerium für Land-und Forstwirtschaft, U.u.W.B., Ed. 2019. Available online: https://wasser.umweltbundesamt.at/h2odb/ (accessed on 1 October 2019).

33. Vreča, P.; Malenšek, N. Slovenian Network of Isotopes in Precipitation (SLONIP)—A review of activities in the period 1981–2015. *Geologija* **2016**, *59*, 67–84. [CrossRef]

34. SLONIP. Slovenian Network of Isotopes in Precipitation. The SLONIP Database. Available online: https://slonip.ijs.si/ (accessed on 1 May 2020).

35. Czuppon, G.; Breuer, H.; Bottyán, E.; Kern, Z.; Simon, G.; Mona, T.; Göndöcs, J. *Role of Meteorological Processes and Isotope Effects on the Stable Isotope Composition of Precipitation Originated from the Mediterranean Region*; Book of Abstracts, National Research and Development Institute for Cryogenics and Isotopic Technologies- ICSI Rm: Válcea, Romania, 2017; p. 64.

36. Vodila, G.; Palcsu, L.; Futó, I.; Szántó, Z. A 9-year record of stable isotope ratios of precipitation in Eastern Hungary: Implications on isotope hydrology and regional palaeoclimatology. *J. Hydrol.* **2011**, *400*, 144–153. [CrossRef]

37. Marković, T.; Karlović, I.; Perčec Tadić, M.; Larva, O. Application of Stable Water Isotopes to Improve Conceptual Model of Alluvial Aquifer in the Varaždin Area. *Water* **2020**, *12*, 379. [CrossRef]

38. IAEA. *IAEA/GNIP precipitation sampling guide V2*; International Atomic Energy Agency: Vienna, Austria, 2 September 2014; p. 19. Available online: http://www-naweb.iaea.org/napc/ih/documents/other/gnip_manual_v2.02_en_hq.pdf (accessed on 15 May 2020).

39. Mann, H.B.; Whitney, D.R. On a Test of Whether one of Two Random Variables is Stochastically Larger than the Other. *Ann. Math. Statist.* **1947**, *18*, 50–60. [CrossRef]

40. Merlivat, L.; Jouzel, J. Global climatic interpretation of the deuterium-oxygen 18 relationship for precipitation. *J. Geophys. Res. Oceans* **1979**, *84*, 5029–5033. [CrossRef]

41. Bershaw, J. Controls on Deuterium Excess across Asia. *Geosciences* **2018**, *8*, 257. [CrossRef]

42. Craig, H. Isotopic Variations in Meteoric Waters. *Science* **1961**, *133*, 1702–1703. [CrossRef] [PubMed]

43. Gat, J. *Isotope Hydrology: A Study of the Water Cycle*; Imperial College Press: London, UK, 2010.

44. Stewart, M.K. Stable isotope fractionation due to evaporation and isotopic exchange of falling waterdrops: Applications to atmospheric processes and evaporation of lakes. *J. Geophys. Res.* **1975**, *80*, 1133–1146. [CrossRef]

45. Moser, H.; Stichler, W. Die Verwendung des Deuterium- und Sauerstoff-18-Gehalts bei hydrologischen Untersuchungen, in English: Application of deuterium and oxygen-18 content measurements in hydrological investigations. *Geol. Bavarica* **1971**, *64*, 7–35.

46. Krajcar Bronić, I.; Horvatinčić, N.; Obelić, B. Two decades of environmental isotope records in Croatia: Reconstruction of the past and prediction of future levels. *Radiocarbon* **1998**, *40*, 399–416. [CrossRef]

47. Vreča, P.; Bronić, I.K.; Horvatinčić, N.; Barešić, J. Isotopic characteristics of precipitation in Slovenia and Croatia: Comparison of continental and maritime stations. *J. Hydrol.* **2006**, *330*, 457–469. [CrossRef]

48. Malík, P.; Michalko, J. Oxygen Isotopes in Different Recession Subregimes of Karst Springs in the Brezovské Karpaty Mts. (Slovakia). *Acta Carsologica* **2010**, *39*. [CrossRef]

49. Holko, L.; Dóša, M.; Michalko, J.; Šanda, M. Isotopes of oxygen-18 and deuterium in precipitation in Slovakia. *J. Hydrol. Hydromech.* **2012**, *60*, 265–276. [CrossRef]

50. Fórizs, I.; Makfalvi, Z.; Deák, J.; Kármán, K.; Vallasek, I.; Süveges, M. Izotópgeokémiai vizsgálatok a Csíki-medence ásványvizeiben/Isotope geochemical investigations of the mineral waters in the Ciuc Basin. *Miskolc. Egy. Közl. Sor. Bány.* **2011**, *81*, 59–67.

51. Salati, E.; Dall'Olio, A.; Matsui, E.; Gat, J.R. Recycling of water in the Amazon Basin: An isotopic study. *Water Resour. Res.* **1979**, *15*, 1250–1258. [CrossRef]

52. Gat, J.R.; Matsui, E. Atmospheric water balance in the Amazon basin: An isotopic evapotranspiration model. *J. Geophys. Res. Atmos.* **1991**, *96*, 13179–13188. [CrossRef]
53. Wang, X.-F.; Yakir, D. Temporal and spatial variations in the oxygen-18 content of leaf water in different plant species. *Plant Cell Environ.* **1995**, *18*, 1377–1385. [CrossRef]